STRONGER THAN A HUNDRED MEN

JOHNS HOPKINS STUDIES IN THE HISTORY OF TECHNOLOGY

general editor: Thomas P. Hughes
advisory editors: Leslie Hannah and Merritt Roe Smith

The Mechanical Engineer in America, 1839–1910: Professional Cultures in Conflict, by Monte Calvert

American Locomotives: An Engineering History, 1830–1880, by John H. White, Jr.

Elmer Sperry: Inventor and Engineer, by Thomas Parke Hughes (Dexter Prize, 1972)

Philadelphia's Philosopher Mechanics: A History of the Franklin Institute, 1824–1865, by Bruce Sinclair (Dexter Prize, 1975)

Images and Enterprise: Technology and the American Photographic Industry, 1839–1925, by Reese V. Jenkins (Dexter Prize, 1978)

The Various and Ingenious Machines of Agostino Ramelli, edited by Eugene S. Ferguson, translated by Martha Teach Gnudi

The American Railroad Passenger Car, New Series, no. 1, by John H. White, Jr.

Neptune's Gift: A History of Common Salt, New Series, no. 2, by Robert P. Multhauf

Electricity before Nationalisation: A Study of the Development of the Electricity Supply Industry in Britain to 1948, New Series, no. 3, by Leslie Hannah

Alexander Holley and the Makers of Steel, New Series, no. 4, by Jeanne McHugh

The Origins of the Turbojet Revolution, New Series, no. 5, by Edward W. Constant II (Dexter Prize, 1982)

Engineers, Managers, and Politicians: The First Fifteen Years of Nationalised Electricity Supply in Britain, New Series, no. 6, by Leslie Hannah

Stronger Than a Hundred Men: A History of the Vertical Water Wheel, New Series, no. 7, by Terry S. Reynolds

*But of all the inconveniences, shortage of water is most to be avoided, . . . because the lifting power of a [water] wheel is much stronger and more certain than that of a hundred men.*
—VANNOCCIO BIRINGUCCIO (1540)

# STRONGER THAN A HUNDRED MEN

## A History of the Vertical Water Wheel

**TERRY S. REYNOLDS**

THE JOHNS HOPKINS UNIVERSITY PRESS
BALTIMORE AND LONDON

The Johns Hopkins University Press, Baltimore, Maryland 21218
The Johns Hopkins Press, Ltd., London

*The Library of Congress has catalogued the
hardcover edition of this book as follows:*

Reynolds, Terry S.
  Stronger than a hundred men.

  (Johns Hopkins studies in the history of
technology; new ser., no. 7)
  Bibliography: pp. 399–430
  Includes index.
  1. Water-wheels—History.  I. Title.  II. Series.
J860.R48      621.2′1′09      82–15346
ISBN 0–8018–2554–7      AACR2

ISBN 10: 0-8018-7248-0
ISBN 13: 978-0-8018-7248-8

*To Ira E. Reynolds and Therasea A. Janzen Reynolds,*
 *my parents, and*
*To Beulah I. Pyle Reynolds, my grandmother,*
 *in partial repayment for the love, guidance, encouragement,*
 *and example they provided in my early years.*

# Contents

# Illustrations

# Tables

# Acknowledgments

To some extent this book began as a doctoral dissertation in the history of science at the University of Kansas, a dissertation which reviewed the relationships between science and water-power technology. The present work, however, is much broader. The bulk of my dissertation research is contained only in a single chapter (Chap. 4), and even that chapter contains additional material and a different organization.

For financial support of my dissertation research I acknowledge the aid of the National Science Foundation Traineeship program and the Woodrow Wilson Dissertation Fellowship program. For guidance and direction in carrying out my dissertation research, I gratefully acknowledge the help of Professor Edward E. Daub. For criticism and suggestions for improvement I am indebted to Professors Jerry Stannard, John Alexander, Nick Willems, and Bernard Gainer, all of the University of Kansas.

Fearing my dissertation covered too narrow a topic to merit publication, I sent Professor Eugene S. Ferguson of the University of Delaware and the Eleutherian-Mills Hagley Foundation a copy and asked him for advice on where to go next. He suggested the need for a comprehensive history of the water wheel. In 1975 or 1976, on the basis of this recommendation, I decided to use my dissertation as a foundation for writing a broader and more comprehensive history of the vertical wheel than that provided by my dissertation.

In carrying out this task I worked largely without outside financial support. I would like to acknowledge, however, the Graduate School Research Committee of the University of Wisconsin for one summer of salary support. For partial relief from teaching duties for one semester, I thank the College of Engineering of the University of Wisconsin and, more particularly, my home department—General Engineering. I also thank Dean W. Robert Marshall of the College of Engineering for providing financial assistance for the photographic work which this volume required.

There are many individuals who assisted or encouraged me in the post-dissertation stage of my writing and research. I owe some acknowledgment to my colleagues in the Department of General Engineering. For the most part they are nonhistorians. But they provided me with a congenial working atmosphere. Ann Morris and Nancy Hansen, the department's secretaries, typed substantial portions of several drafts of this work. Outside my department, thanks are due Karl Willmann of UW–photo extension, who did excellent photographic work for me. Thanks are also due Bert S. Hall of the University of Toronto, who read an earlier draft of this work and provided me with some solid criticisms and suggestions. The librarians of the University of Wisconsin's library system, and particularly those in the Rare Book and Inter-Library Loan departments, likewise have my gratitude for their assistance. Carolyn Moser, copy editor of my manuscript for the Johns Hopkins University Press, did an excellent job of weeding out inconsistencies, detecting organizational flaws, and eliminating repetitious phrases. This work is much better because of her efforts.

Three people should be singled out for special acknowledgment: Edward E. Daub, Edwin T. Layton, and Linda G. Reynolds.

Professor Edward E. Daub, of the Department of General Engineering, University of Wisconsin, and formerly of the Department of History, University of Kansas, did more than merely supervise my dissertation work at the University of Kansas. As a friend and colleague at the University of Wisconsin, he urged me to begin the task of writing a broader history of water power. Without his strong encouragement and support, particularly during the early stages of this work, I seriously doubt whether it would have ever seen the light of day. I am deeply in his debt.

Professor Edwin T. Layton of the University of Minnesota provided me with an extensive critique (50 pages, single-spaced!) of an earlier draft of this work. He also provided me, later, with an even more detailed critique of Chapters 1, 2, and 3. As a result of his suggestions and guidance the first half of this work, in particular, is far better than it would have been otherwise. I owe Ed Layton many thanks for the hours he spent reviewing and criticizing my work and for the encouragement he provided me in the closing stages of this labor.

Finally, I would acknowledge the assistance of my wife, Linda Gail Rainwater Reynolds. Not only did she spend many long and boring hours proofreading all of the drafts of this work, but she also provided me with good meals, companionship, a comfortable home, four sons, and a great deal of love and understanding during the years in which I was involved in researching and writing this book.

To all those who have assisted me in any way, however minor, I express my thanks. This work is better because of you. That it is not better still is my fault, not yours.

# Introduction

Why should one take time to write a comprehensive survey of the history of a prime mover (producer of power) like the vertical water wheel? Two obvious answers are (1) because power technology has played a major role in the evolution of technology and, more generally, of society, and (2) because the vertical water wheel was, for a considerable period, a major element of power technology. Let us consider each of these points in turn.

Prime movers, like the vertical water wheel, have been regarded by some historians as the *central* element in the evolution of technology. For instance, Lewis Mumford, in his classic *Technics and Civilisation* (1934), utilized the idea of characteristic power source and characteristic material to periodize the evolution of technology since 1000 into three broad epochs (eotechnic, paleotechnic, and neotechnic) based, respectively, on water power, coal power, and electric power.[1] A. R. Ubbelohde, in an elementary overview of the history of power technology published in 1955, gave energy a more important role. "Every important technical advance in the past," he declared, has involved "some new phase in the control of energy."[2] And R. J. Forbes, the Dutch historian of technology, argued at about the same time that prime movers were the "keystone of technology," that they determined the size of the units which industry could deal with, the size of the machinery and tools that acted on these units, and the type of products which emerged. Using prime movers as a criterion he distinguished five eras in the history of technology—an era dominated by human muscle, an era when human muscle was supplemented by animal muscle, an era of water power, an era of steam power, and an emerging age of nuclear power.[3]

Others, particularly anthropologists, scientists, and engineers, have pushed the importance of energy further than historians like Mumford, Ubbelohde, and Forbes, asserting that it is not only the "keystone of technology," but that it is also the most important factor in determining the level of culture or civilization attained by a society. For example, in

1

the early 1890s Charles Loring, a prominent American mechanical engineer, pointed to the impact of one particular prime mover, the steam engine, on society. The steam engine, Loring asserted, by liberating man from heavy labor and by increasing productivity, had done more to advance human nature than all the statesmen, monarchs, philosophers, priests, and artists combined.[4] Wilhelm Ostwald, the German philosopher and scientist, declared in 1907 that the "history of civilization" was the "history of man's advancing control over energy."[5] Two decades later his assertion was supported by the anthropologist George MacCurdy, who asserted that the "degree of civilization" of any epoch or people could be measured by that epoch's or that people's "ability to utilize energy for human advancement or needs."[6] And in the past few decades several scientists and engineers have pointed to the close correlation between the standard of living and per capita energy consumption in contemporary societies to argue for the central position of power technology in social development.[7]

One of the most eloquent spokesmen for the importance of power technology to cultural development has been the anthropologist Leslie White. White, in several essays published near the middle of the present century, pointed out that everything a cultural system did, from the basic activities of securing food, clothing, and shelter to the less essential activities such as worship and play, required energy. Thus, he declared, all cultural systems could be reduced to a common denominator—energy. Culture was, therefore, "primarily a mechanism for harnessing energy and of putting it to work in the service of man, and, secondarily, of channelling and regulating his behavior not directly concerned with subsistence and offense and defense." Because everything in cultural systems depended on energy, social systems were determined by technological systems. Religion, philosophy, and the arts simply expressed "experience as it is defined by technology and refracted by social systems." In other words, White contended that culture evolved in its "sociological and ideological aspects . . . as it develops technologically." Culture advanced only as the amount of energy harnessed per capita per year was increased.[8]

Power technology is clearly important to both the evolution of technology and the evolution of culture. But Forbes, Loring, and White, among others, gave it too central a role. In the area of technology, for example, energy is only one of many categories. Others are such things as materials, tools, agriculture, transportation, and communications. All of these areas are mutually interdependent. It may be difficult to conceive of a centralized factory system without the steam engine. But it is just as difficult to conceive of the steam engine's emerging before appropriate metals (cast iron, e.g.) and metal-working machinery (the boring machine, e.g.). All of the different divisions within technology constantly interact with and depend on each other. None can be considered to be always the "keystone" of technology, always more important than the others.

Similarly, energy is not always the primary factor in determining the level of culture. There is, admittedly, a rough correlation between gross national product and per capita energy consumption in contemporary societies. But the level of a culture is based on more than gross national product. And energy consumption is as much a passive follower of social ideals, institutions, and values as a determiner of these ideals, institutions, and values. A society may very well have the technical ability to tap a new source of energy, but refrain for a variety of economic, religious, and philosophical reasons. Lynn White, for instance, has pointed out that both the Islamic and Christian worlds knew how to build windmills, yet only the latter systematically exploited the discovery, a phenomenon White attributes in part to the intellectual climate of Western Christendom.[9] Similarly, contemporary American society has strongly resisted the more extensive introduction of nuclear power. Energy technology is embedded in a social matrix. It influences other elements in the matrix, but it is, in turn, influenced by them. Economic factors, religious convictions, political institutions, and the general economic climate often play just as important a role in determining whether a particular form of energy or power is utilized and how it is utilized as energy technology plays in determining the socioeconomic, political, and intellectual views of a society.

Yet, just because energy technology is embedded in complex and interdependent social and technological matrices and just because it cannot always be singled out as the single most important factor in these matrices does not mean that it is relatively unimportant. It simply means that the claims made for the importance of prime movers to the evolution of technology and society by R. J. Forbes, Leslie White, and others must be moderated. Energy technology may not always be the most important factor in determining what man does technologically and socially, but it often is; and, in the words of Fred Cottrell, "the energy available to man limits what he *can* do and influences what he *will* do."[10] For example, even if Leonardo da Vinci had correctly understood the nature of flight, the absence of a concentrated, mobile power source placed this technological accomplishment beyond the grasp of Renaissance technology. The general mechanization of agriculture and the massive reduction of the proportion of the labor force engaged in agriculture were impossible without a mobile source of mechanical power. The emergence of the modern automobile industry was dependent on a variety of factors, but an appropriate source of power, both for the product itself and for the moving assembly lines used to produce it, was one of the most essential. Ford's River Rouge complex is inconceivable within the context of man-powered, animal-powered, or pre-hydro-electric water-powered technology.

Thus, even though prime movers are not always the central or determining element in technological or social change, they do place definite limits on what can be accomplished and strongly influence what will be undertaken. This suggests that the history of prime movers is at least one of the primary foundations on which a comprehensive history of technology should be built. And this belief provided one of the primary motivations for this study.

Among prime movers, the vertical water wheel was long one of the most important. Invented a century or two before the time of Christ, it was a major turning point in the history of technology. For millenia man had relied on the power of muscles, either his own or those of domesticated animals, for nearly all enterprises, both large and small. The water wheel enabled man, for the first time, to use an inanimate power source for industrial production, and as the use of water power spread in subsequent centuries, it had a major impact on technological and industrial development in several areas:

1. it made possible considerable labor savings in certain industries;
2. it permitted massive increases in production in other industries; and
3. it made certain enterprises possible which would not have been possible at all without the more concentrated energy which the vertical water wheel provided.

The labor-saving potential of the water wheel was considerable. For example, even a small two to three horsepower vertical water wheel could free as many as 30 to 60 men or, more likely, women from the tedious task of grinding grain.[11] However, the amount of flour required by a society and produced by a society at any given time is largely dependent on population, not milling economy. Water power in the flour milling industry permitted labor savings as it gradually came into use in the first millenium of the modern era, but it did not significantly increase flour output.

In other areas, however, the vertical wheel did significantly increase productive capabilities. In its vertical form, the water wheel was a much more concentrated or intensive source of power than any prime mover used previously. (Its horizontal counterpart was not, and hence its impact was largely restricted to saving labor without increasing overall production.) As the possibilities of concentrated, intensive mechanical power were recognized and developed, particularly in the period between c1000 and c1800, the vertical wheel was applied to a wide variety of tasks—to full cloth, to produce hemp, to saw wood, to shape iron, to bore pipes, to crush sugar, to tan leather, and to carry out a large number of other jobs. Large-scale production of many of these goods was uneconomical with animate power. With the vertical wheel much more cloth was fulled, hemp produced, wood sawed, iron shaped, and pipes bored than would have been the case had animate power alone been used.

In some cases the vertical water wheel not only increased production and reduced labor, but also permitted things to be done that would not have been done otherwise. An example is deep-level mining. Simply to drain 48 gallons (182 l) of water per minute from a depth of 97 feet (29.6 m) the Romans at the Rio Tinto mine in Iberia had to use 16 tread-powered water-lifting wheels in sets of two at eight different levels, powered by three eight-hour shifts of 16 men.[12] For depths of over 300 feet (c90 m), where construction of an adit was not feasible, deep-level mining was nearly impossible using animate power. Only water-powered drainage pumps made it possible. Similarly, only water-powered bellows made it possible for Europeans to economically develop high enough temperatures in their blast furnaces to produce cast iron.

Because of its ability to reduce dependence on human labor, to increase productivity, and to carry out tasks nearly impossible with animate power sources, the vertical water wheel became the power source of first choice in European heavy industry by the end of the medieval period. Biringuccio, a mining and metallurgical engineer, wrote in 1540, for example, that "of all the inconveniences, shortage of water is the most to be avoided,... because the lifting power of a [water] wheel is much stronger and more certain than that of a hundred men."[13] Two centuries later Jacob Leupold voiced similar sentiments: "Water-power is to be preferred to everything else if there is enough water and there is sufficient fall. Water gives the smoothest operation. Where water-power is available, other forms of power should not be used."[14] If there was a single key element distinguishing western European technology from the technologies of Islam, Byzantium, India, or even China after around 1200, it was the West's extensive commitment to and use of water power.

For centuries the vertical water wheel was of critical importance not only to Western technology and to the level of Western industrial output, but also to the way Western society evolved. Because water power was not uniformly distributed, it had a major influence on the location of industrial and population centers in Europe. When fulling was mechanized in thirteenth-century England, for example, the wood industry was compelled to relocate in the northern and western parts of the country, where water and waterfalls were available. Many of the southern and eastern English towns which had been important in the medieval cloth trade lost importance and subsequently went into decline.[15] While water power encouraged the growth of mill towns at certain locations, it also placed limits on the size they could attain. The amount of water power available at any particular site was limited; industrial expansion dependent on it could proceed only so far.

There were other ways in which the vertical water wheel influenced Western social and economic development. Because the vertical wheel was relatively expensive, especially if extensive auxiliary works (dams, canals) were required, it tended to restrict entrance to industries that had become dependent on water power. Because the vertical wheel was a concentrated or intensive power source, it contributed to the concentration of workers at specific sites and to the development of the factory system of production. These developments, in turn, contributed to the emergence of new relationships between capital, management, and labor.

Unfortunately, it is difficult to demonstrate quantitatively the emergence of the vertical water wheel as the power source of first choice in Western technology. Reliable statistical data are simply not available prior to the nineteenth century. But a scheme recently suggested by Louis Hunter can be used to lend some support to this theme. Hunter noted that the pattern of a typical river system resembled that of a tree with trunk (main river), limbs (major tributaries), branches (small tributaries and tributaries of tributaries), and small twigs (small creeks and brooks). The smallest elements of the system, brooks and creeks, are like the twigs on a tree. They are more numerous and more widely distributed than the larger elements (the major tributaries and the trunk stream). But because of the small volume of water flowing in them, wheels located where falls are available on creeks are usually able to provide only small amounts of power. The number of potential mill sites declines as one advances from the outlying branches to the central or lower portions of the system, since sudden falls are less common on the plains or in the large valley of the trunk stream than in the foothills which form the headwaters of the system. But the potential power per site is increased considerably because of the enormously increased volume of flow.

Hunter pointed out that one would naturally expect a widely dispersed, largely nonindustrial society or a society with only very small inanimate power requirements and little hydraulic know-how to tap first the smaller and more easily developed water powers available on the streams or brooks in the foothills. Falls would be easier to develop; the volume of water which had to be handled would be small. Only later, as water power became more important in response to growing population, urban concentration on navigable streams, the development of markets, and the extension of mechanization and industrialization, would one expect a society to develop the know-how and have the incentive to use the larger volumes of water on the larger streams where falls were harder to develop. Thus the appearance of hydropower installations on large streams and rivers, according to Hunter, suggests a growing industrial dependence on water power.

Hunter used the trunk-limb-branch schema to sketch the evolution of water power in the United States from subsistence colonial watermills, located on numerous small streams, to the major mid-nineteenth-century textile mills, which, in cities like Lowell, Massachusetts, utilized the power of large trunk rivers.[16]

His device cannot be used as a rigorous tool to document the growing importance of water power to European industry. But because there is an analogy between the emergence of water power in Europe and in early America, Hunter's device can be used as evidence to suggest the growing importance of water power in the West during the medieval and early modern eras. In Europe, as in America, there was a clearly discernable trend towards the use of steadily larger streams and larger quantities of water, a trend which culminated in Europe, as in America, in the harnessing of trunk rivers.

Water power grew to be a central element in Western technology, industry, and society during the medieval and early modern periods primarily because it was a more intensive or concentrated form of power than man or animal. It is for this reason that this study concentrates on the vertical water wheel, instead of other water-powered devices, the horizontal water wheel in particular. For most of the era of water power, the typical horizontal watermill was capable of generating little more power than a donkey or horse, and often not that much. It could be used only for a single task (milling grain), and it was wasteful of water. The horizontal wheel did not make possible the performance of large-scale industrial work, work that would have been either impossible or marginal with the animate power sources that were its alternatives. While it was capable of being developed into a more powerful prime mover, through most of the age of water power, in most of the more industrialized regions of Europe, it was not. It remained largely an item of peasant culture. The vertical water wheel, on the other hand, was usually the best means of economically and reliably securing a relatively large amount of mechanical power from a shaft. It was more easily adapted to tasks other than grain milling. And it was capable of delivering more power from a given volume and fall of water than the horizontal wheel. Consequently it was the vertical water wheel which became the mainstay of European power technology in the medieval period and was to remain the mainstay well into the nineteenth century. The dominance of the vertical wheel in the area of intensive power production, I believe, justifies my decision to concentrate on it rather than on the horizontal water wheel or any of the vertical wheel's other water-powered rivals.

One final point needs to be mentioned. The vertical water wheel does not operate in isolation. It is part of a power system. Dams and reservoirs are required to build up a workable head or fall of water and to help insure regularity of flow. Canals or mill races are needed to lead water from reservoirs or streams to the water wheel. Sluice gates, flumes, and chutes are often used to control and direct the flow of water onto the wheel. Tail races are needed to carry water away from the wheel once it has been used. Gearing, cams, cranks, and shafts transmit the motion of the water wheel to the machinery it activates. The vertical water wheel is, of course, the heart of this system, and for this reason it is the focus of this study. But because it is an intimate part of a larger system, this study does not concentrate on the prime mover to the exclusion of its all important auxiliary works. They, too, play a vital role in the story and are not ignored.

# 1

# Origins:
# The Vertical Water Wheel in
# Antiquity

## TECHNICAL BACKGROUND

The ultimate origin of water power is the sun, for water power, indirectly, is a form of solar power. It is the product of the hydrologic cycle. The water flowing in streams, the source of most water power, is renewed only through the continual evaporation of water by the sun. This process lifts water in the form of vapor from low levels (e.g., the oceans) and returns it in the form of precipitation at higher levels. The water then finds its way back into streams and begins to flow downwards to the oceans, beginning the cycle anew.

The power potential of any given stream results from a combination of meteorological, geological, and topographical conditions. The volume of water available in a stream, for example, is largely dependent on the amount of snowfall or rainfall in an area and on the size of the stream's drainage basin. The fall of water available at any given location is largely set by topography and geology. Volume and fall together determine the power available at any specific location. In mathematical terms, $P = QH$, where $P$ = potential water power available at a site, $Q$ = the volume of weight of water flowing past the site per unit of time, and $H$ = the head or fall which is available or can be developed economically. Thus, a small volume of water with a very high fall can yield as much power as a larger volume with a very small fall.[1]

Depending on the volume of water available at a particular site, its head or fall, and the amount and type (rotary or reciprocating) of power needed, any of several devices can be used to tap a stream's power potential. A wheel is the most frequently used mechanism. Water wheels, however, can come in several different forms. For example, they can be situated in either the horizontal or the vertical plane.[2] Horizontal-plane water wheels can be completely submerged and operated by reaction, or they may have only certain blades struck by the water and operate primarily by impulse. If the wheel is in the vertical plane, it can take

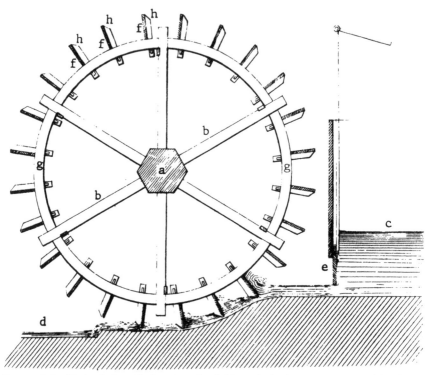

FIGURE 1–1. Undershot vertical water wheel. The principal components are (*a*) axle or shaft; (*b*) spokes or arms; (*c*) head race; (*d*) tail race; (*e*) sluice gate or chute, the device which regulates the admission of water onto the water wheel; (*f*) floats, floatboards, blades, or paddles; (*g*) rim (the circular built-up felloes to which the arms are mortised and the floatboards attached); (*h*) starts or supports, pieces of wood or metal projecting from the rims to which the floatboards or blades are secured.

water from beneath and be operated by impulse (undershot wheel); it can take water from above and be operated by weight or gravity (overshot wheel); it can take water at some intermediate point (breast wheel); it can be operated by reaction (Poncelet wheel); or it can use some combination of the preceding (e.g., an overshot-impact wheel). Water power can even be tapped without the use of a wheel by means of reciprocating devices like the water lever (discussed below), the hydraulic ram, and the water pressure engine.

For the early history of water power only certain of these devices are important: the vertical undershot and overshot wheels, a special type of vertical undershot wheel called the noria, the primitive horizontal impulse wheel, and the water lever. We will look at the operating principles of each of these engines as a necessary background for considering the origins of water power.

Probably the most important of the early engines which utilized water power was the vertical water wheel. Its two basic forms are the undershot and the overshot. The undershot vertical wheel (Fig. 1–1) rotated in the vertical plane and had a horizontal axle. It normally had flat radial blades attached to its periphery and derived its motion from the impact of water flowing under the wheel and against these blades. While capable of working on any convenient stream without mill races (narrow artificial water channels), it worked most effectively in a race and with a stable volume of water running at a fairly high velocity (say, above 5 fps [1.5 mps]). It typically delivered an efficiency of around 15–30%—that is, it was capable of converting 15–30% of the power of the water acting on it into mechanical power at the shaft of the water wheel.

The overshot vertical wheel (Fig. 1–2) was a much more efficient device. Water was fed at the top of the overshot wheel into "buckets" or containers built into the wheel's circumference, and the weight of the impounded water, rather than its impact, turned the wheel. Each "bucket" discharged its water into the tail race at the lower portion of its revolution and ascended empty to repeat the process. The overshot wheel was usually more expensive than the undershot, since a dam and an elevated head race were normally required to build up a large fall of water and to lead the water to the wheel's summit. It was suitable mainly to low water volumes (say 1–10 cfs [0.03–0.3 cms]) and moderately high falls (around 10–40 ft [3–12 m]). Depending on the care taken in building the wheel, it was capable of operating with an efficiency ranging from 50 to 70%. Thus, it could deliver roughly twice as much power as the undershot wheel from an identical volume and fall of water.

While it is conceivable that the undershot and overshot vertical watermills (a vertical water wheel with attached gears and millstones) evolved directly from the rotary quern (hand-powered rotary millstones)[3] and the tread-operated water-lifting wheels,[4] in use shortly before the dawn of the Christian era, it is more likely that their emergence was at least partially influenced by several more primitive devices which tap the power of falling water—the water lever, the noria, and the primitive horizontal watermill.

Perhaps the most primitive water-powered engine is the water lever, sometimes called a spoon tilt-hammer (Fig. 1–3).[5] Operating on the seesaw principle, the water lever utilized the power of falling water, but without the continuous rotary motion of water wheels. One end of a pivoted beam was equipped with a spoon-shaped bucket. On the other end was a hammerlike counterweight used for pounding or crushing. Water was directed into the bucket from a falling stream; the bucket filled, overweighed the hammer, and lifted it. The ascent of the bucket caused the water to spill; the hammer then overbalanced the bucket and fell. The cycle was then repeated to produce a steady pounding action. I know of no tests of this device's efficiency, but I suspect it would be under 10%.

The noria (Fig. 1–4), used for raising water, was a form of undershot water wheel, but it activated no machinery (such as gears or millstones) beyond itself. It was simply a large vertically situated wheel, sometimes

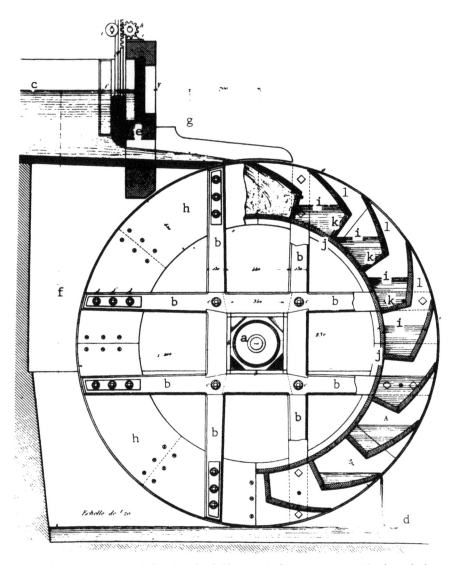

FIGURE 1–2.  Overshot vertical water wheel. The principal components are (*a*) axle or shaft; (*b*) spokes or arms; (*c*) head race (flume); (*d*) tail race; (*e*) penstock or chute; (*f*) head or fall; (*g*) impact or velocity head, the portion of an overshot wheel's fall which provides velocity to the water prior to its entrance to the wheel; (*h*) rims or shrouds, the rings which form the sides of the buckets of overshot wheels (some undershot wheels may also have shrouds); (*i*) buckets, the peripheral compartments used to retain water on the working side of the overshot wheel; (*j*) soal or sole, the boards parallel to the shaft which form the inner enclosure of the buckets and to which the buckets are attached; (*k*) rising boards, boards radial to the shaft in overshot wheels, nailed to the soal and forming the bottom of the buckets; (*l*) bucket boards, boards which form the sloping, as opposed to the radial, component of buckets in some overshot wheels.

FIGURE 1–3. Water lever. This was basically a pivoted beam with a water-retaining compartment on one end and a hammer on the other. When water flowed into the container, that end descended and the hammer was lifted. But as the container descended, the water within it flowed out, and the hammer end fell with sufficient force to pulverize grain.

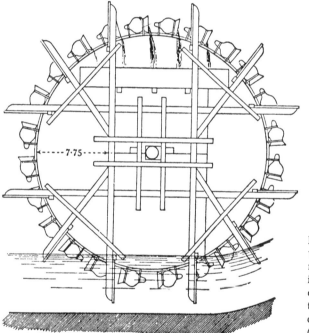

7·75

FIGURE 1–4. The noria. A vertical water wheel without machinery, the noria was powered by water striking its blades. It lifted water with pots or other water-retaining devices attached to its outer rim. The total lift of some norias approached 100 feet (30 m).

as much as 50–80 feet (15–25 m) in diameter, equipped with radial blades which rotated the apparatus as they were impacted by the flowing water in which the lower portion of the wheel was immersed. Buckets or pots of wood, bamboo, or pottery were attached to the rim of the wheel. As the device rotated, they were filled with water at the bottom of the wheel; the water was carried upwards in the buckets and emptied near the top of the wheel into a trough. The buckets were then returned empty to the bottom of the wheel to repeat the process.

The horizontal watermill (Fig. 1–5), sometimes called the Greek or Norse mill, unlike the noria, did activate "machinery:" The upper end of its vertical shaft was connected to the runner of a pair of millstones. The lower end of the same shaft carried a small horizontal "wheel" composed of paddles or blades which were successively impacted by water directed against them by a chute or race. Found most frequently on mountain streams, where the volume of water was quite small but falls high, the horizontal watermill was a relatively inefficient machine (c5–15% efficiency). It was also less sophisticated than the vertical watermill (Fig. 1–6). It did not, for instance, have gearing. So there was no way to alter the plane of rotation of the wheel's motion or to adjust the velocity of the millstones to changing stream conditions.

## ORIGINS OF THE VERTICAL WATER WHEEL

The relationships which existed between primitive water-powered prime movers like the horizontal watermill, the noria, and the water lever, on the one hand, and the more sophisticated vertical watermill, on the other, and even the very origins of water power present a puzzle which is not easily solved. This situation exists primarily because of the sparsity of early literary references to hydraulic devices and the ambiguity of many of the few which are extant.

One possible point of origin for water power is fourth century B.C. India. Joseph Needham noted in 1965 that certain ancient Indian texts dating from around 350 B.C. mentioned a *cakkavaṭṭaka* (turning wheel) which commentaries explained as *arahatta-ghaṭī-yanta* (machine with wheel-pots attached). On this basis he suggested that the machine in question was a noria and that it was the first water-powered prime mover.[6] But the term used in the ancient Indian texts is ambiguous and does not clearly indicate a water-powered device. In fact, as Thorkild Schiøler has noted, it is far more likely that these passages refer to some type of tread- or hand-operated water-lifting device, instead of a water-powered water-lifting wheel.[7]

If Needham is wrong in placing the origins of water-powered wheels in fourth century B.C. India, the next possible point of origin is the Near East around 200 B.C. The surviving manuscripts of the *Pneumatica* of Philo of Byzantium, a Greek technician who probably flourished in the late third or early second century B.C., picture three small overshot and two undershot vertical water wheels (Fig. 1–7). In two cases the small overshot wheels were used merely to produce whistling noises; in the third the overshot wheel only turned. One of Philo's undershot wheels was also very small and simply moved when a stopcock was opened. The other was larger and powered a chain of pots for raising water.[8] But like Needham's early Indian norias, Philo's water wheels are questionable.

FIGURE 1–5. Horizontal watermill. These rather inefficient mills did not utilize gearing and usually required a relatively small volume of water with a modestly high fall. Their chief asset was low first cost.

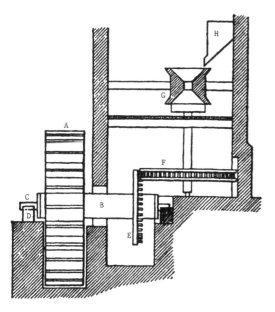

FIGURE 1–6. Vertical watermill. The principal parts are (A) vertical water wheel; (B) axle or shaft; (C) gudgeon, a metal journal mounted in the end of a wooden axle; (D) bearing, the wood or stone block on which the water wheel's axle turned; (E) vertical gear; (F) horizontal gear; (G) millstones; (H) hopper, a device for automatically feeding wheat into the millstones. This drawing shows hourglass-shaped millstones instead of the more usual form.

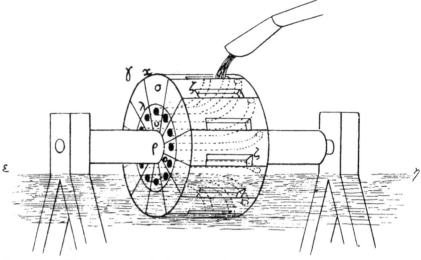

FIGURE 1–7. Water-powered whistling device from Philo. One of the water-powered devices described in surviving manuscripts of Philo of Byzantium. The overshot principle was used to turn this wheel, but its sole function was the production of whistling noises, obtained as water rushed into the revolving chambers of the partially submerged wheel and forced air out the constricted curved tubes. This wheel and the others described in Philo's *Pneumatica* were probably Arabic additions dating from more than a thousand years after Philo's time.

There is a gap of almost two centuries between Philo and the next extant reference to an undershot wheel, and a gap of around four centuries between Philo and the next clear evidence of an overshot wheel. Adding to the suspicion is the fact that the Greek original of the *Pneumatica* is lost. We are dependent on a much later Arabic manuscript for its content. Thus, two of the leading students of early technology, A. G. Drachmann and Thorkild Schiøler, believe that the water wheels in Philo's manuscript were not original, but later additions made by the Arabs.[9]

The next possible references to water-powered devices appear in the first century B.C. writings of Antipater of Thessalonica, Strabo, Lucretius, and Vitruvius. The relevant passages in their works in some cases are vague and do not permit specific identification of the prime mover involved. But all clearly refer to some form of water-powered device.

For example, Strabo in his *Geography* reviewed the spoils which Pompey captured on conquering the last of the Mithridates. Among these was the palace at Cabeira in northern Asia Minor where a "hydralatae" (water grinder) was located. The relevant passage reads: "It was at Cabeira that the palace of Mithridates was built, and also the watermill [*hydraleta*, ὑδραλέτας] and here were the zoological gardens, and, nearby, the hunting grounds, and the mines."[10] If the Cabeira watermill was built with the palace c120 B.C., this passage would contain the earliest definite evidence of the use of water power. But it is probably safe to date the watermill only back to the time that Pompey took the palace (c65 B.C.) or to much later in the century when Strabo visited the area. Exactly what Strabo meant by *hydraleta* is not certain. The device was apparently a water-powered mill, but it could conceivably have been a horizontal watermill or either an overshot or undershot vertical water-mill or a noria.[11] Specific identification is not possible.

Roughly contemporary with Strabo was the poet Antipater, who composed an ode to the watermill. This ode, like the passage in Strabo's *Geography*, clearly refers to a water-powered device, but it, too, furnishes insufficient data to permit identification of the specific type.

> Cease from grinding, ye women who toil at the mill;
> Sleep late, even if the crowing cocks announce the dawn.
> For Demeter has ordered the Nymphs to perform the work of your hands,
> And they, leaping down on the top of the wheel, turn its axle which,
> With its revolving spokes, turns the heavy concave Nisyrian millstones,
> Learning to feast on the products of Demeter without labour.[12]

If the term "leaping down" (κατα) referred to water pouring down the chute of a horizontal mill and the "spokes" (ακτινες) were the crossbars connecting its shaft with the millstones, Antipater's mill could have been a horizontal watermill.[13] But if the spokes refer to the spokes of a water wheel or the teeth of the gears, the water nymphs could conceivably have been leaping down from the head race of a vertical overshot watermill or even, allowing for the liberties poets often take in describing technical details, leaping down the steeply inclined mill race of a vertical under-shot mill.[14]

Only in the works of Lucretius and Vitruvius are water-powered devices described in sufficient length or with sufficient clarity for definite identification. In *De rerum naturae* the poet and philosopher Lucretius (96–55 B.C.) alluded to "rivers which turn wheels and buckets" in explaining the rotation of the heavens.[15] The bucketed noria, which was turned by the stream whose waters it lifted and was commonly found on rivers, was almost certainly the device Lucretius had in mind.[16] Vitruvius, writing around 25 B.C., described in even greater length and with even more detail the operation of both the noria and the vertical undershot watermill in a section dealing with rarely employed machinery:

> Wheels are used in rivers in the same way.... Round the outside, paddles are fixed, and these, when they are acted on by the current of the river, move on and cause the wheel to turn. In this fashion they draw up the water in buckets and carry it to the top without workmen to tread the wheel. Hence, being turned on by the force of the river only, they supply what is required.
>
> Mill wheels are turned on the same principle, except that at one end of the axle a toothed drum is fixed. This is placed vertically on its edge and

turns with the wheel. Adjoining this larger [smaller?] wheel there is a second toothed wheel placed horizontally by which it is gripped. Thus the teeth of the drum which is on the axle, by driving the teeth of the horizontal drum, cause the grindstones to revolve.[17]

Vitruvius' descriptions of the noria and undershot vertical mill are clear and unambiguous. The portion of the passage dealing with the mill's gearing requires some additional comment, however. In the eighteenth century it was the practice to establish gearing ratios so that the millstones turned faster than the wheel by making the vertical gear wheel larger than the horizontal gear wheel. Most early Vitruvian scholars assumed that this is what Vitruvius must have meant and amended the manuscripts to make them read in that manner. Archeological evidence, however (to be reviewed later in this chapter), supports the original version of the Vitruvian text. Roman practice was apparently to use step-down gearing rather than step-up gearing, that is, the stones revolved slower than the wheel.[18]

The earliest Chinese references to water power appear early in the first century A.D. and exhibit many of the same ambiguities as the earliest Western examples. The *Hou Han Shu,* the official history of the Han dynasty, cited Tu Shih, prefect of Nanyang, as inventor of a water-powered bellows for casing iron in 31 A.D.[19] No other technical details are provided. Another early Chinese reference to water power, dating from the same era, merely mentions the use of water power for pounding.[20] In both of these instances a number of devices are possible: the horizontal water wheel, either of the vertical water wheels, and the water lever.

Literary evidence alone is clearly insufficient to establish either the origins of water power or the genealogy of the vertical watermill. Archeology has added only a few pieces to the puzzle. The earliest surviving artifactual evidence of an undershot vertical wheel dates from first century B.C. Italy.[21] The earliest known pictorial representation of this type of wheel is a fifth century A.D. mosaic from the Great Palace of Byzantium (Fig. 1–8). The oldest representation of a noria is a mosaic found at Apamoea in Syria dating from the second century A.D.[22] The earliest definite evidence for the existence of the overshot wheel, literary or archeological, comes from a wall painting in the Roman catacombs dating from the third century A.D. (Fig. 1–9), followed by the remains of a possible overshot complex at Barbegal, France, c300 A.D., and the remains of an Athenian overshot watermill dating from the fifth century.[23] In addition, Danish excavators have unearthed traces of dams, reservoirs, and races of what apparently were two horizontal watermills dating from the beginning of the Christian era or shortly after near Bølle in Jutland.[24] No traces of the other primitive water-powered prime mover, the water lever, have been found. But because this elementary device was made of wood and did not require extensive auxiliary

FIGURE 1–8. Mosaic of a water wheel. Fifth century A.D. mosaic from the Great Palace of Byzantium depicting what appears to be an undershot vertical water wheel. If this interpretation is correct, and the wheel is not a water-lifting wheel of some sort, it is the oldest known representation of a water wheel of this type.

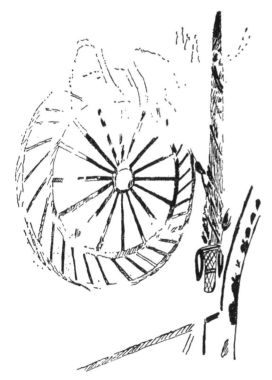

FIGURE 1–9. Drawing of an ancient overshot water wheel. This sketch is based on a Christian wall painting from the Roman catacombs which dates from the third century A.D. It provides the earliest definite evidence of the use of the overshot water wheel.

works like rotary watermills, there is little chance any traces would remain even if it was used.

Based on the surviving evidence, it would appear that the vertical undershot watermill, the horizontal watermill, and the noria appeared almost simultaneously in the Mediterranean world in the first century B.C. and that at approximately the same time some form of water-powered prime mover was developed in China. This coincidence and the paucity and ambiguity of the early evidence have led historians to postulate a variety of different relationships between these devices.

Most of the early scholars who speculated on the origins of water power—Bennett and Elton (1899); Curwen (1944); Forbes (1955–56)—argued that the horizontal and vertical watermills fell within a single sequence of development. Since the horizontal mill is a much more primitive device than the vertical watermill, they assumed that it emerged earlier, a direct offspring of the hand-operated rotary quern, and that the mills mentioned by Strabo and Antipater were of this type. The vertical mill mentioned by Vitruvius they considered to be a descendant of the horizontal mill. Neither Bennett and Elton nor Curwen considered the noria's role in this process, but Forbes did. He believed it was developed after the vertical mill, evolving from it, since the noria "implies the knowledge of the water-wheel as a prime-mover."[25] (See Table 1–1 for schematics of the genealogies postulated by Bennett and Elton, Curwen, and Forbes.)

On the other extreme was Abbot Payson Usher. Accepting the Philonian water wheels as genuine, Usher (1954) argued that the noria, horizontal mill, and vertical mill were invented in that order and were "so distinct that each type should be treated as a separate sequence of inventions."[26] Ludwig Moritz (1958) disagreed with both Usher and earlier scholars. He postulated that the horizontal mill (vertical-axle mill) was an "adaptation for household use" of the earlier, more complicated design of Vitruvius.[27]

Few historians have considered the possibility that the water lever may have entered into the early development of water-powered devices because there is no definite evidence of its existence in either early China or in classical antiquity. The ambiguity of many of the early references to water power, however, leave this possibility open, and at least one scholar has seriously considered the possibility. In the late nineteenth century James Troup (1894) observed water levers in operation in Japan and found that attempts had been made to convert their reciprocal motion to rotary motion by adding a second spoon, counterweight, and lever, and placing these at right angles to the first. Later he found what appeared to be further modifications of the water lever with six or eight spokes, operating in a manner similar to overshot wheels. These observations led Troup to speculate that perhaps the overshot vertical wheel had evolved from the water lever, and the undershot vertical wheel, from the noria.[28]

TABLE 1–1.  THE GENEALOGY OF WATER-POWERED PRIME MOVERS ACCORDING TO VARIOUS AUTHORITIES

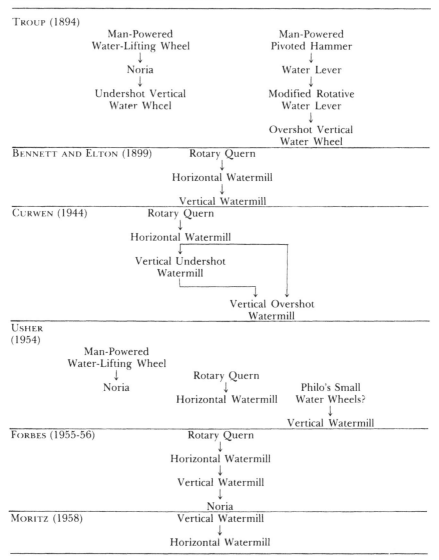

TROUP (1894)

| Man-Powered Water-Lifting Wheel | Man-Powered Pivoted Hammer |
| --- | --- |
| ↓ | ↓ |
| Noria | Water Lever |
| ↓ | ↓ |
| Undershot Vertical Water Wheel | Modified Rotative Water Lever |
|  | ↓ |
|  | Overshot Vertical Water Wheel |

BENNETT AND ELTON (1899)

Rotary Quern
↓
Horizontal Watermill
↓
Vertical Watermill

CURWEN (1944)

Rotary Quern
↓
Horizontal Watermill
↓
Vertical Undershot Watermill
↓
Vertical Overshot Watermill

USHER (1954)

Man-Powered Water-Lifting Wheel
↓
Noria

Rotary Quern
↓
Horizontal Watermill

Philo's Small Water Wheels?
↓
Vertical Watermill

FORBES (1955-56)

Rotary Quern
↓
Horizontal Watermill
↓
Vertical Watermill
↓
Noria

MORITZ (1958)

Vertical Watermill
↓
Horizontal Watermill

Joseph Needham (1965) has also speculated on the water lever's role in the emergence of water power (Table 1–2). Needham argued that the noria and horizontal watermill were separate and unrelated inventions, both because of their distinct structures (one had a vertical, the other a horizontal axle) and different tasks (one ground grain, the other lifted water). Their primitive design and the absence of gearing, Needham felt, made them likely precursors of the more complicated vertical mills.[29] The noria, Needham believed, originated in India, probably

TABLE 1–2. JOSEPH NEEDHAM'S GENEALOGY OF WATER-POWERED PRIME
MOVERS

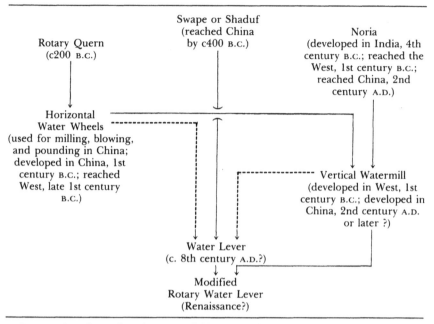

SOURCE: Based on, though not explicitly provided by, Joseph Needham, *Science and Civilisation in China*, vol. 4, pt. 2 (Cambridge, 1965).

around the fifth century B.C. It had spread to the West by the first century B.C. (as evidenced by Lucretius) and had reached China by the second century A.D. In both regions, according to Needham, it was eventually modified into the vertical watermill.[30]

Needham maintained that the horizontal water wheel came from another source and was a Chinese invention of the first century B.C. or earlier, an opinion based on the early Chinese references to water-powered metallurgical devices and the Chinese preference for horizontal wheels in later centuries (Fig. 1–10). Needham speculated that the horizontal wheel had spread from China, had reached Asia Minor by late in the first century B.C. (he felt it was the type of prime mover referred to by Strabo in his *Geography*), and had reached Jutland by the first century A.D.[31]

Because there is no unambiguous evidence to indicate the existence of the water lever before at least the twelfth century A.D., Needham rejected Troup's suggestion that it was a progenitor of the overshot water wheel. He argued, instead, that the rotative water levers observed by Troup were modifications inspired by the overshot wheel, rather than evidence that the water lever had inspired the invention of the overshot wheel.[32]

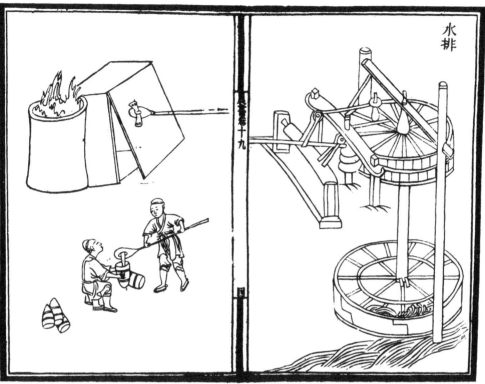

FIGURE 1–10.  Chinese water-powered metallurgical blowing engine, 1313 A.D. This is the earliest extant depiction of a Chinese water-powered bellows. Power from the horizontal water wheel (lower right) was converted from rotary to reciprocating motion for the bellows by means of crank and rocking beam.

The genealogy I have to offer for the evolution of vertical water wheels and for the evolution of other early water-powered engines differs somewhat from those offered by Usher, Forbes, Moritz, Troup, and Needham, and assumes that the course of evolution in the West and in China differed considerably. Let us first look at the probable evolution of water engines in the West (Table 1–3).

I agree with Usher and Needham that the noria and the horizontal water wheel were probably separate and independent inventions. As they have pointed out, the structures and functions of these machines were quite distinct: the vertically situated noria raised water and was probably derivative from earlier tread-operated water-raising machines; the horizontal mill ground grain and was likely an adaptation of the rotary quern. Moreover, they first appeared in widely differing cultures. The first undisputable evidence for the noria comes in the works of Lucretius and Vitruvius; the first indication of the horizontal mill, from Denmark (or possibly China).

Vitruvius clearly described the noria but did not mention the horizon-

TABLE 1–3. THE GENEALOGY OF THE VERTICAL WATER WHEEL: MEDITERRANEAN BASIN

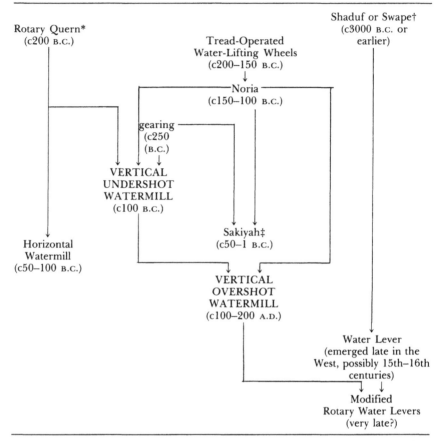

*A set of small millstones; the lower stone is stationary, while the upper stone is balanced on a pivot and rotated by hand with a projecting peg.

†A counterweighted bucket for raising water.

‡A chain of pots or wheel used for raising water and powered through right-angle gearing by man or animal.

tal wheel, suggesting, in the absence of any other solid evidence for the use of horizontal wheels in the Roman world, that he was not familiar with that engine. Arguments for a Greek, Chinese, and "barbarian" origin for the horizontal mill have been made, but no one has amassed sufficient evidence to establish a good case.[33] The best that can be said is that the horizontal watermill was invented probably sometime around the first century B.C. and probably, unlike the noria, outside of the Mediterranean basin.[34]

The noria likely antedated both the horizontal and vertical watermills.[35] Its primitive construction and the absence of gearing or associated machinery certainly seem to make it a predecessor of those mills.

Further, the earliest surviving literary allusion to the noria—Lucretius' use of it in the mid-first century B.C. to illustrate the eternal rotation of the heavens—antedated by 40 or more years both the Danish horizontal mills and the earliest definite allusions to vertical watermills. In fact, the analogy Lucretius drew between the rotation of the heavens and that of the noria suggests that it must already have been a fairly well-known implement. One may even suspect that several earlier references to water-lifting wheels—like a second century B.C. Greek papyrus from Egypt which asks for "Harpaesis and the others who are skilled in working the water wheel"—refer to norias instead of tread- or hand-operated wheels, though, as in the case of Needham's Indian water-lifting wheels, one cannot be sure.[36] I suspect, in any case, that they originated somewhere in the arid regions at the eastern end of the Mediterranean sometime around 150 to 100 B.C.

There are three good reasons for suspecting that the vertical undershot mill (a vertical water wheel plus attached machinery) evolved from the noria. First, as noted already, the surviving literary evidence places the noria slightly earlier than the undershot wheel and in the same cultural area—the Mediterranean basin. Second, Vitruvius, who first described both the noria and the vertical undershot mill in detail, explained the operation of the undershot mill in terms of the noria. This at least suggests that Vitruvius considered the undershot derivative from the noria. Finally, the vertical watermill was a more sophisticated engine than the noria. In addition to the vertical wheel and the horizontal axle it shared with the noria, the vertical undershot mill had right-angle gearing which transformed the vertical-plane rotary motion of the water wheel to the horizontal plane, where it activated the upper of a pair of millstones. The vertical undershot mill can be considered a blend of three earlier inventions—the vertical water-powered wheel of the noria, the rotary millstones of the hand-powered quern, and gearing. Of the three, the water-powered wheel was the latest invention. Rotary querns had emerged around 300 to 200 B.C.[37] Gearing had been developed as early as the third century B.C. in Hellenistic Egypt on small devices. By the first century B.C. the basic operations involved in gearing were well understood, as evidenced by the very complex gearing of certain surviving small devices like the Antikythera mechanism.[38]

Exactly where the vertical undershot mill first emerged is unclear. One possibility would certainly be the Pontus region of Asia Minor, where Strabo, writing shortly before the birth of Christ, mentioned a watermill which may have dated as far back as 65 or 120 B.C. Since norias have a long tradition in nearby Syria, and since Alexandria in Egypt, where gearing first emerged, was not far off, the modification of the noria into the vertical watermill may very well have occurred in the Syria/western Asia Minor area. From there its use would have spread westward—to Greece by Antipater's time, to Rome by Vitruvius'.[39]

The overshot vertical mill apparently emerged later than the noria, the undershot watermill, and the horizontal watermill, since the first certain evidence we have of it dates only from the third century A.D. The assumption made by most authorities is that it was a direct offshoot of the undershot mill. But some direct influence from the older noria, or even the tread-powered water-raising wheel, is probable. These devices can be seen as simply overshot wheels operated in reverse, without machinery. One can readily imagine that the basic idea for the overshot wheel came initially from an incident in which the stream powering a noria was suddenly blocked or the men working a tread-powered wheel ceased working. The weight of the water already collected in the buckets would have reversed the rotation of the engine, temporarily creating an overshot action and possibly keying the process which led to the development of the overshot wheel.

What, then, of the water lever? This machine certainly has all the aspects of primitivity. But there is no evidence that it was used or known in the Western world until well after the medieval period. It is therefore probable that it played no role in the early evolution of water power in the West and that, as Needham suggested, the modified rotative water levers observed by Troup were influenced by the overshot wheel rather than the reverse.

The evolution of the water wheel in China apparently followed quite different lines than in the West. In the West water power was first used to raise water (via the noria), then applied to grind grain, and only much later was it applied to more complicated tasks like blowing and pounding which required devices for transforming rotary motion to reciprocating motion. In China, on the other hand, the early references to water power were invariably to blowing or pounding machinery. Notice of rotary millstones driven by water, for centuries almost the only application of water power in the West (excepting the noria), not only appeared much later, but were much rarer in early Chinese literature.[40]

This divergent course of development can be explained nicely if one supposes that the early Chinese references to water power are to the water lever instead of to the horizontal water wheel (Table 1–4).[41] Both of the earliest Chinese applications of water power, moving bellows and pounding, were processes which required reciprocating motion. The action of the water lever, naturally reciprocating, lends itself to such applications far more readily than the motion of water wheels. Moreover, if the water lever was the device in use, we do not have to postulate, as Needham has, the very early existence in China of complex mechanisms for transforming rotary into rectilinear motion. Needham has claimed that the water lever has neither the force, rhythm, nor speed of action required for operating metallurgical bellows.[42] Although they were probably not as efficient as desired, water levers could conceivably have worked the blowing engines of small furnaces, and Needham himself published an illustration of a European water lever from the eigh-

TABLE 1–4. THE GENEALOGY OF THE VERTICAL WATER WHEEL: CHINA

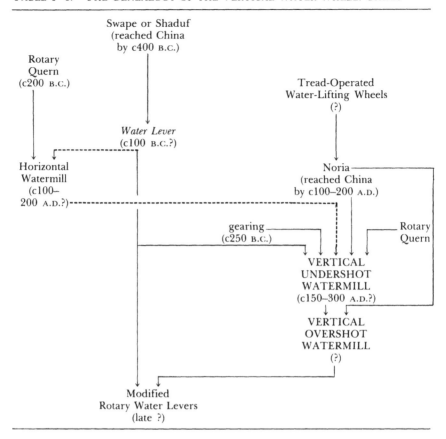

teenth century which was designed to power blowers for furnace and forge.[43]

There are several other pieces of circumstantial evidence that can be used to buttress the argument that the early Chinese references to water power refer to the water lever and not to some form of water wheel. For instance, a passage in the third century A.D. *San Kuo Chih* reads: "Han Chi, when Prefect of Lo-ling, was made Superintendent of Metallurgical Production. The old method was to use horse-power for the blowing-engines, and each picul of refined wrought (iron) took the work of a hundred horses. Man-power was also used, but that too was exceedingly expensive. So Han Chi adapted the furnace bellows to the use of ever-flowing water, and an efficiency three times greater than before was attained."[44] Needham notes that the period referred to by the text is around 238 B.C. and observes that Han Chi was from Nanyang, the city of Tu Shih, the man credited with first applying water power to bellows around 31 A.D. If one assumes, as I have, that Tu Shih's early contribution was to actuate bellows by water lever, then Han Chi's contribution may well have been adapting the bellows to operate from water wheels instead. The notice that he used "ever-flowing water" may lend some support to this interpretation, and the substitution of water wheels for water levers would have given him approximately the threefold improvement in efficiency mentioned. One can even assume that the reason metallurgical production in Han Chi's time was still largely dependent on man and animal power was because the water-lever–activated bellows of Tu Shih were not completely satisfactory and had thus not been adopted to the exclusion of animate power sources. On these assumptions, the water wheel did not come into use in China until around the third century A.D.

There is some additional coincidental evidence that supports this thesis. For example, the third century A.D., the century of Han Chi's improved water-powered bellows, is just when some early Chinese texts place the invention of the watermill (water wheels applied to rotary millstones), attributing it to Chhu Thao (c240–80 A.D.) and Wang Jung (235–306 A.D.).[45]

Moreover, it is also in the third century that Tu Yü is credited with establishing "combined" trip-hammer batteries.[46] If, as I believe, the earliest Chinese references to water-powered trip-hammers refer to devices powered by water levers, the timing of Tu Yü's contribution can be explained. Trip-hammers powered by water levers were necessarily single. The introduction into China in the third century A.D. of the vertical water wheel with its horizontal axle (probably derived from the noria, introduced into China only in the second century A.D.) would have made driving multiple hammers from the same prime mover ("combined trip-hammers") technically possible for the first time.

The near simultaneity of Han Chi's improvements to blowing engines, the introduction of rotary watermilling in China, and the development of the "combined" or multiple trip-hammers in the third century can be explained by assuming that all of these inventions were dependent on the substitution of water wheels for water levers and that the transition from water levers to water wheels occurred in China only sometime around 200 A.D.

Admittedly, my case for the water lever's importance to the early evolution of water power in China is tenuous. The earliest extant illustration of a Chinese water lever dates only from 1313 (Fig. 1–11) and the earliest fairly certain literary allusion to the device dates only from 1145, when Lou Shou, a poet, mentioned "water flowing in and out of the slippery spoon."[47] Moreover, I have relied primarily on Joseph Needham for information on Chinese water power, and he espouses a completely different theory.

But the early Chinese references to water power are so ambiguous that the water lever cannot be completely ruled out. And even though there is no certain reference to water levers before 1145, they had probably been used much earlier. The manner in which Lou Shou mentioned

槽碓

FIGURE 1–11. Chinese water lever, 1313 A.D. Earliest Chinese illustration of the water lever. The most ancient Chinese references to water power may well refer to this device instead of to the horizontal or vertical water wheels.

the water lever in his poem indicates that in his time it was well known. Moreover, several parts of the escapement mechanisms of Chinese water-powered mechanical clocks used the water-lever principle. These date back at least to the eighth and possibly to the second century.[48] All in all, the water lever offers too plausible an explanation for many of the apparent anomalies in the early development of water power in China to be ignored.

If the water lever was the earliest of all water-powered prime movers in China, what influence did it have on the emergence of other water-powered prime movers there? It probably had no influence on the development of the noria, since the noria was apparently imported into China from further west. If the horizontal water wheel was invented outside of China and only imported into China around 100–200 A.D., when it and the vertical water wheel began to replace the water lever, as I believe, its emergence, too, was probably uninfluenced by the water lever. The horizontal wheel may, however, have been independently invented in China, and if that was the case, the water lever probably had some influence on its invention.

The vertical water wheel may have been imported from the West or independently invented in China. If the latter was the case, it probably had many of the same technological progenitors in China as in the West—the noria, rotary milling, gearing. In China, however, it may have had an additional ancestor, the water lever, for undershot wheels with spoon-shaped arms similar to the spoon-shaped arm of the water lever were long popular in the Islamic world.[49]

## DIFFUSION

Data on the diffusion of the vertical water wheel in particular and of water power in general are almost as scant as data about their origins. The very paucity of the data, however, indicates that water power was not at first widely accepted. Vitruvius, in fact, listed the watermill among those "machines which are rarely employed" when he described it late in the first century B.C.,[50] and in all of the pre-Christian period in the West, only Antipater and Strabo, in addition to Vitruvius, mentioned watermills at all.

In the first century A.D. this trend continued. For that century we have only two literary references of water power in the West: a corrupt passage in Pliny (c75 A.D.) noting the use of watermills in Italy and a Talmudic complaint against the operation of watermills on the Sabbath.[51] In addition, the remnants of a vertical undershot wheel dating from late in that century have been uncovered near Pompeii.[52] More definite proof of the relative unimportance of the water wheel comes from the second-century historian Suetonius. Suetonius did not mention watermills, but records that when Caligula seized the animals of the Roman bakeries to haul booty around 40 A.D., famine threatened the city.[53] Had watermills rather than animals been grinding a significant portion of Roman grain, this would not have happened.

In the second century A.D. there are no surviving literary allusions to the water wheel at all, and the same holds true for the third century, though the remains of several mill sites from this period have been unearthed (reviewed later). Surviving evidence indicates that the watermill began to play an important economic and social role only in the fourth century. Very early in that century one of Diocletian's edicts (dated 301 A.D.) listed four types of mills and their value:[54]

| | |
|---|---|
| Horse mill | 1,500 denarii |
| Donkey mill | 1,250 denarii |
| Watermill | 2,000 denarii |
| Handmill | 250 denarii |

Because the watermill appeared third on the list, after the horse and donkey mills, but before the handmill, it may indicate that it was still less important than the first two, but had become more important than the last. Later evidence from the same century attests the increased use of water power both in Rome and in other parts of the Empire. For instance, an epitaph at Sardis from this period mentions a "water wheel engineer" (*magnganeiros hydraleta*), though this could refer to a builder of tread-powered water-raising wheels instead of water-powered ones.[55] Further, according to tradition, the Hellenized Persian Metrodorus introduced the watermill into India during the reign of Constantine early in the fourth century.[56] The *Mosella* of Ausonius (c370 A.D.) refers to watermills for cutting and polishing marble on a tributary of the Moselle, at the very bounds of the Empire, but a number of circumstances make the authenticity of this reference questionable.[57]

In Rome, too, the watermill increased in importance in the fourth century. The chief location of the city's watermills was a region called the Janiculum. Procopius, a sixth-century author, wrote of this area: "Across the Tiber, it happens that there is a great hill where all the mills of the city have been built from of old, because much water is brought by an aqueduct to the crest of the hill, and rushes thence down the incline with great force."[58] Prudentius, around 390 A.D., also mentioned these mills and their importance: "What quarter of the city can endure the dire famine, the gradus being empty? or what, the motion of the mills of the Janiculum being stopped."[59] In 398 an edict of Honorius and Arcadius promised stringent enforcement of laws designed to protect the water supply to these mills, and at approximately the same time the agricultural writer Palladius recommended the construction of water wheels on estates to lessen the load on men and animals.[60] Finally, in 369 the Roman patrician Narsius granted a watermill (*aquimolis*) to the church of San Lorenzo near Subiaco in Latium.[61]

The continued spread of the watermill in the fifth century is evidenced by archeological excavation. A mosaic picturing a water wheel unearthed at the Great Palace of Byzantium, dates from that era (see Fig. 1–8), as do the remains of an overshot mill in the Athenian agora.[62] The latter is especially significant. By the fifth century A.D. Athens was a provincial town of little real consequence. The presence of a mill at such an unimportant locale is witness to the growing use of watermills in the last years of the Roman Empire. Watermills also appear more frequently in the laws and proclamations of the fifth and sixth centuries. In 485 A.D., for example, an edict of the Byzantine emperor Zeno, confirmed by Justinian in 538, prohibited the use of public water for private watermills.[63] And among the *Pandects* of Diocletian there were several relating to watermilling.[64]

The growing importance of the watermill in the period immediately preceding the "dark ages" appears most clearly in Procopius' account of the siege of Rome by the Goths in 536 or 537 A.D. In an attempt to starve the Byzantine general Belisarius into submission, the Goths cut the aqueducts which provided water for the mills of the city. This created a major crisis, indicating how completely, in Rome at least, the watermill had displaced other types of grinding machinery. Belisarius, however, built a new type of undershot mill, the floating, or boat, mill (Fig. 1–12). The wheels on these mills were mounted between two boats anchored several feet apart at a point near a bridge where the current was swift. The Goths responded by floating trees and dead Romans down the river to damage the water wheels. But Belisarius repaired the damages and installed a large chain upstream of the floating mills to protect them from further harm. By these expedients the city was supplied with sufficient flour for the duration of the seige.[65]

Thus, by the fifth or sixth century A.D. the water wheel had gained considerable importance in the classical world, or at least in Rome. This prime mover, however, had been known since at least the first century B.C. Why did it take over five hundred years for its application to become general?

There are a number of factors which may well have combined to retard the diffusion of water power in the Mediterranean world, among them:

1. the attitude of the ruling classes towards nature and towards manual labor;
2. the economic mentality of the ancient world;
3. the Greco-Roman view of the goals and functions of technology;
4. the geography of the Mediterranean basin;
5. the ancient world's labor surplus; and
6. the nascent state of ancient hydraulic technology.

Let us look at each of these in turn.

The mental set of the ruling classes in Greco-Roman antiquity may have been a factor which contributed to the slow diffusion of the vertical water wheel. For instance, one of the underlying attitudes in antiquity was the belief that the gods ruled, appeared, and acted through nature and that human intervention into the natural order of things was improper. This attitude was reflected in the Delphic oracle's response to the Cnidians when they sought approval for their attempts to cut a canal through their peninsula. They were told by the oracle that if Zeus had wanted their peninsula to be an island, he would have made it that way.[66] The reluctance to manipulate nature displayed by this pronouncement could have inhibited the construction of the dams and artificial waterfalls often required for the development of hydropower. Another ancient mental set which may have retarded the diffusion of watermilling was the contempt for manual labor exhibited by the monied classes, a contempt perhaps stimulated by the association of work with slavery. Archimedes, for example, is said by Plutarch to have

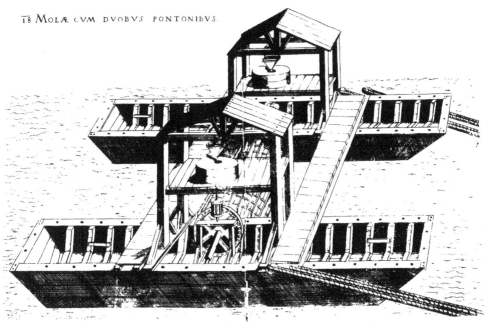

FIGURE 1–12. Boat, or floating, mill. Belisarius used watermills of this type to grind flour when the Goths cut the aqueducts which supplied the Janiculum watermills in Rome in the early sixth century A.D.

regarded "the work of an engineer and every art that ministers to the needs of life as ignoble and vulgar."[67] This contempt for manual labor may well have prevented the classes which had money for investment in mill construction from having much interest in or knowledge of devices like the watermill.

The economic mentality of the ancient world may also have impeded the adoption of devices like the water wheel. The remarks of several ancient agricultural writers, for example, indicate that the prevalent economic philosophy was antagonistic to investment in capital goods. Money not spent (hoarding) was regarded as the best investment. Estate owners were encouraged to buy up more land rather than attempt to cultivate existing land to the fullest extent.[68] With these attitudes it is not surprising that large Roman landowners preferred horse mills, donkey mills, or handmills to watermills, since they cost, according to Diocletian's schedule, respectively 500, 750, and 1,750 denarii less.

How the ancient world viewed the goals and functions of technology may also have played a role in the retarded acceptance of water power. The Greco-Roman world considered aesthetic improvement and the performance of tasks otherwise impossible or nearly impossible to be the proper domain of technology. Emphasis in improving methods of production focused on higher quality.[69] The watermill's potential, however, lay in the area of quantity rather than quality. The product it turned out, flour, was no better, and possibly even worse, than the flour turned out by hand. And the task it performed was not impossible or even nearly impossible to perform without water power, since handmills, donkey mills, and horse mills were available. The watermill's potential throughout most of antiquity lay in its ability to reduce labor and increase productivity, things not high on the value scale in the ancient world.

Although probably not a major factor, the geography of the Mediterranean basin may also have influenced the slow diffusion of vertical watermilling. Most of the rivers which flow into the Mediterranean are small and carry widely varying quantities of water. They dry up in summer, flood in winter. The few rivers carrying enough water to flow year round—for example, the Tiber or the Rhone—were too large to be harnessed with existing technology. This left the ancient hydraulic engineer with two options: (1) construction of an extensive system of auxiliary works (dams and reservoirs) on irregular streams to smooth out their discharge over the year, or (2) use of water from artificial water supply systems (aqueducts) where reservoirs and dams had already been constructed and a substantially even rate of flow already insured. That this latter option was often followed is indicated by the relatively large number of ancient water-power sites which we know were fed by aqueduct water—Venafro, Barbegal, Athens, the Janiculum (the first three sites are discussed in detail later in this chapter). Both options were probably expensive, however, and this expense likely contributed to making the watermill economically uncompetitive with man and animal power, especially in view of the labor surplus which plagued the ancient world for centuries.[70]

The ancient world's labor surplus was probably more important than geography and religious, economic, and philosophical attitudes in impeding the early widespread adoption of water power, for when labor, whether slave or free, is plentiful and cheap, there is often little economic incentive to adopt labor-saving devices like the vertical water wheel.

Two examples from antiquity indicate how labor surplus could negatively affect technological development. First, Suetonius reported that during the reign of Vespasian an engineer offered to bring some large columns to the capitol, apparently through the use of some new machine, for a small fee. Vespasian rewarded the engineer, but did not have the work carried out, saying that he did not wish to take the bread out of "poor people's mouths."[71] Second, there is a story, repeated by a number of Roman writers, that a man invented unbreakable glass. Tiberius was emperor, and the man took the invention to him, hoping for reward. The emperor asked the inventor if anyone else knew about his discovery. When he was assured that no one did, Tiberius ordered the man executed.[72] What both of these anecdotes illustrate is the concern of ancient authorities about the potential effects of technological improvements on the labor supply. In the first case, the device to transport columns would have made Rome's chronic unemployment problems worse by eliminating some of the jobs available to unskilled Roman laborers; in the second case glassmakers would have been put out of work. In both cases the response was to ignore or to suppress technological inventions which might affect the labor situation.[73] The Roman labor surplus probably retarded the introduction of the watermill just as it retarded the introduction of new devices to transport columns and unbreakable glass.

An ancient miller had several options. He could use slave (or cheap free) labor to drive handmills; he could use animal-driven mills; or he could use watermills. According to Diocletian's schedule the initial cost of handmills was far below the other two. But labor (operating) costs would have been higher. The animal mill had higher capital equipment costs, but lower labor costs since, presumably, the output per animal mill was substantially greater than that per handmill and since only one driver was needed for each set of stones turned by horse or donkey. The cost of a watermill, as indicated by Diocletian's schedule, was 30–60% higher than a horse or donkey mill. But its output was probably several times greater per set of millstones, and its labor requirements (only a single person to oversee the entire mill) were likely much lower.

In an era when labor was plentiful and cheap, the best overall investment was probably either the hand- or the animal-powered mill. It was probably impossible in most circumstances in antiquity to make up for the higher initial cost of a watermill (especially if it required auxiliary works) through lower labor (operating) expenditures before its water wheel rotted out or its dam had to be replaced. Moreover, even if we assume, as was likely, that the slave labor for a hand-milling operation and horses for horse milling were quite expensive, slave and animal labor still had advantages over water power. Animals and slaves could be bought and sold easily. They were mobile rather than fixed capital, and this may have been a significant advantage in an era when there were no proper credit instruments. Finally, the power required to adjust production to economic conditions could be more easily regulated by selling or purchasing slaves (or animals) than by buying and selling water wheels.

One other factor, an internal technological one, possibly put the water wheel at an additional disadvantage in competing against other power sources through much of antiquity and impeded its adoption. Early water wheels may have operated poorly because of the nascent state of technological know-how in the area. To document this suspicion we must review what is known about the construction of ancient water wheels.

## CONSTRUCTION

Almost as scarce as information on the genesis and diffusion of the vertical water wheel in antiquity is data on its construction. Extant writings provide practically no information. Most of the available data comes from a handful of archeological finds in Italy, southern France, Britain, and Greece, so let us turn to these next.

The earliest known remains of an ancient water wheel come from Venafro in southern Italy. In 1908 on the Tuliverno River, near Pompeii, the remains of an ancient aqueduct and the imprint of a water wheel were discovered under several feet of lava. The wheel had been covered during the eruption of Vesuvius late in the first century A.D. The porous nature of the lava had permitted the wheel to rot, but the wheel left a hollow impression in the rocks which indicated even the grain of the wood and the nails used to construct the engine. Using this impression the Museo Nazionale in Naples reconstructed it (Fig. 1–13).

The Venafro wheel was an undershot wheel. It had 18 spade-shaped paddles or blades joined by three nails apiece to two parallel rims or shrouds. The wheel was 6.07 feet (1.85 m) in diameter, approximately 1 foot (30 cm) wide, and had an overly large central hub 2.43 feet (74 cm) thick. It was powered by water from an adjacent aqueduct under a fall of around 13 feet (4 m). Luigi Jacono, an Italian engineer, estimated the power of the wheet at around 3 horses. I suspect that 1 to 2 horsepower would be a more realistic figure.[74]

The remains of another Roman undershot mill were uncovered by F. Gerald Simpson in 1907–8 near a weir placed across a small stream, Haltwhistle Burn, in northern England. Simpson found the remnants of an artificial water course leading from the stream and a mill house dated by pottery fragments to the third century A.D. The water course was apparently a mill race which took water from above the weir and led it to an undershot water wheel located at the mill house before returning it to the Burn. While no fragments of the water wheel were found (except perhaps a few nails and pieces of flat iron), fragments of several large millstones (approximately 31 in, or 78 cm, in diameter) confirmed the existence of a water-powered mill. The water course was lined with oak timbers immediately adjacent to the mill building, suggesting that the mill's wheel operated in a close-fitting trough. Some surviving segments of the trough indicated that the water wheel was probably around 14 inches (36 cm) wide. Its diameter was estimated at about 10 to 12 feet (3–3.6 m). The total fall available in the race was around 4.5 feet (1.4 m). About 3 to 3.5 feet (0.9–1.1 m) of this fall were concentrated in the 20 feet (6.1 m) immediately above the wheel, where the race sloped sharply downwards. The output of this wheel was probably between 1 and 2 horsepower, like the Venafro wheel (see Fig. 1–14).[75]

The best preserved overshot mill was that excavated in the Athenian Agora near the Valerian Wall in 1933 by Arthur W. Parsons.[76] Coins and pottery placed the construction of the mill in the reign of Leo I (457–74 A.D.) and its destruction in the late sixth century A.D. As at Venafro, water was supplied to this mill by aqueduct. Unlike Venafro, no impression of the water wheel survived. The wheel, however, was

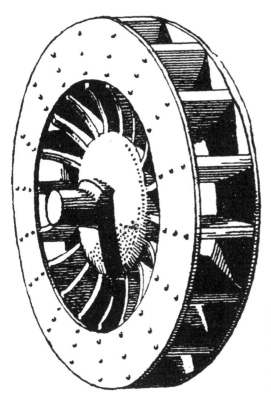

FIGURE 1–13. Venafro water wheel. Reconstruction of an undershot Roman water wheel based on impressions left in the lava that had covered it in the first century A.D. The wheel was found at Venafro, near Pompeii.

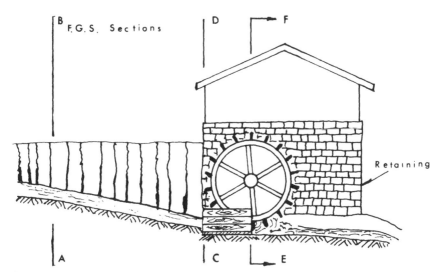

FIGURE 1–14. Roman watermill at Haltwhistle Burn (reconstruction). The undershot wheel at this site in northern England apparently rotated in a tight-fitting wooden trough.

situated in a pit 18.04 feet (5.50 m) long, 3.61 feet (1.10 m) wide, and 13.78 feet (4.20 m) deep. This pit and the race which led to it did survive and offered considerable evidence of the nature and size of the wheel used. The slotlike shape of the wheel pit, for instance, made it obvious that the mill's power source was a vertical wheel. And the way lime deposits had formed on the wall revealed "at once," according to the excavator, that the wheel was overshot, that it had received the water at the upper left-hand portion of its circumference and turned counterclockwise: "Only if this were the case could the [lime] deposit have formed [in the wheel pit] as it has, most thickly and evenly along the left perimeter of the wheel and down to the floor of the race, and again in great, irregular blobs at the lower right where the last drops were kicked off."[77]

The engineer who constructed the edifice blundered slightly. The axis of the flume or head race was slightly out of line with the pit. In order to make the wheel take the water as squarely as possible, he had to set it at a slight angle to the pit walls. As a result the rim of the wheel turned very close to the east side of the pit, leaving a series of concentric grooves in the heavy lime incrustations there. This enabled Parsons to establish rather exactly the location of the center of the wheel and its diameter, as well as other construction details.

The Athens wheel, like the Venafro wheel, had rims attached outside the spokes with nails. It was larger than the Venafro wheel: 10.63 feet (3.24 m) high, with a probable width of 1.77 feet (54 cm). Although the wheel was overshot, the designer used a combination gravity-impact design, that is, a wheel powered by both the impulse and the weight of the water. The gradient indicated by the floor tiles of the mill race where they were preserved some 50 feet (15 m) to the south of the pit would have brought the water to the wheel at a height of around 4.6 feet (1.4 m) above the wheel. Lime deposits indicated that the water was led from that height to the wheel by a steep chute (see Fig. 1–15). My calculations, based on an estimate of the quantity of water available to the mill, indicate that the wheel may have had an output of between 2 and 4 horsepower.

Impressions left on the stone and brick of the gear pit also permitted the reconstruction of the gearing and axle of this mill. The axle was a single beam around 11.5 feet (3.5 m) long, with a diameter of around 0.66 feet (20 cm). It was tapered where it rode on its bearings and strengthened by a 0.13 foot (4 cm) wide metal collar. The bearings were wooden blocks set in stone sockets. The gearing ratio, by modern standards at least, was rather unusual. The horizontal gear (4.46 ft, or 1.36 m, in diameter) was larger than the vertical gear (3.64 ft, or 1.11 m, in diameter), so that the millstones revolved at a slower rate than the wheel. The high velocity given the wheel by the large impact head might explain this anomaly.

The most impressive archeological evidence for the application of

FIGURE 1–15. Roman watermill in the Athenian Agora (reconstruction). This fifth-century watermill used reduction gearing, probably because of the high velocity given the water wheel by water falling down the steeply sloping chute.

water power in antiquity comes from southern France, at Barbegal.[78] On the summit of a hill near Barbegal, 6 miles (9.7 km) from Arles, the channel of an ancient aqueduct was split into two branches. These branches descended a slope of around 30° with a total fall of 61 feet (18.6 m). Excavations made in the late 1930s revealed that both of the branches descended the hill in eight steps or tiers (Fig. 1–16). Each of the 16 tiers had a constant fall (8.53 ft, or 2.60 m) and a constant width (7.22 ft, or 2.20 m), though their length was somewhat variable (15.58–19.68 ft, or 4.75–6.0 m). As work progressed it became clear that the edifice was not some type of storage reservior, as previously believed, but a large water-powered milling establishment. Coins and potsherds at the site, as well as the methods used to build the aqueduct, dated the mill to the late third or early fourth century A.D.

No trace of the wheels which powered the flour factory at Barbegal remained, not even imprints in the lime incrustations. Nonetheless, some suggestion of their size and type was inferred from the remains of the races and falls. The terraced arrangement suggested that vertical overshot wheels were used. Benoit, who published the first reports on the excavation, believed that the site once contained 16 of them, one for each terrace, each powering a set of millstones (Fig. 1–17). The forma-

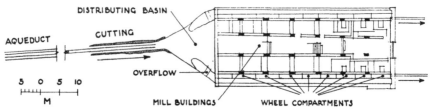

FIGURE 1–16. Ground plan of the Roman flour factory at Barbegal. The channel of an aqueduct was divided into two channels and led down a slope in two sets of eight terraced falls, providing the means for operating 16 water wheels, probably overshot.

FIGURE 1–17. Probable appearance of the Roman flour factory at Barbegal, c300 A.D.

tion of lime deposits on the side walls of several of the races permitted him to estimate the diameter of the wheels at 7.22 feet (2.20 m) and their width at around 2.30 feet (0.70 m). The floor of each race was horizontal, and incrustations indicated that a chute led the water from each tail race to the wheel at the next lowest level.

Benoit's analysis of the Barbegal wheels as overshot has been generally accepted. But it is not completely clear from the evidence he records that this was the case. Indeed, the concavity of the lime deposits reported at the bottom of the races would seem to indicate a current far swifter than that usually found in overshot races. Thus, while overshot wheels may have been used at Barbegal, there is a distinct possibility that the wheels could have been undershot, with water directed on them by steeply inclined wooden chutes.[79]

Sagui has estimated the total output of the Barbegal factory at 28 tons daily and the power of each of the Barbegal wheels at around 10 horse-

| | Venafro | Haltwhistle | Athens | Barbegal |
|---|---|---|---|---|
| Date excavated | 1908 | 1907–8 | 1933 | 1937–38 |
| Century constructed | 1st A.D. | 3rd A.D. | 5th A.D. | 3rd/4th A.D. |
| Type of wheel | undershot | undershot | overshot | overshot? |
| Number of wheels | 1 | 1 | 1 | 16 |
| Height | 6.07 ft. | c12 ft. | 10.63 ft. | 7.22 ft. |
| | (1.85 m) | (c3.7 m) | (3.24 m) | (2.20 m) |
| Width | 0.978 ft. | c1.17 ft. | 1.77 ft. | 2.30 ft. |
| | (0.298 m) | (c0.36 m) | (0.54 m) | (0.70 m) |
| Water supply | | | | |
|   Source of water | aqueduct | mill race | aqueduct | aqueduct |
|   Available head | 13.0 ft. | c4.5 ft. | 16.4 ft. | 8.53 ft. |
| | (4.0 m) | (c1.4 m) | (5.0 m) | (2.60 m) |
|   Velocity or | | | | |
|     impact head | 13.0 ft. | c4.5 ft. | 4.60 ft. | 0.98 ft.? |
| | (4.0 m) | (c1.4 m) | (1.40 m) | (0.30 m)? |
| Estimated power | | | | |
|   output | 1–2 hp | 1–2 hp | 2–4 hp | 4–8 hp |

power.[80] I suspect their output was more on the order of 4–8 horsepower. Since the population of Roman Arles was small and since the output of the flour mills was much larger than the potential consumption of flour in the region, the establishment was probably used to supply the needs of the Roman army stationed in the region. The Venafro, Haltwhistle Burn, Athens, and Barbegal wheels are compared in Table 1–5.

To these major archeological finds only a few fragments can be added. Remnants of races and millstones, though not of the wheels powering the stones, have been found at the Janiculum in Rome.[81] About all that can be determined from these finds is that the mills were probably undershot and used aqueduct water. Traces of possible watermills have also been uncovered at several locations near Hadrian's Wall in Britain besides the Haltwhistle Burn site, already noted. At Chesters Bridge a covered water course 130 feet (40 m) long may have led water to an undershot water wheel.[82] A similar water channel was uncovered at Willowford.[83] Both of these sites were dated to the second or third century A.D. The fifth-century mosaic picturing a watermill at the Great Palace of Byzantium has already been mentioned.

While archeological excavations have yielded only a little information on the way ancient water-powered wheels were constructed, some additional information can be garnered from the closely related field of water-raising wheels, where construction techniques were probably similar. As with water-powered wheels, extant literature provides us with little in the way of construction details. Vitruvius is the major exception. In discussing tread-operated water lifting wheels he noted that the axles were cylindrical and constructed either by lathe or by hand and that the ends of the axles were hooped with iron bands. He also noted that their buckets were made watertight with pitch and wax.[84]

Artifacts of water-lifting wheels are more helpful. The remains of a number of tread-operated wheels have been discovered in Roman mines, 30 in the southern Spanish mine of Rio Tinto alone. The remains of these devices indicate that both wooden dowels (tree nails) and iron nails were used in construction. The hubs of the wheels were sometimes solid oak, sometimes composite, with inner and outer discs made from oak planks strapped and pinned together like the staves of a barrel. The spokes, buckets, and rims of the wheels were made of pine. Bronze axles were used on several, and most seem to have turned on oak bearings.[85]

Although the evidence is fragmentary, it is possible by combining the extant literary and archeological evidence on ancient water-powered and water-lifting wheels to reach some tentative conclusions regarding their construction. The wheels were almost certainly made largely of wood, joined with either tree nails or iron nails. If vertical water-powered wheels were constructed like water-raising wheels, and this is likely, their hubs were of oak, while their spokes, rims, and paddles were of pine. The axles were probably constructed from oak, reinforced with metal hoops at the point where they were tapered to ride on oak or stone bearings.

Ancient overshot or gravity wheels apparently made use of only the simplest form of bucket—the inclined bucket. This type of bucket was formed by boards set between rims and inclined upward on the side the water was received (the catacomb painting). These buckets, however, were probably made watertight, as ancient water-lifting buckets were, with pitch and wax (Vitruvius) or, possibly, with a mastic made of linseed oil and manganese dioxide (evidence at Barbegal). The need for a well-directed water supply for overshot wheels was recognized. At both the Athens and Barbegal sites wooden chutes were apparently used to direct water more smoothly onto the wheels. And at both of these sites the delivery chutes may have been equipped with sliding hatches to allow the water to drop to the rear of the wheels so they could be stopped for inspection, repair, or rest.

Ancient hydraulic engineers clearly recognized that water could produce power either by its impulse (Venafro, Haltwhistle Burn, Vitruvius), by its weight (the catacomb painting, Barbegal?, Athens), or by a combination of both (Athens). They also recognized that maximum impact occurred when the water struck the wheel perpendicularly in a confined channel. This is evidenced by the race at Haltwhistle Burn and by the situation of the Athens wheel at an angle to its pit so that it could receive the perpendicular impact of the water.

The Barbegal site is clearly the one that shows off the abilities of ancient water power engineers at their best. It would have been impossible to make use of the volume and head of water available there with a single wheel. So the water channel was split in two, and two sets of eight terraced falls were constructed. Moreover, the installation was placed at a point on the hill where a pocket of earth permitted the deepening of the lowest terraces and the construction of the last tier partly in subsoil. This added more than 3 feet (1 m) to the available fall. The 16 wheels placed on the slope made much better use of the available power than any single wheel possibly could have.

But there is scattered evidence of serious deficiencies in ancient water power technology, deficiencies due, no doubt, to the nascent state of experience in the area. For instance, the hub of the Venafro wheel, the only ancient water wheel where an impression of the hub has survived, was very large and heavy, so large and heavy that the wheel's effective operation may have been impaired. The very large hub suggests that ancient methods of joining the spokes to the axles of water wheels may have been deficient. If so, the output and reliability of ancient watermills would have been low.

Possibly related to this problem was the use of rims on undershot wheels. If the two examples of undershot construction from antiquity where the presence or absence of rims can be determined (Venafro and the Byzantine mosaic) were typical, ancient undershot wheels were normally equipped with rims. Rims on undershot wheels have the advantage of giving a wheel greater stability, but the disadvantage of impeding the exit of water from the wheel after impact. There are two possible explanations for the use of rims on ancient undershot wheels. First, if linking spokes to axles was a problem in antiquity, as posited above, the greater stability provided by the use of rims would be important. Second, early undershot wheels may have been equipped with rims in imitation of their predecessors, the compartmentalized, tread-operated water-lifting wheels, without realization that the freer exit of water after impact permitted by rimless wheels was important to efficient operation.

The nascent state of technological know-how on water power is suggested, also, by the available evidence on ancient overshot wheels. The catacomb painting indicates that ancient overshot wheels made use of inclined buckets, a rather inefficient form which does not retain water well on the lower portion of its revolution. Moreover, ancient millwrights did not recognize that gravity was more efficient than impulse as a motive force. In the overshot wheel at Athens, for instance, the water should, for maximum efficiency, have been given a very moderate fall before it impacted the wheel (say around 6 in, or 15 cm). But the impact head was around 4.6 feet (1.4 m), 30% of the total fall. The higher impact head gave the wheel a higher rotational velocity but reduced its power output.

High rotational velocities, in fact, may have been common to Roman watermills, so common that step-down gearing was used in milling, instead of the more conventional step-up gearing of later times. There is evidence, for example, that, correctly translated, the watermill described by Vitruvius used step-down gearing. The remnants of the mill at

Athens definitely yielded evidence of step-down gearing. And the Venafro wheel apparently rotated at very high velocities. It is easy to confuse velocity with power,[86] and this confusion may well have been common among Greek and Roman water-power technicians, seriously diminishing the effectiveness of the installations they erected.

The large hub on the Venafro wheel, the rims apparently used on ancient undershot water wheels, the mis-setting of the Athens wheel, the employment of inefficient inclined buckets on overshot wheels, and the use of excessive rotational velocities indicate that Roman technological knowledge in the area of hydropower may have been seriously incomplete. In conjunction with prevailing attitudes, economic conditions in the ancient world, and the geography of the Mediterranean basin, this may well have contributed to the slow adoption of water power.

A water wheel, of course, does not operate in isolation. It must have supporting facilities, such as dams, reservoirs, and mill races, especially in areas, like the Mediterranean basin, where stream-flow conditions are sporadic. Unfortunately, other than the Jutland excavation, which uncovered remnants of dams, reservoirs, and races used with horizontal wheels, little evidence has been discovered of hydropower dams and reservoirs in the ancient world. At first glance, one might be tempted to suggest (1) that this was because Roman hydraulic engineers did not know how to dam, store, and direct natural streams and (2) that their body of technical know-how was even more deficient in the area of auxiliary hydropower elements than in the construction of the prime mover itself.

These presumptions would be false. Roman engineers certainly had the ability to construct storage dams and reservoirs. They erected a number of very large storage dams for water supply in Spain, for example.[87] And the Roman aqueduct system was, in effect, a combination dam-reservoir-canal system which could be, and sometimes was, adapted to the production of water power. In fact, most of the watermills whose remains have been discovered—the mills at Venafro, Barbegal, Athens, and the Janiculum—were dependent on aqueduct water.

Why was ancient watermilling so strongly dependent on aqueduct supplies? The primary reason was probably economic. The construction of a dam, storage reservoir, and canal network sufficient to build up a fair head of water and insure regularity of flow in the small and flashy streams typical of the Mediterranean basin would have required a large investment of capital. An aqueduct provided a ready-made fall and reasonably steady year-round flow of water. Even if watermills using aqueducts were required to pay a rental fee, and even if their use of water was sometimes restricted by the needs of the urban population in summer months, the use of aqueducts was probably a better option in much of the Roman era than construction of an independent water regulation system. Aqueduct supplies, moreover, had another advantage. Because they had been designed for urban water supply, they were usually close to cities, the natural market for goods produced by a mill, an important factor in Greco-Roman antiquity, when overland freight rates were high.

The use of aqueducts for watermills can be regarded as a logical first step in the emergence of water power. The aqueduct was, in effect, a small, regularly flowing artificial stream. Because the volume of water it could provide water wheels was small and because its flow was already regulated, the skill and know-how required of the infant technology of water power did not have to be high for success to be achieved in the construction of watermills dependent on aqueduct water. Moreover, in the centuries immediately following the introduction of the water wheel, the demand for water power was slight. The economic advantages of man and animal power were too great. Aqueducts were quite sufficient to supply this rather small and occasional demand. They would become inadequate, forcing watermillers to turn to more powerful and more troublesome natural streams, only as the conditions which had long retarded the wider adoption of the water wheel began to disappear.

At least some of these hindrances began to disappear in the last century or two of the Roman era. For example, Christianity, a religion whose attitude towards nature was much more aggressive and radically different from that of most other religions, replaced the paganism and animism previously dominant in the Empire. As the Empire ceased to expand, the supply of new slaves dwindled, and slavery became an increasingly less significant factor in the economy. The decline of slavery and a generally declining population slowly transformed classical antiquity's labor surplus into a labor shortage. This labor shortage was quite serious by the fourth century. It led the anonymous author of the military treatise *De rebus bellicus* to urge the emperor to introduce a number of new labor-saving war machines into the Roman army.[88] It also had its effect on the emergence of water power, for the Roman agricultural writer Palladius in the fourth century recommended the installation of watermills on country estates because of his concern about the growing shortage of labor.[89]

The centuries that saw social and economic conditions begin to make water power a viable alternative in certain areas to man and animal power also provide evidence suggesting that the Romans had overcome earlier deficiencies and developed considerable sophistication in water-power engineering. This increased sophistication in late Roman hydro-power technology may be signaled by the possible use of water power for cutting marble on the Moselle in the late fourth century.[90] It is definitely indicated by other evidence, such as the use of double water channels in eight tiers to power 16 wheels at the Barbegal plant, c300 A.D. Another indication of growing Roman sophistication in water wheel design comes from the anonymous author of *De rebus bellicus* late in the fourth century. He suggested reversing the operation of the undershot water wheel. Instead of using the motion of water against this wheel to produce power, he designed a system which used ox power, through gearing, to provide power to the wheel, turning it against water to propel naval vessels.[91] Finally, Belisarius' boat mills on the Tiber c536–37 A.D. indicate that Roman (or Byzantine) engineering was sufficiently advanced to begin to tackle the problem of tapping the power of large streams.

Thus, by the time the Roman Empire completely collapsed in the West, the foundations had been laid for the more extensive and diversified use of water power.

# 2

# Diffusion and Diversification: The Water Wheel in the Medieval Period, c500 to c1500

In 1940, in an essay entitled "Technology and Invention in the Middle Ages," Lynn White suggested that the West during the medieval period had experienced a power revolution.[1] Briefly in this essay, and more extensively in later works, he argued that the medieval epoch was "decisive" in the history of attempts to tap the forces of nature and that, particularly from the eleventh century on, there was a rapid replacement of human by nonhuman energy sources "whenever great quantities of power were needed or where the required motion was so simple and monotonous that a man could be replaced by a mechanism." "The chief glory of the later Middle Ages," he concluded, "was not its cathedrals or its epics or its scholasticism: it was the building for the first time in history of a complex civilization which rested not on the backs of sweating slaves, or coolies but primarily on non-human power."[2]

White supported this contention by outlining medieval activities in a number of areas involving the production, control, and use of power. He pointed out, for example, that the efficiency of horse power was considerably augmented by medieval innovations. The introduction of the nailed horseshoe and the padded horse collar around the ninth century quadrupled the horse's effectiveness as a prime mover. Another element of the medieval power revolution was the interest shown in the force of expanding vapors and gases. Medieval technicians attempted to tap the power of steam, hot air, and compressed air and made considerable progress in the development of apparatus to use gunpowder. Medieval European craftsmen also endeavored to make use of the power of gravity. They were successful in tapping this means of storing energy, as White noted, in the weight-driven mechanical clock and in the trebuchet, the characteristic medieval projectile-throwing engine. Another element of the medieval power revolution was the windmill. Invented in the twelfth century, the Western vertical windmill (a windmill whose blades rotate in the vertical plane) quickly became a major prime mover in regions where winds were relatively dependable. Already by the thirteenth century 120 wind-powered mills could be found around Ypers in Flanders, and by the early fourteenth century a British monastic chronicler was complaining that the search for timbers for windmill vanes was a major cause of deforestation.

White, however, properly gave primacy of place in the medieval power revolution to the water wheel. One of the ways in which the water wheel was important was as a parent technology. Very likely the invention of the Western vertical windmill was inspired by the widely used vertical watermill, for the Western windmill adopted the vertical watermill's horizontal axle, gearing, and machinery. The vertical windmill differed from its parent primarily in having its blades adapted to air instead of to water.[3] The water wheel, however, was important as more than just a parent.

Of all the inanimate sources of power developed in the Middle Ages, the water wheel was, by far, the most important. In the centuries following the demise of the Roman Empire in the West, the water wheel spread from a few small pockets to practically every region of the European continent. In antiquity the watermill had usually been tied to aqueducts; in the medieval period water wheels were put into operation on streams of every size, from small mountain brooks to substantial rivers; they were even put to work on tidal inlets. In antiquity water wheels had been used primarily for flour milling. By 1500 they were being applied to a host of different industrial processes, some on a very substantial scale. For a variety of reasons medieval European society enthusiastically adopted water power and incorporated it into the predominant feudal-manorial system. If the ancient world gave birth to the vertical water wheel and nurtured the earliest stages of its growth, it was the medieval West that brought it through adolescence and into adulthood.

In the sections which follow I will review in more detail the general trends in the medieval development of hydropower intimated in the preceding paragraph: (1) geographical expansion, (2) numerical expansion, (3) development of power on large streams, and (4) industrial diversification. I will then examine surviving evidence on the construction of medieval water wheels, review how the watermill was incorporated into medieval European society, and discuss the social factors which encouraged the unprecedented growth of water power in medieval Europe. Finally, to set European achievements in hydropower into a wider context, I will summarize the accomplishments of China, Islam, and Byzantium in developing water power during the same era.


THE SPREAD OF WATER POWER

As we have seen, in classical antiquity waterpower was not extensively used. During the Middle Ages, however, the use of waterpower spread to nearly every corner of Europe, and it did this in spite of barbarian incursions and the general political and economic uncertainties of the era.[4]

This diffusion of the water wheel seems to have followed a definite pattern. Watermilling apparently survived the collapse of the Roman Empire in a few pockets in southern France and Italy, definitely around urban areas like Rome, and perhaps around a few monastic centers. From these pockets the vertical wheel diffused outwards. This process of diffusion accelerated during the eighth and ninth centuries, primarily within the bounds of the Frankish Empire and probably because of the relative peace and stability which came with the growth of Frankish power. But in the twelfth century the watermill spread well beyond the limits of the by now defunct Frankish kingdom.

Data are so sparse, however, that a second possible pattern of diffusion cannot be ruled out. If Anders Jespersen is right in dating the remnants of a horizontal milling establishment in Jutland (Denmark) to around the first century A.D., there may have been a parallel milling tradition using the horizontal water wheel outside of the Roman Empire, and this milling tradition could have spread southwards and eastwards at the same time that the Roman vertical milling tradition was spreading north, east, and west out of the Mediterranean basin. This theory, however, has serious deficiencies. Other than Jespersen's discovery, there is no evidence for a "barbarian" watermilling tradition. And, sparse as the documentary evidence is, it is fairly consistent in pointing to a steady northward diffusion of watermilling from surviving pockets in the Roman heartland. Let us turn now to this data.

In Italy and France the barbarian invasions of the fourth, fifth, and sixth centuries apparently did not completely destroy the few pockets where watermills were already in use. In Italy, for example, laws relating to watermills were enacted by Theodoric the Great (475–526), king of the Ostrogoths, and watermills are mentioned by Cassidorius (d. 575), one of his officials.[5] There are indications that a watermill was one of the first facilities built by Saint Columban (d. 615) at Bobio, along with other evidence of watermilling from the same era.[6] Further, Paul Aebischer, who studied the occurrence of Latin terms for "watermill" in medieval Italian documents, found that by the eighth century watermills were known in Tuscany, Latium, and Lombardy, and by the ninth in the provinces of Emilia, Piedmont-Liguria, and Compania.[7] By the mid-tenth century they were commonplace through most of Italy.[8]

France presents a similar picture. Watermilling apparently survived the barbarian incursions there as in Italy. Gregory of Tours (538–94) noted a watermill at Dijon and a monastic mill at Loches,[9] and the poet Venantius Fortunatus, at about the same time, wrote of a watermill on the Moselle.[10] Moreover, watermills were regulated by sixth-century Visigothic and Salic law codes.[11] Under Carologinian rule the watermill spread to all parts of France. For example, in the ninth century, documents indicate it at the abbeys of Saint Bertin (Pas-de-Calais), Gellone

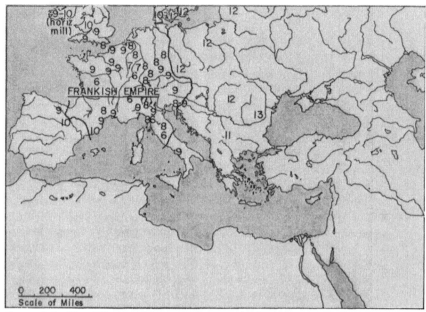

FIGURE 2–1. The diffusion of the vertical water wheel, c500 to c1300. The numerals indicate the century in which water wheels are documented in a given region.

(Hérault), Corbie (Somme), and Saint-Germain des Prés (Paris), Narbonne (Aude), and elsewhere.[12] From the mid-tenth century on, according to Blaine, "the appearance of mills in French cartularies and other relevant documents is the rule rather than the exception."[13]

From southern France and Italy the watermill spread relatively quickly to neighboring regions. The earliest known Swiss mill, one near Geneva, dates from the sixth century.[14] Water wheels were in use by as early as the middle of the seventh century in modern Belgium and by the eighth century in Holland.[15] Watermills were known in western Germany by the seventh century (Trier).[16] In the eighth century they appeared in Swabian and Alaman law codes[17] and were in operation in central and eastern Germany.[18] The earliest Austrian watermill dates from the ninth century.[19] The Alpine Slavs were also apparently using watermills by the eighth or ninth century.[20]

The map of Figure 2–1 indicates a very close correlation between the boundaries of the Frankish Empire and those of the water wheel around 800 to 850. In fact, the only non-Frankish areas where there is firm evidence of the use of water wheels prior to 900 are areas immediately adjacent to the Frankish domain, like Kent in England and extreme northern Spain. This suggests that the early diffusion of the watermill from its bases in France and Italy was made possible, at least in part, by Frankish conquests and settlements and by the relative peace imposed by the Franks over large portions of early medieval Europe.[21]

Outside of the Frankish realm the watermill was adopted more slowly if surviving documents present an accurate picture. In Jutland, for example, other than the archeological evidence for horizontal mills dating from around the time of Christ (Chap. 1), there is no other indication of the use of water power until approximately 1000 A.D.[22] For the main Danish island of Sjaelland, the earliest surviving evidence of the use of water power comes only from 1140; for the Scandinavian peninsula the earliest evidence also comes from the mid-twelfth century.[23] At about the same time the watermill first reached Poland. The earliest documents mentioning the use of water power there date from 1145 and 1149.[24] In Bohemia and on the Baltic coast watermills also first appear in twelfth-century documents, and the same applies to Transylvanian Rumania.[25] The watermill had reached the Balkans somewhat earlier, however, for in 1098 the Crusaders burned several floating mills at Niš, in the extreme eastern part of modern Yugoslavia.[26] By the thirteenth century the watermill was common in Poland,[27] and was known south of the Carpathians in Rumania[28] and in both northeast and southwest Russia.[29]

The diffusion of water power was apparently somewhat faster west and south of the Frankish kingdom. For Britain there is no evidence of watermills between the collapse of the Roman Empire and the eighth century, when a document of King Ethelbert of Kent (762) mentions a watermill.[30] But by the ninth century the watermill was definitely known over wide areas of England,[31] and by the tenth century it was in use in Wales.[32] Irish watermilling was under way by at least the ninth or tenth century, with strong indications of earlier use.[33] But in Ireland, the horizontal wheel, not the vertical wheel predominated. The earliest known Spanish watermills also date from the ninth and tenth centuries, at Santander, Barcelona, and Burgos.[34]

In summary, the surviving evidence on the diffusion of water power in medieval Europe seems to indicate that it spread outward from a few surviving pockets in Italy and southern France. The relative rapidity of this diffusion supports White's thesis of a medieval power revolution. In the period of approximately 600 years between the invention of the water wheel and the complete collapse of the Roman Empire in the West (c100 B.C. to c500 A.D.) the use of water power had apparently not spread beyond a few sharply restricted areas. In the 700 years between 500 and 1200 A.D., in spite of long periods of political and economic chaos, the water wheel spread over the entire European continent. By the thirteenth century it was known and in use from Spain to Sweden, from Britain to Bulgaria, from Rome to Russia.

## NUMERICAL GROWTH

In the medieval period the use of water power not only spread over the entire continent; it increased quantitatively as well. There were certainly many more water wheels in use c1500 than there had been c500. Here too, however, evidence is scrappy. Most extant medieval documents do not provide reliable statistical data, and there have been few detailed regional studies of even the medieval documents which are extant. But there are some exceptions, and these documents and studies lend considerable support to White's contention that medieval Europe experienced a major power revolution.

One of the most striking indications of this power revolution is provided by the Domesday Book (1080–86), the results of a census ordered by William the Conqueror shortly after the Norman conquest of England. Among the items enumerated in the census were mills. Domesday listed 5,624 watermills in 3,000 different locations. As Figure 2–2 indicates, most small streams in southern and eastern England were covered with mills. In many areas they were placed less than a mile (1.6 km) apart; in some areas there were as many as 30 mills in 10 miles (16 km). Over all of England there was an average of one watermill for every 50 households.[35] And, since there is no reason to believe that England was technologically ahead of the continent in the eleventh century, it is quite possible that watermills were in even heavier use by that date in some parts of Europe.

There are some estimates and local figures for the use of water power in the medieval period on the Continent, though no data as broad or as comprehensive as that provided by Domesday for England. Toulouse around 1100, for example, had 60 floating mills. These were replaced in the twelfth century by 43 fixed mills.[36] Paris at the beginning of the fourteenth century had around 70 watermills on about one mile (1.6 km) of the Seine.[37] In the sixteenth century from Poitiers to the confluence of the Saint-Benoit River with the Vienne River there were 24 mills, and the Abbey of Saint-Maxient in Poitou possessed 20.[38] Dembińska estimated that Poland in the late medieval period had approximately 1,000 watermills.[39]

These figures do not indicate the rate of expansion, but some local French figures do. Only two mills are mentioned as being on the banks of the Robec, a minor tributary which joined the Seine at Rouen, in tenth-century documents, but five in twelfth, and ten in thirteenth. Twelve watermills were in operation there by c1300.[40] On the outskirts of Rouen the number of mills grew from one in the eleventh century to five in the twelfth and six in the thirteenth century.[41] In the Forez district there was only a single watermill in the early twelfth century, but nearly 80 by the thirteenth; in Aube, 14 mills were mentioned in extant records from the eleventh century, but 60 in records from the twelfth century, and nearly 200 watermills were in the region by the early part of the thirteenth century.[42] In Picardy, Duby reported 40 watermills by 1080, 80 watermills by 1125, and 245 by 1175.[43] The area around Troyes had 11 watermills on the Seine and Meldancon rivers in the late twelfth century, but by the late fifteenth century the district had 20 corn

FIGURE 2–2. Map of England with location of Domesday watermills.

mills, 14 paper mills, 2 tanning mills, 4 fulling mills, and 1 cloth mill, for a total of 41.[44] The Aubette River had one lone mill in the eleventh century; there were three by the twelfth and six by the thirteenth century.[45] This data is summarized in Table 2–1.

These data suggest that the greatest numerical expansion of the water wheel in medieval Europe took place between the twelfth and thirteenth centuries, say 1150 to 1250, an era which a few other, less quantitative studies of specific regions in France (Dauphiné, Provence, Yonne) also point to as a period of rapid expansion for watermills.[46] Europe in this century enjoyed a long period of peace and prosperity. Population grew

| | Aube | Aubette | Forez | Picardy | Robec (at Rouen) | Rouen (outskirts) | Troyes |
|---|---|---|---|---|---|---|---|
| | | | | Number of watermills | | | |
| 10th century | | | | | 2 | | |
| 11th century | 14 | 1 | | 40 (1080) | | 1 | |
| 12th century | 60 | 3 | 1 | 80 (1125) | 5 | 5 | 11 |
| 13th century | 200 | 6 | 80 | 245 (1180) | 10 | 6 | |
| 14th century | | | | | 12 (early) | | |
| 15th century | | | | | | | 41 |

by approximately 35%, the largest increase for any century in the medieval period.[47] The increase demands of a prosperous and growing people may well have stimulated a significant increase in the utilization of water power in that epoch.

Unfortunately several of the critical parameters needed to evaluate quantitatively the full extent of medieval Europe's accomplishments in the utilization of water power are missing. Despite the statistical data from Domesday and the scattered data from a few detailed studies of French provinces, we have no way of knowing the total number of water wheels in operation at the beginning and end of the medieval period. We have no way of determining the total horsepower they generated. Similarly, we have no solid data to establish just how much power could be concentrated for industrial purposes at a single water-power site at either the beginning or the end of the epoch.

Yet, when compared to classical antiquity, the numerical growth of watermilling implied by the fragmentary data we do have is striking. For instance, in just the Aube region of France, in just the eleventh century, we can identify 14 different watermills, more than we can identify in all of the Roman Empire in the entire Greco-Roman period. And England in the late eleventh century had almost 6,000 watermills. The scattered figures available on hydropower in Europe in the medieval period may not provide adequate quantitative data and may in part reflect only the survival of a larger number of documents from later centuries instead of the growing use of water power. But the figures which are available certainly hint of a medieval revolution in the use of water power, even though they are not sufficient to prove it conclusively.

## FROM AQUEDUCTS TO RIVERS

A stronger demonstration of the growth of hydropower in medieval Europe comes from evidence that medieval hydraulic engineers made extensive and widespread efforts to harness the water power of medium to large streams.

This demonstration is based on the analogy between the development of water power in early America and in medieval Europe. Louis Hunter, in reviewing the history of water power in America, found that most early American watermills were located primarily on small headwaters and had only a very small power output. As the American economy grew and became more dependent on water power, American millwrights and engineers began to construct mills on larger streams and on small rivers, where more power could be developed. This trend eventually culminated in the harnessing of major rivers, like the Connecticut and the Merrimack, in the mid-nineteenth century.[48] Data on the growth of water power is not as abundant for medieval Europe as for early America, but an analogous trend is clearly visible. During the medieval period water wheels moved from aqueducts, where they had often been confined in antiquity, and from small brooks to larger streams and eventually to rivers as European hydraulic engineers modified the vertical water wheel and developed appropriate auxiliary works, like hydropower dams and power canals. Thus, the widespread and successful efforts made by European technicians to adapt the vertical water wheel to operation on large streams indicates, by analogy to developments in America, a growing industrial dependence on water power.

In the late Roman and early medieval period efforts were already underway to modify the vertical undershot wheel for operation on large and troublesome streams. In its primitive form the undershot wheel was inefficient and cumbersome. Rising water drowned the machine; falling water levels left it high and dry. Often the banks of a river were unsuited for the construction of undershot mills, either because of urban congestion or because of topography. And all to frequently the flow of water was not rapid enough in large streams to deliver sufficient power to the wheel. In other words, the use of conventional undershot wheels on large streams was plagued with problems. It is thus not surprising that its use was often confined to small, artificial streams (aqueducts) during antiquity.

What ultimately emerged from medieval attempts to adapt the undershot wheel to large, natural streams was the hydropower dam, a major medieval technological achievement and one which made it possible for medieval engineers to begin to tap the power of fairly large rivers. It is not likely, however, that the hydropower river dam emerged fully developed on the European scene. Instead, it probably evolved in three distinct stages:

1. Late Roman or early medieval technicians mounted the primitive undershot wheel on boats, creating the boat mill, which could operate continuously in spite of the variable flow conditions of large natural streams;

2. Medieval technicians found that certain man-made structures, notably bridges, modified the natural flow of streams in a manner

le chastel du kahaire    le chite de babylone

h gardin du baume    h fleuue du fuce z les molins qui sui

FIGURE 2–3. "Mills of Babylon" (boat, or floating, mills) from a fourteenth-century manuscript.

advantageous to power production and began to take advantage of these modifications with mills anchored near bridges or attached to bridges (bridge mills);

3. Medieval technicians began to make deliberate efforts to modify flow conditions in streams specifically for power production purposes. In this stage preexisting European irrigation or water-supply dams probably contributed to the emergence of the necessary conceptions and technology.

The boat mill was the earliest modification of the conventional vertical wheel to enable medieval Europeans to tap the power of rivers with some degree of economy and reliability. A boat mill, of course, had no trouble adapting itself to changes in water level or crowded bank conditions. And the low velocity of large rivers on which it was situated could frequently be augmented by mooring the mill at a point where the stream was swifter than usual because of narrow channels formed by islands or sandbars. A better option was to moor the mill between the piers of a bridge or just beyond a bridge. Mooring boat mills close to bridges was a very popular practice in the medieval period. A stream's velocity was frequently high near bridges, since much of the channel's width was obstructed by the cumbersome, thick piers and short arch spans of medieval bridges.

The first record of the floating or boat mill comes from the Gothic siege of Rome in 536–37 A.D. (see Chap. 1).[49] There was a continuous tradition of boat-milling in medieval Europe from this date on. It was probably mills of the floating type that were noted by Marius, bishop of Avenches, in 563, when he cited a bridge with mills at Geneva.[50] The *Chronicles* of Saint Martin of Trier in the mid-eleventh century mentioned a barge mill.[51] And the Crusaders burned seven floating mills at Niš in 1098.[52] The earliest medieval illustration of a boat mill comes

FIGURE 2–4. Boat, or floating, mills on the Rhone at Lyons, c1550. These mills were moored to the shore in a narrow channel formed by an island in the river where the current was swifter than in open stream.

from the fourteenth-century French manuscript the *Romain d'Alexandre*, which depicts a pair, calling them "Mills of Babylon" (Fig. 2–3). Toulouse in the late twelfth century had 60. Paris at one time had 70 on about one mile (1.6 km) of the Seine.[53] Boat mills had become so numerous by 1570 that they were being cited as serious obstructions to river traffic in French documents.[54] (See Fig. 2–4.)

Two varieties of boat or floating mill developed in the medieval West. On the Seine single-hull mills with a wheel on each side dominated. In most other areas of Europe floating mills had double hulls with a single,

FIGURE 2–5. Bridge mills under the Grand Pont in Paris, from a French manuscript of 1317.

large wheel suspended between. The single-wheel, double-hulled mill was the more efficient and powerful of the two. The double hull channeled water onto the wheel and allowed the use of very large wheels, often around 25 feet (8 m) wide. The two wheels of the single-hull mill received less impulse from the water since they were not enclosed in a channel, and since the prow of the hull divided the stream, what impact they did receive was oblique and therefore less effective. Without a channel, the single-hull mill could have no control gates and no way to regulate the amount of water impinging on its wheels. To stop the mill a friction brake, expensive and often dangerous to handle, had to be used. Finally, the single-hull mill was less stable than the double-hull mill owing to its narrower base.[55]

While useful, the boat mill, in either single- or double-hull form, was only a partial solution to the problem of adapting the undershot wheel to large rivers. Because of the inherent instability of their situation (on water), the stones of boat mills were never set as well as on fixed mills, and the grain they ground was of lower quality. Their power output and efficiency were low. Access was difficult. They were, in addition, a constant hazard to navigation, especially since their mooring ropes were sometimes cut by ice in cold weather, setting them adrift and out of control.

The practice of anchoring or mooring boat mills near or in the openings of bridges, however, probably provided the stimulus for a more sophisticated medieval attempt to adapt the vertical water wheel to large rivers—the bridge mill, the second step in the evolution of European hydropower dams. A bridge mill is a mill whose entire edifice—wheel, mill house, gearing, stones, etc.—is directly connected in some way to the superstructure of a bridge. Marius' "bridge with mills" at Geneva, c563, may have been a mill of this type, although it was probably simply a bridge with boat mills anchored nearby. The earliest definite evidence of a bridge mill comes, not from Christian Europe, but from Moslem Spain. The Arabic geographer, al-Idrisi, in the mid-twelfth century described a bridge mill at Córdoba which had several sets of wheels and stones.[56] Shortly after al-Idrisi's report, however, there is evidence of similar structures farther north. In 1175, for example, a place named Moulin-du-Pont existed in the Côte-d'Or region of France, suggesting a bridge mill.[57] A French manuscript of 1317 pictured three such mills under the arches of a Paris bridge, each having one axle set on a masonry pier of the bridge, the other set on pilings which also supported the mill house (Fig. 2–5). From the reign of Louis XI (1461–83) there is a plan for the bridge mills at Corbeil (Seine-et-Oisne). The plan depicts three water wheels under three arches, with sluice gates to control the flow of water (Fig. 2–6).[58]

Bridge mills, unlike boat mills, required some mechanism for adjusting the position of the wheel to changing water levels. The extant medieval illustrations of bridge mills do not provide sufficient detail to establish exactly what type of mechanism was used. But there are drawings of bridge mills from shortly after 1600 which do indicate such devices. Frequently the mill was suspended from the bridge by chains which could be adjusted to changing flow conditions. In view of the simplicity of this and other methods (see Fig. 2–7), we may suppose that similar devices were also in use in medieval times.[59]

It is hard to distinguish between bridge and boat mills in medieval

FIGURE 2–6. Plan of the bridge mills at Corbeil, France, fifteenth century. Visible in the drawings are the sluice gates used to control the flow of water in the channels beneath the bridge's piers.

documents, for most refer merely to a mill "in," "under," or "below" a bridge. But the frequency with which mills came to be associated with bridges from the eleventh century on indicates that medieval hydro-technicians fully appreciated the advantages to be gained by locating under or near bridges. Not only was the velocity of the stream greatly accelerated, but there were savings on labor, materials, and capital, since

FIGURE 2–7. Bridge mill from Zonca, 1607. The vertical water wheel of this establishment was used to drive a grindstone (far left), to turn millstones, and to activate a trip hammer (lower right). The height of the wheel (mill) was adjusted by means of the windlasses at *H* and the chains leading from them to the mill. Mills of this type were sometimes called hanging or suspended mills.

mill races and other auxiliary works to channel or direct water were automatically provided by the bridge. How fully the advantages of such sites were appreciated is evidenced by the actions of Parisian millers in the thirteenth century. The Grand Pont, a masonry bridge under which a number of mills were situated, was destroyed in 1296. The owners of the mills formerly situated there built another bridge (this one of timber) simply to facilitate the operation of their mills. By 1323 there were 13 mills under the Grand Pont.[60]

The hydropower dam and its adjunct, the power canal, presumably evolved from the boat mill and the bridge mill. The boat mill had been an attempt to adapt the vertical wheel to the natural flow of streams. The bridge mill had been a further step in that direction, but it had also demonstrated that natural conditions of flow could be significantly improved by human artifices. In one sense, the medieval river bridge, when combined with either boat or bridge mill, was a primitive form of hydropower dam, an intermediate step between the unassisted boat mill and the full-scale hydropower dam. Intentionally designed hydropower dams and the power canals frequently associated with them, however, went a step further. The boat mill and, to a lesser extent, the bridge mill had adapted the water wheel to natural stream conditions. The hydropower dam did the reverse. It adapted the stream to the water wheel.

Water-power dams were used by medieval European engineers for three primary purposes: (1) storage, (2) development of head or fall, and (3) diversion. The water backed up and stored behind a dam permitted the owner of a water wheel using that water to even out natural flow conditions. If, for example, a stream provided only enough water to operate a wheel at capacity for 12 hours, he could use the dam to store up water in the reservoir overnight, permitting full operation of the wheel the following day. The water backed up behind a dam also increased the head or fall available at the site. In hilly regions, where gradients were steep and valleys narrow, it was frequently possible to develop sufficient fall at the dam site so that a water wheel could be located on or adjacent to the dam. In wide, flat valleys, on the other hand, high dams would have backed waters up and flooded too much land. Here, a lower dam was used to divert water into a mill race, or power canal. The canal led water along an almost level course to some point further downstream, where, because of the greater gradient of the stream, a higher fall could be developed. Finally, a hydropower dam could be used to divert water into a mill race, or power canal, which would lead water to a more convenient location. For instance, if the banks of a stream were marshy and subject to flooding, or if the location of a dam was too far removed from access to transportation, the combination of dam and canal could carry water to a point where it was possible to construct mill buildings economically and carry out industrial operations.

Extant data on medieval hydropower dams is scattered but sufficient to suggest widespread use. We do not know when the first hydropower dam was constructed in Europe, but the abbey of St. Albans in Hertfordshire, founded in 793, may have used an old Roman causeway for a mill dam as early as the eighth, ninth, or tenth century.[61] Forbes reported that a weir (a small overflow dam) was built across the Leck c1000 to supply Augsburg watermills.[62] Fountains Abbey in Yorkshire, founded c1150, used a weir on the River Skell to divert water to a watermill, and the monks of Dale in Derbyshire in 1278 set up a weir on the Derwent for their cornmills, cutting off Derby from access to the Trent River and the sea for centuries.[63] In 1171 the count of Toulouse allowed the bishop of Cavaillon to divert water from the Durance River by means of a dam to operate flour mills.[64] Just a few years later, in 1179, work began on a dam on the River Ticino in northern Italy, near the town of Oleggio. This dam diverted water into a canal, where it was used for power, as well as irrigation and, ultimately, navigation.[65] In 1191 the River Drac, near Grenoble, was dammed to produce water power.[66] Around 1200 the town of Gournay, near Beauvais, France, had a dam which operated watermills, for Philip Augustus was able to hasten the surrender of the city during a seige by demolishing the dam which fed the grain mills of the town.[67] In the county of Toulouse, by the late twelfth century, mill dams had become numerous enough to interfere with each other, leading to legal disputes.[68] And Gille has claimed that in certain regions of medieval France, where the current of streams was small or irregular, so many dams were built and reservoirs formed that the whole nature of the landscape was changed.[69]

Power canals are usually associated with water-power dams. As with dams, information about medieval power canals is fragmentary but sufficient to suggest widespread use. The Saxons before the end of the ninth century had cut a channel about 0.7 miles (1.1 km) long, 20 feet (6 m) wide, and 13 feet (4 m) deep through a loop of the Thames to provide power for three vertical wheels.[70] The monastery of Clairvaux around 1136 had a mill race 2.2 miles (3.5 km) long which led water from the river Aube to the abbey, where it was used to supplement the power provided by two small streams which flowed naturally through the buildings.[71] At Obazine, in France, the Cistercians dug a canal almost a mile (1.6 km) long through solid rock to secure an adequate power supply.[72] Shorter mill races, or power canals, could be cited.[73]

Urban areas near medium to large rivers sometimes constructed networks of canals for developing water power. The city of Troyes, for instance, constructed a number of canals to divert water from the Seine for both drinking and power between the twelfth and fourteenth centuries. By 1500, the city had an extensive system of canals and branch canals which ultimately powered 25 mills.[74] Nor was Troyes unique. According to Fernand Braudel, a number of other medieval cities similarly modified the flow of adjacent rivers for power production, including Nuremberg, Rheims, Colmar, Marne, and Bar-le-Duc.[75] Bologna by the mid-1500s was using a sluice from the Reno to provide water for mills making copper pots and weapons, for mills pounding herbs and materials for dyes, for mills polishing arms and sharpening various implements, and for mills sawing planks.[76] Filarete, in designing his ideal city, Sforzinda, equipped it with steeply flowing canals which were to conduct water to mills for grinding grain, hammering iron and copper, polishing arms, sharpening knives, fulling cloth, and making paper, among other tasks.[77]

The emergence of the hydropower dam and the power canal (mill race or mill leat) permitted European water-powered industry to migrate from the aqueducts and small headwaters where it had been confined in antiquity to larger streams where more power could be developed. This movement toward the use of larger streams is most apparent in medieval England. The distribution of mills in the Domesday census of the late eleventh century indicates that the use of watermills had already spread from very small upland streams to the larger tributaries of navigable rivers and perhaps to a few of the smaller rivers. But very few mills seem to have been in operation on the lower, navigable portions of Britain's larger rivers (see Fig. 2–2).[78] This situation had changed by late in the medieval period. For example, navigation on the River Exe was ruined in 1313 by construction of a mill dam.[79] In reviewing medieval documents relating to milling in Britain, Bennett and Elton found that by the fourteenth century the construction of leats (mill races or power canals), dams, and weirs or causeways to supply water to mills located on navigable streams had "extended to exceedingly mischievous proportions." Legal actions resulted. In 1352 Parliament passed an act directing that all causeways established since the time of Edward I (1272–1307) be destroyed. In 1399 another act repeated and confirmed that of 1352, directing that weirs and causeways be surveyed "to the end that those which have been much enhanced [raised] since their erection, shall be amended to the old time level." From 1300 on, according to Bennett and Elton, there was a tremendous increase in the number of laws and judicial proceedings relating to impediments to navigation caused by milling establishments, as well as disputes over water rights.[80] They concluded:

> Mills were increasing in size; ... causeways for heading up the entire upper reaches of a river, and providing a heavier flow through the mill-race, were in universal demand. The character of these causeways, and their success in effectually blocking up a stream may be seen even yet at Chester, where the great curved stone structure in the bed of the river Dee, raising the water level of the water behind it three or four feet for the purpose of driving the Chester mills, has existed almost since the period of Domesday; and, despite many a struggle, and more than one Parliamentary order to destroy it, remains to the present time.[81]

The surge of legislation in Britain in the late medieval period relating to conflicts between navigation and milling interests provides compelling evidence for the increased use of large streams and rivers for water-power development.

The best single example of the high level of medieval European hydro-power engineering, however, comes not from Britain but from France—the damming of the Garonne at Toulouse in the twelfth century. The Garonne at Toulouse is a relatively large and deep river, subject to violent flooding on occasions. Thus, early in the medieval period attempts to utilize its power focused on adapting watermills to the stream, rather than vice versa. Toulouse's flour requirements were supplied, up until the twelfth century, by 60 floating mills. These caused problems. They interfered with navigation, and during floods they often broke their moorings and smashed boats and docks. To remedy this situation the owners of the floating mills erected three dams across the Garonne on which 43 fixed mills were erected. (These mills may originally have used vertical water wheels, but by the seventeenth century they were using horizontal wheels, at least for flour milling.) The Bazacle dam, the largest of the three, was in operation by 1177. It was more than 1,300 feet (400 m) long and was slanted diagonally to the course of the river. Built by ramming thousands of oak piles approximately 20 feet (6 m) long into the river bed in two parallel palisades and filling the space between with earth, wood, gravel, and boulders, this dam, on completion, was probably the largest dam in the world. It survived the Garonne's floods until 1709, and was quickly replaced by another dam when it finally failed.[82]

If the analogy between the development of water power in early America and in medieval Europe referred to at the beginning of this section is correct, then the appearance of large milling complexes like that on the Garonne in Toulouse and the appearance of hydropower dams on navigable rivers all over western Europe indicates a very substantial growth in industrial dependence on water power during the Middle Ages. By this view the late medieval hydropower complexes at the Grand Pont in Paris or on the Garonne in Toulouse can be considered to be the close relatives of the larger and more sophisticated water-power complexes at Lowell and Lawrence, Massachusetts, in the United States in the mid-nineteenth century.

At Toulouse, in fact, we find that the growth in the scale of industrial enterprise occasioned by the construction of Garonne dams gave rise to the same divisions between capital and labor which were to be characteristic of the water-powered and stream-powered factories of the later British and American industrial revolutions. Even before the construction of the dams on the Garonne, ownership of the 60 floating mills which satisfied the region's grain needs was divided in shares, or *uchaus,* worth one-eighth of a mill. These shares could be bought and sold, and like stocks today, their value fluctuated. Their exact worth at any specific time depended on economic conditions, the state of the mill, conditions on the river, and speculative activities. But a substantial portion of the shares were held by the millers who operated the mills. When the decision was made to dam the Garonne in the twelfth century, the shareholders who owned the floating mills agreed to share the cost of constructing the dams and the profits garnered from the sale of fishing rights in proportion to their holdings. The cost of building and maintaining the dams and prosecuting lawsuits, as well as occasional economic fluctuations, encouraged shareholders with marginal resources to sell their holdings. The high rate of return which the shares normally brought encouraged wealthy Toulouse investors to purchase shares, or *uchaus,* when they became available. As a result, by the thirteenth century, ownership of the mills at Toulouse had passed entirely into the hands of stockholders who were neither millers nor millwrights, but merely wealthy citizens, capitalists making profits from an investment instead of their own labor. The millers were no longer owners or part owners of the mills they operated, only employees. By the late fourteenth century stockholders associated with specific dams, for example, the Bazacle, had even pooled their shares to form mini-corporations. The owner of a share no longer owned part of specific mill, but a share in the Société du Bazacle, which owned a dam, a reservoir, associated fishing rights, and a number of mills, and which held annual meetings of stockholders where accounts were rendered and managers elected.[83] Thus, the growth of water-powered industry at Toulouse contributed to the creation and growth of a division between capital and labor similar to that found on a larger scale in the "industrial revolution" of the late eighteenth and early nineteenth centuries.

While the emergence of hydropower dams, power canals, and large-scale water-powered enterprises like that at the Bazacle in Toulouse provide the strongest indications of a medieval revolution in water power, some additional evidence comes from another medieval development in water-power technology—the tide mill. In some regions of Europe the combination of bridge and mill or dam and mill was insufficient to provide a reliable source of water power. The streams which flowed into low-lying coastal regions, for instance, were often too broad or too deep to be bridged or dammed and too slow to provide much power without these artifices. Other streams disappeared into marshy estuaries. In medieval Europe the vertical water wheel penetrated those areas in the form of the tide mill.

Tide mills could be constructed in many ways. The earliest may have been merely anchored or moored boat mills. When the tide came in, their wheels would be pushed in one direction; when the tide went out, in the other. But the form of the tide mill that came to be most widely applied was more complex. It required impounding the water of incoming tides by means of dams and gates placed across the mouth of a river, creek, or bay. The gates would be open to an incoming tide, but closed when the tide went out, creating large tidal reservoirs. The water from these reservoirs would then be released through a mill race onto the blades of a water wheel. Tide mills were not as convenient as conven-

FIGURE 2–8. Tide mill from Taccola, c1430. This is the earliest extant illustration of a tide mill. As the tide came in, the sluice gate shown in the upper portion of the drawing was raised. But when the tide began to go out, this gate was closed to create a tidal pond. Water from this pond was released through a mill race onto the water wheel.

tional watermills. They could be used only six to ten hours a day, and the hours when they could be operated varied from day to day as the tides changed. But where conventional mills could not be used, they were applied, and they did have the advantage of never freezing over.

References to tide mills appear first in Europe around Venice in 1044 and 1078. These mills were probably of the floating variety, if they were in fact tide mills. The more usual arrangement which utilized dams and tidal reservoirs probably went into use at about the same time. The Domesday Book, for example, mentions a mill at Dover as causing "disaster to vessels by the great disturbance of the sea." A tidal mill with impounding reservoir was probably involved, though the meaning of the passage is not at all clear. Tide mills were in use on the Adour River near Bayonne, southern France, between 1124 and 1133, and in Essex in 1135. They were used in Brittany by 1182. By 1300 there were at least 38 tide mills in operation in Britain, and by 1600 there were at least 89. The earliest extant illustration of a tide mill comes from the notebooks of the fourteenth-century Italian engineer Taccola (Fig. 2–8).[84]

The medieval tide mill did not exert a major influence on the development of hydropower. Like the boat and bridge mill, it was a modest improvement on the conventional undershot mill, and it probably

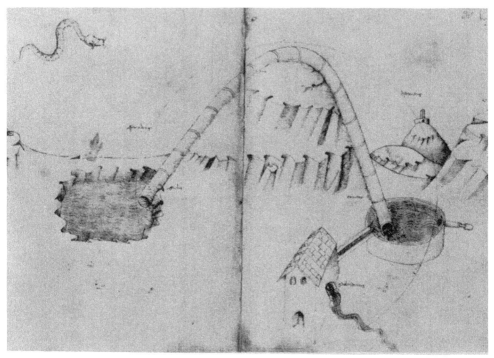

FIGURE 2–9. Siphon line from Taccola, c1430. Taccola designed this line to deliver water across a hill or mountain to a reservoir for powering a watermill. This project was probably only an idea since a siphon could not lift water higher than about 32 feet (c10 m). But it illustrates the increased interest in artificially contrived systems for delivering water to watermills towards the end of the medieval era.

tapped an even smaller portion of the energy available to it. Nonetheless, the development of the tide mill was significant. It permitted water power to penetrate into areas which had not used it before, and it may indicate that medieval Europe was, as Lynn White has postulated, a power-hungry society attempting to mobilize all possible energy sources.[85]

Finally, there is other, scattered evidence which suggests that Europe's power needs by the end of the medieval period had grown to the point where very unusual solutions to water-power shortages could be considered. For example, Taccola, in the fifteenth century, suggested using a siphon line to bring water over a mountain to a mill site (Fig. 2–9). In another instance, where no fall of water was available, he suggested using a hand-powered pumping complex to supply water to an elevated reservoir which powered a mill.[86]

In summary, although extant evidence is scattered, what is available indicates a general trend in the medieval use of water power away from primary dependence on aqueduct water and the natural flow of very small streams towards the use of progressively larger streams and water volumes. The first step in tapping these larger streams was the boat mill, an adaptation of the conventional undershot water wheel to existing flow conditions. The boat mill alone often produced little more power than a small horizontal mill on a mountain brook. But the boat mill was soon placed adjacent to a bridge and evolved into the bridge mill, the bridge serving as a proto-dam, increasing the working head available and permitting the development of a complex of several mills on the same site. This evolution eventually culminated at Toulouse, Chester, and other points in the construction of large dams or weirs across navigable rivers and at Troyes, Bologna, and elsewhere in the use of networks of canals paralleling rivers. By the close of the Middle Ages watermills were in use on streams of every type. They dammed up the rivers of medieval man; they were on the banks of his brooks and creeks, in the middle of his rivers, under his bridges, and along his coastlines. They impeded navigation and created streams (in the form of mill races and power canals) and lakes (in the form of storage reservoirs behind water-power dams) where none had existed before.

If medieval European developments were analagous to early American developments, the steadily growing use of watermills on progressively larger streams certainly implies, if only roughly, that industrial dependence on water power climbed significantly during the course of the Middle Ages. This trend, in conjunction with the available quantitative information on medieval water wheels and the evidence of their widespread geographic dispersion, provides considerable support for White's contention that the medieval epoch experienced a revolution in the use of nonhuman power sources. The available data on medieval industrial uses of water power, reviewed below, provides even stronger support for this thesis.

## FROM FLOUR MILL TO INDUSTRIAL PRIME MOVER

In the medieval application of water power to industrial processes we find the roots of a tradition that was to continue without a significant break to the heavily mechanized cotton textile factories of Britain's industrial revolution.[87] Through all of antiquity and on into the early Middle Ages almost the only work to which the force of falling water was applied was grinding wheat. This was always to be one of its more important functions. Buy by the tenth century, European technicians had begun to adapt the vertical water wheel to other tasks. By the sixteenth century, in addition to flour mills, there were hydropowered mills for smelting, forging, sharpening, rolling, slitting, polishing, grinding, and shaping metals. Water wheels were available for hoisting materials and for crushing ores. There were mills for making beer, olive oil, poppy oil, mustard, coins, and wire. Water wheels were used in the preparation of pigment, paper, hemp, and tanning bark, and for fulling, sawing wood, boring pipes, and ventilating mines. The replacement of human power

Table 2–2.   Medieval Applications of the Vertical Water Wheel

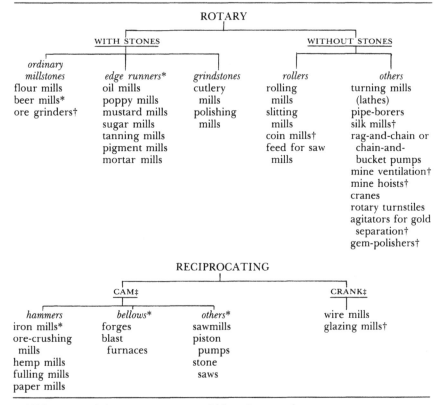

ROTARY

| WITH STONES | | | WITHOUT STONES | |
|---|---|---|---|---|
| *ordinary millstones* | *edge runners** | *grindstones* | *rollers* | *others* |
| flour mills | oil mills | cutlery | rolling | turning mills |
| beer mills* | poppy mills | mills | mills | (lathes) |
| ore grinders† | mustard mills | polishing | slitting | pipe-borers |
| | sugar mills | mills | mills | silk mills† |
| | tanning mills | | coin mills† | rag-and-chain or |
| | pigment mills | | feed for saw | chain-and- |
| | mortar mills | | mills | bucket pumps |
| | | | | mine ventilation† |
| | | | | mine hoists† |
| | | | | cranes |
| | | | | rotary turnstiles |
| | | | | agitators for gold |
| | | | | separation† |
| | | | | gem-polishers† |

RECIPROCATING

| CAM‡ | | | CRANK‡ |
|---|---|---|---|
| *hammers* | *bellows** | *others** | wire mills |
| iron mills* | forges | sawmills | glazing mills† |
| ore-crushing | blast | piston | |
| mills | furnaces | pumps | |
| hemp mills | | stone | |
| fulling mills | | saws | |
| paper mills | | | |

*Other transmission arrangements could have been used.
†First definite evidence of this application of water power comes only after 1500, but so soon afterwards (1500–1560) that it may have initially occurred prior to that date.
‡By the sixteenth century many of the processes previously carried out using the cam were adapted to the crank (e.g., pumping with the piston pump, bellows operation, sawmilling).

by inanimate power for industrial production, therefore, did not begin in Britain's cotton textile mills. The roots of this movement were firmly embedded in the soil of medieval Europe. Table 2–2 summarizes the medieval achievement in hydropower diversification, an achievement which will be discussed in more detail in the following pages.

Perhaps the earliest use of water power, other than for grain milling and water raising, was in the preparation of malt for beer mash. Although it is possible that other mechanisms may have been involved in this process (see the discussion of the cam-activated hydraulic trip-hammer below), most early beer mills probably resembled conventional flour mills (Fig. 2–10).[88] Their millstones may have been a bit different because the substance they pulverized was malt instead of wheat, but there were probably few other distinctions. Beer mills initially appear in docu-

FIGURE 2–10. Flour mill powered by undershot water wheel, 1588. Medieval beer mills had the same general configuration as flour mills, but ground malt instead of wheat.

ments dating from the ninth century in France (Picardy, 861),[89] and have a continuous tradition from the tenth century on.[90] Water-powered, horizontally situated millstones, similar to those used in flour and beer mills, were also being used to grind tin and gold ore to powder by the sixteenth century.[91] They may have been used to prepare tanning bark as well, although other arrangements are more likely (see below).

The oil mill (used to produce olive oil) needed stones that would crush rather than grind. The solution of this problem was the edge-runner, or edge-roller, mill. It usually consisted of one or a pair of broad stone

FIGURE 2-11. Edge-runner mill from Zonca, 1607. Edge rollers crushed rather than ground and were applied in the medieval period to the production of oil, sugar, pigments, tannin, and mortar. They could be equipped with either one edge runner, as in the mill pictured here, or two.

wheels, situated vertically rather than horizontally, and connected by a short axle or shaft to a rotating vertical drive shaft. The rotation of the drive shaft caused the wheel(s) to travel in a circular path, crushing the olives laid in their way (Fig. 2–11). Although this arrangement, powered by men or animals, existed in classical times, and even though the continuous circular motion of the vertical water wheel was well suited for the task, it was apparently not adapted to the task until after 1000. Water-

FIGURE 2–12. Water-powered sugar mill, c1580. One of the earliest extant illustrations of a water-powered sugar mill. The vertical water wheel can be seen at the extreme left part of the engraving. This sugar mill used conventional millstones, but later mills tended to rely on edge-runner stones for crushing sugar cane.

powered oil mills first appear in documents from southeastern France (Dauphiné) in the eleventh century. Forbes reports one in Province in 1101, and some additional references, rare and scattered, come from the thirteenth and fourteenth centuries.[92]

Edge-roller mills could be used in the preparation of other materials besides olive oil—mustard, poppy oil, sugar, pigments or dyes, mortar. In the thirteenth century a mill for crushing mustard seed was reported in the Forez district of central France, and a mill for producing oil from the poppy, in the Nord.[93] Sugar mills, likely using the edge roller and possibly water powered, appeared in Norman Sicily in 1176.[94] The earliest extant engraving of a water-powered sugar mill, shown in Figure 2–12, dates from around 1580. There is some evidence for the existence of pigment mills, used in the preparation of dyes in the medieval period. They were apparently a rather late application of water power, for the earliest are mentioned in documents dating only from the mid-fourteenth century in the textile-producing areas of northern France at Hesdin (Pas-de-Calais) and Peronne (Somme). The earliest known non-French pigment mill comes only in 1532.[95] There is, finally, a medieval document dating from 1321 from Augsburg referring to a mill for pre-

FIGURE 2–13. A cutlery (grinding and polishing) mill from Böckler, 1661. The grindstones are indicated by G. Some of the utensils sharpened in this mill are shown scattered on the floor. The grinder lay on the inclined plane near the lower grindstone while working.

paring mortar.[96] Later, after 1500, the edge-runner mill was applied to the preparation of snuff, pipe clay, china, cement, and gunpowder.[97]

Another medieval industrial application of hydropower was to the tanning, or bark, mill. The object of the tanning, or bark, mill was to reduce bark (usually oak) to as small and as uniform a size as possible for

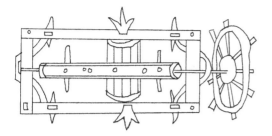

FIGURE 2–14. Fifteenth-century pipe-boring machine. This sketch by an anonymous "Hussite" engineer seems to show a vertical water wheel (on the right) boring a pipe.

FIGURE 2–15. Seventeenth-century pipe-boring plant. The various forms of cutting tools used in this device are shown lying on the ground at *K*.

the leaching process which would extract tannin. The extant documents do not make it clear whether vertical stamps, horizontal millstones, or edge runners were used. There is good evidence to indicate that probably several of these methods were applied.[98] But edge rollers were probably the most common. The use of water power to rotate stones or lift hammers for the preparation of tanning agents dates back to the early twelfth century. The first firm reference to a tanning mill comes from Charment, near Paris, in 1138. By the end of the twelfth century the tanning industry in northern France and northern Italy had become dependent on water power, and by the second half of the thirteenth century tanning mills were in use throughout much of Europe.[99]

FIGURE 2–16. Eighteenth-century engraving of a rolling and slitting mill. The rolling mill is situated just behind the man in the center of the illustration, while the slitting wheels are directly in front of him.

Mills for sharpening and polishing blades, armor, and cutlery also used stones, but grindstones, instead of the horizontal millstones or the edge-runner stones used by flour, oil, and bark mills. The grindstones were either attached directly to the water wheel's axle or took their motion from the main shaft via gearing or belting (see Fig. 2–13). The earliest extant citation of a water-powered cutlery mill comes from 1204 at Evereux (Normandy). By the end of the thirteenth century, sharpening and polishing mills were in extensive use in France. By the middle of the thirteenth century they were being built over the rest of Europe as well.[100]

The continuous rotary motion produced by the vertical water wheel was applied to both working wood and rolling and cutting metals during the medieval period, as well as to milling and grinding. Although Woodbury found the earliest example of a water-powered wood lathe only in 1590, Blaine believes that they came earlier, citing several documents from the fourteenth and fifteenth centuries from Dauphiné in southeastern France.[101] Pipes were usually made of wood before the nineteenth century, and an anonymous engineer illustrated a pipe-boring machine at Nuremberg around 1480 which appears to be driven by a vertical water wheel (Fig. 2–14).[102] Hydropowered pipe-boring machines were definitely in use by the seventeenth century (Fig. 2-15).

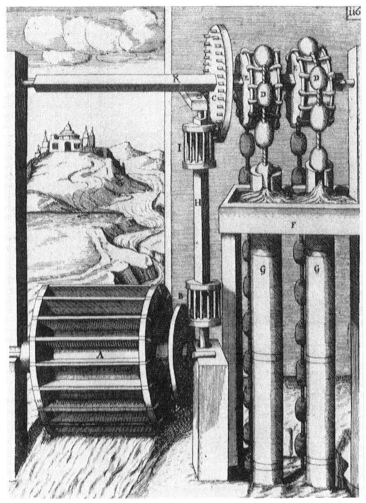

FIGURE 2–17. Water-powered rag-and-chain pumps from Böckler, 1661. Similar pumps were pictured by Agricola in 1556 and were in use by the early fourteenth century for mine drainage in central Europe.

A very late medieval application of the vertical water wheel was in the rolling and slitting, or cutting, mill (Fig. 2–16). These mills had pairs of revolving cylinders through which metal was passed either to form sheets (rolling) or to be cut into strips or rods (slitting). The earliest firm evidence of a cutting mill dates from 1443 at Raveau, Nivernais, France. The next evidence is from 1532 in Germany. The diffusion of this innovation was rather slow. In England rolling mills were not known until the seventeenth century, though cutting mills were in use before the end of the sixteenth.[103] Coin mills were very closely related to rolling mills. In coin mills two water-powered rotating cylinders, upon which the desired design was engraved, were fed heated blanks. The first

FIGURE 2–18. Water-powered ventilating fan for mines, Agricola, 1556. Paddles were rotated in the cylinder *A* by the overshot wheel, *G*, taking in air at *B* and forcing it into the mine through the wooden conduit situated near the lower right portion of the picture.

FIGURE 2–19. Triple suction pump operated by overshot water wheel, Agricola, 1556. The piston rods of these pumps were lifted by the cams situated along the axle, *B*, of the water wheel.

water-activated coin mills of which we have certain knowledge were put into operation in Paris in 1551 and 1552, but coin mills powered by water may well have been constructed in Germany a few years before the beginning of the sixteenth century.[104] The Paris coin mill replaced a gem-polishing mill, presumably also water-powered, which had been in operation since 1534.[105]

The rotary motion of the water wheel may also have been used in the medieval period for other tasks. There is evidence, dating from four-teenth-century Moravia (Iglau, 1315), as well as from Silesia and eastern Germany, of water-powered chain-of-bucket or rag-and-chain pumps for draining mines (Fig. 2–17).[106] Water-powered ventilating apparatus (Fig. 2–18), mine-hoisting machinery, and piston pumps (Fig. 2–19) are pictured in Agricola's *De re metallica* of 1556, and it is possible that some of these applications of water power were a century or more old, even though there is no definite evidence.[107] For silk-twisting mills the same applies. While there is no proof of water power being applied to silk until 1607, when Zonca illustrated a silk mill (Fig. 2–20), there are documents which hint at possible water-powered silk mills in fourteenth- and fifteenth-century Florence and Lucca,[108] and Leonardo da Vinci (c1480–1500) has silk spinning on a list of activities that could be powered by water if the Arno river were modified.[109]

Finally, there are several direct applications of the rotary motion of water wheels that were conceived in the medieval period, though not necessarily carried out in practice. A thirteenth-century German poem (*Wigalois*, by Wirnt von Grafenberg) described a water-activated rotary turnstile equipped with knives to keep unwanted intruders from passing through a doorway.[110] Filarete, the Italian architect, described an ideal city in the fourteenth century which used water-powered cranes or hoists to unload ships.[111] Taccola, in the same century, planned to use water wheels to pull boats upstream (Fig. 2–21).

Probably more important in the long run than the expansion of the vertical water wheel's role in industries requiring rotary motion was its application to those requiring reciprocating motion. This necessitated resolution of the delicate problem of transforming circular motion to reciprocating motion. Ultimately, medieval technicians devised two solutions—the camshaft and the crankshaft. The cam was the first of the two to emerge and was long the more widely applied. Only very late in the medieval period were the crank and crankshaft applied to water-powered systems.

The cam, like many revolutionary innovations, is a simple device. Basically it is a small projection fixed on an axle. It was not a medieval innovation. The Alexandrian mechanicians, Hero in particular, had employed it on automata in antiquity. But, like the water wheel, the cam was an invention that came into full practical application only in the Middle Ages.[112] The earliest applications of the cam in water-powered systems were in conjunction with mallets or pounders. There were two basic

Figure 2–20. Silk-spinning mill from Zonca, 1607. The spindles (not shown) were fixed on the "garland" (the device on the left side of the illustration). Although there is no definite evidence that hydropowered silk-spinning mills like this were in operation before 1500, there are strong indications that they were.

FIGURE 2–21. Device to enable a boat to ascend upstream through the use of water power. The impact of the water flowing downstream struck the blades of the water wheels, causing them to rotate and wind in a rope connected to some fixed object upstream, thus pulling the boat forward. This scheme appeared in the notebooks of Taccola in the fourteenth century. Although attempted on several occasions later, it never proved to be a practical success.

mechanisms for utilizing the cam for pounding—the vertical stamp and the recumbent trip-hammer. In the vertical stamp, cams fixed at intervals along the main axle of a water wheel were brought by the rotation of the axle against similar projections on the vertical shaft of a hammer, stamp, or mallet (see Fig. 2–27). The cams, in turning, lifted the stamps until their rotation took them out of contact with the vertical shafts. The weight of the hammers then caused them to fall, producing the desired impact. The recumbent trip-hammer operated in a similar manner. Again, the cam's rotation against a hammer or shaft raised the pounding implement and then released it (see Figs. 2–23 and 2–24 for examples). But in this system the shaft of the mallet or hammer was situated horizontally rather than vertically and was pivoted.[113]

Both of these systems were certainly in use by the fifteenth century, for both appear in the notebooks of the anonymous "Hussite Engineer."[114] In the sixteenth century Agricola pictured mainly vertical stamps; Olaus Magnus drew recumbent hammers.[115] Which system was most heavily used in medieval Europe is impossible to determine, since surviving documents referring to hammer or stamp mills do not differentiate between the two, and there are no extant illustrations of stamp mills before 1400.

The first industry to use the hydraulic hammer may have been brewing. A plan of the monastery of St. Gall dating from around 820 has two enclosures on the outer edge of the monastic complex. One of these contains two circles and is labelled *molae,* the other contains two circles with hammerlike objects attached ( ⌐◠○ ) and is labelled *pilae.* Situated just inside the monastic complex from these enclosures are, respectively, the monastic bakery and the monastic brewery.[116] Blaine believes that the circles in the enclosure adjacent to the bakery represent water wheels, especially since water-powered grain milling had a long tradition in the area and the term *molae* was frequently used in medieval documents to mean watermill. The adjacent pilae, he believes, were water-powered trip-hammers used to crush malt for beer mash.[117] The sketch of the pilae could be interpreted as a water wheel and recumbent trip hammer viewed from the side, a reasonable assumption since medieval artistic convention was to depict objects in the view that best revealed their operation. However, the circles of the molae could represent millstones instead of water wheels, and in later periods conventional millstones rather than trip-hammers were used to prepare malt for beer, so the early use of hydraulic trip-hammers in the preparation of beer must remain highly conjectural.

While the early application of hydraulic hammers to beer production is questionable, their early application in the fulling and hemp industries is beyond dispute. The fuller took freshly woven cloth and subjected it to a pummeling action either in or out of water. This permanently altered the texture of the cloth by thickening and shrinking the material and by felting the fibers so that they became intimately bound together. At the same time the cloth was also scoured and cleaned, removing the oil with which the wool had been impregnated before spinning to strengthen the warp threads. Originally this process involved walking on the cloth or manual beating. The cam and hydraulic hammer allowed the fuller to substitute water-activated wooden hammers for human arms and feet. The earliest generally accepted evidence of a water-powered fulling mill comes from Lodi, near Milan, around 1008. Fulling mills were shortly after in operation in other parts of Alpine Italy and France. From there they spread quickly. By the eleventh century fulling mills were in operation in northern France and throughout northern Italy.[118] During that century and the following one, according to Bautier, the water-powered fulling mill revolutionized the French textile industry, forcing it to center in areas where water power was available and eventually leading small fullers to resist mechanization.[119] By the twelfth century mechanized fulling had begun in Scandinavia and in England. Carus-Wilson's study of the fulling mill in England indicates that it was the primary factor behind the decay of a number of once famous, but water-power-deficient clothing-making cities on the eastern plain. Although the first English fulling mill dated only from 1185, there were at least 130 of them by 1327.[120] By the twelfth or thirteenth century water-powered

FIGURE 2–22. One of the earliest extant illustrations of a fulling mill, Zonca, 1607. The water wheel, *I*, the camshaft, *G*, the cams, *H*, and the recumbent trip-hammers, *A*, *B*, are visible. The earliest fulling mills may have used vertical trip-hammers instead of recumbent hammers.

fulling mills were in operation on the Rhine, and by 1212 they had reached Poland.[121] Among the earliest illustrations of water-powered fulling mills is that of Zonca, early in the seventeenth century (Fig. 2–22).

Hemp mills are very similar to fulling mills and also applied hydraulic trip-hammers quite early. The only major difference between the two mills lay in the task to which the trip-hammers were applied. Fulling mills were designed to cleanse, thicken, and strengthen newly woven cloth. Hemp mills used their stocks or mallets to break up the woody tissues of dried hemp or flax stalks and free the fibers for use in the manufacture of ropes and cords. The earliest records of hemp mills come from Dauphiné around 990 and from other points of Alpine France in the early eleventh century. By the twelfth or thirteenth century water-powered hemp mills were found scattered throughout France; by the fourteenth century, through most of Europe.[122]

Paper mills also employed hydraulic trip-hammers. Here they were used to beat rags to pulp (see Fig. 2–23). The mechanization of this

FIGURE 2–23. Water-powered paper mill. The trip-hammers were used to beat rags into pulp. The pulp, after being transferred to a vat, was dipped out in a mold, pressed, and dried, producing paper.

industry was of major importance, for it made paper cheap and common for the first time. It has often been assumed that the water-powered paper mill entered Europe through Moslem Spain in the mid-twelfth century, but there is no indication that the mills used by the Moslems were anything but hand- or animal-powered. The earliest certain example of a water-powered paper mill comes from Italy at Fabriano in 1276, and shortly after that, from Christian Spain. By the end of the thirteenth century paper mills driven by water were common throughout northern Italy. In the 1300s they spread to France. The first German paper mill is recorded in 1390. By 1400 paper mills were common in France and during the fifteenth century were widely diffused among the German

TABLE 2–3.  SMELTING WORKS IN THE SIEGEN AREA, 1311–1505

| | 1311 | 1417 | 1444 | 1463 | 1492 | 1505 |
|---|---|---|---|---|---|---|
| Smelting works | ? | 19 | 26 | 25 | 18 | 16 |
| Bloomeries | 1 | ... | ... | ... | ... | ... |
| Pig-iron works | ... | ... | 2 | 4 | 11 | 13 |
| Forges making steel in | | | | | | |
| hearth fires | ? | 1 | 8 | 11 | 9 | 15 |
| Total | ? | 20 | 36 | 40 | 38 | 44 |
| Total using | | | | | | |
| water power | 1 | 6 | 24 | 30 | 38 | 44 |
| Disused | ? | ... | 3 | ... | ... | ... |

SOURCE: R. J. Forbes, "Metallurgy," in Charles Singer et al., eds., *A History of Technology*, 2 (Oxford, 1956): 74.

states. By 1500 they had reached Austria, Poland, Bohemia, and Britain, and before the end of the sixteenth century, Scandinavia.[123]

The introduction of the cam not only revolutionized the fulling, hemp, and paper industries, but the all-important iron industry as well. Early methods of smelting iron ore did not achieve high enough temperatures to produce a completely liquified metal. What emerged from the small furnaces then in use was a spongelike mass called the bloom. This was consolidated by extensive hammering. The advent of the hydraulic trip-hammer naturally made this aspect of the iron industry subject to mechanization. When, later in the Middle Ages, blast furnaces served by water-powered bellows became operative, it was possible to produce large quantities of iron in its molten state and strain off the slag. But the product, cast iron, was brittle and unsuitable for many purposes. To produce quality wrought iron, the cast or pig iron was submitted to a process of reheating and hammering until conversion to wrought iron was affected and additional impurities removed. Here too, medieval engineers combined the wheel, cam, and trip-hammer to advantage. By the late Middle Ages running water had become a critical factor in the location of iron works, as Table 2–3, covering smelting works in the Siegen area of Germany, indicates. The association of iron works with hydropower, begun in the late Middle Ages, was to remain an almost invariable aspect of the industry until the closing years of the eighteenth century.[124]

Early documents mentioning iron mills do not make it clear whether water wheels were being applied to power hammers or bellows, or whether, if bellows, they were used with the bellows of the furnace or forge. Probably the first iron mills merely had mechanized hammers, similar to fulling or hemp mills, and only later added the more complicated systems required to power bellows.

The first positive allusion to an iron mill comes from southern Scandinavia in 1197 or 1224.[125] But there are good reasons for assuming that iron mills existed several centuries earlier. There was, for example,

FIGURE 2–24. A fifteenth-century hammer forge. Iron mills similar to this one may have been in operation as early as 1025 and were definitely in use by 1200 or shortly after.

a place named Smidimulni ("Smith Mill") in southern Germany around 1024, suggesting that water-powered hammers were already in use. Two of the English Domesday mills paid their rents in blooms of iron, again suggesting the application of water power to metallurgy. Similar evidence indicates eleventh- or twelfth-century iron mills in France, Italy, and Spain. Thus, as Blaine has noted, while there is no definitive documentary proof of hydropowered iron mills before the Scandinavian example of around 1200, there are impressive indications that they were in use much earlier. In any case, by the thirteenth and fourteenth centuries, hydraulic trip-hammers for treating iron were in use from Spain to Scandinavia and from Britain to Bohemia.[126] One of the earliest extant illustrations of a hammer forge is shown in Figure 2–24.

Water-powered bellows for forges, blast furnaces, and ventilation seem to have come later than water-powered trip-hammers. While some of the early iron mills may have had hydraulic bellows, as well as trip-hammers, there is no evidence which indicates the application of water power to bellows until 1214 at a monastic iron mill at Trent, northern Italy. Hydraulic bellows did not become common until after 1300.[127] Water wheels activating forge bellows were depicted by Taccola several times in the first half of the fifteenth century (see Fig. 2–25).

The introduction of water power to bellows revolutionized the iron industry when those bellows were moved from the forge to the furnace.

FIGURE 2–25.  Forge bellows activated by overshot water wheel and camshaft from Taccola, c1449.

Before the introduction of the water-powered blast furnace, both the quantity and quality of iron produced were low and costs were high. The larger and hotter furnaces which water power made possible enabled the medieval iron masters to liquify the metal, remove impurities easier, and cut costs. It was probably the single most important step in the transition from a technology based on wood to one based on metal. The cam, again, was the basic device which made this application of water power possible (see Fig. 2–26). The first solid evidence for a blast furnace activated by a water wheel comes from Liège, Belgium, in 1384. But the innovation spread quickly, and by the fifteenth century water-powered blast furnaces were known in Germany, Italy, France, and England.[128]

Another aspect of medieval metallurgy which became dependent on water power was ore stamping. Ore stamping mills used trip-hammers

FIGURE 2–26. Water-powered blast furnace from Agricola, 1556. A portion of the water wheel can be seen at the extreme right. The bellows were cam-activated. As the cam rotated, it closed the bellows, forcing air through the furnace. They were reopened by means of a counterweighted lever arm after the cam had rotated out of contact.

similar to those used in fulling, hemp, and paper mills to break ore into small pieces in preparation for reduction in the blast furnace. Documents dating back to 1135 and 1175 from the monastery of Admont in Styria may refer to hydraulic ore stamps. They were definitely in use shortly after 1300.[129] Agricola very clearly illustrated a set in operation in 1556 (Fig. 2–27).[130]

The ability of medieval hydraulic engineers to devise mechanisms for directly utilizing a water wheel's rotary motion while at the same time converting a portion of it to reciprocating motion can be seen in the case of the sawmill. A sawmill requires reciprocating motion to move the saw back and forth. It also requires a device for feeding timber into the saw, and this often utilized the rotation of the main axle. As early as 1235

FIGURE 2–27. Ore stamps with vertical trip-hammers, Agricola, 1556. These were used to prepare ores for reduction. Vertical trip-hammers of this type were possibly used in most medieval stamp mills for fulling, hemp, paper, and iron production.

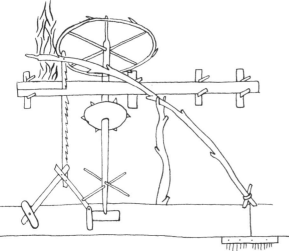

FIGURE 2–28. Water-powered sawmill of Villard de Honnecourt, c1235. The sticklike projections on the axle are cams and were used to depress the saw blade. A spring-pole returned it to position for the next stroke. The toothed wheel shown in the center of the axle fed the timber continuously into the saw.

Villard de Honnecourt sketched a mill of this type (Fig. 2–28).[131] Villard's apparatus had an unsupported blade drawn downwards by the action of cams projecting from the axle of a water wheel and returned to its original position by a flexible pole spring. At the same time the rotation of the axle of the water wheel was used as a feed mechanism.

Water-powered saws for cutting stone may have been in use by the tenth century, but the earliest known certain reference to a sawmill

FIGURE 2–29. Seventeenth-century sawmill. This mill used a crank instead of cams to produce reciprocating motion; feed was provided by the weights *A* and *B*.

comes from Normandy in 1204. The hydraulic saw became common in northern France and Switzerland in the thirteenth century and throughout France and southern Germany in the fourteenth. But it did not emerge in Scandinavia or England until the sixteenth or seventeenth century.[132] In certain areas—for example, southeastern France, around Vizille—so much timber was cut for transportation to sawmills that local officials considered the timber trade one of the primary factors behind deforestation in the area. In 1304 the Dauphin ordered that all sawmills established near or in the forests of the Prémol Mountains be removed in an attempt to alleviate the situation.[133]

The alternative to the cam as a means of converting rotary to reciprocating motion was the crank. Although apparently known in ancient China, it appeared in the West only in the Middle Ages and seems to have found application in a few hydropowered systems shortly before

FIGURE 2–30. Water-activated bellows in a forge hearth from Ramelli, 1588. Note the use of a crank in this system as opposed to the cams used in the Taccola drawing of Fig. 2–25. Vertical rods from the rocking arms (B, C, D) raised the upper leaf of the bellows, which was lowered by the rope connected to the upper leaf and to the weighted, hinged flap beneath the lower leaf.

1500. A number of cam-activated systems were either then or shortly afterwards converted to the crank. Cranks, for example, provided the double-acting reciprocal motion needed for sawmills (Fig. 2–29) and metallurgical bellows (Fig. 2–30).[134] It was also applied to piston pumps, where it was much better suited than the cam (Fig. 2–31).

The crank was also applied to several systems where the cam was apparently never utilized. One example was the wire mill. In the wire

FIGURE 2–31. Water-powered piston force pumps using crankshaft and rocking beam, Ramelli, 1588.

mill the linear force of the crank was used to pull wires through draw plates. Water-powered wire pullers may have been in operation as early as 1351, when Augsburg city records refer to a wire mill. But the first certain evidence, a watercolor by Dürer of the exterior of a wire mill situated beside a stream, comes only from 1489. The wire-drawing mechanism itself, along with its linkage to the crankshaft and water wheel, was very clearly depicted by Biringuccio in 1540 (Fig. 2–32).[135]

The crank was also utilized in mills which ground and mixed the colors used for pottery glazes (the glazing mill). As a sixteenth-century illustration of one of those mills indicates (Fig. 2–33), a small millstone

FIGURE 2–32. Wire mill from Biringuccio, 1540. As the water wheel rotated the crankshaft to the left, the wire was gripped and pulled through a draw plate.

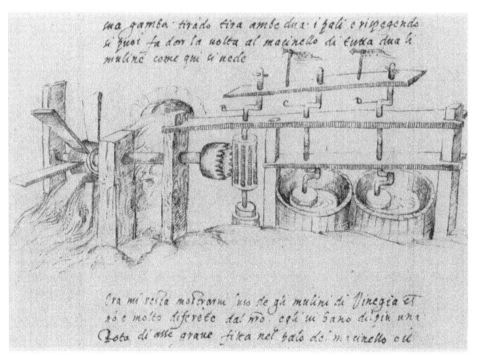

FIGURE 2–33. Glazing mill used at Foligno, central Italy, early sixteenth century. The materials used in producing the glazes for pottery were ground and mixed by the motion of a small millstone over a large one placed in the bottom of a container shaped like a half-barrel.

TABLE 2–4. EMERGENCE OF NEW INDUSTRIAL APPLICATIONS OF THE
VERTICAL WATER WHEEL TO c1550

| Type of mill | Date of first occurrence | Location |
|---|---|---|
| Beer | 861 | NW France |
| Hemp | 990 | SE France |
| Fulling | 1008 | N Italy |
| | 820? | Switzerland? |
| Iron | 1025? | S Germany |
| | 1197 | S Sweden |
| Oil | c1100 | SE France |
| Ore-stamping | 1135 | N Italy |
| Tanning | 1138 | NW France |
| Sugar | 1176? | Sicily? |
| Cutlery (grinding and polishing) | 1204 | NW France |
| Saw | 1204 | NW France |
| Mechanical bellows | 1214 | Styria |
| Mustard | c1250 | SE France |
| Poppy | 1251 | NW France |
| Paper | 1276 | N Italy |
| Mine-pumping (chain type) | 1315 | Moravia |
| Mortar | 1321 | S Germany |
| Turning (lathes) | 1347 | SE France |
| Pigment (paint) | 1348 | NW France |
| Blast furnace | 1384 | Belgium |
| Pipe-boring | c1480 | S Germany |
| Rolling and slitting | 1443 | Cent. France |
| Wire | 1351? | S Germany |
| | 1489 | S Germany |
| Gem-polishing | 1534 | NW France |
| Coin | 1551 | NW France |

SOURCE: Bradford B. Blaine, "The Application of Water-Power to Industry during the Middle Ages" (Ph.D. diss., University of California, Los Angeles, 1966).

was rotated by means of a crank over the surface of a large stone. Glazing mills were first mentioned by Cipriano Piccolpasso shortly before 1560, but in terms suggesting that they were widely known and possibly had a tradition long enough to place their initial use 50 to 100 years earlier.[136]

As Table 2–4 indicates, the industrial diversification of the water wheel, reviewed in the preceding pages, occurred in two main centers. One was in the Alpine regions of southeastern France, northern Italy, Switzerland, and southern Germany. Eight of the major new medieval applications of water power—hemp mills, fulling mills; paper, oil, iron, and wire mills; hydraulically operated ore stamps; and metallurgical bellows—first emerged in or near this area. Seven of these processes definitely depended on the conversion of rotary to reciprocating motion (hemp, fulling, paper, iron, and wire mills; ore stamps; and bellows), and the remaining one (oil mills) may have. By way of contrast, in the other center of innovative activity—northern France and adjacent areas—most of the new uses of the water wheel depended on rotary motion. This region was responsible for the introduction of tanning, beer, cutlery, saw, pigment, rolling, and coin mills, plus water-activated blast furnaces. Of these eight, six utilized rotary motion, with only the sawmill and blast furnace clearly dependent on reciprocating motion.

In the early 1930s John U. Nef published a comprehensive study, *The Rise of the British Coal Industry*. He concluded from this study that the traditional "industrial revolution" of the late eighteenth century was not really a sharp break with the past. He found that the transition from wood to coal, considered to be one of the central elements of the industrial revolution, had begun on a significant scale in Britain as early as the reigns of Queen Elizabeth and her two Stuart successors, and that long before the nineteenth century coal had had a major impact on British economic and industrial life.

Nef extended his studies to other industries in subsequent years. He reviewed the growth of mining and metallurgy and the growth of the salt, glass, gunpowder, and textile industries, for example. He found in these industries, as well as in coal mining, a significant increase in production and an emerging capitalistic organization, particularly in the period c1570 to c1640. Nef therefore concluded: "The rise of industrialism in Great Britain can be more properly regarded as a long process stretching back to the middle of the sixteenth century and coming down to the final triumph of the industrial state towards the end of the nineteenth, than as a sudden phenomenon associated with the late eighteenth and early nineteenth centuries."[137]

I have no serious objection to Nef's emphasis on the continuity of the process of industrialization in Britain from c1650 on. But I believe that if we view western Europe as a whole, the beginnings of the "long process" of industrialization should be pushed even further back, back to the late Middle Ages, when European engineers began to aggressively expand the applications of water power in industrial processes.

We can see in this expansion, for example, the seedbed from which the modern factory system ultimately emerged. The modern factory which first emerged in Britain's cotton textile industry and then spread in the nineteenth century to other industries and other countries, had four basic characteristics: (1) mechanization, (2) concentration of a large number of workers under one roof, (3) the imposition of work discipline governing hours and tasks, and (4) devotion to a mass market.[138] There can be no question that the industrial diversification of water power in the medieval period made a very substantial contribution towards the emergence of the first of these characteristics—industrial mechanization. But in the other areas, as well, the medieval water-powered industrial plant contained the germ of the nineteenth-century factory. The introduction of hydropowered dams and canals and more complex and expensive water-powered equipment like hydraulic stamps or the larger bellows used in blast furnaces, for example, increased capital requirements and generally increased the scale of industrial undertakings. These developments, in turn, encouraged the growth of outside capital investment and, ultimately, encouraged the imposition of work discipline on employees who no longer owned the equipment with which they worked, the concentration of a larger work force at a single site, and production for larger markets. At Toulouse, for instance, the construction of large river dams for hydropower clearly encouraged the concentration of works at a single site (e.g., the Bazacle dam), if not under a single roof, and fostered divisions between capital and labor that prepared the way for imposition of work discipline. The monastic complexes, discussed in more detail later in this chapter, made heavy use of water power, imposed a very rigid work discipline governing hours and tasks on initiates, and concentrated a large number of workers under a single roof. Finally, while most medieval watermills probably produced only for the limited local market, the larger mills in certain industries, like the iron and fulling industries, certainly passed beyond this stage. The quantity of iron produced by blast furnaces and hydraulic trip-hammers was much larger than a local market could absorb. Iron mills probably served regional if not national or mass markets.

It would be an exaggeration to assert that the modern factory first emerged in the medieval period. But it is not an exaggeration to make the more limited claim that the roots from which the modern factory system emerged were quite deeply embedded in the Middle Ages and that there were no really sharp breaks between the water-powered fulling and iron mills of the late Middle Ages and the textile mills of Strutt and Arkwright, only a series of rather modest, incremental shifts.

The tremendous expansion in the water wheel's industrial applications in the medieval period provides strong additional evidence for White's postulated "power revolution." In antiquity the impact of water power on a typical Roman could have been only slight: it might have produced his flour or irrigated his crops. But that is all. By the close of the medieval period, however, the water wheel had been applied to literally dozens of different processes. The house medieval man lived in might have been made of wood sawed at a hydropowered sawmill; the bowls he ate from, turned on a water-powered lathe; the pipes that his water flowed from, bored by water power. The flour he ate was probably ground at a watermill; the oil he put on his bread could have been crushed from olives by water wheel. The leather of the shoes he put on his feet and the textiles he wore on his back could have been produced, in part, by water-powered tanning and fulling mills. The iron of his tools could have been mined with the aid of water-powered drainage pumps, ventilating fans, and hoisting devices; was probably smelted in a furnace with water-activated bellows; and was probably forged with hydraulic stamps and bellows. If he was a clerk, the paper he wrote on was, most likely, the product of a water-powered mill. If he was a soldier, his armour and weapons might have been polished and sharpened by stones turned by vertical water wheels. While probably no single individual in the Middle Ages was affected by water power in all of these ways, most were affected in at least several. Water power had become, by the sixteenth century, and probably much earlier, an inescapable part of daily life in the West.

## THE ICONOGRAPHY OF THE MEDIEVAL VERTICAL WATER WHEEL

The medieval revolution in the use of hydropower, and particularly the European success in adapting the vertical water wheel to a wide variety of industrial processes and to a variety of stream conditions, implies the development of a large body of know-how in the area of water power. Unfortunately, most medieval documents provide little precise information on the nature or extent of this knowledge. Most medieval documents are legal rather than technical and refer only to a "watermill," "iron mill," or "fulling mill" on such and such a stream or estate, paying an annual rent of so much. From a legal standpoint the construction details of dam, canal, reservoir, or even wheel were usually unimportant and thus not mentioned.

Information on the design and construction of medieval watermills must therefore be based on a handful of medieval drawings, especially since the archeology of medieval milling is largely untouched. But even in dealing with the extant drawings considerable caution is necessary. Realism was not always the primary concern of medieval artistry, and good perspective techniques were late in emerging. Moreover, the earliest surviving medieval illustrations of water power come only from the twelfth century.

In surviving medieval illustrations, however, it is usually possible to judge whether a wheel was vertical or horizontal and, if the former, whether it was undershot or overshot. Thus, these drawings do give some indication of the wheel type which was preferred late in the medieval period. Based strictly on a count of medieval illustrations, Usher declared that the undershot wheel was the most common form of medieval water wheel.[139] His conclusion has been supported by Maria Dembińska, who found that undershot wheels were more frequently employed than overshot or horizontal wheels in medieval Poland.[140]

The very earliest surviving medieval reproductions of the watermill depict only vertical undershot wheels. Among the oldest, for instance, is a twelfth-century manuscript titled *L'image du monde* by Gautier of Metz. It has a very rude sketch of a medieval town with a mill on its outskirts. Although the figure is small and crude, the mill was clearly vertical and undershot.[141] Slightly later comes a south German manuscript, *Hortus*

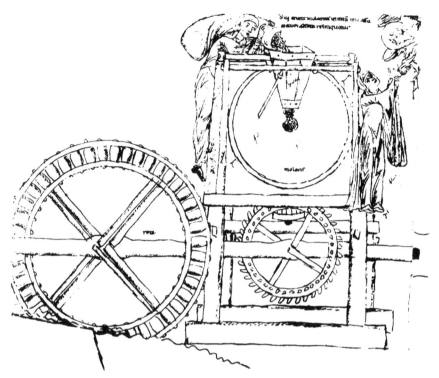

FIGURE 2–34. Watermill from the *Hortus Deliciarum* of the Abbess Herrad of Landsberg, c1190. This is one of the earliest surviving illustrations of a watermill.

*deliciarum,* of the Abbess Harrad of Landsberg. It contained a much larger and more detailed sketch showing, in addition to an undershot wheel operating in a confined channel, the mill's gearing arrangement (Fig. 2–34). A thirteenth-century miniature, from the *Veil rentier* of Audenarde, has what appears to be an external view of an undershot mill supplied with water by a very steeply inclined head race (Fig. 2–35),[142] while a Spanish reliquary from around 1300 pictures an undershot mill grinding grain with its wheel operating in a confined race (Fig. 2–36). The wheel which powered Villard de Honnecourt's sawmill (c1235) seems also to have been vertical and undershot (see Fig. 2–28). There are, in addition, a number of undershot illustrations from the fourteenth and fifteenth centuries.[143]

Little can be determined from these drawings beyond the fact that the wheels were undershot. Presumably they had radial blades (i.e., blades arranged as rays projecting from the wheel's center), though the blades of Honnecourt's wheel appear to have been inclined. Unlike ancient water wheels, most of the undershot wheels of medieval illustrations are without shrouding (rims), suggesting that medieval technicians may have recognized the superior performance of unshrouded undershot wheels or may have developed better methods than ancient engineers

FIGURE 2–35. Vertical water wheel from the *Veil rentier,* c1275. If the vertical wheel shown is, as some believe, a breast wheel, it is the earliest evidence we have of this type of wheel.

FIGURE 2–36. Vertical undershot wheel from a Spanish reliquary of the late thirteenth century.

for linking wheel to shaft, obviating the need for rims to strengthen the wheel.

Illustrations of medieval overshot wheels exist, but later than the earliest undershot illustrations and not earlier than the thirteenth and fourteenth centuries. Then several crude drawings of overshot wheels appear in works like the *Dresdener Bilderhandschrift des Sachsenspiegels,* c1350 (Fig. 2–37).[144] The *Luttrell Psalter,* dating from around 1340, shows an overshot mill fed by water from a dammed mill pond (Fig. 2–38). And a design of the German technician Conrad Kyeser, c1405, indicates very clearly a small chute leading water onto the summit of an overshot wheel (Fig. 2–39). Overshot water wheel illustrations are more frequent in the

FIGURE 2–37. Overshot water wheel from the *Dresdener Bilderhandschrift des Sachsenspiegels*, c1350. One of the oldest surviving drawings of an overshot water wheel.

fifteenth century, as evidenced by the notebooks of Italian engineers like Taccola and Francesco di Georgio Martini (Fig. 2–40).[145] There may be a correlation between the development of hydropower dams and the greater frequency of overshot wheel drawings in the late medieval period. The greater availability of artificially contrived (and hence higher) falls would have given European hydraulic engineers more opportunities to make use of the more efficient and presumably more powerful overshot wheel, which required water supplied over its summit.

As with undershot wheels, it is difficult to determine many technical details from surviving overshot illustrations. We cannot even be sure that the early examples of overshot wheels had buckets, for those sketched in the *Luttrell Psalter* and Kyeser's *Bellifortis* do not clearly show them. On the basis of the drawings alone, one could assume that these were overshot-impact wheels, wheels which had blades instead of buckets and which differed from the undershot wheel only in the point at which water was applied. Although not the most common type of overshot wheel, some wheels of this type were used as late as the eighteenth century. Where buckets are clearly indicated, it seems that the inclined bucket was the preferred medieval style. It was the easiest to construct, since it was made simply by inclining undershot-type blades upward

FIGURE 2–38. Overshot water wheel supplied with water from a dammed mill pond, *Luttrell Psalter*, c1340.

FIGURE 2–39. Overshot water wheel from Conrad Kyeser's *Bellifortis*, c1405.

towards the point where water was received on the wheel and enclosing the sides of the wheel with rims. The more efficient elbow bucket, which later became popular, may also have been available. Although no medieval water-powered wheel is shown with elbow buckets, Kyeser (c1405) drew them on a water-raising wheel (Fig. 2–41).

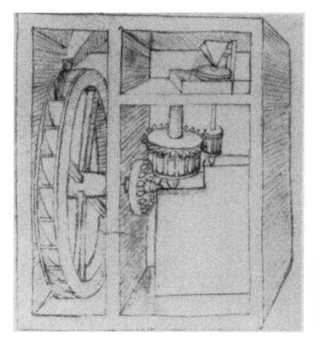

FIGURE 2–40. A late-fifteenth-century flour mill with overshot water wheel from Francesco di Giorgio Martini. Note Martini's use of a tapered spout to deliver water to the wheel.

There are two basic ways to attach a water wheel, overshot or undershot, to its axle or drive shaft—the compass arm method and the clasp arm method. Examples of the former may be seen in Figures 2–22 to 2–25, and elsewhere; examples of the latter in Figures 2–18, 2–19, and 2–31. In the compass arrangement the arms or spokes of the wheel are mortised into the shaft of the wheel. In the clasp arrangement a portion of the axle is squared off and several parallel pairs of spokes arranged so that they "clasp" this part of the shaft. Wedges are used to insure a tight fit. Both methods had shortcomings. A serious disadvantage of the compass arm wheel was that the mortises weakened the axle and allowed water to penetrate to the heart of the timber, so that the axle rotted more quickly. Many millwrights thus preferred the clasp arm arrangement. But clasp arm wheels were not particularly strong at the point where the spokes were wedged tightly against the axle. If the surviving medieval water wheel illustrations are technically correct and representative of practice, the compass arm construction was the more common.[146]

Because of the paucity of available information it is difficult to determine with any precision just how medieval vertical water wheels were superior to ancient water wheels, or if they were at all. But available information does indicate some possible areas of improvement—the use of unshrouded undershot wheels, the development of better methods of linking spokes to shaft, the introduction of more efficient buckets on overshot wheels, and the increased use of the more efficient overshot wheel late in the Middle Ages.

FIGURE 2–41. Water-raising wheel with elbow buckets, Kyeser, c1405. This is the earliest evidence I have been able to locate of the use of elbow buckets. Most water wheels in the medieval period seem to have depended on inclined buckets (see Figs. 2–18, 2–19, and 2–27 for illustrations of inclined buckets).

What of the other form of water-powered prime mover available in the era—the horizontal water wheel?

## THE HORIZONTAL WATER WHEEL AND THE SOCIAL SETTING OF WATER POWER

The relationship of the horizontal water wheel to the vertical wheel and the position of this implement in the medieval water-power revolution present a major puzzle. One set of data seems to indicate that the horizontal wheel was little used in the West; the other set hints that in many areas of Europe it was extensively applied.

The distribution of horizontal mills, for instance, suggests widespread application. Horizontal mills, at one time or another, were found in Norway, Ireland, southern France, Spain, Portugal, the Alpine regions of Austria and Italy, Rumania, Yugoslavia, and Greece.[147] The horizontal wheel seems to have been almost the only type used in grist mills in certain areas of the Austrian Alps in the sixteenth century and around Pistoia in Tuscany in the fourteenth century.[148] The very low rent paid by many of the watermills in the Domesday survey suggests that small horizontal mills may even have prevailed in eleventh-century England.[149] Moreover, some authorities have suggested that horizontal water wheels may have been applied in Europe to more than just grist milling. A blast furnace described by the fifteenth-century architect-engineer Filarete may have been driven by a horizontal wheel.[150] And the engineer Juanelo Turriano, in an unpublished fifteenth-century manuscript, pictured horizontal water wheels carrying out a variety of industrial duties.[151] The simple horizontal mill, furthermore, had clear advantages over its vertical rival. It was cheaper to build; it could be erected much faster; and some could even be moved easily if the stream on which they were located played out. Their simplicity made them easier to operate and repair than the vertical mill. For a peasant household, the cost advantages of a simple horizontal mill over the larger, more expensive, and more complex vertical mill were obvious and should have contributed to its extensive utilization throughout the medieval period.

But this picture of actual or potential widespread use is clouded by other data. The horizontal mill (also called the Norse mill) may not have been known in Norway until the thirteenth century and did not become common there until the fourteenth.[152] The first clear description of the horizontal grist mill comes only from around 900 A.D.,[153] almost a thousand years after Vitruvius described the vertical mill. The earliest medieval illustration of a watermill dates from around the mid-twelfth century, and it is vertical. From around 1150 to after 1400, in fact, only vertical water wheels are pictured in extant drawings.[154] In the machine books of the Renaissance vertical wheels are pictured much more often. In Ramelli's *Diverse et artificiose machine* of 1588, for example, illustrations of vertical wheels outnumber those of horizontal wheels four to one. Agricola's *De re metallica* of 1556 probably reflected contemporary practice more closely than Ramelli. It pictured a large number of vertical wheels in use in mining activities, but no horizontal wheels. Travellers who saw horizontal wheels in the sixteenth and seventeenth centuries commented on their novelty, in itself an indication that they were rare in most of western Europe by this time.[155] And Jacob Leupold, the German technical writer, wondered in the early eighteenth century why the horizontal wheel was so rare in view of its advantages in price and simplicity over the vertical wheel.[156] Finally, Anders Jesperson in this century observed: "From my study of literature on this point [the comparative price for building and operating vertical and horizontal watermills] I am fairly satisfied that the vertical mill could offer no serious competition to the horizontal, yet by 1600 AD the vertical mill had superseded the horizontal in the majority of the areas."[157]

The best reasoned attempt to ferret out the horizontal mill's role in the expansion of hydropower in the Middle Ages has come from Abbot Payson Usher. He postulated that the horizontal grist mill was the predominant type in Europe between the fifth and ninth centuries and that it was displaced by the vertical wheel only after that period.[158] Usher's theory is largely based on the early Irish and Teutonic law codes. He argues as follows. The *Senchus Mor,* an Irish law code dating back at least to the ninth or tenth century A.D. and probably earlier, lists the parts of a watermill in terms that make it certain that a horizontal mill is being described. Roughly contemporary with the *Senchus Mor* was another primary text of early Irish law, the *Book of Aicill.* In the latter there is a long section describing responsibilities for accidents at the mill. The passages suggest that the mill was provided by the owner for the use of others, that there was free access to the mill, that it was operated intermittently, and that there was no class of professional millers. Each party ground his own grain at the mill.[159] Provisions in the early Teutonic law codes seem to indicate that similar conditions prevailed at Continental mills. Hence, argues Usher, it is likely that these codes were also referring to the horizontal mill.[160] And since Teutonic codes were compiled between the fifth and ninth centuries, the horizontal mill was probably the dominant form in early medieval Europe.

The professional miller, restricted access to the mill, regular rather than intermittent use, and the imposition of the obligation to use the lord's mill were all features of feudal law. These features began to appear only in the ninth century, for instance in the statutes of Adalhard, abbot of Corbie, first issued in 822.[161] Within the next century or two these new legal arrangements for watermilling were consolidated, signaling, to Usher, the transition from horizontal to vertical watermilling.[162] Thus, by the time the first extant illustrations of watermills appear in the twelfth century, the vertical water wheel was dominant and the absence of the horizontal mill illustrations explicable. Moreover, since the European horizontal mill was primarily a grist mill, Usher's hypothesis would explain in part why the new applications of water power began to emerge only around 1000 A.D., for it was only then that the more adaptable vertical water wheel began to replace the horizontal.

Usher's edifice, however, rests on some very shaky assumptions. For example, Usher assumes that small mills, free access to the mill, and the absence of a professional milling class are characteristic only of horizontal mills and that the professional miller, restricted access to the mill, the feudal milling monopoly, and regular rather than intermittent use are characteristic of vertical watermilling. But these distinctions are not necessarily valid. The grain mills of late medieval Pistoia, in northern Italy, were horizontal, yet Muendel indicates that many of them were operated, presumably with regularity, by a class of professional millers.[163] Moreover, there is no logical reason why vertical mills could not be small, have free access, and be operated irregularly without professional millers. The changes which led to the emergence of large mills, the professional miller, and the obligation to grind at the lord's mill may have been just other aspects of the general tightening up of feudal authority in the period 800 to 1000 and may have been applicable to whatever mills were in operation, vertical or horizontal.

Contrary to Usher, I suspect that in the early medieval period neither the vertical nor the horizontal wheel was dominant over Europe as a whole. Each probably had pockets where is was dominant, areas where topographic or economic conditions gave it an advantage. The horizontal mill, for instance, was probably used more extensively by small, independent freeholders, especially in sparsely populated areas, in areas where commercial considerations of scale and efficiency were not important, and in areas where transportation facilities were poor. These areas often had the mountainous or hilly topography for which the horizontal mill was suited. The larger (usually) vertical mill, on the other hand, was probably the more frequent type in areas where the demand for flour was greater—in more urbanized areas, in monastic centers, in the more densely populated lowlands on large estates. Since superior water transportation routes and dense populations tended to occur more frequently in the foothills and plains, where water volumes were larger but water velocity was slower, there was a need not only for mills with larger capacity, but for mills with gearing (i.e., vertical mills) so that the slow velocity of the water and water wheel could be stepped up to the 120 revolutions per minute needed for grain milling.

In areas between the two extremes outlined above, vertical and horizontal watermilling traditions probably existed side by side, the form of prime mover utilized depending on the individual's or group's needs. The remnants of a Saxon watermilling establishment of the mid-ninth century, for example, indicated the use of both vertical and horizontal wheels.[164] In the sixteenth-century Austrian Alps, in the Gastein Valley, watermills used to supply urban areas and mining camps were vertical; those used to satisfy the needs of individual peasants or freeholders were horizontal.[165]

Why, then, was the vertical wheel able to oust the horizontal wheel from most of western Europe by the end of the medieval period? We do not have clear descriptions of the horizontal mills used in the early medieval period, but if we assume that they were similar to the primitive horizontal wheels used in the backwaters of Europe in the nineteenth and twentieth centuries, their eclipse presents no mystery. In the first place, the primitive horizontal wheel was inefficient. It could transfer only about 5 to 15% of the power of the water acting against its wheel to the shaft. The vertical undershot wheel, on the other hand, delivered an efficiency of 20 to 30% and the vertical overshot wheel, 50 to 60%. The primitive horizontal wheel was usually very small, often capable of generating no more than 0.5 to 1.0 horsepower. A vertical water wheel typically generated four to ten times more power. The conditions under which the primitive horizontal wheel operated effectively were limited. It was usually dependent on very small volumes of water with fairly high falls. The vertical wheel, in one or another of its forms (overshot or undershot), could handle a wide variety of stream conditions—large volumes or small volumes, high falls or low falls. Moreover, the vertical mill's gearing gave the millwright the opportunity to adjust the speed of his millstones to specific stream conditions, an option not available with primitive horizontal mills. Finally, the vertical water wheel was almost certainly the better design for most industrial applications of water power, for it was easier to adapt to tasks involving pounding or rotary motion in the vertical plane. Thus, as water power was expanded from grain milling to other tasks, the vertical wheel naturally came to the fore.[166] Even in areas where horizontal wheels continued to be used extensively in the medieval period, like Pistoia in Italy, vertical wheels dominated industrial (non–grain milling) applications of water power.[167]

But technological superiority alone cannot explain the all-but-complete dominance assumed by the vertical watermill in much of western Europe, for it displaced the horizontal mill even in many areas where water was plentiful, power demand low because of sparse population, and the grinding of grain the chief application for water power, areas where, it would seem, the horizontal mill's low first cost and ease of operation should have continued to give it an economic advantage. The incorporation of the watermill into the manorial system, as Usher suggests, probably provides the best explanation for this phenomenon.

When central authority broke down in the ninth and tenth centuries, the emerging feudal lords successfully assumed exclusive jurisdiction over a number of areas where previously their authority had been limited. One of these areas was water rights. In substantial areas of western Europe the feudal aristocracy utilized this authority to impose the "soke," a milling monopoly which compelled all tenants to grind grain only at the lord's watermill. This monopoly was one of the most widespread and was fairly general, though not universal, in England, France, and Germany.[168]

Since horizontal watermills were primarily suitable for small power units and capable of grinding the grain of only a few families, they were both technically and economically unsuitable in a situation calling for a centralized, monopolistic mill to serve not only the lord's family, but also large numbers of retainers and serfs. The larger vertical mill was much more appropriate. It is even possible that the peasant-owned horizontal watermills of earlier centuries were outlawed or destroyed to make the manorial mill monopoly effective. We lack concrete evidence of this, but there are cases of the destruction or outlawing of peasant-owned handmills.[169] It is therefore at least possible that peasant-owned horizontal mills suffered a similar fate. The hypothesis that consolidation of the feudal system and the incorporation of the watermill into that system drove the primitive horizontal mill out of much of Europe is supported by the fact that primitive horizontal mills tended to survive longest primarily in those areas where feudal authority was weakest—the Alps, the Shetlands, Norway, the Balkans, and Ireland, for instance.[170]

The position of the horizontal mill may also have been weakened by the growth of urban centers, the revival of trade, and the beginnings of a capitalist market system in the second half of the Middle Ages. In a subsistence peasant agricultural economy the primitive horizontal mill had one great advantage over the vertical mill—lower first cost. It obtained this advantage at the price of low power output. But low first cost declined in importance when Europe in the eleventh and twelfth centuries began to grow prosperous and develop cooperative financial arrangements which made it easier to raise capital. At the same time greater power output grew in importance as the growth of cities and trade increased the size of markets and the emergence of a capitalist economy increased the importance of profit. A vertical watermill was usually able to grind more grain, serve a larger market, and, thus, bring in a larger profit than the typically small horizontal mill. It enjoyed, in addition, certain economies of scale. Thus the higher power output of the vertical wheel became steadily more important than the low first cost of the horizontal wheel, leading to the horizontal wheel's slow disappearance from much of western Europe. Only during the Renaissance, when horizontal wheels were modified by the addition of curved blades, improved nozzles, and tight-fitting masonry conduits, did they become competitive with vertical wheels in the development of power in large units.[171]

Thus a combination of factors regulated the horizontal mill to a minor role in European society and in the medieval European economy. The primitive horizontal mill's technical deficiencies were important. It was inherently inefficient and had a small power output. It was simply not well suited to the intensive development of power, nor to industrial applications involving crushing, blowing, lifting, or pounding. But social factors also contributed to the dominance of the vertical wheel, especially in the area of grain milling. The manorial milling monpoly and the emergence of the merchant mill in urban areas gave an economic advantage to large mills with a large power output sufficient to satisfy the needs of many families—that is, to the vertical watermill. Thus, by the late medieval period the horizontal wheel had been relegated to the backwaters of Europe, where it largely remained an item of subsistence peasant agriculture. The vertical wheel, in the meantime, became the workhorse of European industry.[172]

The substitution of larger vertical water wheels for primitive horizontal water wheels, in conjunction with other developments—such as the wide geographical dispersion of the watermill, the substantial number of watermills known to have been in operation in medieval Europe, the general tendency of medieval hydropower engineers to tap progressively larger streams, and the development of a multitude of new industrial applications of water power—provides a large foundation of data which supports White's contention that there was a power revolution in medieval Europe, a revolution which saw the rapid replacement of human by nonhuman power sources.

Why was the medieval West so much more responsive to water power than classical antiquity and, as we shall see later in this chapter, contemporary cultures to the south and east? This question is a difficult one to answer, but part of the answer surely lies in the predominance of Christianity during the Western Middle Ages. Christianity was one of the most distinctive elements of medieval Western culture; it pervaded practically every aspect of medieval life and thought. It most certainly had, in one way or another, directly or indirectly, an important influence on the adoption and diffusion of water power in medieval Europe. Nowhere is this influence seen more clearly than in the Western monastic tradition.[173]

Two elements in the Western monastic tradition made water power important to European monasteries: (1) a belief in the dignity of manual labor, an idea foreign to most of the prevailing philosophies of antiquity, and (2) the desire for self-sufficiency as a means of isolating the monastery from wordly corruption. These ideals were introduced into Western monasticism by Saint Benedict. Reacting against the excesses of oriental asceticism and austerity and the disorganization of the Western monastic movement, Saint Benedict established early in the sixth century a set of rather rigid rules for monastic brotherhoods, rules which quickly gained widespread acceptance in Europe. Benedict incorporated into his rules a belief in the dignity of manual labor, an assertion that work was worship. Benedictine monks were therefore required to devote certain rigidly regulated periods of time to manual labor, as well as to reading and study and to spiritual duties like prayer and meditation.[174] The equal weight given the physical, intellectual, and spiritual elements of Benedict's program, coupled with the Benedictine inclination towards self-sufficiency, an inclination prompted both by desire for isolation from worldly influences and by political and economic conditions in early medieval Europe, provided a solid incentive for monastic use of water power. Only by adopting water power for tedious and very time-consuming tasks like flour production could Benedict's program be carried out.

Watermills thus early emerged as an essential element of monastic life in the West. Saint Benedict himself may have strongly encouraged their use, for his rule specifies: "The monastery should, if possible, be so arranged that all necessary things, such as water, mill [*molendinum*], garden, and various crafts may be within the enclosure, so that the monks may not be compelled to wander outside it."[175] The term *molendinum,* or "mill," does not have to mean watermill. But in the Latin West in the Middle Ages the term was used so frequently in conjunction with streams of water that it is clear that *molendinum* generally did mean watermill.[176] Whether Benedict meant "watermill" or simply "mill," leaving the power source to individual discretion, most monasteries eventually came to depend on water power. For example, Luckhurst, who studied water power in English monasteries, concluded that "the watermill was an integral part of those monastic houses which lay near a suitable water supply."[177]

Certain monastic brotherhoods, notably the Cistercians, who concentrated on developing or reclaiming underpopulated waste areas, depended quite heavily on water power.[178] Founded early in the twelfth century, the Cistercian order had over 500 monasteries by the opening of the thirteenth century. Most of these seem to have owned watermills. Constance Berman, for instance, found that almost all of the Cistercian abbeys in the county of Toulouse in the twelfth and thirteenth centuries acquired watermills.[179] And David Williams found that practically all Welsh Cistercian monasteries had watermills.[180]

While most Western monasteries had at least one watermill, many had more. An abbey in the Pas-de-Calais had three watermills by 855, while the abbey of Corbie (Somme) had mills with up to six wheels in 822.[181] In 845 the monastery of Montier-en-Der (Haute Marne) owned 11 watermills on the river Voire, while the monastery of Saint-Germain de Prés owned 59.[182] The abbey at Silvanès in southern France, founded in 1136, was small and never particularly important. Yet documents dating from the period 1150 to 1170 indicate that it already had four or five watermills, used both for flour milling and fulling, as well as two dams which were at least partially used for power purposes.[183] The Welsh Cistercian abbey of Whitland, founded in 1140, owned ten watermills.[184] In the county of Toulouse, in southwest France, the monastery of Grandselve owned 16 watermills by around 1250, Gimont owned 14, and Bennecombe and Berdoues owned 12 each.[185]

The mills erected or purchased by monastic brotherhoods were not only flour mills. Donkin found that 28 of the English Cistercian houses owned at least one fulling mill, and many owned more. Newminster in northeastern England owned five.[186] The earliest water-powered hemp mills, beer mills, tanning mills, hammer mills, and ore stamps of which we have knowledge were monastic mills, and the Cistercians, in particular, played a very active role in the medieval iron industry.[187]

Moreover, the water-power systems installed at some abbeys were quite sophisticated. At Royaumont, near Paris, monks constructed a structure over a small river, creating a tunnel 105 feet (32 m) long and 7.71 feet (2.35 m) wide. The river flowed through this tunnel, activating vertical water wheels which ground grain, tanned, fulled, and worked iron. At the abbey of Vaux de Gernay, also near Paris, monks erected a similar structure.[188] At Fontenay in Burgandy a long structure containing water wheels was built alongside a river.[189]

Monastic enthusiasm for water power is perhaps best exemplified by the case of Clairvaux, in France. Clairvaux was rebuilt in 1136 by Saint Bernard, founder of the Cistercian order. A contemporary of Saint

Bernard, Abbot Arnold of Bonneval, described the rebuilding of Clairvaux in the twelfth century without mentioning the church. He gloated, instead, over the abbey's use of water power. His description of Clairvaux's water system is worth quoting:

> Entering the Abbey under the boundary wall, which like a janitor allows it to pass, the stream hurls itself impetuously at the mill where in a welter of movement it strains itself, first to crush the wheat beneath the weight of the millstones, then to shake the fine sieve which separates flour from bran.
>
> Already it has reached the next building; it replenishes the vats... to prepare beer for the monks....
>
> The stream does not yet consider itself discharged. The fullers established near the mill beckon to it. In the mill it had been occupied in preparing food for the brethren; it is therefore only right that it should now look to their clothing. It never shrinks back or refuses to do anything that is asked of it. One by one it lifts and drops the heavy pestels, the fullers' great wooden hammers. When it has spun the shaft as fast as any wheel can move, it disappears in a foaming frenzy....
>
> Leaving here it enters the tannery, where in preparing the leather for the shoes of the monks it exercises as much exertion as diligence; then it dissolves in a host of streamlets and proceeds along its appointed course to the duties laid down for it, looking out all the time for affairs requiring its attention, whatever they might be, such as cooking, sieving, turning, grinding, watering or washing, never refusing its assistance in any task. At last, in case it receives any reward for work which it has not done, it carries away the waste and leaves everywhere spotless.[190]

Arnold's pride in the abbey's ultilization of water power technology is clear. He was not alone. Another monastic observer of Clairvaux showed the same enthusiasm.[191]

It was installations like that at Clairvaux which permitted the monastic brotherhoods in the West to be completely self-sufficient, while still allowing their initiates the appointed periods for labor, study, prayer, meditation, and the church offices.

In addition to contributing directly to the diffusion of water power in medieval Europe, Western monasticism probably also made an indirect contribution. In the Middle Ages technology was diffused primarily through the movement of craftsmen and artisans. The construction of new monasteries with water-power apparatus thus diffused knowledge of water power, for it brought lay craftsmen in new areas into contact with advanced elements of water-power technology.

The successful use of watermills on monastic precincts probably also provided considerable encouragement to lay lords to construct mills. The feudal aristocracy faced many of the same problems as monasteries. They, too, sought to be self-sufficient. They, too, had to support large numbers of retainers and agricultural servants and provide them with the necessities of life while providing periods for time-consuming pursuits like training men to fight mounted on horseback.[192] Unfortunately, there is no definative evidence to prove that lay lords adopted water-mills as a result of monastic influence. But evidence is strong in other areas—e.g., sheep farming—that lay lords adopted monastic practices they considered advantageous, so it is likely that they imitated monasteries in the area of watermilling as well.[193]

While Latin Christianity had its strongest impact on the growth of water power through its monasteries, there are several more subtle ways in which it could have exerted an influence favorable to the growth of water power. For example, as noted in the previous chapter, the Christian view of nature was more aggressive than the animalistic religions of antiquity. This more aggressive attitude may have contributed to the creation of a "cultural climate" which made men more willing to undertake massive modifications to the natural flow of streams. The greater dignity accorded manual labor by medieval Christendom may also have contributed to the creation of a better cultural climate for technological development.[194] Finally, Christian compassion may have had an occasional role in the adoption of water power, for it led a sixth-century abbot to build a watermill in order to free his monks from the monotony of grinding wheat by hand.[195]

The program of Western monasticism quite clearly played an important tole in the acceptance and diffusion of water power in medieval Europe. More generally, Christianity may have created an environment favorable to its wider use. But other factors also contributed to the medieval revolution in water power, among them:

1. early medieval Europe's labor scarcity,
2. the geography of western Europe, and
3. the emergence of social classes with an economic interest in the construction and operation of watermills.

Late in the Roman era, as we have seen already, Europe had already begun to experience problems with a shortage of labor, particularly agricultural labor. This problem grew worse in the early medieval period. The population of west and central Europe, declining already in the Roman period, declined further, dropping, according to one estimate, from nine million to five and one half million people between 500 and 650 A.D.[196] In this context a labor-saving device like the water wheel assumed more importance than it had in much of Greco-Roman antiquity, when labor was plentiful. Even a primitive horizontal watermill could save about one man day per week for every group of five adults, an important gain if labor was in short supply and if the gain could be achieved cheaply (as it could with the primitive horizontal mill). The labor saved was probably primarily the labor of women, for women traditionally had the task of grinding grain for their families' meals, a monotonous task which required two hours with the hand quern per day. The time gained from the use of watermills would allow women to work more intensively in the gardens or in the fields and provided more time for work on other tasks, like textile production, during the winter months.[197] It is thus quite likely that the demographic crisis of early medieval Europe provided an important stimulus to the more rapid adoption and diffusion of water power.

Europe's labor scarcity may also have been responsible for creating the advantageous cultural climate which Lynn White has considered to be a key to Western medieval technological advance. One could argue, for instance, that the Latin Church's emphasis on work as worship and the elevated position it accorded manual labor were initially rooted in and responses to the economic realities of severe labor shortages.

Geographical circumstances, however, probably also contributed to the medieval West's increased use of water power vis-à-vis both classical civilizations and contemporary civilizations. Much of the area under the sway of the Egyptian, Chinese, Roman, and Islamic civilizations was either arid or had rainfall of a highly seasonal nature. Often there were few streams suitable for easy development of water power. Moreover, the limited supplies of water power available had other uses, notably irrigation and transportation. In these civilizations the use of water power for irrigation and transportation not only antedated but was usually considered to be more vital than the use of water for mechanical power. Thus, even though the water wheel was occasionally used in Egypt, China, the Roman Empire, and Islam, its expansion was, in part, limited by the priority given to navigation and irrigation for the use of water.

Western Europe, on the other hand, was generally well watered. There was no need for irrigation, and there were an abundance of small to medium-sized streams, with relatively constant year-round flow, suitable for water-power development. The construction of watermills on many streams even antedated the emergence of large-scale trade and hence navigation. As trade grew in western Europe, navigation and milling interests often came into conflict, but in the West the needs of navigation were never able to firmly and definitely establish priority over the needs of power production. Irrigation and navigation were therefore, as a result of the geography of western Europe, not the barriers to the expansion of water power in the West that they were in many other civilizations.

One final factor which may have contributed to the impressive growth of water power in the medieval West was the existence of powerful social classes with an interest in its development. As we have seen, in large parts of Europe the Western feudal aristocracy, both military and clerical, successfully imposed the "soke," or manorial milling monopoly, on their peasants after the ninth or tenth century. The successful imposition of this monopoly and the profits it brought the aristocracy very likely contributed to the expanded use of water power. In fact, the profits gained from the soke were apparently so attractive that in some areas of Europe it was extended to fulling, tanning, and other processes as they began to exploit water power. Carus-Wilson, for example, found that the high profits obtainable from the manorial milling monopoly provided the original incentive for the development of the water-powered fulling industry in England.[198] Towards the end of the medieval period, with the revival of urban life and trade, yet another influential social group emerged which had an interest in the application of water power—the merchant class. Most other cultures, both previous and contemporary, did not develop powerful social classes with vested interests in the expansion of water power. Their aristocracies were quite different from the aristocracy created by the Western feudal system, and their merchants (as well as their craftsmen) were largely powerless and often severely restricted in their activities by governmental fiat.

In conclusion, the West's comparatively rapid development of water power in the Middle Ages was probably due to a combination of factors. The general cultural environment of medieval Europe may have been important. The Roman Catholic Church's attitude towards manual labor and towards nature, for example, was favorable to technological development generally and to the use of water power specifically. More important, however, was the Western monastic movement, with its emphasis on self-sufficiency and work as worship. It spread knowledge of water-power technology all over Europe and provided an example to lay lords. Perhaps as important as monasticism in encouraging the use of water power in early medieval Europe was a severe labor shortage. This shortage not only promoted the adoption of labor-saving devices like the water wheel, but it may have influenced the Latin Church's attitudes towards craftmanship, labor, and nature. An additional bias towards the use of water power was provided by Europe's geography, a geography which provided it with abundant rainfall (hence no conflict with irrigation interests) and numerous streams which flowed at a relatively constant volume (compared to the streams further south). Finally, the way the watermill was incorporated, first, into feudal/manorial society and, later, into the emerging capitalist/mercantile economy gave powerful social classes (the landed aristocracy, in the first case; the rising merchant class, in the second case) a vested interest in the use and development of water power. The support of these classes insured the continued expansion of water power in Europe, even after the severe labor shortages that had early encouraged its use had disappeared.

## WATER POWER IN CHINA, ISLAM, AND BYZANTIUM

The West's achievements in hydropower utilization during the medieval period can best be comprehended by comparing them to the accomplishments of several of the other great cultures of the same era—China, Islam, and Byzantium.

The early history of water power in China, as noted in the previous chapter, is confused by uncertainties over the nature of the water-

FIGURE 2–42. Battery of Chinese hydraulic trip-hammers, seventeenth century. Mills using vertical water wheels and batteries of recumbent trip-hammers like these were common in the Chinese medieval period.

powered prime mover involved in early Chinese references to water power. Even if, as I have postulated, the water wheel did not begin to replace the water level until around the third century A.D., it is clear that by the Mongol era (1280–1386), it was predominant. Of the two basic forms of water wheel—vertical and horizontal—the horizontal seems to have been the preferred type at least from 1313, when extant illustrations of water-powered systems began to appear.[199] But vertical wheels were also used extensively, especially for work involving the use of hydraulic trip-hammers (Fig. 2–42).[200]

In size medieval Chinese watermills were probably comparable to European mills. For instance, Kao Li-Shih in 748 A.D. owned a mill with five water wheels which ground 300 bushels a day. Assuming that the mill operated ten hours a day, each of its wheels would have had an output of around 6 horsepower, or if 12 hours a day, 5 horsepower. In the fourteenth century Wang Chên reported Chinese mills large enough to meet the daily needs of a thousand families.[201]

In water-power diversification China early led the West. The evidence presented by Needham indicates that by the fall of the Han dynasty (220 A.D.) the Chinese were already using water power for metallurgical bellows and dehusking rice, in addition to raising water and grinding grain. Vertical water wheels may also have been in use by this time for trip-hammers in forges and for providing motion to astronomical equipment, a task to which they were certainly being applied by the twelfth century.[202]

Diversification of the water wheel's functions continued during the Chinese medieval period. But the rate was slower than Europe's, especially after 1000 A.D. Hydraulic trip-hammers had found use in crushing mica and other minerals used in Taoist drugs by the eighth century A.D.[203] The water wheel was applied to driving fans in the imperial palace in the same century, and to grinding tea by the eleventh.[204] In the fourteenth century there is evidence of water-powered textile mills for spinning hemp and ramie.[205] But the earliest Chinese sawmills come only from 1627, the earliest Chinese paper mills from only around 1570, and mills for winding silk from cocoons from 1780.[206] In the seventeenth century Sung Ying-Hsing described a watermill that would simultaneously grind flour, pound rice, and lift water for irrigation.[207] While the list of Chinese applications of water power is moderately impressive, it does not compare to the European accomplishment, particularly in the period from the eleventh to sixteenth centuries.

While watermills did play some role in Chinese life, they seem never to have occupied the position of importance they did in the West. Why was this? One possible explanation is population. In China manpower was more abundant than in the West. Hence the need for resorting to hydropower was correspondingly less. But other factors contributed. As we have seen, in the West, power usually took first place on the list of priorities for the use of water, despite the occasional complaints of navigation interests. This priority was insured by the monopoly which the Western feudal nobility was able to enforce over water and milling rights, for this gave the ruling class a vested interest in the use of water power that for centuries no other group was able to challenge effectively. In China it was otherwise. Chinese interest in waterways centered on transportation, irrigation, and flood control. And when these needs conflicted with water power in China, it was usually the latter that gave way. In 721, for example, Li Yuan-Hung was allowed to destroy a number of watermills on the grounds that they jeopardized irrigation projects. In 764 Li Hsi-Yün demolished 70 watermills belonging to wealthy merchants and Buddhist abbeys on similar grounds. In 778 eighty more watermills were torn down for the same reason. One of the first actions of Fang-I when be became governor of Fuchow in 653 was the repair of the moat surrounding the city and the removal of watermills from the area. He also heavily taxed the remaining watermills to feed the poor. The Thang ordinances (737 A.D.) make it clear that the use of water power came second to transportation and irrigation in China. Eventually, in the tenth century, all watermilling was put under the control of a bureaucratic commission to keep it subservient to general water conservancy programs.[208] The Confucian bureaucracy in China thus provided a check to those very groups—merchants, nobility, and the monasteries—which were the most active in the expansion, diversification, and use of water power in the West.

Unfortunately, there is no survey of technology in Islam to compare to Joseph Needham's volumes on China. Hence, evidence for the use of water power in the Moslem world is scattered, and no clear picture has emerged. It is not even certain when the Moslems began to use the water wheel. It may have been due to Byzantine influences, for several Moslem writers mention a Byzantine ambassador to Baghdad who, around 775 A.D., constructed a large watermilling complex for the Caliph.[209] The complex, however, may have been built at an even earlier date, the work of Nestorian Christians.[210] And there is no good reason to suppose that water wheels were not at work in the Near East at a much earlier date, since, as mentioned in Chapter 1, watermills appeared early in that area and could easily have been inherited by the Moslems.

Whenever watermilling was introduced, it is clear from scattered remarks in the works of various medieval Moslem geographers and historians that by the eleventh or twelfth centuries it was known from one end of the Islamic world to the other. Watermills were reported in Spain,[211] Palestine,[212] Syria,[213] Mesopotamia,[214] Azerbaidzhan (modern U.S.S.R.),[215] both western[216] and eastern[217] Persia, Turkistan,[218] and within the boundaries of modern Afganistan[219] and Pakistan.[220] The noria, too, was widespread, particularly in Syria and Spain.[221]

The density of watermills in the Moslem world is difficult to determine. There are a few remarks here and there which indicate some heavy local concentrations, but no solid statistics. Istakhri, writing in 951, for example, commented that one of Baghdad's canals, the Isa, was not navigable for large boats on account of its weirs, dams, and watermills.[222] Al-Idrisi, writing in 1154 of Damascus, commented that within the city "there are many mills, on the streams."[223] An eleventh-century author noted of Antioch, also, that there were "many mills" along the banks of the Orontes,[224] while Yakut in 1225 noted that the small city of Imm, between Aleppo and Antioch, had a spring and "all round it were mills."[225] In the tenth century Ibn Hawkal described Jiruft in southeastern Iran as a great city, noting sufficient water power for turning from 20 to 50 mill wheels.[226] Balkh, in modern Afganistan, had in the tenth century ten watermills turned by a small river.[227] Samarkand was reported to have many water wheels.[228] Finally, large numbers of water wheels were reported on several occasions in the Nishapur area (extreme northeastern Iran). In 1340 Mustawfi commented on a stream which in spring had sufficient water to turn 20 mills in the course of 20 leagues, but at other seasons was dry. He also noted of the Nishapur

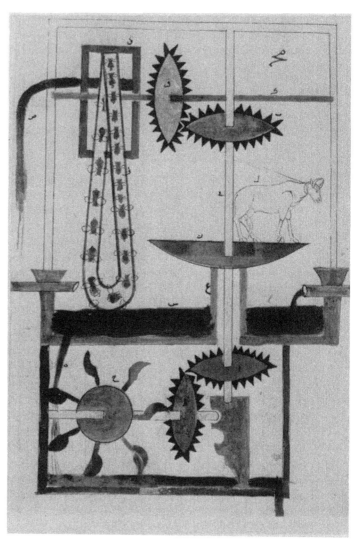

FIGURE 2–43. Water-raising system from al-Jazari, c1205. Note the water wheel at the lower left with its spoon-shaped paddles or buckets. Wheels of this type, operated both undershot and overshot, appear in several medieval Arabic manuscripts.

River that it rose in the mountains 2 leagues to the east of the city, but had sufficient current to turn 40 mills before it reached the town.[229] Earlier, around 960 A.D., Muqaddasi had reported that the river turned 70 mills.[230]

Exactly what type of watermills these were—vertical or horizontal—is impossible to determine, since most references to watermills in medieval Moslem documents, like medieval European documents, merely mention mills, with no technical details. The Moslem world certainly knew both types. Al-Jazari, who wrote a treatise on mechanical devices in the twelfth century, described horizontal vaned wheels, undershot paddle wheels, and both undershot and overshot spoon wheels (see Fig. 2–43).[231] The boat, bridge, and tide mills of the Arab world, as well as its norias, certainly had the vertical configuration. And this may well have been the dominant form, for al-Quazimi noted that a wheel on Majorca was worked as an overshot wheel when the amount of water available was too small for it to be worked as "common" mills were.[232] On the other hand, Donald Hill, in his translation of al-Jazari's treatise, indicated that the wheel type most familiar to that author was a horizontal vaned wheel (similar in appearance to an American windmill laid on its side), called by some Arabic writers a "Byzantine mill."[233] Different portions of the Moslem world may have preferred different wheels, but until more detailed studies are made of water power in the Moslem world the question of which form of wheel was preferred where must remain in abeyance.

Perhaps because water was a much rarer commodity in Islam than in Europe the Moslem world more quickly developed some artificial contrivances for adapting streams of water to the water wheel and the water wheel to unfavorable stream conditions. The earliest known bridge mill, as already mentioned, was one at Cordoba, described by al-Idrisi in the twelfth century.[234] The earliest Moslem tide mill of which we have knowledge was one located on a canal at Basra in southern Iraq. It was clearly described around 960 A.D. by al-Muqaddasi.[235] It anticipates the earliest known European tide mill by a century and suggests that the tide mill may well have been a Moslem contribution to European technology. Other than this isolated example, however, there is little other indication of Islamic application of tide mills.

The Moslems seem to have made somewhat wider use of dams for hydropower than bridge or tide mills. The earliest solid evidence for the use of dams for water power in Islam comes quite early, from al-Muqaddasi, who described around 960 a dam at Band-e-Amir, in Fars province in southern Iran. This dam was designed primarily for irrigation, but its water also powered water-lifting wheels and watermills.[236] Norman Smith says there were other Moslem hydropower dams on the Karun, Kur, Helmund, and Oxus rivers in the East and on the Guadalquivir in Spain.[237] He also notes the Pul-i-Bulaiti dam on the Ab-i-Gargar in southwestern Iran, built with tunnels cut through the rock on either side of the main channel to serve watermills.[238]

The early development of tide mills, bridge mills, and hydropower dams in medieval Islam certainly indicates that Moslem technicians, and particularly those at the eastern (Iran) and western (Spain) extremes of medieval Islam, were fully capable of building water-powered industrial plants and suggests that the use of water power in Islam may have been far more extensive, at least in some regions, than commonly supposed.

But the use of water power in the Islamic world seems to have been restricted to two tasks—grinding grain and raising water.[239] There have been a few authorities who have suggested a slightly more diversified role for water power in Islam, but largely on inadequate evidence. Norman Smith, for instance, in his *History of Dams* mentions a Persian dam called "Fullers' Dam" and a canal at Samarkand called "Fullers' River," and suggests that they were used to drive water wheels for fulling mills.[240] But the fulling process requires a large quantity of water simply for washing and rinsing. In the absence of any other firm evidence of water-power diversification in the medieval Moslem world, we can only assume that this, and not the presence of water-powered fulling mills, accounts for the names.

It is fairly evident that while watermills were used in many portions of Islam, they never came to mean to Islamic civilization anything like what they meant to Europe. To identify why this was the case would take an intensive study of Moslem use of water power, Moslem water laws, economic conditions of Islam, Moslem technology, and the like. Most of these studies have not yet been made. It is at least possible that the key factor was the relative scarcity of water in Islam. But this does not provide a completely satisfactory explanation. Willocks, writing around 1900, noted that in the Egyptian irrigation system there were many regulators near important towns where a relatively constant fall and volume of water could have been used to develop water power. Yet the only mills powered by water in all of Egypt were confined to the Fayoum district.[241]

It is also possible that the deprivations of the Mongols in the thirteenth century and of Timur in the fourteenth century set back a promising Moslem beginning in the use of waterpower at the eastern end of Islam. There may be some significance, for example, to the fact that a pre-Mongol author reported 70 watermills in Nishapur, a city sacked by the Mongols in 1221, while later reporters (for example, Mustawfi in the fourteenth century) indicated only 40 in the area.[242] The large number of canals which were destroyed or silted up during the thirteenth and fourteenth centuries may well have set back the use of hydropower in the eastern portions of Islam by centuries.

Other factors may have entered into Islam's poor showing in the use of water power vis-à-vis the West in the late medieval period. The Moslem industrial tradition, with its emphasis on hand-wrought goods manufactured with close attention to detail, may have contributed to the neglect of industrial applications of water power. It has also been suggested that Moslem water laws, which grew out of principles laid down in the Koran, inhibited things, since these laws declared that no one could own or monopolize a river. Slavery and the abundance of cheap labor may have contributed, as well as Islamic laws and customs discouraging the rise of a manufacturing class that could have made personal profits exploiting water power. Finally, the Moslem world's near reverence for water could have deterred the use of water for power. When asked why they had no mill on a nearby brook, the citizens of Ijli, Morocco, replied:

"How could we compel the sweet water to turn a mill?"[243] Some combination of factors like these probably led to the Islamic world's failure to apply and diversify water power on the same scale as had the West.[244] In brief, medieval Islam, like medieval China, anticipated some of the medieval West's accomplishments in water power. But, like China, Islam did not push the use of water power to the limits that the West did.

For the history of water power in the Byzantine Empire there is even less information than for Islam. That the Eastern Orthodox cultural sphere failed to keep up with western Europe is indicated, however, by a letter written in 1444 by a Greek scholar, Cardinal Bessarion, to Constantine Palaeologos, despot of the autonomous Byzantine province of Morea. Bessarion, born, raised, and educated in the East, had been sent to Italy with a delegation charged with negotiating the unification of the eastern and western branches of Catholicism. He became a strong advocate of such a union and by the 1440s was, in effect, in self-imposed exile in the West because of his beliefs. Bessarion, however, was deeply interested in reviving Greek power, and he saw Morea as a possible base for that revival. In his letter he urged Constantine to take advantage of the West's mechanical advances. Bessarion was particularly impressed with the West's use of water power to eliminate hand labor, especially in sawmills and iron mills. The whole tone of Bessarion's letter suggests that in the Eastern Roman Empire, as in China and medieval Islam, water power was applied on a much more limited scale than in the West.[245]

# 3

# Continuity:
# The Traditional Vertical Water Wheel at Its Pinnacle, c1500 to c1750

## WATER POWER AND THE GROWTH OF MECHANIZATION

*Introduction*

In his classic analysis of the rise of the British coal industry, as well as in subsequent publications dealing with economic and technological developments between 1500 and 1800, John U. Nef opposed the prevalent idea that an industrial revolution had suddenly occurred in the late eighteenth century in Britain. He suggested, instead, that there was an earlier industrial revolution in Britain in the sixteenth and early seventeenth centuries and that the later industrial revolution had its roots in this earlier period.[1]

In recent years some historians have found Nef's thesis of an early industrial revolution wanting. His estimates of coal output and his use of percentage rates of growth, for example, have been criticized. He has also been accused of over-stressing exceptional examples of large works to demonstrate the early emergence of large-scale capitalist organization in industry.[2] Thus, Nef's early industrial revolution may well turn out to be an exaggeration. But even if this is so, Nef's work remains important. His emphasis on the essential continuity of industrial and technological growth and his demonstration that the industrial revolution was not a sudden phenomenon, but one with roots extending back into earlier centuries, seems to be as solid as ever.

The history of water power in the period between the fifteenth and nineteenth centuries, for example, supports Nef's theme of continuity in industrial growth. In this era many of the trends which first emerged in water-power technology in the medieval period continued without abatement. The number of watermills steadily increased; European engineers expanded their efforts to harness larger streams and concentrate more power at specific sites, as well as to tap more systematically the power of limited regions; the role of water power in industry grew, culminating in the large water-powered cotton mills of Strutt and Arkwright in the 1770s. There was, in other words, no sharp discontinuity between the water-powered monastic milling complexes of the medieval period and the early mechanized factories of late-eighteenth-century Britain or nineteenth-century France, Germany, and the United States.

### Numerical Growth

Although the paucity of available statistical data does not permit me to document exactly the steady and uninterrupted growth of water power in Europe in the period from around 1500 to 1750 or 1800, scattered data dealing with limited areas suggest that the replacement of animate with inanimate power was a process well underway before the time of the industrial revolution.

In Britain, for instance, the Domesday survey of the late eleventh century indicated approximately 5,624 watermills. By the eighteenth century there were between 10,000 and 20,000 watermills in Britain.[3] Moreover, certain areas of Britain were, by the 1700s, beginning to experience problems with mill crowding. By 1600 or 1700 there were nine mills on 8 miles (13.6 km) of the Ecclesbourne, a small tributary of the Derwent in Derbyshire. These mills were used for lead mining and smelting and for grinding dyes and crushing bones for fertilizer.[4] Around 1700 there were 60 watermills on 3 miles (5 km) of the Mersey below Manchester.[5] The extensive development of water power in the Sheffield area began only around 1500. By the end of the 1700s there were over 100 watermills on about 20 miles (c35 km) of stream in the area, or five watermills for every mile (three per km) of stream (Fig. 3–1).[6] In Westmorland the number of watermills increased from around 25 in the thirteenth century to 90 by around 1800.[7] As a result of the steady growth in the use of water wheels in Britain, practically all of the good water-power sites around industrial towns were occupied some years before the time of Arkwright.[8]

Tide mills seem to have shared in the steady numerical expansion of watermills in Britain. In Devon and Cornwall, where data are available, there were only 5 tide mills reported in the fourteenth century, but 9 in the sixteenth, 11 in the seventeenth, 14 in the eighteenth, and 25 in the nineteenth century.[9]

France presents a similar picture. There are no data on the number of watermills in France during the medieval period, but data from the seventeenth century indicates the considerable importance of water power in French industry. Vauban, the French military engineer, estimated in 1694 that France had 80,000 flour mills, 15,000 industrial mills (fulling, oil, hemp, paper, saw, etc.), and 500 iron mills and metallurgical works.[10] Vauban's estimates included windmills as well as watermills. Assuming that around 20% of the flour mills were windmills, France had around 80,000 watermills in 1700. These watermills, particularly the 15,000 industrial watermills and the 500 water-powered metallurgical plants, provided the French kingdom with a substantial basis for

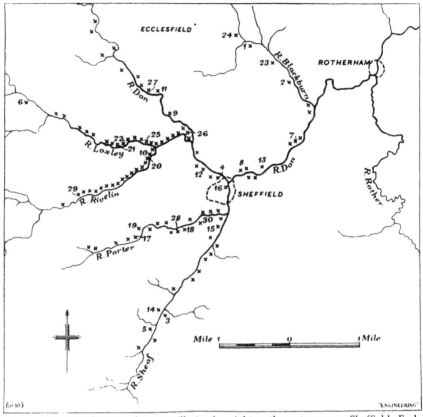

FIGURE 3–1. Distribution of watermills in the eighteenth century near Sheffield. Each cross indicates the position of a mill. Mill crowding along certain stretches of stream is evident.

industrialization. They no doubt played an important role in the revival of the French economy in the eighteenth century, a revival which enabled the French to match England's rate of industrial growth between 1725 and 1785 and to reduce substantially the lead which Britain had acquired in industrialization.[11]

As in England, the growing use of waterpower in France had produced regions where watermills were crowded quite close together, a problem which worried Bélidor in 1737.[12] By the early modern period there were 300 watermills around Ambert in central France.[13] Ferrendier noted that the Furan River in 1753 had more than 250 watermills in a distance of around 25 miles (40 km), including flour, saw, paper, polishing, gunpowder, and hammer mills. At Vienne, in eastern France, on 3 miles (5 km) of stream there were more than 100 water wheels, almost one every 150 feet (45 m). In Picardy, Ferrendier reported 100 watermills on the Thirain River, 51 on the Brèche, 60 on the Authie, 50 on the Aa, 40 on the Escaut, 34 on the Selle, and 130 on the Bresle, all rather small streams.[14] "Over all the territory of France," he concluded, "there was not a river which did not drive a mill."[15]

There are fragmentary data on boat milling in France which indicate that boat mills shared in the steady growth of water power in France. For instance, Ferrendier reported that in 1493 there were 17 boat mills on the Rhone at Lyons. Their number had grown to 20 by 1516 and to 27 by 1817. Other French rivers also had boat mills, although there are insufficient data to trace numerical increases over time. In 1758 there were 15 boat mills around Agen, 4 at Le Réole, and 3 at Gironde. In 1776 there were 20 on the Garonne near St. Macaire and 21 others on the River Dordogne near its confluence with the Vézère.[16] Even though boat mills went into decline as river navigation increased, some mills of this type survived well into the 1800s and 1900s. Vienna, Austria, for example, still had 55 around 1870.[17] In the early twentieth century boat mills could still be found at Verona (northern Italy) and Brataslava (Czechoslovakia) and in Rumania.[18]

Even on the frontiers of Europe the watermill had come into extensive use by the end of the 1700s, if not earlier. Late in the eighteenth century Austrian-occupied Poland had an area of about 50,000 square miles (c140,000 km²), a population of around two million, and 5,243 water-mills.[19] A census of part of the Ukraine dating from 1666 and covering the catchment areas of the northern tributaries of the Dneiper from the Sula to the Vorskla rivers indicated 50 dams and 300 water wheels. One of the 14 rivers covered by the census, the Udai, had 72 wheels. A later document (1744) referred to 1,273 watermills owned by 314 cossack families.[20]

While many of the figures reviewed above deal with scattered territories only for one specific year, they are sufficient to suggest that the use of water power (and hence the substitution of inanimate for animate power) begun during the medieval period continued with little abatement between 1500 and the late eighteenth century and had reached very high levels (approaching saturation) on certain European streams. Unfortunately the fragmentary nature of this data does not let us do more.

Nevertheless, the continuity of the shift to inanimate power in European industry and the steady expansion of mechanization in European industry are also indicated by nonquantitative data, notably (1) a discernible trend towards the more intensive use of larger amounts of power through extended use of hydropower dams, reservoirs, and canals; and (2) a steady expansion of the water wheel's role in industry. We will review these developments in the next two sections.

### Dams, Canals, and Reservoirs

Water-power dams were used in the early modern period for basically the same reasons technicians had used them in the medieval period. They were used in conjunction with storage reservoirs to even out the

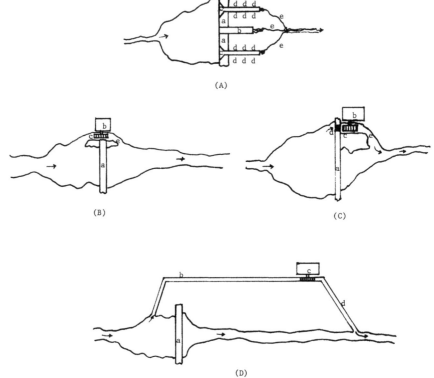

(A)

(B)                                          (C)

(D)

FIGURE 3–2. Options for locating dams, canals, and water wheels. (*A*) Sketch of the dam
and races at the Ekaterinburg plant, Urals, in the 1730s (*a*, dam; *b*, waste water discharge;
*c*, working races; *d*, shops utilizing power from the working races; *e*, tail race). (*B* and *C*)
Watermills operating adjacent to a weir or small dam (*a*, weir or dam; *b*, mill building; *c*,
water wheel; *d*, sluice gate; *e*, tail race). (*D*) Use of a weir and leat or canal to divert water to
another location (*a*, weir or dam; *b*, leat or canal; *c*, mill building and water wheel; *d*, tail
race).

flow of streams and to insure that water was available when it was
needed, thus permitting regular rather than intermittent operation of
watermills. They were used to increase the fall of the water backed up
behind them, thus increasing the amount of power that could be developed at a given site. Finally, in conjunction with a mill leat or power
canal, they were used to divert and convey water to locations where more
power could be developed or where use of water power was more convenient.[21] (See Fig. 3–2.)

Because dams, as well as reservoirs and power canals, promoted regularity in production, permitted significant amounts of inanimate power
to be developed at a specific site, and required heavy capital investment—all hallmarks of an industrial rather than a subsistence, agricultural economy—their growing popularity between the fifteenth century and the late eighteenth century provides further evidence of the

steadily rising dependence of European industry on water power in the centuries which preceded the industrial revolution.

As early as the sixteenth century scattered evidence suggests that the use of dams, reservoirs, and power canals had become almost commonplace in some regions of western Europe. For example, the English writer Fitzherbert commented in 1539 in describing "ground mills" (i.e., undershot mills): "Commonly these mills are not set upon the large streams of great rivers, but a great part of the water is converted out of the great stream by a mill flume made with man's hand and brought to a site where the mill is and held up by setting a yere [a weir—a small overflow dam] over the stream made of timber, stone, or both."[22] In other words, English millwrights "commonly" tapped the power of large streams by diverting water (probably through the use of a diversion dam)[23] into a power canal or mill leat, carrying that water to a more convenient site, where construction of a dam and reservoir could provide more power with better regulation.

Later, in the same century, William Camden noted the extensive use of hydropower reservoirs in the Sussex iron industry: "[Sussex] is full of Iron-mines all over . . . ; many streams are drawn into one chanel [*sic*], and a great deal of meadow-ground is turned into Ponds and Pools, for the driving of Mills . . . ; which, beating with hammers upon iron, fill the neighborhood round about, night and day, with continual noise."[24] And Frank Nixon noted a "great boldness" in the construction of weirs in Derbyshire between the sixteenth and mid-eighteenth centuries in his review of the industrial archeology of that county.[25]

Nor was England unique in its growing use of water-power dams, canals, and reservoirs in the era. Jacques Levainville claimed that most of the ponds of Brittany were artificial, constructed to provide power for the French iron industry, and Ferrendier noted that in some areas of France by this time water-powered forges had become so numerous that dams, mill races, and reservoirs were situated one after the other, almost without break on some streams, each forge using the water of the forge above it.[26] Around Auvergne, in central France, the regularity of the flow of area streams made dams with storage reservoirs unnecessary, but diversion dams and long power canals were used quite extensively. In the Lagat valley, for instance, the termination of one power canal was followed, a few yards downstream, by the entrance to another. So closely did power canals and mills follow each other that very few feet of fall on the stream went unutilized.[27] Finally, the French engineer Henri Pitot noted in the 1720s that water-powered engines were most commonly operated from reservoirs where the water level was built up artificially.[28]

In Germany, too, dams, reservoirs, and power canals were in frequent use by the sixteenth and seventeenth centuries. Agricola, for example, noted in 1556 that when no stream could be diverted to the top of a vertical water wheel, Saxon miners usually collected water in large reservoirs, presumably through the use of a system of dams and canals, and

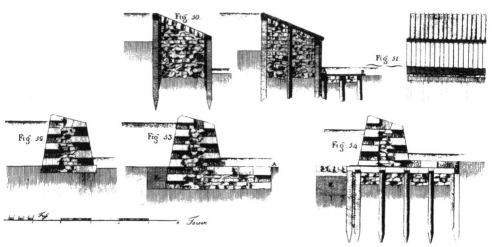

FIGURE 3–3. Some examples of small masonry overflow dams. Dams of this type were used for hydropower development over much of Europe between 1500 and 1750.

then directed it by means of sluice gates against the blades of undershot water wheels.[29]

Most of the hydropower dams familiar to Fitzherbert, Camden, Pitot, Agricola, and their contemporaries were probably, like most hydropower dams in the medieval period, small, unsophisticated structures (Fig. 3–3).[30] Typically, they were overflow weirs, no more than 10 feet (c3 m) high and usually under 100 feet (c30 m) long. They were dependent on their bulk or weight for stability (i.e., they were gravity dams), although wood piles were often used to anchor them firmly to the subsoil. Some were situated perpendicularly to the flow of a stream, but the more common practice was to place the dam obliquely across the stream to gain greater length for the water to run over. This enabled it to handle flood water better and decreased the danger which the overflow posed to the foundations. In the eighteenth century overflow V-shaped dams (the vertex of the V pointing upstream) and arch dams began to come into use as the size of the streams being confined increased.

Although all-masonry construction would have provided greater durability and stability to overflow-gravity dams, most water-power dams of this period in western Europe (as well as in Russia and America) were built up from a combination of wood, earth, and stone (composite dams). Although less stable and durable than masonry construction, composite dams used easily obtainable materials and were therefore cheaper to erect. There were a legion of possible material combinations for composite dams. On the larger ones, for example, European technicians frequently utilized a clay core with either a masonry or a battered earth and masonry covering. The cores of the much more numerous small dams were usually built up from timber cribs, bound together with spikes or dowels, and filled with the most convenient available mate-

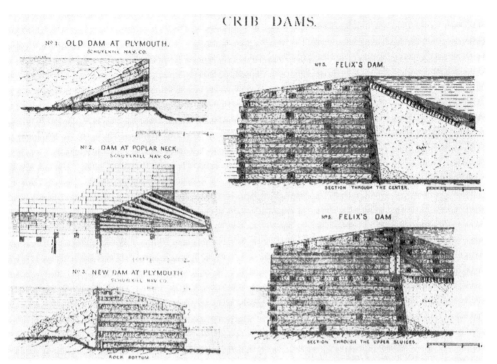

CRIB DAMS.

Nº 1. OLD DAM AT PLYMOUTH.
SCHUYLKILL NAV. CO.

Nº 2. DAM AT POPLAR NECK.
SCHUYLKILL NAV CO

Nº 3. NEW DAM AT PLYMOUTH
SCHUYLKILL NAV CO

ROCK BOTTOM

Nº 5. FELIX'S DAM.

SECTION THROUGH THE CENTER.

Nº 5. FELIX'S DAM

SECTION THROUGH THE UPPER SLUICES.

FIGURE 3–4. Examples of composite hydropower dams. Dams of this type were frequently used to develop water power in much of Europe and America between 1500 and 1750. The sloping water face was usually covered with stones or heavy clay to prevent washing, while the air face was often sloped so that the overflow would not undermine the foundations. All of the dams pictured used timber frames with stone or earth fill.

rials—stones, earth, rock, brush, clay, or some combination. Laborers then piled earth and rock over the cribs and sloped these materials up to the crest until the structure was completed. (See Fig. 3–4.)

European millwrights and engineers had to take particular care to protect the water faces and foundations of composite dams. The water face was subject to erosion from current and wave action; the foundation of a dam could be undermined by water pouring over the crest of the structure. In most instances the protection provided the water face consisted of a layer of compacted sod, clay, timber, or stone. Protecting the base of a composite dam was more difficult. European dam builders most frequently combatted the threat of undermining by designing dams with an inclined slope of stone, timber, or compacted clay on the downstream side (air face). To further alleviate the impact of the falling water, particularly on larger dams, they sometimes erected a very low weir a short distance downstream. This weir penned a foot or two (c0.3 to 0.6 m) of water against the foot of the main dam, so the overflow struck standing water and lost its force. In areas where water economy was a critical issue, of course, water was not allowed to freely flow over

TABLE 3–1. SOME EXAMPLES OF EARLY LARGE HYDROPOWER DAMS

| Dam | Location (date built) | Height | Length |
|---|---|---|---|
| Cento | Savio R., near Cesena, Italy (c1450) | 19 ft (5.8 m) | 234 ft (71.3 m) |
| Chalviri | nr. Potosi, Bolivia (c1575) | 28 ft (8.5 m) | 740 ft (226 m) |
| Oderteich | Oberharz region, Germany (1714–21) | 60 ft (18 m) | 475 ft (145 m) |
| Ekaterinburg | Iset River, Urals, Russia (1721–23) | 22 ft (6.5 m) | 650 ft (209 m) |
| Bedia | Ibaizabal River, nr. Bedia, Spain (c1730) | 12 ft (3.7 m) | 175 ft (53.3 m) |
| Albera de Feria | Almendralejo, Spain (1747) | 64 ft (19.5 m) | 400 ft (122 m) |
| Zmeinogorsk | Zmeevka River, nr. Zmeinogorsk, Siberia (1780s) | 60 ft (18 m) | 500 ft (158 m) |
| Typical small European water-power dam | Western Europe (1500–1750) | c10 ft (c3 m) | c50 ft (c15 m) |

the crest of a dam. The dam was built high enough to prevent this, and water was drawn out, instead, through small gate-controlled flumes. In the Harz mining region of Germany, for example, practically no overflow dams were used. The exit of water from storage reservoirs was carefully controlled by sluice gates built into or adjacent to the dams.[31]

While dams constructed with some combination of wood, stone, and earth were the most common type of hydropower dam in early modern Europe, all-masonry dams were sometimes used, especially on larger streams. In the Harz, for example, most dams were fabricated from stone only.[32] Many of the weirs on British rivers were all-stone in construction—for example the seventeenth-century dam on the River Bain that supplied the Low Mill in Bainbridge, Yorkshire, with an operating head of around 13 feet (3.9 m) and the three large weirs erected on the River Derwent in Derby.[33]

In the fifteenth, sixteenth, seventeenth, and eighteenth centuries large hydropower dams became steadily more common (see Table 3–1). One of the early large hydropower dams was a composite structure—the Cento dam on the Savio River, near Cesena, in Italy. Built around 1450, Cento was a gravity dam of brick and wood. It was 234 feet (71.3 m) long at the crest, with a vertical water face some 19 feet (5.8 m) high and an air face which sloped downwards for more than 40 feet (12.2 m) before dropping vertically about 5 feet (1.5 m) to the level of the foundations. The Cento dam had a mill built on one of its abutments and a long canal, carried on an embankment of bricks, which provided power to other mills further downstream.[34]

The first large dam ever built in Germany was the all-masonry Oderteich dam in the Harz mining area. Erected between 1714 and 1721,

Oderteich had a central core of granite rock between two outer walls of granite block roughly cut to shape and fitted together. The spaces between the blocks were packed with earth and moss. Oderteich was around 475 feet (145 m) long, 60 feet (18 m) high, and around 155 feet (47 m) thick at base. The pond formed by the dam was almost a mile (1.7 km) long, and the canals leading from this dam eventually drove 47 water wheels.[35]

Several large water-power dams were erected in Spain between 1500 and 1750. In fact, the earliest known examples of both multiple-arch and modern buttress dams emerged there. One of the pioneer arch dam builders was Don Pedro Bernardo Villarreal, a Basque nobleman. In a work published in 1736 he described five arch dams built to his specifications for the development of water power. The largest was located at Bedia on the Rio Ibaizabal, a tributary of the Rio Nervio. It was a five-arch structure, approximately 175 feet (53 m) long and 12 feet (3.7 m) high, reinforced by four large buttresses. The dam, still standing, diverts water into a canal on one bank of the river to power a local mill.[36]

The first modern buttress dam was the rubble-masonry Almendralejo dam, built in 1747 in northern Spain. Originally this dam was some 64 feet (19.5 m) high, although it has subsequently been raised. Its crest length is around 400 feet (122 m). Shortly after Almendralejo was completed, several of its buttresses were roofed over to house a mill, and the overflow diverted to a special side spillway. These modifications made the Almendralejo dam the first to contain a water-powered device actually within the body of the structure.[37]

Bedia and Almendralejo were not the only large Spanish hydropower dams built between 1500 and 1750 or 1800. Garcia-Diego's study of 23 pre-1800 dams in Extremadura ranging from 13 feet (4 m) to 79 feet (24 m) in height and 246 feet (75 m) to 860 feet (262 m) in length indicated that 18 had at least one mill and that supplying water to mills was the predominant function of these dams.[38]

Some of the largest hydropower dams of the period preceding the industrial revolution were constructed in Russia. The Ekaterinburg plant in the Urals, built between 1721 and 1723, is an example. The main dam at this plant was built up from wood cribs and clay and was 650 feet (209 m) long and 21.3 feet (6.5 m) high. To supplement the water supplies stored by this dam during dry seasons, several additional storage dams were later constructed slightly upstream of the Ekaterinburg dam. Unlike most western European power dams, Ekaterinburg was not an overflow structure. Surplus water was discharged from a centrally-located sluice (Fig. 3–2A). Two races, also built into the dam, provided water to water wheels which produced power for a large number of adjacent workshops. The whole complex eventually contained 50 water wheels which drove 22 hammers, 107 bellows, 10 wire-drawing mills, several mint machines, a drill, and a roller.[39] The total output of the water wheels in the system was probably around 200 to 500 hp.

There were other large Russian hydropower dams besides Ekaterin-burg. But the largest of those built before 1800 was that designed by K. D. Frolov in the Altai region of Siberia at Zmeinogorsk in the 1780s. The dam, which still exists, was placed across a deep ravine of the Zmeevka River. It was 518 feet (158 m) long at crest, 282 feet (86 m) long at base, and 59 feet (18 m) high, with a volume of around 30,000 cubic yards (c23,000 m³). It created a reservoir with a surface area of around 2.4 square miles (6.5 km²). A canal led water from the reservoir to four large overshot vertical wheels which sawed wood, hoisted ore, and pumped water from nearby mines. The Zmeinogorsk reservoir was probably the largest artificial lake impounded solely for water-power requirements until well into the nineteenth century.[40]

Many medieval and early modern hydropower dams, both large and small, required no long power canals, and had water wheels located either on or immediately adjacent to them; these included the Ekaterin-burg dam in Russia, the Almendralejo in Spain, the Garonne dam at Toulouse, and the dam at Bainbridge in Yorkshire. But there were two sets of circumstances, as noted in the previous chapter, where this ar-rangement was inadequate, and a power canal (mill leat) of at least moderate length was needed: (1) if sufficient fall could not be developed at the dam site, and (2) if the dam location was inconvenient (e.g., owing to insufficient area for mill buildings, isolation from established trans-portation routes, or distance from where the power was needed).

Let us look first at the problem of insufficient fall. This problem might be caused, in part, by the economic limits placed on the height of a dam by topography. In flat areas, for example, a dam of even moderate height would flood enormous amounts of land and might have to be unreasonably long to prevent water from flowing around it. But the problem was serious even in topographically suitable areas, because of mill crowding, increased demand for power, and streamside property owners. For instance, in a populated river valley with a moderate gra-dient a millwright might be able to build or raise a weir to a height of no more than 6 feet (c1.8 m) at a given location because a higher weir would flood too much farmland or back water up against a dam further up-stream (a frequent problem from the sixteenth century on). If 9 feet (c2.7 m) of fall were needed to meet the mill owner's power needs, the 3 feet (c0.9 m) additional fall required could be developed only through the use of a mill leat (sometimes called a head race or power canal).

There was widespread agreement among technical writers between 1500 and 1750 that a mill leat or head race constructed to develop additional fall should have only a very slight slope. The basic rule was to use only as much slope as was necessary to insure flow into and along the channel. This usually amounted to something on the order of an inch or two (c2.5–5.0 cm) of drop for every 400 to 500 feet (c120–50 m) of run.[41] Thus, if the stream on which the hypothetical weir mentioned in the preceding paragraph stood dropped at the rate of 1 foot (c0.3 m) every 500 feet (150 m), the millwright could construct a leat or power canal from the mill dam in a generally downstream direction and, by intercepting the main stream 1,500 to 1,600 feet (c450–550 m) downstream, gain the extra 3 feet (c0.9 m) of fall needed.

Because mills with leats were usually able to tap more power than mills located immediately adjacent to dams, they tended to emerge later and only as power demands grew. This at least was the case in Gloucester, where Jennifer Tann made a detailed study of mill siting practices. But already by the sixteenth century, mills with leats were common there. Most of the mills in Kingswood parish, Gloucestershire, for example, utilized leats to increase operating heads along to the gentle gradient of the Little Avon River valley.[42] Some of the Spanish mills observed by Garcia-Diego utilized the combination of dam and power canal for the same reason.[43]

In other cases the primary purpose of a mill leat or power canal was to conduct water to a more convenient location, instead of increasing operating head. (In many cases, of course, a leat could perform both functions.) Mines, for example, usually could not be moved to points where water power was naturally available, so water had to be brought to the mine head to operate water wheels there. The Zmeinogorsk system in Russian Siberia, mentioned above, was constructed for this reason. German mining engineers in 1587 constructed a canal almost 4 miles (c6.5 km) long to bring water to operate a reversible water wheel used to hoist materials at the Heiliger Geist shaft of the Röhrerbühel mine.[44] Similar dam-reservoir-canal systems were erected to bring water to water-powered pumps operating at mine heads in British coal, lead, and iron mines in the sixteenth, seventeenth, and eighteenth centuries.[45]

Accompanying the growing use of power canals and large hydro-power dams from the sixteenth through the eighteenth centuries was the growth of networks which utilized multiple dams, reservoirs, and power canals to effectively tap the water-power resources of entire regions.

Perhaps the most systematic attempt to exploit the hydropower re-sources of a region during this period was in the Harz mining region of Germany, an area of between 9 and 12 miles (c15–20 km) in radius. Around 1550 German mining engineers began to develop a complex set of dams, water-storage reservoirs, and water canals in the Harz Mountains as mineral demands forced mines deeper and deeper and accelerated the need for power. By the early nineteenth century they had erected about 60 hydropower dams and reservoirs within a radius of 2.5 miles (4 km) of Clausthal, the center of the upper Harz mining district alone. These dams, which varied from 20 to 50 feet (c6–15 m) in height, were linked to each other and to an assortment of water-powered installations by an extensive system of surface, elevated, and subterranean power canals which were typically around 3 feet (c1 m) wide and 3 feet (c1 m) deep. Most of the surface conduits in the system were lined with

TABLE 3–2.  USE OF WATER POWER IN THE HARZ MINING REGION OF
GERMANY, c1800

| | Number of installations | Water wheels | |
| | | Number | Use |
|---|---|---|---|
| Mines | 61 | 92 | 55 for pumps<br>37 for drawing engines |
| Mills for the mechanical preparation of ores (stamps and washers) | 56 | 56 | 52 for stamp mills<br>4 for ore washers |
| Foundries and metallurgical furnaces for the production of lead, silver, and copper | 97 | 77 | 77 for blowing engines |

SOURCE: A. M. Héron de Villefosse, *De la richesse minérale* . . ., vol. 1 (Paris, 1819), tables 1, 2, and 3, facing p. 102.

sandstone to inhibit leakage and covered on top with a layer of timber to reduce losses due to evaporation and to reduce the possibility of blockage by debris. All-timber flumes were used for the elevated portions of the system. The subterranean channels were constructed from masonry and provided with arched roofs. The Harz by 1800 had almost 120 miles (190 km) of these canals. The longest of the channels, the Dammgraben, was 17.9 miles (28.8 km) long. The Harz's networks of dams, reservoirs, and canals provided water to 225 water wheels which powered mine pumps, drawing engines, ore washers, and ore stamps, concentrated mainly in the area immediately around Clausthal, and powered blowing engines for lead, silver, and copper foundries scattered all over the Harz region (see Table 3–2).[46] The aggregate power generated by the system probably exceeded 1,000 horses.

Rivaling the Harz region in the magnitude of its water power network was the dam and reservoir complex erected by the Spanish in the rich silver-mining district around Potosi in Bolivian Andes. The Conquistadores began exploiting the silver deposits in the area around 1545. By 1566 the rich surface veins had been exhausted. In order to work the poorer ores, Spanish engineers were compelled to turn water-powered crushing machinery. This presented a major problem, since the nearest permanent river in the generally dry area was nearly 9 miles (15 km) away. To provide supplies of water for crushing or stamp mills in the Potosi area, Spanish engineers began in 1573 to construct a series of dams and reservoirs to impound summer rains and a series of canals to lead the water thus collected to the stamp mills. By 1621 they had erected 32 dams, most situated in the glacial troughs in the region which radiated toward Potosi. These dams provided a total storage capacity of around 5,850 acre feet (c6 million m³). The larger dams in the system were about 25 feet (8 m) high and varied from 210 to 1,640 feet (64–500

m) in length. The most impressive was a masonry and clay dam 28 feet (8.5 m) high and 740 feet (225 m) long; it had a reservoir with a storage capacity of 2,200 acre feet (c2.7 million m³), almost half the total of the entire system.

The dams were located at varying distances from the city, so Spanish engineers designed a series of canals which fed water from those further out to those nearer Potosi and then to the 132 ore-crushing mills located adjacent to or in the city itself. The main channel which supplied water to the mills was 26 feet (8 m) wide and 3 miles (5 km) long. It had a total fall of around 1,950 feet (594 m) and an estimated average flow of around 9 second-feet (250 l/sec.). The power actually developed by the system has been estimated at around 600 horses.[47]

The magnitude of the Harz and Potosi systems was exceptional. The concept was not. Similar arrangements on a somewhat smaller scale were put in use between 1500 and 1750 in other areas—for example, in the Schemnitz mining district of Hungary (Slovakia) and at Kongsberg in Norway.[48]

The water wheels used in conjunction with hydropower dams throughout most of Europe were usually vertical wheels. But successful attempts were also made between the late Middle Ages and the eighteenth century to adapt the primitive horizontal water wheel to hydropower dams for more intensive development of water power. These adaptations of the horizontal wheel took two primary forms. One type of horizontal mill (moulin à trompe; regolfo mill) utilized an enclosed masonry conduit, usually of circular or pyramidal shape, to lead water under a moderately high head to a spout which directed an isolated jet of water against a horizontal wheel with curved blades or a helical rotor. These mills developed an efficiency of around 30%. The second type of horizontal mill (moulin à cuve; tub mill) had a horizontal wheel with curved blades or a helical rotor placed at the bottom of a close-fitting masonry cylinder. This wheel, unlike the wheel of the moulin à trompe, was completely submerged by the water which moved through the cylinder and powered it. Moulins à cuves were in use at the Bazacle dam at Toulouse by the seventeenth century. They operated at an efficiency of around 10 to 20%. Horizontal wheels of both types were used mainly for flour milling and were confined largely to southern Europe (Italy, Spain, southern France).[49] But the evolution of the horizontal wheel between the late medieval period and the eighteenth century from a primitive implement of peasant culture used primarily on small mountain brooks with only the most primitive auxiliary works to an engine frequently associated with hydropower dams on the rivers of southern Europe and capable of delivering significant amounts of power roughly parallels the evolution of the more widely used vertical wheel.

As noted in Chapter II, Louis Hunter found, in dealing with the emergence of water power in America, that the slow emergence of an industrial economy could be traced, in the absence of statistical data, by following the movement of water power from small upland streams, where power could be developed only in small units, to progressively larger streams capable of supporting progressively larger establishments and producing more power. There is little solid statistical data on the growth of water power in Europe prior to the nineteenth century. But, by analogy to Hunter's findings for America, the continued spread of hydropower dams in Europe between 1500 and 1800, the emergence of large hydropower dams, and the development of extensive reservoir, dam, and canal networks like that at the Harz suggest, along with the bits and pieces of statistical data available, that the use of inanimate power in large units was under way in certain parts of Europe well before the industrial revolution.

*Industrial Diversification*

Another gauge of the growth of mechanization and inanimate (usually water) power in European industry in the centuries preceding the industrial revolution is the continuing expansion of the water wheel's role in industry. As noted in the previous chapter, the vertical water wheel had begun to play an important role in some industries in the medieval period. By 1500 European engineers had recognized that, in principle, water power could be applied to almost any industrial process, and they had made considerable progress in expanding the realm of water power in industry. For example, around 1500 Leonardo da Vinci drew up a long list of industrial activities that might be expected to use water power if the Arno River were improved. This list included sawmills, fulling mills, paper mills, hydraulic trip-hammers, flour mills, and mills for burnishing arms, grinding and sharpening knives, manufacturing gunpowder and saltpetre, spinning silk, weaving ribbons, and shaping jasper and prophyry vases with water-activated lathes.[50] Between 1500 and 1750 water power continued to grow in importance in industries which had long had some reliance on it, like silk, flour milling, paper, and mining, while penetrating completely new industries like glass, lead processing, and linen.

The textile industry provides a good example of an industry which had already begun to use water power in the medieval period and which became steadily more dependent on it between 1500 and 1750. Water power had first entered the textile industry with the fulling mill, introduced around 1000 and used to finish wool cloth. By 1500 fulling was largely dominated by water-powered establishments in much of western Europe.[51] The importance of water power to the textile industry had been further extended with the introduction of the water-powered silk mill towards the end of the medieval period. This industry's use of water power also grew. By 1700 there were over 100 water-driven silk mills in operation around Venice and in the Po River valley alone.[52] Some of the early silk mills were quite large. For instance, Thomas Lombe's mill, erected on an island in the Derwent near Derby c1720, was five to six stories high and 500 feet (152 m) long. It employed 300 mill hands and was a clear forerunner of the water-powered cotton textile mills of the late eighteenth century.[53]

Late in the seventeenth century and early in the eighteenth, water power was adapted to the production of linen with the introduction of scutching, washing, and beetling mills. At the scutching mill water-powered wooden blades, similar to fan blades, beat the pith and stem of rotted flax plants away from linen fibers. Linen washing mills used rubbing boards of corrugated wood. These were driven back and forth by water wheel and cam or crank as wet, soaped linen was drawn through. Beetling mills were a modification of the water-powered fulling mill long used in the woolen textile industry. In a beetling mill linen was slowly drawn over a large wooden roller or stone beam while smooth wooden hammers rapidly beat on it, toughening the weave and giving the linen a slight sheen or gloss which completed its preparation for sale. Water-powered scutching, washing, and beetling mills were developed and widely applied in Ireland, the center of the linen industry in the late seventeenth and early eighteenth centuries. Linen mills were usually associated with bleaching greens, and in Ulster alone more than 200 bleaching greens with associated water-powered mills were established between 1700 and 1760. These mills transformed what had been a small-scale, completely manual industry into a "nascent mechanized industry."[54]

The industry par excellence of the traditional industrial revolution was the cotton textile industry. It was in that industry that spinning and weaving were first mechanized, that Watt's double-acting, rotative steam engine first found widespread application, and that the modern factory system first fully emerged. Unlike the production of wool, silk, and linen, the production of cotton textiles was not mechanized in any stage prior to 1750. Nonetheless, the mechanization of the cotton textile industry is best viewed, not as a radical break with the past, but as the culmination of a process that had begun with the introduction of the water-powered fulling mill around 1000 and the water-powered silk mill some centuries later, for like these, the early cotton textile mill was water-powered. It is true that water power was applied to wool and silk in only a single stage of the productive process. But even before the mechanization of cotton textile production, European technicians had successfully used water to mechanize several stages of linen production. Seen in this light the cotton textile mills of the industrial revolution were simply the culmination of an evolutionary process which had been gradually extending the role of inanimate power and mechanization in textile production for centuries before the time of Strutt and Arkwright.

In the food processing industry, like the textile industry, water power grew in importance between 1500 and 1750. In the medieval period water power had been used to grind grain to flour, to prepare malt for beer making, to grind pepper and mustard for use as condiments, and to

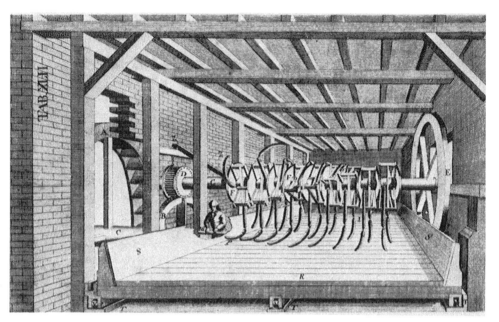

FIGURE 3–5. A water-powered threshing machine, 1735. The shrouded, but unsoaled water wheel to the left (*A*), rotated the shaft (*G*), on which a number of flails were mounted. The wheat was placed beneath the flails on table *R*.

produce olive oil and sugar. In the mid-sixteenth century the water-powered rice dehusking mill (long used by the Chinese) first appeared in Europe in the manuscripts of the Italian engineer Juanelo Turriano; Turriano also sketched a vertical water wheel operating a wedge press for extracting beeswax.[55] The water-powered threshing machine, first illustrated by Johannes Beyer in 1735 (Fig. 3–5), is yet another example of the growth of mechanization and the extended use of inanimate power in European food processing before the industrial revolution.[56]

Probably the most widely applied water-powered innovation in the food processing industry in this period was the mechanical bolter, used in flour mills to automatically sift flour. The bolter was basically a sheet or roll of wire mesh or cloth (most often canvas or linen, but sometimes silk or another fabric). The flour produced by the mill was fed through or over the device, which was shaken by a mechanism (several were possible) taking power from the drive train leading from the water wheel to the millstones.[57] The earliest known reference to the combination of millstone and mechanical bolter occurs in one of Leonardo da Vinci's notebooks c1500.[58] But the prime mover is not visible in da Vinci's drawing. The first clear evidence of a water-powered bolter comes from Ramelli in 1588.[59] The bolter depicted by Böckler and illustrated in Figure 3–6 indicates just how simple the device could sometimes be. Use of the bolter spread rapidly. According to Howell, after 1730 most mills of any size had them, and by the mid-eighteenth century there were

FIGURE 3–6. Overshot watermill with mechanical bolter from Böckler, 1661. The bolter (*H*) is a cylindrical piece of cloth which is shaken by a rod jogged by the lantern gear (*D*).

some areas of Europe—for example, the Nord region of France—where practically no grist mill was without a mechanical bolter.[60]

In the paper industry, the use of water power also grew steadily, stimulated in large part by the rising demand for paper which followed the introduction of the printing press. This demand had led as early as the late fifteenth or early sixteenth century to the conversion of numerous water-powered flour mills to paper mills in some parts of France.[61] In England there was similar growth. There were 38 water-powered paper mills in England in the first half of the seventeenth century, but 106 by 1690, around 200 by 1710, nearly 350 by 1763, and about 400 by 1800.[62]

The most important technical innovations in paper-making in the period were quickly adapted to water power. Probably the most significant of these was the "hollander." Developed shortly before 1700, the hollander was a cylinder covered with iron blades. Set on one side of an oval tub and rotated by a vertical water wheel, it macerated rags in the pulp production process much more effectively than the stamp hammers used for the process since the medieval period.[63] Water power was also extended to the finishing process on fine papers in the early eighteenth century with the pressing or glazing mill. This device used one or more water-activated light hammers to rapidly pound sheets of paper to give their surface a smoother finish.[64]

The introduction of the hollander and the glazing hammer, and the growing demand for paper, steadily increased the scale of paper-making. While the typical mill remained small, some examples of factory-scale production had emerged before the mid-eighteenth century. For example, John Spilman's paper mill at Dartford, Kent, employed scores of workers in the late sixteenth century.[65] The largest paper mill in the Auvergne region of France, the Grand Rive mill, had seven water wheels, 38 sets of stamps, and four vats in 1676.[66] The great French paper mill (l'Anglée) at Montargis, south of Paris, utilized a diversion from the River Loing for power. The main body of this factory was 450 feet (137 m) long, and in the factory itself there was substantial division of labor.[67]

The background for the expanded role of water power in mining and metallurgy between 1500 and 1750 was a growing demand for metals, both precious and common, a demand stimulated by a growing taste for luxury and ostentation among Europe's expanding wealthy classes and by the rise of large military establishments in the emerging centralized nation states. The wealthy classes, encouraged by the secular spirit of the Renaissance, wished to display their wealth with gold and silver ornamentation, with silver utensils and plate, with bronze and lead statuary. At the same time, the financial demands of the larger and more expensive military establishments and bureaucracies which the absolute monarchs of the emerging nation states used to support and extend their power led most rulers to encourage strongly the mining of precious metals in their realms. The emergence of large military establishments likewise led to the expanded use of iron, copper, and tin, since significant quantities of these metals were needed to manufacture cannon, cannon balls, muskets, and bayonets.

Mining activities had slowly been depleting easily accessible supplies of metals through the medieval period. The accelerated demands of the period from 1500 to 1750 speeded up this process. By the sixteenth or seventeenth century European engineers were being increasingly compelled to resort to deep-level mining. As early as the late sixteenth century, for instance, drainage problems involved in deep-level mining had created a crisis in the mining area around Liège in Belgium,[68] and by 1615 one of the shafts at the Kitzbühel mine in the Tyrol was nearly 3,000 feet (c900 m) deep.[69] Mining at depths of even 100 feet (30 m) below the surface or below an adit (water drainage tunnel) created serious drainage and hoisting problems and required mining engineers to turn increasingly to water-powered equipment.

Water-powered mining equipment had been known to a limited extent in late medieval mines. But by the mid-sixteenth century it had become common, as the large number of water-powered mine pumps mentioned in Agricola's treatise of 1556 suggests.[70] By around 1550

FIGURE 3–7. The stagenkunst system. Networks of pivoted field rods like these were used to transmit power from water wheels to mine heads sometimes as far as several miles (3–4 km) away by 1700 in central Europe.

water-powered drainage and hoisting equipment had substantially replaced man- and animal-powered devices at the Joachimsthal and Schwaz mines in central Europe and at the Kitzbühel mine in the Tyrol.[71] Also in the sixteenth century animal-powered whims were replaced by water-powered whims in the Bohemian tin mines and elsewhere.[72] Around 1700 the lead and silver mines of Hugh Myddleton near Keswick were drained by water power.[73] The lead mines in Britain's Pennines switched to water power in the 1730s when horse whims failed to keep the mines dry.[74] And by the mid to late eighteenth century most of the deeper collieries in Britain had turned to water-powered pumps.[75] In general, by 1750 most large mining enterprises were making use of water-powered equipment if water power could economically be provided at or near mine heads.

Moreover, certain European mines were making extraordinary efforts to insure that water power could be delivered to mine heads. One of the major new technological innovations of the period was the stagenkunst, a method of transmitting mechanical power which increased the geographical range that could be served by a water-power site. This system employed long lines of pivoted field rods, connected by rigid linkages, to transmit reciprocating motion from the crank of a water wheel shaft to pumps or drawing engines located up to several miles away (see Fig. 3–7).[76] An invention of the late sixteenth century, by 1590 or 1600 stagenkunst more than a mile (1.6 km) in length were in use, and by

1700 the system was being extensively applied in central European mines.[77]

By the 1700s stagenkunst were also in use in Sweden. The Swedish engineer Christopher Polhem, for example, used field rods at the Bisperg mines to transmit power from a water wheel to mine pumps more than 1.5 miles (2.5 km) away.[78] Polhem further extended the domain of water-powered mechanization in Swedish mines by designing and building, around 1700, a water-powered machine which hoisted ore out of a mine shaft in buckets, carried the buckets to a dump, emptied them, and automatically returned the empty buckets to the mine.[79]

Animal power and even human power were, of course, not completely replaced in all European mines in the period, even for jobs like mine drainage. Many mines were shallow and small and did not require the more concentrated power available from a vertical water wheel. In other cases it was not possible to secure a satisfactory quantity or fall of water economically at or near the mine head; and mines, unlike fulling hammers, silk twisting machinery, or millstones, could not be moved to where water power was available. But the role of water power in mining in Europe clearly underwent substantial expansion between 1500 and 1750. The growth of deep-level mining, the expanded use of water-powered equipment, and the development of power transmission systems like the stagenkunst permitted European miners to meet the continent's growing demand for metals and reflect the growth of inanimate power and mechanization in European mining.

The growth of dependence on water power also had another effect; it encouraged the emergence of capitalist organization in the mining industry. In fact, in no other European industry did early forms of capitalism emerge as early or develop as quickly as in mining. The individual miner or small cooperative mining company did not have the capital resources to install the expensive (often water-powered) mine drainage, haulage, and ventilation equipment required for deep-level mining. Thus, as early as 1500 limited stock companies and even international corporations had begun to appear in rudimentary form in the European mining industry.[80]

European metallurgy, like European mining, had been partially mechanized in the medieval period. By 1500 water power was being used to activate the hammers which prepared ore for the smelter, to power the bellows which helped reduce ore to metal, and to lift the hammers which pounded metal into desired forms. Probably under the stimulus of military demands, the contribution of water power to European metallurgy steadily expanded during the next 250 years. For example, the water-powered blast furnace was introduced into Britain late in the fifteenth century. By 1653 there were 73 blast furnaces in the country.[81] Scattered data indicate that in the Bohemian iron industry the introduction of water-powered bellows and hammers doubled the output of iron per man between 1575 and 1747.[82] Tin processing was

FIGURE 3–8. Machine for boring and grinding the barrels of muskets. In this device four horizontal wooden barrel drills were placed abreast and rotated by water power. During drilling the barrels were placed on a bed and fed into the drills by lever. A shorter shaft to the right of the drill beds carried a grindstone used for external grinding of barrels.

usually conducted without the use of water power in the medieval period. But in Britain by the late sixteenth century, water power was being extensively used for smelting and stamping tin.[83] In both the iron and tin industry, the introduction of larger, more expensive water-powered equipment led to the progressive displacement of the small craftsman by the capitalist employer.[84]

Military demands contributed not only to the expansion of water power in metallurgical processes, where it had been first applied in the medieval period; it also contributed to the development of new applications for the vertical water wheel. For example, between 1500 and 1750 water power was used for the first time to bore the barrels of muskets and cannons (Fig. 3–8)[85] and to drive metal lathes which turned the trunnions of cannons.[86] Another new water-powered device of the period was the rotary ore washer. By the mid-eighteenth century some very elaborate water-powered ore washing installations had been put into operation in the French iron industry (see Fig. 3–9).[87]

Other extensions of the use of waterpower in metallurgy in the sixteenth, seventeenth, and early eighteenth centuries were probably spurred more by the desire for wealth and luxuries than by military requirements. The mechanization of gold processing, for example, was probably stimulated by the growing demand for that metal for coinage, for ornamentation, and for decoration. Agricola in 1556 depicted a

FIGURE 3–9. Mill for washing ore. This plant was designed in the eighteenth century by the French forgemasters Ruel de Chaville and Ruel de Bellisle. Ore from iron mines was placed in the circular vats ($\check{K}$) where it was mixed with water by iron harrows (*I*). The overflow from the vats was led to the tail race along the channel *P*. When the vat began to overflow, its contents were released by sluice gate (*Q*) into the concave trough (*S*), where a vertical paddle wheel (*T*) rotated. The paddles on the wheel pushed the washed ore out at *X*. The overflow water from *S* was fed into channel *P* and conducted to the tail race. The overshot water wheel (*B*) powered both the rotating harrows and the paddle wheels.

mechanized, water-powered gold processing plant (Fig. 3–10). In the plant three distinct industrial processes were powered by the same water wheel. By means of cams on the left side of its axle the water wheel operated ore stamps which reduced gold-bearing quartz rock to grain size. These grains were then ground to powder by millstones geared into the right side of the same axle. The powder fell from the mill into a series of buckets where it was mixed with mercury and water by agitators rotated by the same axle. The gold analgamated with the mercury; the soil, dust, and unwanted rock fragments associated with it were floated away.[88]

The desires of Europe's affluent classes also contributed to the extension of the role of water power in metallurgy outside the precious metal area, for example, in lead metallurgy. Due to the small demand for lead in the medieval period, lead processing had generally been carried out

FIGURE 3–10. Water-powered gold-processing plant from Agricola, 1556. The overshot wheel (A), by means of its axle (B), drove vertical stamps (C, at the left), millstones (K), and agitators (P, P, P) which stirred mercury and water in the vats (O, O, O). The stamps reduced the gold-containing quartz to grains, the millstones ground the grains to powder, and the agitators separated the gold powder from the waste materials.

without mechanical aids. By the sixteenth century, however, lead sheet for roofing, lead pipes, and lead garden statuary became popular among the wealthy. The increased demand for lead led to the introduction of the water-powered stamp mill for preparing lead ore, the water-powered blast furnace for smelting lead ore, and the water-powered rolling mill for producing lead sheet (Fig. 3–11).[89]

Another industry which saw an extended role for water power was the gunpowder industry. The triumph of musket and cannon over bow and trebuchet, the growing scale of war, and the rise of the centralized nation state created closer links between military power and manufacturing potential. The size of the demand for gunpowder and the vital interest of governments in both the volume and quality of production tended, both directly and indirectly, to promote the use of water power.

FIGURE 3–11. Water-powered mill for rolling lead. The soaled and shrouded undershot wheel on the right provided the power for this mill. The lead rollers are on the left at *BB*, with the space between them adjustable. The lever *T* controls the sliding switch lock *E*. In the position shown the sliding lock has engaged *F* so that power from the water wheel is transmitted directly through the axle to *D* and thence to the rollers. If the lever were lowered, power would be transmitted to *D* only through gears *F, L, K,* and *dd,* and the rotation of the rollers *BB* would be reversed.

The increased rate of production possible with water-activated machinery provided a direct stimulus for the use of water power. Indirectly, the interest of governments in gunpowder production tended to concentrate production (sometimes under direct state sponsorship). And concentration into large units inclined the gunpowder industry further towards water power, since larger production units could afford the expense of installing a water-power system and profit more from it than small units.

Gunpowder required three basic ingredients—charcoal, sulphur, and saltpeter. If these were not already in granular or powdered form, they had to be reduced. By the seventeenth century this was commonly being

FIGURE 3–12. Water-powered vertical trip-hammer mill for reducing charcoal to powder for gunpowder manufacture. The operation of the plant is clear: wheel *A* turns axle *B*, which by means of cams *C, C* lifts and drops the hammers *D, D*.

done by water-powered stamp mills similar to those used in the fulling, hemp, and iron industries (Fig. 3–12). In the gunpowder mill proper either stamps (Fig. 3–13) or edge-runner stones (Fig. 3–14), powered by a vertical water wheel, were used to pound or crush the charcoal, sulphur, and saltpeter into a progressively finer and more homogeneous mixture.[90]

According to Oscar Guttmann, gunpowder mills existed as early as 1340 at Augsburg in Germany, but whether the early gunpowder mills were water-powered or not is uncertain.[91] Gunpowder manufacturing was dangerous, and it may long have been carried out by hand. By the sixteenth century, however, most gunpowder was produced with the aid of water power. By 1692 there were 22 gunpowder mills, with 829

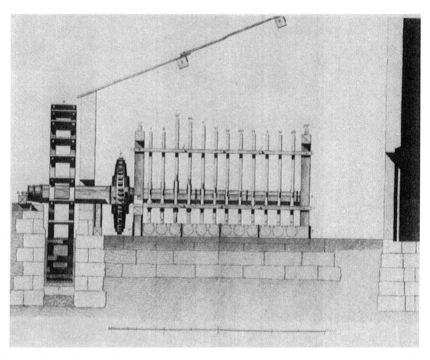

FIGURE 3–13. Battery of trip-hammers in one of the French government's gunpowder mills at Essonnes. The drawing gives a side view of the works, showing the undershot water wheel and a row of the hammers. The ingredients for gunpowder and a little water were placed in the spherical containers beneath the hammers to be pulverized and mixed. After being pounded for about an hour in one container, they were scraped into an adjacent compartment and the process resumed. After 20 or more hours of this, the powder was ready for drying and graining.

stamps, and an annual production of 2.31 million pounds (1.05 million kg) in France.[92] Many earlier French gunpowder mills had been quite small, but by 1700 or shortly thereafter, under the sponsorship of the French government, some very large establishments were built. The gunpowder mills at La Fére in Picardy and at Essonnes, south of Paris, for example, were very large, approaching in scale the cotton textile mills of the early industrial revolution. Some of the British gunpowder mill factories were also quite large by the mid-eighteenth century. The Royal Gunpowder Factory at Faversham in Kent, for example, had 11 watermills and 6 horsemills by 1760.[93]

An example of an industry where water power had not been put to use at all before 1500 but into which it steadily penetrated after that date is glass manufacturing. The demand for increased production came largely from the rich and near rich, instead of from the military, as was the case with gunpowder.

The heart of the glass-making process was a furnace. Because a thermal rather than a mechanical process was at the center of glass produc-

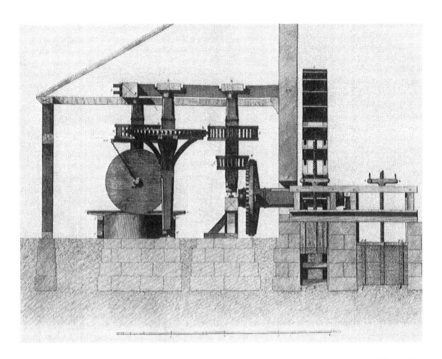

FIGURE 3–14. An edge-roller mill at Essonnes for pulverizing gunpowder. Edge-rollers eventually replaced vertical stamps in gunpowder mills since they posed less danger of explosion.

tion, water power was never able to become a dominant factor in the glass industry. But other processes associated with glass production required mechanical power, and in the course of the seventeenth and eighteenth centuries, glass makers began to use water to provide this power, especially as the scale of the industry grew under state encouragement and support. For example, in the mid-eighteenth century the scientist-engineer Lomonosov (1711–65), with the support of the Russian government, constructed a glass factory where the basic ingredients of glass—sand, potash or soda, and lime—were stamped, mixed, and ground by water power.[94] The French glass factory of Saint-Louis in Lorraine also had a watermill for crushing the materials used in producing glass.[95] Engraving finished glass was another element of glass production partially mechanized through the use of water power prior to 1750. As early as 1690 water power was used to engrave glass in Silesia.[96]

Water power was also sometimes applied to the grinding and polishing of cast plate glass by the mid-eighteenth century. Cast glass was an important French invention dating from 1688. Before 1688 flat or plate glass could be produced only by skilled glass blowers. There were few of these, and plate glass was therefore very expensive. Cast plate glass did not require skilled glass blowers. But unlike blown plate glass, it was not transparent after casting. It was badly marked, dulled, and scarred by

FIGURE 3–15. Machine to polish glass with water power in the factory of St. Ildefonse, near Madrid, mid-eighteenth century. This is a simplified drawing of the plant. Only the buffers on the lower right (*T, T, T*) are shown. In the original machine sets of buffers were located on the opposite side of the water wheel on the same level, and above the water wheel on level *L*. The buffers were moved back and forth over plate glass placed beneath them by means of the crank located at *d*. The crank moved the vertical beam (*i*) back and forth; it in turn moved the horizontal beam (*II*) which was tied to the beam *LL-KK*, on which the buffers were mounted.

the casting process and had to be ground and polished to give it transparency. Once these steps were carried out, however, cast plate glass was of comparable quality to blown plate glass and much cheaper. There was therefore a heavy demand from affluent Europeans for cast plate glass. To speed up the slow and tedious process of grinding and polishing, water-activated grinders and polishers were introduced in some eighteenth-century glass plants, like the Spanish factory near Madrid illustrated in Figure 3–15.[97] The French glass factory at Saint-Quirin had three water-powered machines for polishing its plate glass by the 1770s.[98]

Thus, the role of water power in glass production steadily expanded in the seventeenth and eighteenth centuries. According to Walter Scoville, who studied the French glass industry, many glass factories built watermills during the eighteenth century.[99] In some cases they directly served the glass-making process as indicated above. In other cases they indirectly served it by providing services to the industrial villages that grew up around the large-scale glass factories subsidized or otherwise supported by the French monarchy. The glass factory at Saint-Louis in Lorraine, for example, was practically a self-contained mill village. It had a power canal approximately 425 feet (c260 m) long and 4 feet (c2.5 m) wide. This canal supplied power to a mill which crushed the materials used in making glass. It also supplied power to a sawmill which provided timber for use at the site, and a grist mill, which provided flour for the factory hands and their families.[100]

Glass was not the only new industry to adopt water power between 1500 and 1750. There were a host of others. Water wheels were being used by the mid-eighteenth century to manufacture snuff,[101] to smash bones for fertilizer,[102] to crush chalk for whitewash,[103] to pulverize flint for pottery glazes,[104] and, possibly, to drive piles.[105] They were also being used to power butter churns,[106] polish mosaics,[107] manufacture bricks,[108] and produce music (Fig. 3–16).[109] Attempts were even made to produce perpetual motion (Fig. 3–17),[110] and to drive boats upstream using water wheels.[111]

By the sixteenth or seventeenth century water power had definitely become a major ingredient of Europe's emerging industrial economy. Although man power and animal power continued in heavy use because water power was not always economically available and because all processes were not easily mechanized, water was generally recognized as the preferred prime mover. Biringuccio noted in 1540, for instance, that whatever could be done with man power could also be done "much more easily with water."[112] In commenting on the factors to be considered in establishing a metallurgical plant, he observed:

> But of all the inconveniences, shortage of water is most to be avoided, for it is a material of the utmost importance in such work because wheels and other ingenious machines are driven by its power and weight. It can easily raise up large and powerful bellows ... ; and it causes the heaviest hammers to strike, mills to turn ... , for it would be almost impossible to arrive in any other way at the same desired ends because the lifting power of a wheel is much stronger and more certain than that of a hundred men.[113]

He also commented:

> Certainly this [the vertical water wheel as applied to the blast furnace] is a very useful thing as well as an ingenious one, for this wheel is a strong worker that endures much labor and never tires.... Surely it is impossible to work without it. If you did, you would wear out an infinite number of men.[114]

Less than a century later Jacob Leupold noted that water power was to be preferred to everything else if there was enough water and sufficient fall. "Water," he declared, "gives the smoothest operation. Where water power is available, other forms of power should not be used."[115]

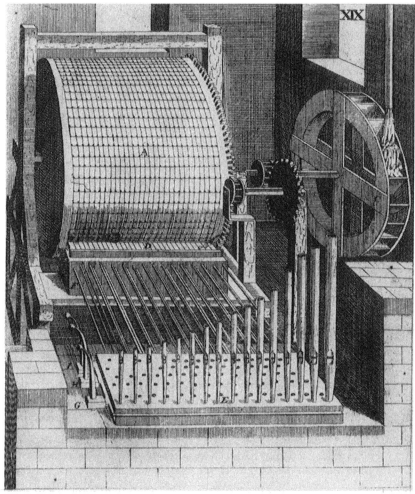

FIGURE 3–16. Musical automata powered by a small overshot water wheel, 1659.

Biringuccio and Leupold were not the only technicians to favor the use of water power over other sources of power. Around 1700 John Mortimer reviewed the various power sources available for milling and concluded: "Water is... maintained with least cost, and being the most certain and most advantageous hath gained the Preheminence [sic]."[116] And John Robison, the Scottish natural philosopher, observed towards the end of the eighteenth century: "Water is the more common power, and indeed the best, as being the most common and equable."[117]

The very high value that Europeans placed on water power is reflected not only by the scattered comments made by technicians and engineers. It is reflected, as well, by the avidity with which European colonials transferred hydropower technology to their new homes. Hunt-

Within the figure: 130, L, I, K, N, M, H, A, B, G, E, F, D, C, P, Q, O O, R

*Bulgen oder Schöpff*   *Eÿmer.*

FIGURE 3–17. Perpetual motion system utilizing overshot water wheel. The water wheel, *C*, through the gearing network *D, E, F, G, H, I*, drove a chain-of-buckets pump (*K*) which lifted water to reservoir *A*, from which the water wheel derived its power. The designer of this system, besides producing perpetual motion, hoped to drive a set of vertical stamps (*M, N*).

er, for example, found, in reviewing water power in colonial America, that watermills were regarded as a prime necessity by early colonists.[118] Hence, the resort to water power usually came quickly after settlement. The first permanent English settlement in North America was at Jamestown, Virginia, in 1607. Early in that settlement's history the Virginia company instructed its governor to build watermills on every plantation.[119] By 1649 Virginia had five watermills.[120] Maryland had a watermill in 1634, the very year it was first settled, and Swedish authorities responsible for settlements on the Delaware in the 1640s made the erection of watermills one of their first concerns.[121] The colony of Massachusetts, first settled in 1620, had a watermill at Dorchester by 1633, and mills at Roxbury, Lynn, and Watertown by 1635.[122] These were all flour mills. But according to one authority, the Piscataqua River at South Windham, Maine, was dammed for a sawmill as early as 1623.[123] In 1646, on the Saugus River, Massachusetts built an iron mill, complete with water-activated trip-hammers, blast furnace bellows, rollers, and slitters.[124] By 1700 there were few New England villages without a watermill.

The English colonies in North America were not unique. The transplantation of the watermill occurred with equal rapidity in other European colonies of the period. The Dutch, for example, began colonizing southern Africa in 1652. By 1659 the colony had a vertical watermill.[125] The French attempted to transplant Europe's seignorial system, including the manorial watermilling monopoly, to their colonies in Canada. French Canadian law encouraged seigneurs to construct mills by requiring those who intended to enforce banal rights to erect a mill within a year of assuming their estates. Most seigneurs met the requirement. French Canada had 44 gristmills by 1688 and 120 by 1739, most of them water-powered.[126]

Water power played a prominent role in certain Spanish and Russian colonization activities. The water-power system at Potosi in the Bolivian Andes has already been mentioned.[127] The Potosi mines' output added enormously to Spanish wealth in the sixteenth and seventeenth centuries, and the volume of metal yielded by that water-powered complex ultimately contributed to the near total destruction of Europe's silver mining industry. The Russian sphere of colonization was in the Urals and Siberia. Many of the Russian activities in these areas focused on the development of mineral resources, part of the effort of Peter the Great and his successors to Westernize and industrialize Russia. Since the labor supply in these areas was limited, mechanized plants, such as those already reviewed at Ekaterinburg and Zmeinogorsk, played a prominent role in this effort.[128]

*Summary*
There is no way to document precisely the steady expansion of mechanization and the growing use of water power in European industry in

the period between 1500 and the beginnings of the industrial revolution around 1770 and 1780. But the data outlined in the preceding sections largely supports the picture of continuity in industrial development posited by John Nef.

The gradual and continuous growth of mechanization and of the use of inanimate power during these two and a half or three centuries is suggested in part by the scattered and fragmentary quantitative data available, and particularly by strong evidence of mill saturation on some streams. This picture is also strongly supported by the trend towards tapping the power of progressively larger streams and towards developing power in more concentrated and larger units, a trend most discernible in the growth of hydropower dams. As we have seen, hydropower dams were apparently unknown in antiquity but began to be used in the medieval period. By the sixteenth or seventeenth century small waterpower dams, reservoirs, and canals had clearly become quite commonplace in certain areas of Europe. Moreover, between 1500 and 1750 or 1800 a number of very large hydropower dams were built. In some cases, the amount of power developed by these dams and their associated works was significant and far in excess of the amount of power developed in the early water- or steam-powered cotton textile mills of Britain. The Oderteich dam in Germany, for example, provided power to 47 water wheels, which probably developed a total of 200 to 400 horsepower. The Ekaterinburg plant in Russia developed at least 200–300 horsepower at the dam site itself. Moreover, the extended use of networks of dams, reservoirs, and canals permitted European engineers to develop effectively the water-power potential of entire areas. The Harz system may have developed as much as 1,000 horsepower; the Potosi system in Bolivia, around 600 horsepower.

Further evidence of the shift of European industry to inanimate power before the industrial revolution comes from the steady expansion of the role of water power in industry. Initially limited to milling flour, the water wheel had been applied in the medieval period to a host of other operations. This expansion of role, as we have seen, continued. Between 1500 and 1750 water-powered mechanical devices played a steadily larger role in mining, metallurgy, food processing, gunpowder and glass manufacturing, and a host of other areas. And while the cotton textile industry did not use water power and was largely unmechanized prior to the 1770s and 1780s, the role of water-powered, mechanized equipment had been steadily growing in the wool, silk, and linen textile industries for decades or even centuries.

By the late eighteenth century, the period in which the traditional industrial revolution began, mechanization through the use of an inanimate power source, the vertical water wheel, had become an important, if not a vital factor in European economic life. The high value placed on water power in Europe is confirmed by the statements of European technicians like Biringuccio and Leupold and by the avidity with which

European colonists transferred and applied hydropower technology in their new homes. The use of water power had become so much a part of European life by the eighteenth century that life without it was apparently difficult to conceive or tolerate.

Bowden, Karpovich, and Usher in their economic history of modern Europe declared: "The outstanding feature... which distinguishes the period since the industrial revolution from the ages preceding it, is the extent of the use of machines as substitutes for manual labor."[129] While there is some validity in this claim, the use of machines as substitutes for manual labor is clearly not the best way to distinguish the period before the industrial revolution from the period after. The replacement of manual labor with power machines in the cotton mills of Britain in the late eighteenth century and in mills and factories of all types all over the Western world in the nineteenth century was not really a radical break with the past. In many areas of Europe, in a large number of industries, the substitution of water-powered machinery for manual labor was a process well under way before the industrial revolution. The mechanized factories of the nineteenth century were simply the culmination of a process whose roots went unbroken back to medieval Europe.


THE TRADITIONAL WOODEN VERTICAL WATER WHEEL

Although there is little detailed information on the construction, size, or power output of vertical water wheels before 1500, there is a growing volume of data after that date. Several factors were responsible. One was the growing importance of water power, documented in the first portion of this chapter. As European industry grew more dependent on the vertical wheel and as hydropower installations grew larger, the vertical water wheel naturally attracted more attention.

A more important factor, however, was the printing press. The press's ability to quickly turn out thousands of identical copies of books eventually increased the quantity of written material in circulation in practically every area of knowledge. When it was applied in the 1500s to the production of technical works, it had a major impact on both the quantity and the quality of the data in circulation. The most important qualitative advance was in illustrations. Illustrations were a vital part of technical communication, and they were far more likely than text to be distorted by the hand-copying methods that had been used. In fact, the pictorial record is often more extensive and more important than the written record for understanding the status of water-power technology between 1500 and 1750. The Renaissance theaters of machines, for example, tended to be very rich in illustrations, but poor in text. Ramelli's *Diverse et artificiose machine,* to provide just one instance, had more than 60 excellent and detailed illustrations of water-powered machinery. But the text accompanying the illustrations was uninformative,

TABLE 3–3. WORKS CONTAINING INFORMATION ABOUT THE DESIGN AND
CONSTRUCTION OF THE TRADITIONAL WOODEN VERTICAL WATER WHEEL,
c1500–c1900

| | |
|---|---|
| 1539 | Anthony Fitzherbert, *Surveying* |
| 1540 | Vannoccio Biringuccio, *Pirotechnia* |
| 1556 | Georgius Agricola, *De re metallica* |
| 1579 | Jacques Besson, *Theatre des instrumens mathématiques et méchaniques* |
| 1588 | Agostino Ramelli, *Le diverse et artificiose machine* |
| c1595 | Fausto Veranzio, *Machinae novae* |
| 1607 | Vittorio Zonca, *Novo teatro di machine et edificii per varie et sicure operationi* |
| 1612–14 | Heinrich Zeising, *Theatri machinarvm* |
| 1615 | Salomon de Caus, *Les raisons des forces mouvantes* |
| 1629 | Giovanni Branca, *La machine* |
| 1633 | John Bate, *Mysteries of Nature and Art* |
| 1644 | Isaak de Caus, *New and Rare Inventions of Water-Works* |
| 1661 | Georg Böckler, *Theatrum machinarum novum* |
| 1724 | Jacob Leupold, *Theatrum machinarum generale* |
| 1725 | Jacob Leupold, *Theatri machinarum hydraulicarum* |
| 1729 | Stephen Switzer, *Hydrostaticks and Hydraulicks* |
| 1731–32 | Henry Beighton, "Description of the Water-Works at London-Bridge" |
| 1735 | Johannes Beyer, *Theatrum machinarum molarium* (in the Leupold *Theatrum machinarum universale* series, vol. 10; also reprinted in 1767) |
| 1737–39 | Bernard Forest de Bélidor, *Architecture hydraulique*, vols. 1 and 2 |
| 1744 | John T. Desaguliers, *A Course of Experimental Philosophy*, vol. 2 |
| 1751–77 | Denis Diderot and Jean d'Alembert, eds., *Encyclopédie; ou, Dictionnaire raisonné des sciences*, 17 vols. text, 11 vols. plates with *Supplement*, 4 vols. text, 1 vol. plates (water wheels are described or depicted in conjunction with a large number of industrial processes scattered through the set) |
| 1771 | M. Malouin, "L'art de meunier," in *Description des arts et métiers*, vol. 1 |
| 1775 | Edme Beguillet, *Manuel du meunier* |
| 1782–91 | *Encyclopédie méthodique: Arts et métiers mécaniques*, 8 vols. text, 8 vols. plates (water wheels are described or depicted in conjunction with a number of industrial plants through the set) |
| 1795 | Oliver Evans, *The Young Mill-Wright and Miller's Guide* |
| 1797 | John Robison, articles in the 3rd ed. of the *Encyclopaedia Britannica* ("Mechanics" and "Water-Works") |
| 1811 | Robertson Buchanan, *Practical Essays on Mill Work and Other Machinery*, 2nd ed. (1823) and 3rd ed. (1841) |
| 1819 | "Water" in Abraham Rees, ed., *The Cyclopaedia* . . . , vol. 38 |
| 1829 | Zachariah Allen, *Science of Mechanics* |
| 1830 | John Nicholson, *The Millwright's Guide* |
| 1849 | David Scott, *The Engineer and Machinist's Assistant*, 2nd ed. (1856) |
| 1853 | William Hughes, *The American Miller, and Millwright's Assistant* |
| 1882 | David Craik, *Practical American Millwright* |
| 1882 | Robert Grimshaw, *The Miller, Millwright, and Millfurnisher* |
| 1893 | Joseph Frizell, "Old-Time Water-Wheels" |

invariably taking the form: "Water wheel *A* turns axle *B*, which turns gear *C*, which meshes with gear *D*" and so on.

Because the pictorial record is often more extensive and informative than the written record, I have relied heavily on it in attempting to determine the prevalent design patterns and methods of assembly used for vertical water wheels between 1500 and 1750. But written material cannot be neglected. I have drawn on a number of works dating from 1500 to as late as 1900 in attempting to supplement the pictorial record (Table 3–3). In drawing on materials after 1750 to describe construction methods and techniques used before 1750 I do not believe I have strayed far into error. Millwrighting, like many traditional, wood-oriented crafts, changed slowly through the centuries. The methods used to construct wooden water wheels c1900 probably differed very little from those used c1500.

*Construction*

The traditional vertical water wheel (whether undershot or overshot) had five parts: (1) the axle, (2) the arms or spokes, (3) the rims and/or shrouds, (4) the soal, and (5) the floatboards or bucketboards (see Figs. 1–1 and 1–2). All of these parts were made primarily of wood. This is indicated by the illustrations in the works of Agricola (1556), Böckler (1661), Ramelli (1588), and other technicians of the period. It is also supported by archeological evidence. During excavations at the site of the seventeenth-century American iron works on the Saugus River in Massachusetts, for example, a segment of one of its water wheels was uncovered. It was almost totally wood. Even the shrouds were pegged with wooden dowels, instead of nails. The only iron used was in the bands that reinforced the axle.[130] The wheel of the Wakefield mill in Britain in 1754 was similarly built. Iron was used only for the axle journals and, in the form of small plates, for tying the shroud segments of the wheel together.[131] As late as 1882 David Craik noted that while timber was cheap, iron was "a precious metal" and that, therefore, a good millwright "must exert his ingenuity to construct a mill that will do the required work with the greatest possible economy of iron. Mills are sometimes thus built almost entirely of wood, without bands or bolts, or even nails, with scarcely anything metallic."[132]

Information on the specific type of wood used in traditional vertical water wheels is scattered. It seems clear, however, that oak was the most popular and that, in many cases, the entire water wheel was built with it. For the axle, in particular, oak was usually used, although white pine was recommended by Zachariah Allen in 1829 and Weisbach noted that deal and pitch pine were sometimes used.[133] Tamarac, black ash, and pine were mentioned as alternatives by Craik in 1882.[134] Other woods—such as elm, beech, ash, or redwood fir—were sometimes used for the arms, rims, soaling, or floatboards of vertical wheels. Beguillet, for example, noted in 1775 that the floats on French wheels were ordinarily made of soft woods, and he recommended elm.[135] Elm was often favored for parts exposed to water because of its water-resistant qualities. When elm was wet, a slimy coating formed on it which protected the wood from decay. In the United States pine and cypress were sometimes used for floats for the same reason. The average life expectancy of a wheel made of wood was about ten years, though some repairs undoubtedly would have to be made earlier than that.[136]

Probably the most important single part of the traditional vertical water wheel was its axle. It was certainly the most expensive. Straight oak

TABLE 3–4.  MATERIAL REQUIRED TO BUILD AN OVERSHOT WATER WHEEL
(18 feet in diameter, 2 feet 2 inches wide, including axle and large cog wheel)

Wood:   1 shaft, 18 ft long × 2 ft diameter
        8 arms, 18 ft long × 3 in wide × 9 in deep
        16 shrouds, 8.5 ft long × 2 in thick × 8 in deep
        16 face boards, 8 ft long × 1 in thick × 9 in deep
        56 bucket boards, 2 ft 4 in long × 17 in wide
        140 ft boards for soaling
        3 arms for cog wheel, 9 ft × 4 in × 14 in
        16 cants, 6 ft long × 4 in × 17 in

Metal:  2 gudgeons, 2 ft 2 in long; neck 4.25 in long × 3 in diameter
        2 bands, 19 in inside diameter × 0.75 in thick × 3 in wide
        2 bands, 20.5 in inside diameter × 0.5 in thick × 2.5 in wide
        2 bands, 23 in inside diameter × 0.5 in thick × 2.5 in wide

SOURCE: Oliver Evans, *The Young Mill-Wright and Miller's Guide,* 4th ed. (Philadelphia, 1821), pp. 344–45.

trunks that could be cut down to the 15 to 24 inch (c40–60 cm) diameter commonly employed on water wheels between 1500 and 1750 were hard to find. And the axle was critical, since it had to be strong enough to support the entire weight of the wheel, typically between 5,000 and 10,000 pounds (c2,000–4,000 kg), and the weight or force of the water, as well as transmit power to the attached machinery.[137]

Because of its critical role and its cost, the axle was the one part of the traditional vertical water wheel which almost invariably was strengthened with metal work (see Table 3–4). The friction between the wood axle and its bearings would quickly have worn down the axle, so very early vertical wheels were probably equipped with iron gudgeons or journals (Fig. 3–18). One of the most widely used forms of gudgeon in later years was the wing gudgeon pictured in Figure 3–19. The butt ends of an oak axle would be mortised so that the wings could be inserted, leaving the cylindrical part of the gudgeon, commonly between 1.5 and 3 inches (c4–8 cm) in diameter and 3 to 6 inches (c8–15 cm) long, projecting from the end of the axle. It was on the cylindrical projection that the wheel turned. To insure that the gudgeons remained firmly in place and to reinforce the weakened ends of the axle, several iron bands would be installed near the ends of the axle where the gudgeon had been mortised. These bands were typically around 4 inches (10 cm) wide and from 0.5 to 1.0 inches (1.2–2.5 cm) thick. They were pounded on hot, so that in cooling the bands would contract and tighten. Any remaining play between gudgeon and axle was taken up by inserting keys and wedges next to the wings of the gudgeons and by driving iron wedges into the butt ends of the axle.[138]

The bearings on which the gudgeons ran were made of hardwood, stone, brass, or iron. Fitzherbert noted in 1539, for example, that mills commonly ran on stone or brass bearings.[139] The larger wheels found in mining and other heavy industries probably used iron, for Agricola in

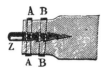 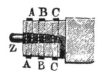  

FIGURE 3–18. Forms of gudgeons commonly used on wooden vertical water wheels.

FIGURE 3–19. Wing gudgeon. Note the iron rings (*r, r*) used to reinforce the axle at the mortise. The wheel rotated on the cylindrical part (*A*) of the gudgeon.

1556 reported the use of iron bearings in German mining districts.[140] The friction of the gudgeon on the bearing was considerable and generated sufficient heat to shorten the life of both gudgeon and bearing if cooling and lubrication were not utilized. The favored lubricants were rendered animal fats, olive and sperm oils, tallow, and sometimes powdered black lead. Where cooling was needed, water was dripped very slowly on the bearings. Evans in 1795 and Craik in 1882 recommended that the flow rate be sufficient to maintain evaporation but not allow water to remain long on the bearing, since it would destroy the polish made by the grease.[141]

A second primary part of the traditional vertical water wheel was its arms or spokes. These varied in size. Typically they might be either 3-by-9-inch (7.6 × 22.9 cm) or 4-by-6-inch (10.2 × 15.2 cm) planks.[142] They were linked to the axle in two basic ways, as noted in the last chapter—compass linkage or clasp linkage.

The clasp technique required that the axle be squared off at the point of contact with the arms. Sets of spokes or arms ran from one side of the wheel to the other in parallel pairs and were spaced so that they "clasped" the squared portion of the axle between them. Where sets crossed each other, they were bolted together. The connection between spokes and axle was stiffened with wedges, which were also used to adjust the wheel if it were out of true. Clasp arms were used in Europe between 1500 and 1750, but they were apparently not as popular as compass arms. My count of water wheel illustrations in books published between 1500 and 1750 indicates a strong preference for compass arms. In some areas of Europe, like Portugal, clasp-arm wheels were hardly ever used.[143] And, according to Frizell, they were rare in the United States.[144] The difficulty of tightly binding wooden clasp arms to an axle and keeping them from working loose in service probably accounts for the preference for the compass arm construction.

When compass arms were used, the point where the wheel joined the

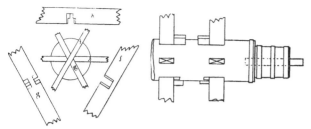

FIGURE 3–20. Notching arrangement for a compass-arm water wheel. For a wheel with six spokes (three arms) the centers of the arms (*f, g, h*) would be notched to approximately two-thirds of their thickness in the manner shown. The arms were then placed in three mortises cut completely through the axle of the water wheel and linked.

axle could be left round, but was often shaped hexagonally or octagonally depending on whether six or eight spokes were used. Mortises were cut completely through the axle at equal distances around the circumference. Then arms running from one side of the wheel to the other were notched so that they could be locked as they passed through the axle. If three arms (six spokes) were used, each of the arms was notched to ⅔ of its thickness at center (see Fig. 3–20); if four arms (eight spokes) were to be inserted (and this was usually the maximum), each arm was notched to ¾ of its thickness. In order to insert the arms in the axle and lock them, the axle mortises had to be larger than the arms. The spaces left after the arms had been inserted and linked were closed by oak keys, by wedges, or by a combination of the two.

Both compass and clasp designs provided a strong and reliable means for transmitting motion from the wheel to the axle as long as the wheel's diameter did not exceed 15 feet (c4.5 m). Above 15 feet simple clasp and compass arms were insufficient and had to be reinforced with diagonal supports. These supports ran either between the different sets of arms or from the arms to the rims. (See Figs. 3–21 and 3–22.)

A third primary part of a vertical water wheel was its rims or shrouds. These were built up from planks around 2 to 4 inches (c5–10 cm) thick, cut and assembled to form a ring. The separate planks which formed these rings can be seen clearly in Figures 3–10, 3–24, and 3–29, for example. After the arms of a water wheel had been prepared, but before they were inserted into the axle, they were laid out on the ground or on a stand, and the planks intended for the rims or shrouds were laid out around the ends of the spokes. The periphery of the wheel and the joints between the planks were marked off and the planks cut to size for assembly. There were several methods of joining the segments that composed the rim or shroud. The butt ends of adjacent planks were sometimes cut to half thickness and overlapped; sometimes they were scarfed or checked and counter-checked; sometimes they were joined by mortise and tenon, or by joggle. Frequently, the rims or shrouds of water wheels were built up from two or more segmented plank rings connected so that the joints of one ring fell in the middle of the segments of the other.

There were three basic methods of joining the rims or shrouds to the spokes: (1) the rims or shrouds of a wheel could be mortised to receive

FIGURE 3–21. West Indian sugar mill powered by a reinforced compass-arm water wheel. The use of vertically situated roller presses instead of edge-runner stones, regular mill-stones, or vertical stamps in sugar mills was another innovation of the 1500–1750 period.

FIGURE 3–22. French watermill with reinforced clasp arms. The man is raising a sluice gate to allow water to flow beneath the undershot water wheel. Note the curved race beneath the water wheel. Sometimes a vertical ledge or an inclined plane was used instead.

the arms (e.g., Figs. 2–31 and 3–31); (2) the arms could be notched and the rims or shrouds set in the notches (Figs. 2–29, 3–10, 3–29); or (3) the arms could simply be pinned to the sides of the rims or shrouds (Figs. 1–2, 2–18, 3–16). Mortising was the method most frequently depicted in illustrations between 1500 and 1750 and was probably the preferred practice because of the stronger joint it provided.

The fourth primary part of the traditional water wheel was its soal, a cylindrical layer of planks which linked the two rims or shrouds. Attaching the soal to the rims was a relatively routine matter. Planks of around 1 to 2 inches (2.5–5 cm) thick were simply pinned by dowels or nails to the bottom of the rims or shrouds, forming a circular drum which would retain water if the wheel were overshot or would prevent water from splashing over the blades or floatboards into the interior of the wheel if it were undershot. In addition, the soal provided the wheel with increased stability by binding the rims of the wheel rigidly together. Some undershot wheels, however, were constructed without a soal.

The fifth primary part of the traditional wooden water wheel was its floatboards (undershot) or bucketboards (overshot). Several methods were used to attach these to the soal or rim. Most frequently, in undershot wheels at least, mortises were cut into the rims of the wheel, starts were inserted in the mortises, and the blades were pinned to the starts (Figs. 2–10, 2–31, 3–13, and 3–14, e.g.). In most overshot wheels, and in some undershot, notches were cut in the shrouds and the floats inserted and pinned (Figs. 2–17 and 3–31, e.g.). Other methods were sometimes used. If the soal of the wheel was thick enough, the floats might be set into notches there, projecting radially without starts or shroud support (see Fig. 2–30, e.g.). And if there were any fear that the floats were not held solidly, circular hoops (Figs. 2–11, 2–20, 3–25) or intermediate rings or rims (Fig. 3–31) could be installed to reinforce them.[145]

The order in which the various parts of a traditional vertical water wheel were shaped and assembled probably varied. The order of assembly given in Table 3–5 is based on information provided by a French author in 1765, by an American millwright in 1795, and by a Russian engineer in 1819. It is supplemented by the comments of two late-nineteenth-century American millwrights. Undershot and overshot wheels were constructed in much the same way, so the order of assembly outlined in Table 3–5 probably applied to both.

The axles, spokes or arms, rims, and often the shrouds and soaling of undershot and overshot wheels were identical. But because water acted on the two wheels in different ways—by impulse on the undershot wheel, by weight on the overshot—there were differences, particularly in (1) the design of the boards which received the water and (2) the auxiliary works which fed water onto the wheel. We will look at these differences in turn.

The undershot wheel had simple blades or floatboards. Between 1500 and 1750 they were typically from 1 to 4 feet (0.3–1.2 m) long, 1 foot (0.3

TABLE 3–5.  ORDER OF ASSEMBLY FOR A TRADITIONAL VERTICAL WATER WHEEL

---

1. Construct a work stand. It should consist of a center post (representing the axle or shaft of the water wheel) and a number of stakes (equal to the number of spokes which the wheel will have) spaced at uniform intervals around the center post and at a distance approximately equal to the water wheel's intended radius.
2. Lay out the arms or spokes of the water wheel so that their ends lie on the stakes and their centers on the center post. Notch the arms as they will be joined in the axle of the wheel.
3. Lay out the planks for the rims and/or shrouds on the ends of the arms or spokes. Mark the intended periphery of the water wheel on the planks and trim them so that when joined they will form a circular, segmented ring.
4. Make the cuts and notches necessary to join the rim and/or shroud to the spokes and temporarily assemble the rim or shroud and join it to the spokes.
5. Mark off the notches or mortises needed to fix the floatboards to the rims or shrouds and make the appropriate cuts.
6. Prepare the axle, making the mortises needed to insert and link the arms of the wheel in the axle. Install the gudgeons in the axle or shaft.
7. Set the axle or shaft up in its working position.
8. Install the arms of the wheel in the axle or shaft, fix firmly, and align.
9. Install the rim or shroud segments on the arms or spokes of the wheel and align. Permanently link the rim or shroud segments to the arms.
10. Line the inside of the wheel with planks (i.e., install the soal if the wheel is to have one).
11. Permanently fix the floatboards in place.

---

SOURCES: Oliver Evans, *The Young Mill-Wright and Miller's Guide* (Philadelphia, 1795), pp. 315–19; "Pompe," *Encyclopédie; ou, Dictionnaire raisonné des sciences*, ed. Denis Diderot and Jean Lerond d'Alembert, 13 (1765): 10–11; David Craik, *The Practical American Millwright and Miller* (Philadelphia, 1882), pp. 105–14; Robert Grimshaw, *The Miller, Millwright, and Millfurnisher* (New York, 1882), pp. 506–7; and V. V. Danilevskii, *History of Hydroengineering in Russia before the Nineteenth Century* (Jerusalem, 1968), pp. 178–79. Danilevskii based his account on P. Asonov, *Sistematicheskoe opisanie gornogo i zavodskogo proizbodsva Zlatoustovskogo savoda* [Systematic description of the metallurgical and plant production of the Zlatoustovskii plant], a manuscript dating from 1819. Danilevskii says, p. 183, n. 6: "Although this description was written in 1819, it describes the technique of manufacturing water-wheels according to experience at the turn of the 18th century."

m) deep, and around 2 inches (5.1 cm) thick. Theoretically the blades could be either radial or inclined, but in practice only radial blades were used. They were distributed uniformly around the periphery of the wheel so that at least two, if not three or four, were always immersed in the water. Fitzherbert in 1539 recommended that at least three be always submerged, and illustrations of undershot wheels in the era indicate that this practice was widely followed.[146]

Shrouding and soaling were optional on undershot wheels. Some wheels had both (Figs. 2–13, 2–17, 3–11). Some had only one (see, for example, Figs. 2–15 and 3–5 for a wheel with a shroud but no soal, and Fig. 3–25 for a wheel with a soal but no shroud). Some had neither (Figs. 3–12, 3–13, 3–14, 3–33). Both shrouding and soaling gave the wheel generally, and the floats in particular, greater overall solidity, and shrouding may have been an aid in guiding water onto the blades. But both added to the cost of a wheel, impeded the exit of water from the wheel after impact, and hindered the wheel in driving out backwater when partially flooded. Some millwrights also felt that the water trapped in the spaces between the blades in a shrouded and soaled wheel would, due to the vacuum formed, create a back pressure on the wheel and diminish its power.[147] The unshrouded undershot wheel was thus preferred to the shrouded wheel, and between two-thirds and three-fourths of the undershot wheels pictured in works dealing with water power between 1500 and 1750 were unshrouded. With soaling the issue was closer. Slightly more wheels appear with soals than without prior to 1700, but after 1700 the unsoaled undershot wheel seems to have become more popular. Almost all of the undershot wheels pictured in Bélidor's *Architecture hydraulique* (vols. 1–2, 1737–39) and Beyer's volume of Leupold's *Theatrum machinarm universale* series (vol. 10, first published 1735), for example, were unsoaled.

The analogue of the undershot wheel's blades or floatboards was the overshot wheel's buckets. The soaling of the overshot wheel formed the bottom of the overshot's bucket, the shrouds formed two of the sides, and the inclined bucketboards set between the shrouds formed the other two. Commonly the buckets of overshot wheels (and hence the shrouds as well) were between 8 and 24 inches (0.2–0.6 m) deep, and the bucketboards were spaced about 12 to 18 inches (0.3–0.45 m) apart.

The primary problem encountered in the design of the buckets of overshot wheels was preventing water from spilling from them before the bottom of their revolution. Spillage could occur very early in the downward arc on the loaded side of an overshot wheel, as Figure 3–23 illustrates (see also Fig. 3–27). This problem was widely recognized in the period, and several attempts were made to alleviate it through better bucket design. The usual design applied throughout the Renaissance and on up to the eighteenth century was the inclined bucket. Basically it consisted of floatboards, much like those on undershot wheels, set between shrouds at an angle to the radius of the wheel (Fig. 3–23 or 3–27). The elbow bucket (Fig. 3–26), developed by the early fifteenth century if not earlier, retained water better, but was rarely used before 1700. Only in the 1730s is there firm evidence that it had begun to replace the inclined bucket extensively. As with the rise in popularity of the unsoaled undershot wheel, Bélidor's *Architecture hydraulique* of 1737–39 and Beyer's volume in the Leupold series signal the change. While some of the buckets pictured in Beyer and Bélidor retain the inclined design, elbows appear with equal frequency. By the late eighteenth century elbow buckets were dominant. In areas like America, however, where water economy was not a major concern, the less efficient, but easier to construct, inclined buckets continued to be popular. The American millwright Oliver Evans, for example, depicted nothing but inclined buckets in his *Mill-Wright and Miller's Guide.*

Besides redesigning the buckets, there were several other attempts to alleviate the spillage problem. Leonardo da Vinci, in one of his man-

FIGURE 3–23. Overshot water wheel with inclined buckets, 1795. Note the spilling from the buckets before the bottom of their revolution. This represented a serious loss of power and eventually led to the widespread use of elbow buckets, which retained water better.

uscripts, suggested pivoted buckets. These would have largely eliminated spillage, but they required a device to tilt and empty the buckets at the bottom of the wheel. Leonardo planned a ratchet mechanism to do this, but the cost and complexity of his system, combined with heavy power losses due to friction, probably kept it from ever coming into use.[148] Another solution was to enclose the periphery of the overshot wheel in a close-fitting casing, or breast. Leonardo also suggested this.[149] But it, too, never came into wide use. The basic idea of using a casing to inhibit water spillage and confine water to a wheel, however, eventually led to the development of the breast wheel. Since the breast, or breastshot, wheel played a very minor role in vertical wheel technology before

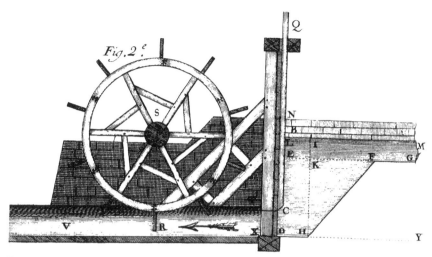

FIGURE 3–24. Use of an undershot water wheel in conjunction with a mill dam. The water was allowed to spurt through a sluice gate located at the bottom of the dam onto the water wheel. This was the usual manner of operating an undershot wheel when a fall of 5 to 6 feet (c1.5–2.0 m) was available.

1750, only to emerge as one of the major forms of vertical wheel after that date, I will postpone any discussion of its development until Chapter 5. The chain-of-bucket and chain-of-plug engines developed between 1630 and 1750 and reviewed later in this chapter can be seen as another attempt, albeit rather radical (and largely unsuccessful), to solve the spillage problem in overshot wheels.

A water wheel does not operate in isolation. It requires auxiliary works to collect and store water and to build up a head (dams and reservoirs), to lead water to the mill site (canals), to direct water onto the wheel (head races, flumes, sluice gates, penstocks), and to carry water away from the wheel (tail races). Dams, reservoirs, and canals were reviewed earlier in this chapter. We shall now consider the auxiliary works which fed water onto water wheels and led the water away after use. Because vertical undershot and overshot wheels operated under different principles, their auxiliary works also differed.

The undershot mill was sometimes called a "ground shot" mill because the leat, or head race, which carried water to it normally rested in or on the ground. By the Renaissance it had become standard practice to constrict the race at the point where the undershot wheel rotated to bring the maximum amount of water at the maximum velocity against the blades. Usually less than 3 inches (c7.6 cm) separated the blades of an undershot wheel from both the bottom and sides of the channel or trough in which it rotated. Where a fall of 3 to 6 feet (c1–2 m) could be developed, there were two options open to the millwright. First, he could use a dam and release the water on the wheel from beneath a vertically lifting sluice gate near its bottom (Fig. 3–24). In this case the water wheel

FIGURE 3–25. Pumping engine built at Nymphenbourg by le Comte de Wahl, director of works for the elector of Bavaria. The water was led along the canal Q and then dropped along an inclined plane onto the blades of the undershot wheel. This was an alternate method for utilizing high heads with undershot wheels.

was usually placed fairly close to the gate, say within 3 to 6 feet (c1–2 m). This was the more usual arrangement and the more efficient. But a second arrangement was possible. This involved leading the water along an almost level channel until just before it reached the wheel and then dropping it against the bottom of the wheel with a sharp incline, as in Figure 3–25.

One of the key design problems with undershot wheels, as previously noted, was getting the water away from the wheel as quickly as possible after impact to avoid wasting a part of the wheel's power in pushing dead water. While the absence of soaling and shrouding and the use of large blades facilitated this task, the key element was the trough or channel (the tail race) behind the wheel. Traditionally, the millwright provided a drop of between 3 and 9 inches (7.6–23 cm) in the tail race

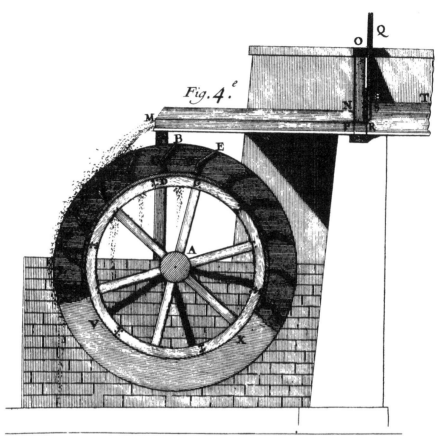

FIGURE 3–26. Typical eighteenth-century overshot water wheel, from Bélidor, 1737. The use of only a small velocity head (c10 in, or 25 cm) on overshot wheels was usual in this period. Note the use of elbow rather than inclined buckets.

directly behind the wheel to facilitate the exit of water. This drop could be designed in several ways. A vertical ledge could fall off that distance, or the channel could be curved (Fig. 3–22) or sharply inclined downward. Another technique used to facilitate drainage was widening the trough or channel behind the wheel. At many mills where drainage and back water were serious problems, further steps were taken. An apron of timber, stone, or masonry was built to carry water from the wheel to a point 15 to 30 feet (c4–10 m) downstream. This apron was usually sloped to drop off between 6 and 18 inches (15–45 cm) in that distance. Thus, the drop between the water wheel and the tail race in traditional vertical watermills could be as great as 1 to 2 feet (0.3–0.6 m).[150]

The canals, flumes, races, or launders which led water onto overshot water wheels tended to be smaller than those conducting water to undershot wheels. And, because of the nature of the instrument they served, they were usually elevated instead of built on or into the ground. There

FIGURE 3–27. Overshot watermill for grinding corn at Nun-Eaton in Warwickshire, England, 1744. The Nun-Eaton wheel is a good example of the mixed-overshot pattern. The wheel was 16 feet (4.9 m) in diameter and was fed water through a trough or race (*YZ*) from a reservoir (*X*). The surface of the reservoir was 7.5 feet (2.3 m) above the top of the wheel, so that around 32% of the total fall was devoted to impulse. Water was fed into the trough by means of a small gate located in the penstock (*ab*) and raised by a lever (*e*).

were two basic patterns for directing water into overshot wheels between 1500 and 1750, as well as two subdivisions within one of these patterns. They are summarized below:

1. The gravity-overshot wheel pattern. Water was carried in an almost level race over the wheel and fed into the buckets from a moderate height (say, less than 3 feet, or 1 meter), so that the water had little impact against the wheel. (See Fig. 3–26.)
2. The mixed-overshot wheel pattern. Water was fed onto the wheel under a substantial head (say, 3 feet or greater), so that a considerable portion of the wheel's power came from the impact of the water. There were two subpatterns in this group.

FIGURE 3-28. Flour mill complex from Zonca, 1607. Three of the overshot wheels in this complex receive water in the Alpine fashion, with its distinctive steeply inclined chutes. Enclosed spouts, like that of the small wheel in the lower center, were rarely used in this period.

(a) The normal mixed-overshot pattern. A head of water was built up in a penstock situated very near or immediately above the wheel, with the water being fed onto the wheel under pressure either through a nearly horizontal flume slightly higher than the wheel or through short inclined shuttles. (Fig. 3–27.)

(b) The Alpine wheel. Water was carried in an almost level race to a point some distance above the wheel, and then dropped on the wheel by a steeply inclined chute. (Fig. 3–28.)

Most mills which used overshot wheels used the gravity-overshot pattern, probably because of its higher efficiency. But the pattern selected often depended on the machinery that the wheel was to drive. In sawmills, for example, high speed was often more important than efficiency, so wheels of the Alpine pattern (2b) were used.[151]

The flow of water from the head race into the chute which led to an overshot wheel was usually controlled by a vertically lifting sluice gate similar to those which regulated the entrance of water to undershot wheels. The chutes or flumes leading to the wheels could be designed in several ways. The Alpine wheel, for example, commonly had an open chute, but one which tapered toward the end (Fig. 3–28). An enclosed nozzle was sometimes used on both Alpine and normal mixed-overshot water wheels (Figs. 3–6, 3–17, 3–28 at bottom). The most common arrangement, however, was that depicted in Figures 3–9, 3–15, and elsewhere, where water was led along a rectangular open chute of uniform width and dropped out of the end onto the buckets of the water wheel. On rare occasions the water might be directed onto the wheel from notches cut in the bottom of one of these chutes, instead of from an open end (Fig. 3–27), but this method achieved wide popularity only later.[152]

Whether a millwright chose an undershot or an overshot wheel seems to have depended largely on the terrain. Scattered remarks from a number of sources between 1600 and 1800 suggest that undershot wheels were usually preferred in low, flat regions where water was plentiful, while overshot wheels were preferred in hilly areas where high heads could easily be developed.[153] The breaking point was somewhere between 6 and 10 feet (2–3 m). Below that, an undershot wheel would almost always be built; above that, an overshot. In 1754, in reviewing the use of water power in iron works, for example, the French forge master Bochu specified 9 feet (c3 m) as the critical height.[154]

*Size and Power*

The dimensions given a vertical water wheel, whether undershot or overshot, varied with the task it was to perform, the quantity of water available, and the head of water that could be developed. Nonetheless, it is possible to gain some idea of the dimensions of a "typical" vertical water wheel between 1500 and 1750 or 1800. Scattered remarks, for instance, indicate that vertical wheels were commonly around 10 to 15 feet (3.0–4.6 m) in diameter (see Table 3–6).

An average or typical width is much harder to establish, for few writers mention any one width or narrow range of widths as common. But illustrations from the period indicate that most water wheels were rather narrow relative to their height. If 10 to 15 feet (3.0–4.6 m) was the typical height, then the typical width was between 1 and 4 feet (0.3–1.2 m), but with frequent exceptions.

Estimates of the average power output of vertical water wheels before 1750 have ranged from a fraction of a horsepower all the way up to 50 horsepower. On the high side of the spectrum is Leslie Aitchison. Aitchison suggested that early medieval mills probably generated around 8 horsepower, while in the thirteenth and fourteenth centuries "a single mill would generate from forty to sixty horsepower."[155] But Aitchison cited no specific evidence to support this estimate. Bern Dibner, in describing the wheels illustrated by Agricola in *De re metallica*, declared: "By constant improvement, these sources of primary power had reached

TABLE 3–6. USUAL DIAMETER OF VERTICAL WATER WHEELS, c1500–c1800

| Authority | Usual diameter | Comments |
|---|---|---|
| Biringuccio, 1540 | 11.5 to 15.5 ft | large overshot wheels |
| Pryce, 1778 (referring to wheels built c1700) | 12 to 15 ft | Cornish mine wheels |
| Leupold, 1724 | 12 to 16 ft | |
| Beyer, 1735 | 14 to 16 ft | undershot wheels |
| Bélidor, 1737 | 12 to 18 ft | ordinary flour mills |
| Ferguson, 1764 | 17 to 18 ft | common breast wheels |
| Beguillet, 1775 | 10 to 12 ft | wheels of flour mills |
| Encyclopédie méthodique, 1788 | 10 to 12 ft | wheels of flour mills |
| Robison, 1797 | 12 to 18 ft | |

SOURCES: Vannoccio Biringuccio, *Pirotechnia* (1540), transl. C. Smith and M. Gnudi (Cambridge, Mass., 1966), p. 301; William Pryce, *Mineralogia Cornubiensis* . . . (London, 1778), p. 307; Bernard Forest de Bélidor, *Architecture hydraulique*, 1 (Paris, 1737): 287; Johann Beyer, *Theatrum machinarum molarium* . . . (Dresden, 1767 [originally published in 1735]), pp. 13–15, 23–28; James Ferguson, *Lectures on Select Subjects* . . . (London, 1764), p. 45; Edme Beguillet, *Manuel de meunier et du charpentier des moulins* . . . (Paris, 1775), pp. 23–24; "Meunier," *Encyclopédie méthodique: Arts et métiers mécaniques*, 5 (Paris, 1788): 42; [John Robison], "Mechanics," *Encyclopaedia Britannica*, 3rd ed., 10 (Edinburgh, 1797): 760.

a top limit of about ten horsepower. It required another century, the late 1600's, for water-wheels to attain a power of twenty horsepower each."[156] In stark contrast to the estimates of Aitchison and Dibner are several statements made in Ernest Straker's *Wealden Iron*. Straker quoted entries dating from 1744 and 1754 from the diary of John Fuller. These referred to a lack of water for the blowing engines at iron foundries in southern England and mentioned the use of man-powered treadwheels and capstans to operate the blowers.[157] Paul Wilson, citing these passages, commented:

> The power developed by many of these early waterwheels was remarkably small. It is unlikely that more than two or three men could tread a wheel at the same time, and even if they worked in relays, it is doubtful that they could maintain, continuously, more than 0.2 or 0.3 horsepower. Probably the blast was less satisfactory than that obtained when the wheel was driven by water, but even so it appears likely that large numbers of early industrial waterwheels developed well under 1 brake horsepower.[158]

Other estimates, between these two extremes, have come from Paul Gille (5 hp), A. P. Usher (2–5 hp), M. Ferrendier (7.5 hp), and Carlo Cipolla (1–3.5 hp).[159] (See Table 3–7.)

Usher provided the best documentary support for his estimate. On the basis of studies around 1800 which indicated that it took approximately 1 horsepower to grind 1 bushel of wheat per hour,[160] he noted: "It is, of course, especially difficult to form any notions of the potentialities of water wheels, but a number of references by Evans [1795] and Smeaton [c1760–c1790] indicate clearly that flour mills run by water

TABLE 3–7. ESTIMATES OF THE POWER OUTPUT OF TRADITIONAL VERTICAL WATER WHEELS

| Authority | Estimate | Comments |
|---|---|---|
| Aitchison | 40–60 hp | by the 13th to 14th centuries; it is not clear whether Aitchison intended this to be an average or the upper limit |
| Dibner | 10 hp | top limit in the 1500s |
| | 20 hp | attainable by 1700; not clear whether Dibner intended this to be the upper limit or the average power |
| Ferrendier | 7.5 hp | average power |
| Gille | 5 hp | average power; Gille estimated 10 hp as the top limit |
| Usher | 2–5 hp | average power |
| Cipolla | 1–3.5 hp | average power, 13th century |
| Wilson | less than 1 hp | "large number of early industrial wheels" |

power seldom ground more than five or less than two bushels of wheat per hour."[161] This suggested to Usher 2 to 5 horsepower as typical for watermills in the late eighteenth century. This figure would seem to apply to earlier periods as well, for Samuel Hartlib, writing in 1651, noted that English watermills, as a whole, could not be expected to grind above four bushels of wheat per hour, suggesting an output of 4 horsepower.[162]

There are a few works between 1700 and 1800 which do provide sufficient quantitative data about specific water wheels to allow a reasonable estimate of their power output for comparison with the estimates of Usher and others. Table 3–8 reproduces this data along with my estimates of their horsepower. In Table 3–9 I have utilized quantitative data provided by Héron de Villefosse on the water wheels used in the mining districts of central Europe around 1800 to reach an estimate of their power. While there is no assurance that the wheels in these tables are representative, the data about them provide the only solid quantitative information available for calculating the power of traditional wooden water wheels. On the basis of this data I have concluded that Usher's estimate of 2 to 5 horsepower is probably a little low. The average output of wooden water wheels by the eighteenth century was more likely in the neighborhood of 5 to 7 horspower at the shaft. The estimates of Dibner and Aitchison (20 to 60 horsepower) probably reflect the upper limits of the technology.

These values may seem rather insignificant when compared to modern power units. But it must be remembered that they represented a concentrated and powerful source of energy in their own day. Since a water wheel could operate 24 hours a day, a 5 to 7 horsepower wheel could, theoretically, produce an amount of power equivalent to that of 20 to 30 horses or 150 to 200 men.

FIGURE 3–29. Mine hoist with reversible water wheel from Agricola, 1556. The water wheel on this hoist was partitioned into two parts, *G* and *H*. The buckets on each side of the partition were inclined in opposite directions so that the man at *O*, by lowering either the spout *E* or the spout *F*, could determine the direction in which the wheel rotated. The right side of the axle of this wheel (not shown) was equipped with a friction brake of the type shown in Fig. 3–30. This wheel, according to Agricola, was 36 feet (c11 m) in diameter and was the largest type then in use.

While the average water wheel may have been 10 to 15 feet (3–4.5 m) in diameter and 1 to 4 feet (0.3–1.2 m) wide, and while it may have generated 5 to 7 horsepower, much larger and more powerful wheels were sometimes used, especially in heavy industries like mining and water supply. Taller wheels were reported, for example, by Agricola in 1556. He noted that the water wheels used to power rag-and-chain pumps of the type pictured in Figure 2–17 were commonly 24 feet (7.1 m) in diameter, and often 30 feet (8.85 m) in diameter.[163] The largest wheel he pictured was the overshot wheel of Figure 3–29. It was 36 feet (11 m) in diameter, must have been at least 4 to 5 feet (1.2–1.5 m) wide, and may well have generated on the order of 30 to 40 horsepower. It was unique in several respects. It had two sets of buckets, inclined in opposite directions. These allowed the operator of the wheel to power the wheel in either direction. While most wheels were stopped merely by cutting off the water supply, the momentum built up by the motion of this wheel was such that it was equipped with a friction brake similar to that pictured in Figure 3–30.[164]

TABLE 3–8. AN ESTIMATE OF THE POWER OF SPECIFIC WATER WHEELS IN THE PERIOD c1700–c1800

| (I) Wheel (source)[1] | (II) Type | (III) D | (IV) W | (V) Dimen. of Blds/Orifice |
|---|---|---|---|---|
| La Fére flour mill (a) | US | 16* | | |
| Mont Royal flour mill (a) | US | 14* | | |
| La Fére powder mill (a) | US | 17* | | |
| La Fére sawmill (a) | OSI | 10.5* | | |
| Nymphenbourg pumping engine (a) | US | 25 | 5.0 | |
| Pont Neuf water works (a) | US | 20 | 18.0 | 18 × 4 |
| Notre Dame water works (a) | US | 20 | 18.0 | 18 × 3 |
| Nun-Eaton flour mill (b) | OS | 16 | 1.5 | |
| Lord Pembroke's wheel (b) | US | 12 | 2.58 | 2.58 × 0.83 |
| Lord Tinley's wheel (b) | US | 30 | 1.5 | 1.5 × 1.0 |
| London Bridge Water Works (b,c) | US | 20 | 14.0 | 14 × 1.5 |
| Marly Water Works (b,a) | US | 36 | 7.5 | 7.5 × 4.0 |
| "a good mill" (d) | | | | |
| "a bad mill" (d) | | | | |
| "an undershot mill" (e) | US | 14 | 1.3 | |
| "an overshot mill" (e) | OS | 9 | 3.5 | |
| "a corn mill" (f) | US | 12 | 3.17 | |
| "a paper mill" (f) | B | 8 | 8.25 | |
| Leith sawmill (g) | US | 4.5 | 3.17 | 3.17 × 1.5 |
| Dartford sawmill (g) | US | 16 | 4.5 | |
| Osburn oil mill (g) | OS? | 14 | 2.5 | |
| P. Dawson's engine (g) | OS | 22 | 4.0 | |
| Ficket paper mill (g) | B? | 13.5 | 3.58 | |
| Tamworth paper mill (g) | US | 15 | 3.0 | |
| Elford paper mill (g) | US? | 14 | 3.75 | |
| West Bromwich forge mill (g) | B | 14 | 2.83 | |
| Bromwich slitting mill (g) | OS? | 18 | 4.38 | |
| Huslet slitting mill (g) | US? | 12 | 3.83 | |
| Huslet fulling mill (g) | OS? | 16 | 2.0 | 1.85 × 0.5 |
| Tamworth scouring mill (g) | US | 14 | 2.5 | |
| Musselburgh fulling mill (g) | US? | 13 | 2.33 | 1.75 × 2.33 |
| Braunston fulling mill (g) | OS? | 14.5 | 3.0 | 3.08 × 0.17 |
| Oakenrod water wheel (h) | B | 15.67 | 2.0 | 2.17 × 0.58 |
| Oakenrod water wheel #2 (h) | B | 15.67 | 2.25 | 2.25 × 0.42 |
| Stanton, Del., wheel (i) | OS | 18 | | |
| Stanton, De., wheel #2 (i) | OS | 18 | | |
| Bush, Md., wheel (i) | OS | 16.4 | | |
| Alexandria, Va., wheel (i) | OS | 19.3 | | |
| Bush, Md., wheel #2 (i) | OS | 16.4 | | |
| wheel, location not given (i) | OS | 11 | | |

[1] For explanation of column headings see Key to Column Headings, following.

| (VI) A | (VII) V | (VIII) H | (IX) h | (X) Q | (XI) n | (XII) Bush. | (XIII) Est. hp |
|---|---|---|---|---|---|---|---|
| 2.17* | 17.28 | 5.0 | 5.0 | 37.5* | .3 | 6.25 | 7.0 |
| 12.25* | 9.0 | 1.34* | 1.34* | 110.25 | .25 | | 4.6 |
| 2.5* | 20.1 | 6.72 | 6.72 | 50.25* | .25 | | 10.5 |
| 2.33* | 19.8 | 7.0 | | 15.17 | .3 | | 4.0 |
| 10.0 | 2.0 | 12.0 | | 20.0* | .25 | | 7.4 |
| 72.0* | 6.17 | 0.63* | 0.63* | 444.24* | .25 | | 8.7 |
| 54.0* | 8.75 | 1.27* | 1.27* | 472.50* | .25 | | 18.6 |
| 0.15 | 22.0* | 23.5 | 7.5 | 3.3 | .55 | 2.5 | 4.9*(3.0) |
| 2.15* | 14.9 | 5.0 | | 32.0* | .25 | | 4.6 |
| 1.5* | 21.0 | 7.0 | 7.0 | 31.5* | .25 | | 6.3 |
| 21.0* | 11.4 | 2.0 | 2.0 | 239.4* | .25 | | 13.6 |
| 30.0* | 13.5* | 3.0 | 3.0 | 403.8* | .25 | | 37.6 |
| | | | | | | 5.5 | 8.2 |
| | | | | | | 2.0 | 3.0 |
| 1.18 | 17.95* | 5.33 | 5.33 | 21.2* | .25 | 2.0 | 3.5*(2.2) |
| 0.81 | 7.8* | 11.0? | 1.0? | 6.32* | .6 | 5.0 | 5.2*(3.2) |
| 1.6 | 19.2 | 6.0 | 6.0 | 30.7* | .25 | 5.0+ | 5.2*(3.2) |
| 1.4 | 11.3 | 8.1 | 2.0* | 15.8* | .5 | | 7.3*(4.5) |
| 4.75* | 6.9* | 3.58 | 0.75? | 32.8* | .25 | | 3.3 |
| | | 2.75 | 2.25 | 50.0 | .25 | | 3.9 |
| | | 15.5? | 1.5 | 3.24 | .6 | | 3.4 |
| 1.5 | 2.9 | 19.5 | | 4.35* | .6 | | 5.8*(3.6) |
| | | 5.0 | 1.4 | 21.5 | .5 | | 6.1 |
| | 16.3 | 5.5 | 4.15 | 40.8 | .25 | | 6.4 |
| | 15.54 | 4.58 | | 36.7 | .25 | | 4.8 |
| | 20.3 | 6.7 | 3.3 | 22.4 | .5 | | 8.6 |
| | 1.26 | 20.3? | 2.3 | 6.8 | .6 | | 9.4 |
| | 16.7 | 5.08 | 3.08 | 93.0 | .25 | | 13.5 |
| 0.92* | 5.7* | 17.5 | 0.5 | 5.2* | .6 | | 6.2*(3.9) |
| | | 3.67 | | 41.8 | .25 | | 4.4 |
| 4.1* | 11.3* | 3.75 | 2.0 | 46.3* | .25 | | 5.0*(3.1) |
| 0.52* | 8.0* | 15.5? | 1.0? | 4.2* | .6 | | 4.5*(2.8) |
| 1.26* | 11.3* | 8.75 | 2.0 | 14.24* | .5 | | 7.1*(4.6) |
| 0.95* | 16.0* | 8.75 | 4.0 | 15.2* | .45 | | 6.8*(4.2) |
| 0.385 | 12.9 | 20.0 | 2.67 | 3.8 | .6 | 3.5 | 5.2*(4.2) |
| 0.325 | 11.17 | 19.2 | 1.9 | 3.5 | .6 | 2.5 | 4.6*(3.0) |
| 0.345 | 15.79 | 17.8 | 3.83 | 5.18 | .6 | 3.75 | 6.3*(4.1) |
| | 16.2 | 21.5 | 4.0 | 3.57 | .6 | 4.5 | 5.25 |
| 0.425 | 14.96 | 17.8 | 3.5 | 6.16 | .6 | | 7.5*(4.8) |
| 0.567 | 14.0 | 12.5 | 3.0 | 7.6 | .6 | 3.5 | 6.5*(4.2) |

| | | | | | | | |
|---|---|---|---|---|---|---|---|
| Average horsepower for 40 wheels | | | | | | | 7.4 |
| Average using lower values | | | | | | | 6.6 |
| Average excluding Marly wheel | | | | | | | 6.5 |

(continued)

TABLE 3–8 (*cont.*)
KEY TO COLUMN HEADINGS

| Column I | Wheel (source): |
|---|---|

Column I     Wheel (source):
a = Bélidor, 1737–39     f = Anonymous, 1775
b = Desaguliers, 1744     g = Rennie, 1845 [c1770–1800]
c = Beighton, 1731–32     h = Ewart, 1791 [from Hills]
d = Malouin, 1771     i = Evans, 1795
e = *Ency. méthodique*, 1788

Column II     Type: US = undershot; B = breast; OS = overshot; OSI = overshot impulse

Column III     $D$ = diameter (in British feet for British wheels; in French feet [pieds] for French wheels; 1 Fr. ft = 1.07 Br. ft); an asterisk indicates that the diameter was measured from the center of the wheel's blades instead of from its periphery

Column IV     $W$ = width of wheel (in British feet for British wheels; French feet for French wheels)

Column V     Dimen. of Blds/Orifice = dimensions of the blades of the wheel or the orifice providing water to the wheel; used to determine the value of $A$ (Column VI)

Column VI     $A$ = area of the section of the stream of water acting on the wheel; assumed to be equal to the area of the surface of a blade on undershot wheels or the area of an orifice spouting water on the wheel; the asterisk indicates the value given is the area of a blade or floatboard

Column VII     $V$ = velocity of the water impacting the wheel; the asterisk indicates that $V$ was calculated with the equation $V = \sqrt{2gh}$; no asterisk indicates the velocity was given by the authority (Column I) describing the wheel

Column VIII     $H$ = total head or fall of water available; the asterisk indicates $H$ was calculated from the velocity of the water (Column VII) using the equation $H = V^2/2g$

Column IX     $h$ = impulse or velocity head; the asterisk indicates that $h$ was calculated from the equation $h = V^2/2g$; no asterisk indicates that this datum was provided by the authority describing the wheel

Column X     $Q$ = quantity of water in cubic feet per second delivered to the wheel (cubic British feet for British wheels; cubic French feet for French wheels); an asterisk indicates the volume was computed using the equation $Q = AV$; no asterisk indicates the figure was provided by the authority who described the wheel

Column XI     $n$ = assumed efficiency of the water wheel (generally 25% for undershot, 50% for breast, 60% for overshot)

Column XII     bush. = grinding capacity of the mill in bushels of wheat ground per hour; according to Whitham, *Water Rights Determination*, pp. 2–14, it takes between 1 and 2 hp to grind 1 bushel of wheat per hour; these figures can be used as a check on my calculated horsepower ratings

Column XIII     Est. hp = estimated horsepower of the water wheel, calculated with the equation $hp = 62.5\ HQn/550$ (appropriately modified for calculating wheels with French dimensions); the figure in parenthesis following estimates with an asterisk is the calculated horsepower when $Q$ (Column X) was calculated using the equation $Q = 0.62AV$, instead of $Q = AV$, to take into account possible contraction of a spouting stream of water

Agricola's 36-foot (11-m) diameter wheel did not represent the upper limits of traditional vertical wheel technology. Pryce in 1770 recommended wheels of 38-foot (12.2-m) diameters for Cornish mine engines, and he reported seeing wheels of 48 feet (14.6 m).[165] Smeaton in the 1750s saw wheels 45 and 48 feet (13.7 and 14.6 m) in diameter in Cornwall.[166] The water wheel built by the Swedish engineer Christo-

TABLE 3–9. AN ESTIMATE OF THE POWER OUTPUT OF CENTRAL EUROPEAN VERTICAL WATER WHEELS USED IN MINING c1800 BASED ON HÉRON DE VILLEFOSSE'S REPORT

| | Quantity of water used in ft.³/min.* (Héron) | Diameter of water wheel (ft.)* (Héron) | Estimated head (ft.), assuming 1.5 ft. above diam. | Estimated output, assuming use of overshot wheels and efficiency of 60%† |
|---|---|---|---|---|
| Water wheels used to power mine pumps, Harz region | 231.4 | 26.7 | 28.2 | 7.4 hp |
| Water wheels used to power hoisting machines, Harz region | 174 | 26.7 | 28.2 | 5.6 hp |
| Water wheels used to power 6-hammer stamp mills, Harz region | 313.2 | 13.4 | 14.9 | 5.3 hp |
| Water wheels used to power 9-hammer stamp mills, Harz region | 400.2 | 13.4 | 14.9 | 6.8 hp |
| Water wheels used for drainage, Freyberg area, Saxony | 86 | 41.8 | 43.3 | 4.2 hp |

*Héron's values have been converted to English measure: 1 Harz ft = 0.955 English ft; 1 Saxon ft = 0.95 English ft.

†Some of these wheels, notably those of the stamp mills, may have been undershot wheels instead of overshot. If this was the case, their output cannot be estimated, since the approximate head of water under which they operated cannot be estimated from their diameter, as with overshot wheels.

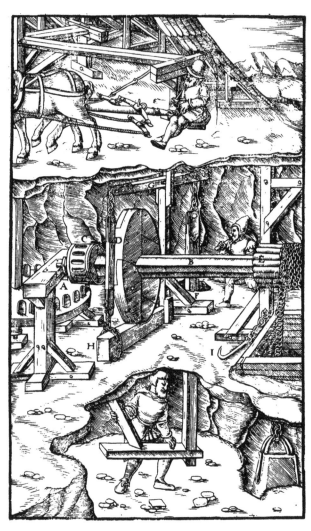

FIGURE 3–30. Friction brake of the type used on large German mine wheels. The man in the lower center of the woodcut could stop the ascent of the loaded bucket (lower right) by sitting on the horizontal beam. This would depress the right end of the pivoted lever $G$, and would raise the beam $H$ against the friction drum $D$.

pher Polhem for the King Charles-Gustav mine in the early eighteenth century was 48 feet (14.6 m) in diameter.[167] The Russian engineer K. P. Frolov built overshot wheels 52.5 feet (16 m) and 55.8 feet (17 m) in diameter to pump out the Zmeinogorski mines,[168] and in 1797 John Robison reported seeing a wheel 58 feet (17.7 m) in diameter.[169]

While traditional vertical water wheels could be constructed as tall as 60 feet (18.3 m), 40 feet (12.2 m) was the effective limit. Above that there were difficulties. Robison, for example, noted that tall wheels were "of very difficult construction and extremely apt to warp and go out of shape by their weight."[170] And the great British engineer John Smeaton noted that when water wheels exceeded 40 feet in diameter they became "so heavy, unwieldy, and expensive, as on those accounts to be inelligible [*sic*]; and they rarely exceed fifty feet [15.2 m]."[171]

Wheels of larger than average diameter were used extensively in the

Plate XXIX.

FIGURE 3–31. One of the four water wheels at the London Bridge Water Works. This wheel was built by George Sorocold, c1700. It was 20 feet (6.1 m) in diameter and had 26 blades, 14 feet (4.3 m) long, reinforced by four rims or rings. It drove eight pumps by means of crankshafts and rocking beams and could raise more than 2,500 tons of water 120 feet (36.6 m) high per day. The water wheel could be raised or lowered as the level of the Thames rose or fell because the axle was mounted on pivoted levers.

mining industry, for very high falls could often be secured in the mountainous areas where mines were commonly found. Wheels with larger than average widths could be found most commonly on large streams with low falls. Some of the best-known examples were used for water supply. The London Bridge Water Works, for example, had four large wheels installed under the arches of a bridge. The wheels (Fig. 3–31), built around 1700, were 14 feet (4.3 m) wide.[172] The wheels used at the Paris pumping stations at Pont-Neuf (la Samaratine) (Fig. 3–32) and Notre Dame were 19 feet (5.8 m) wide.[173] Just as 40 feet (12.2 m) represented the usual top limit for the diameter of wooden water wheels, 20 to 25 feet (6.1 to 7.6 m) seems to have represented the upper limit for the width.[174] (See Table 3–10.)

As the scale of projects that European engineers were able to tackle grew larger, so did their need for increased amounts of power. Since there were practical limits to the dimensions and hence to the power of traditional vertical water wheels, these needs often had to be met by using several wheels and linking them together. One such assemblage of water wheels was the Tower Engine of Cornwall. It consisted of ten overshot water wheels, each 20 feet (6.1 m) in diameter, placed one

*Fig.1.ʳᵉ*

FIGURE 3–32. The Paris water works called la Samáritaine. Located under the second arch of Pont Neuf, these works were erected c1600, but restored and redesigned by Bélidor in 1714. The wheel was about 20 feet (6.1 m) in diameter by 18 feet (5.5 m) wide. It powered pumps which delivered water to royal palaces.

above the other so that the tail race of one wheel served as the head race for the next. A building in the form of a tower contained these wheels. Each wheel was equipped with two cranks. The cranks of the ten wheels were linked by means of two long rods which acted on two large pumps.[175] Unfortunately there is not enough data to determine the overall output of this pumping complex, but it was probably in the neighborhood of 50 to 100 horsepower. A similar, but smaller, assemblage of water wheels was used at the Blackett mine (two water wheels) and at the Allenheads mine (four water wheels) in the Pennine lead-mining district in the early eighteenth century.[176]

Probably the best-known assembly of water wheels in the period between 1500 and 1750 was the Marley plant, built in the 1680s to provide water to the fountains, gardens, and palaces at Versailles, Marly, and Trianon for Louis XIV. The Marly plant was erected on an arm of the

TABLE 3–10. AVERAGE AND UPPER LIMITS OF THE DIMENSIONS AND POWER OUTPUT OF TRADITIONAL WOODEN VERTICAL WATER WHEELS

|  | Average | Upper limits* |
|---|---|---|
| Height | c12 ft (3.6 m) | c60 ft (18 m) |
| Width | c2.5 ft (0.75 m) | c25 ft (7.5 m) |
| Power | 5 to 7 hp | c40 hp |

*While it was possible for a single water wheel to have the "average" dimensions given in this table, no single water wheel was built with the maximum attainable height and width. Such a wheel would have broken under its own weight or have been so severely warped as to be useless.

Seine. To divert water onto the water wheels and build up a respectable head a dam was built across the arm and a canal constructed around the dam for navigation. Water was directed into 14 races which led to 14 undershot water wheels distributed in three lines across the river: seven were in the first line, six in the second, and only one in the third. The water wheels were large—36 feet (11 m) in diameter and 4.5 feet (1.4 m) wide (Fig. 3–33).

The Marly wheels drove 64 pumps at river level by means of cranks and rocker arms. These pumps lifted water from the Seine to a reservoir 150 feet (48.7 m) higher. Then, by means of a system of field rods (cranks, rocker arms, connecting rods) extending over 600 feet (194 m), the wheels drove a second group of 49 pumps, which lifted water from the first reservoir an additional 175 feet (56.8 m) to a second reservoir. Through more cranks, connecting rods, rocking beams, and chains extending 1,944 feet (631 m) from the river, the wheels drove 78 more pumps, which lifted the water through an additional 177 feet (57.5 m) to an aqueduct, which led the water to Marly, Versailles, and Trianon. Thus 14 water wheels drove a total of 221 pumps, operating at three different levels, and lifted water some 502 feet (163 m) above river level.[177] (See Figs. 3–34 and 3–35.)

While Marly was widely celebrated because of the magnitude of the installation and the complexity of the transmission apparatus, its design was inefficient. Contemporary descriptions indicate that the entire complex, even under the most favorable conditions, delivered no more than 150 effective horsepower, or around 11 horsepower per wheel. This figure, however, reflects the cumbersome and inefficient nature of the pumps and power transmission apparatus. The Marly wheels probably produced around 30 to 40 horsepower each at the shaft.

It is clear that European millwrights and engineers had developed sufficient know-how by the mid-eighteenth century to tap falls as high as 40 to 50 feet (12.2 to 15.2 m), to handle volumes up to at least 500 second-feet (c14 m³/sec), and to develop power outputs of up to 40 horsepower. They had, in other words, the capability of dealing with the volumes and falls commonly available on most European streams and of delivering units of power appropriate to the scale of most contemporary

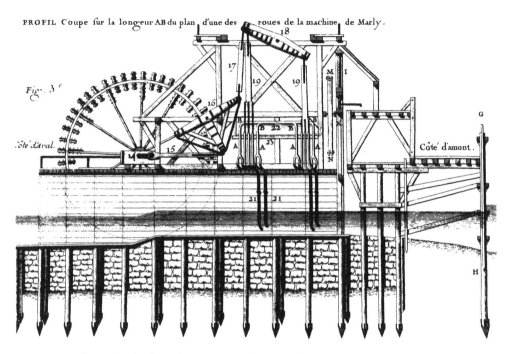

PROFIL Coupe fur la long-eur AB du plan d'une des roues de la machine, de Marly.

18

Fig. 3.

17

19    19

16

M

I

B

B  22  B

Côte d'aval.

15

A    A  23  A    A

N

Côte d'amont.

G

21  21

H

FIGURE 3–33. One of the 14 water wheels of the Marly plant. Each of the Marly wheels operated in its own closely fitted channel and was 36 feet (11 m) in diameter by 7.5 feet (2.3 m) wide. The entrance of the water to the channel was controlled by the usual sluice or penstock gate. The Marly wheels drove a set of pumps at river level as well as vertical rocking beams and other linkages which powered pumps at reservoirs located at considerable distances from the river. The illustration here shows the drive to the river-level pumps from the crank, 14, through the rocking beams, 16 and 18.

industrial demands. By using a number of wheels at the same site, European technicians were even able to develop higher falls, larger volumes, and greater outputs, though systems which required coupling water wheels probably sharply raised costs and decreased reliability.

Despite these accomplishments, the vertical water wheel was still noticeably deficient in several respects. One of the primary technical deficiencies of the traditional undershot wheel was its limited adaptability to changing water levels (e.g., flood conditions). One of the primary technical deficiencies of the traditional overshot wheel was its unsuitability to falls above 40 to 50 feet (12.2 to 15.2 m), especially if the available volume of water was small. Much of the inventive activity in water-power technology between 1500 and 1750 focused on attempts to resolve these two problems.

### Innovation

The undershot wheel's limited ability to adapt to changing water levels was a problem which had already been attacked by medieval millwrights. Their inventions—in particular, the boat mill and the suspended

FIGURE 3–34. Overview of the works at Marly from a German illustration of 1725. This complex, built by Rennequin Sualem of Liège in the 1680s, lifted water from the Seine, through two intermediate reservoirs and pumping stations (32 and 34) to an aqueduct 502 feet (153 m) above the river and almost three-fourths of a mile (cl.2 (cl.2 km) away. The aqueduct can be seen in the extreme upper left portion of the illustration.

(bridge) mill—were partial solutions to this problem. Both boat and suspended mills continued to be used in the Renaissance and for some centuries after because of their effectiveness in this area.

Of the two, the suspended mill probably underwent the greatest amount of additional alteration. This type of mill, it will be recalled, required some mechanism to alter the level of the wheel as the water level changed. This machinery could be very simple, as the chain and capstan arrangement depicted by Zonca in 1607 and illustrated in Figure 2–7 of the last chapter indicates. By the sixteenth century, however, more sophisticated arrangements had been invented to handle larger water wheels. Ramelli, for example, pictured a suspended mill on pilings in a river (Fig. 3–36). Four long wooden screws mounted at the corners of the edifice raised or lowered vertical shafts attached to the beams that carried the shaft of the water wheel. In this design, as in most western European mills of the suspended type, the mill house was located directly above the water wheel.[178] A lesser-known variant of the suspended mill was the Rumanian "Alvan" design, introduced in the nineteenth century. Here, a light wheel was suspended by means of ropes and poles at the side of the mill, instead of beneath it, and raised or lowered by means of capstans and pulleys.[179]

Figure 3–35. The Marly installation in the early eighteenth century.

Boat and suspended mills, however, were far from ideal solutions to the problem of changing water levels. The boat mill could float on high water and keep its wheel(s) in effective operation, but it was inherently unstable, inefficient, and often a serious navigational hazard. Suspended mills could adjust their wheels to changing water levels. But the equipment necessary to raise or lower the wheel made them expensive to construct and limited the size of their water wheels. Moreover, in most cases, the mill house had to be built above the wheel, increasing capital costs. European engineers, therefore, continued to seek other solutions to the problem of changing water levels.

Although high-water levels flooded fixed wheels and seriously impeded their operation, there were some positive aspects to the situation. The increased stream flow meant that there was more power available. If this power could be tapped, it could be used to drive out the back water (the water at the rear of the wheel impeding its motion) and allow mills to operate at their normal output. One mid-eighteenth-century solution to the problem of flooded wheels involved the use of sliding floatboards. For normal flow conditions an undershot wheel would have only the usual fixed blades, normally around a foot (c0.3 m) deep. But at high water, when the lower portion of the wheel was flooded, sliding float-

FIGURE 3–36. A suspended watermill from Ramelli, 1588. The water wheel, *B*, could be raised or lowered, depending on the water level, by means of four screws (*I* and *K* are visible) mounted at the corners of the edifice. The screws adjusted the height of the vertical beams *D*, *H*, *L*, *G*. The horizontal beams on which the axle of the water wheel rode were linked to these vertical beams. The rectangular flat plate mounted on *S* in front of the water wheel provided some control over the water flowing against the wheel.

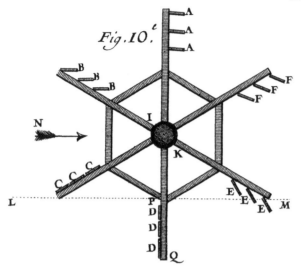

*Fig. 10.*

FIGURE 3–37. Undershot water wheel designed by Gosset and de la Deville in the 1730s to rotate when completely submerged. *M* indicates the normal level of the stream; *GH* indicates the water level when the wheel was completely flooded. The floats of the wheel were hinged so that at the bottom of the wheel (*D,D,D*) they presented their faces to the stream, while at the top (*A,A,A*) they presented only their edges. Because flowing water would have a greater impact at the bottom of the wheel than at the top, it would turn even when fully submerged.

boards would be inserted above the fixed floatboards, doubling or tripling the size of the blades. This increased the amount of water effectively impacting against the wheel and augmented the wheel's power, enabling it to drive out the back water that was impeding its operation. This system had possibilities, but it was slow to catch on. In 1775, for example, one writer observed: "This method is yet uncommon, though a great improvement." He explained the reluctance to adopt it, noting that many people seriously doubted that a back water of 2 to 3 feet (c0.3–0.9 m) could really effectively be removed.[180] The cost of fitting an undershot wheel with sliding blades may also have contributed to the neglect.

In certain tidal estuaries the rise and fall of the water level was so great that a wheel was needed which would operate not only when partially submerged, but also when completely submerged. Around 1730 there was an attempt to adapt the vertical undershot wheel to this condition. The wheel, pictured in Figure 3–37, was the work of two French inventors, Gosset and de la Deville of Laon. Their wheel had hinged floatboards arranged so that at the bottom of the wheel a blade presented its face to the flow of the water, but as it rotated towards the top it presented only its edge. Thus flowing water would have a greater effect on the lower than the upper part of the wheel, and the wheel would rotate even when completely submerged. The French engineer Bélidor reported in 1739 that he had been present at the first trial of this wheel in Paris and that it operated as well as could be desired.[181] In practice the device seems to have found little application, for after Bélidor there are no reports of its use. The small amount of power it generated when submerged was probably the reason.

Most inventive efforts to adapt the vertical undershot-impact wheel to various stream conditions focused on large streams whose water levels varied sharply. New efforts to adapt the vertical overshot-gravity wheel, on the other hand, focused on streams with very high falls and very low volumes. The traditional form of overshot wheel could not deal effectively with falls above 40 to 60 feet (c12–18 m), since wheels of even that diameter were unstable. It was also difficult for normal overshot wheels to operate with very small volumes of water, since too much of an already precious commodity was lost in feeding the water into the buckets and in spillage from the buckets during rotation.

No minor modification of the traditional overshot wheel was sufficient to meet these difficulties. Hence, efforts to solve the problem of efficiently using high falls and low volumes required major modifications to the overshot wheel, modifications which involved, in effect, detaching its buckets. The resulting engine, commonly called a bucket engine or a chain-and-bucket engine, can be considered a water wheel with suspended rather than attached buckets. Chain-and-bucket engines reversed the operation of the bucket hoist. Instead of applying power through a shaft to chains or ropes and buckets to lift water, water was allowed to fall while contained in a bucket or buckets attached by rope or chain to a wheel and axle to produce power. Two water-bucket engines were described by Caspar Schott in 1667. One consisted of two buckets, one at each end of a rope which led over a wheel or pulley. When one bucket was filled with water and allowed to fall, it would raise a smaller quantity of water in the other bucket. The other machine, attributed by Schott to Geronomo Finugio in 1616 and a European version of the primitive water lever, had a bucket on one end of a lever. When filled with water its weight pushed the lever down, working a pump attached to the other end. Whether either of Schott's machines was ever built is not clear.[182]

In 1668, however, a French inventor, Francini, designed a chain-of-buckets engine which was put into operation. Francini's engine had two endless chains of buckets, both mounted on a common wheel and axle (Fig. 3–38). The buckets of one chain made use of a moderate fall of water to drive the other chain of buckets, which lifted water to the top of the machine. France's finance minister, Colbert, was sufficiently impressed with Francini's designs to order one of the engines constructed in the garden of the king's library, where it lifted water from a small spring to a reservoir which powered a fountain in the middle of the garden.[183]

Both bucket-and-pulley engines like Schott's and chain-of-buckets engines like Francini's were erected in England in the early eighteenth century. For example, George Gerves, a carpenter, built a modified version of Schott's engine for Sir John Chester of Chichester. He substituted a differential pulley for Schott's simply pulley and made the machine self-acting. Francini's chain-of-bucket design was apparently reinvented by Costar in England for pumping out mines in Devon and

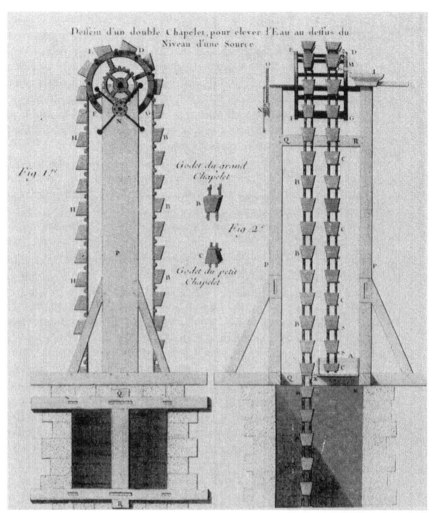

FIGURE 3-38. Chain-of-buckets engine, Francini, 1668. Water from a small spring at level *Q* was fed into the buckets on the left side from the reservoir at *A* and fell in them to the bottom of a cistern. This provided power, through the wheel *EFGD*, to the smaller chain of buckets on the right, which lifted water from the reservoir at *A* to *I*.

Cornwall shortly before 1700.[184] An engine which utilized a chain-of-plugs instead of a chain-of-buckets arrangement was described by Leupold in 1725.[185]

Water-bucket engines were apparently able to tap small volumes with high falls efficiently. But despite their promise, these engines never seem to have been widely used save in a few mining districts. There were several practical problems which proved insurmountable. First, their power output was very small in comparison to their cost. Second, both chain-of-bucket and chain-of-plug engines had so many parts in constant motion that breakdowns were frequent and maintenance expenses high. Sooner or later a chain link would break, usually on the loaded side of the engine near the top, and the whole assembly would come crashing down, causing extensive damage.[186] Thus, while these engines were occasionally built and operated with moderate success, they never provided a serious challenge to the dominance of the standard vertical overshot water wheel.

## THE MILLWRIGHT BEFORE 1750

Despite the growing importance of water power to the economic and industrial development of the Western world, surprisingly little is known of the background and training of the men who, before the late eighteenth century, invented devices like the bucket engine or who designed and erected the hydropower dams and built the vertical water wheels.[187] Largely this is because millwrights were part of a craft tradition where information and instruction were primarily passed on by word of mouth rather than through the written word.[188]

There is, however, one brief description of millwrighting dating from 1747:

> The Trade is a Branch of Carpentry (with some assistance from the Smith) but rather heavier work, yet very ingenious, to understand and perform which well a person ought to have a good Turn of Mind for Mechanics, at least to have some knowledge in Arithmetic, in which a lad ought to be instructed before he goes to learn this Art; for there is a great variety in Mills, as well as in the Structure and Workmanship of them, some being worked by Horses, some by Wind, others by Water shooting over and others by its running under. And why not in Time by Fire too, as well as Engines? They take an Apprentice £5 to £10, work from six to six, and pay a journeyman 12/– to 15/– a week; but £50 or £100 worth of timber and £50 to spare will make a Master of him.[189]

The most complete description of the millwright is that of William Fairbairn (1789–1874), the great British mechanical engineer, who was apprenticed to a millwright and practiced the trade for some years. Around 1840 he described the traditional millwright in this manner:

> The millwright in former days was to a great extent the sole representative of mechanical art, and was looked upon as the authority in all applications of wind and water, under whatever conditions they were used, as a motive power for the purposes of manufacture. He was the engineer of the district in which he lived, a kind of jack-of-all-trades, who could with equal facility work at a lathe, the anvil, or the carpenter's bench. In country districts, far removed from towns, he had to exercise all these professions, ... and ... went about the country from mill to mill.
> Thus the millwright of the last century was an itinerant engineer and mechanic of high reputation.[190]

According to Fairbairn, the millwright was "a fair arithmetician, knew something of geometry, leveling, and mensuration, and in some cases possessed a very competent knowledge of practical mechanics":

> He could calculate the velocities, strength, and power of machines: could draw in plan and section, and could construct buildings, conduits, or watercourses, in all the forms and under all the conditions required in his professional practice.[191]

As far as education, Fairbairn noted:

> Their education and habits were those of the times in which they lived. There were then no schools for the working classes but those of the parish, nor any libraries or mechanics' institutes; and after the usual course of reading, writing, and accounts, the millwright was thrown upon his own resources in the attainment of the knowledge which might aid him in his profession.[192]

The craft itself was learned via the apprenticeship system, with knowledge being passed from the master almost entirely by example and the spoken word. Books and formal education played no important role in the process.

Fairbairn's description of the millwrights with whom he was acquainted is probably a fair estimate of those who practiced the craft in the late eighteenth and early nineteenth centuries.[193] But we must be wary of applying it uncritically to all millwrights before 1750. One may question, for example, whether the typical millwright of the seventeenth century was able to "calculate the velocities, strength, and power of machines," since the first sophisticated, mathematical analysis of the operation of the vertical water wheel was published only in the early years of the eighteenth century.

I think it possible that some of the best millwrights and some of the engineers associated with large-scale water power projects did, prior to the eighteenth century, make some type of crude quantitative measurements with wood shavings or blades of grass to determine water velocity. These may also have recognized, even without formal scientific proof, that a stream's power potential depended on the product of volume and fall. And using this knowledge some may have made some crude calculations to determine whether their wheels would develop power sufficient for their needs. Unfortunately, however, this must remain speculation, for sufficient data are not available to prove this.

While the best millwrights and engineers *may* have gathered some quantitative data and made some crude calculations of power output to assist them in design, the typical millwright prior to 1750 apparently did not. John T. Desaguliers complained in 1744 that Britain was "over run" with "Engineers (not Ingenieurs of a proper Education for the science)," particularly in the area of "Water-Works."

> Almost all the ... Mill-Wrights now set up for [claim to be] Engineers; tho' I hardly believe there are two of them who know how to measure the

Quantity of Water requir'd to turn an Undershot, or an Overshot, or a Breast-Mill. They only judge of a Stream by the Eye, and he that has the most practice is likely to succeed the best.[194]

In another place Desaguliers discussed the work of a designer and builder of water wheels, Holland, whom he credited with a "very great Genius for Mechanics." But Holland, he added,

[was not] able either to calculate the Force of a Power, or apply it to the best Advantage. In that he went by guess-work, as most of those Persons do who put up Engines for raising Water: for Example, he did not know how to measure the Quantity of Water he was to make use of, or how to proportion the number of Ladle-Boards or Floats in his Wheels: He knew that a great deal of Friction was to be given to produce a very high Jet with but little Intermission of Variation in the Height; but provided he thought he had Water enough, and that he gave Friction enough, he was satisfied. As for the rest, he was very exact and skilful.[195]

Desagulier's claim that mill builders, for the most part, did not calculate the dimensions of their wheels or rely on theoretical guidance is confirmed by several other eighteenth-century writers. An anonymous author in 1775 noted: "The extensive use of water for mills and other machines is generally known; but the manner of constructing such machines, and their application of the water to produce the greatest effect, is not sufficiently understood and attended to; the construction of them being, for the most part, left to persons not well skilled in the principles of mechanics and force of running water."[196] A French theoretician, Fabre, observed in 1783 that most mill builders were "not able to handle even the simplest algebraic calculations."[197] And Antoine Parent, around 1700, asserted that many of the most skilled mill builders in his time often succeeded only by employing as much force for a single machine as would have been needed to power several similar ones.[198]

It seems probable, then, that most millwrights between 1500 and at least 1700, with perhaps some exceptions, had little or no knowledge of mathematics above the simple arithmetic necessary for establishing gearing ratios and estimating costs. They did not know how to "calculate the velocities, strength, and power of machines." Neither did they approach the problems involved in water-power development with such quantitative questions as, What is the velocity of the stream in feet per second? What is its fall in feet? What will be the impacting force of the water against the wheel in pounds? What will the wheel's efficiency be? Instead, the traditional millwright probably relied on a combination of intuition, past experience, judgment by eye, and observations of other machines. The questions he asked were comparative, not quantitative: Is the water supply more or less than I had when I last worked on a job like this? Is the fall of water higher or lower than it should be? Should the wheel go slower or faster? Is the water here faster or slower than at that mill I saw a couple of miles upstream? And the standards by which he judged his wheel designs were purely normative, that is, he judged his work against what was typical of the time, instead of against an ideal standard. Was the wheel well built and of good materials? Would it perform its function with some margin for emergencies? Did it appear ingenious enough to augment the reputation of its builder.[199]

There are a few scattered passages in works between 1500 and 1750 that support this view of the traditional millwright's methodology. Ramelli, for example, in his 1588 theater of machines gave no specific directions for adjusting the size of a water wheel to its task; he gave only the general advice that it was always necessary to proportion the size of a machine to the motive force operating it, presumably leaving the details to the millwright's intuition and past experience.[200] In the section on watermills in Fitzherbert's *Surverying* of 1539 there are a number of practical rules to guide the design of water wheels. For example, Fitzherbert noted: "And the lengar that the ladell [blade] is, the better it is, so that it have sufficient water. And that the mylner nedeth not drawe up his gate too hyghe, as if the ladell be shorte, for the ebber [thinner] the water is the swyfter it is."[201] And again: "The greater compasse the whele is, the lesse water will dryue [drive] it, but it wyll nat go so oft about as a lyttell whele wyll do."[202] Similar in tone are the remarks of R. D'Acres, a mid-seventeenth century English writer on water supply. Referring to overshot wheels he declared:

> How much greater your fall is, so much lesser water spend to move your work. Therefore are wheels generally made large to receive this great fall; if their bigness exceed their fall of water, they will have more strength, but then what is gained in strength is lost in time, what you gain in the hundred you lose in the shear; if you make them much lesse, the violence of the fall of the water will be such, that it will not be contained in the buckets, but be lost.[203]

What is significant in both the Fitzherbert and D'Acres' passages is their terminology. Their terms are all comparative: greater, lesser; longer, shorter; more, less. Nowhere is there any indication of exact measurement or calculation.

To sum up, the methods most millwrights used to design and build water wheels prior to the eighteenth century were not abstract. They judged the velocity, head, and flow of water by eye, dimensions by feel. Based on these judgments and on their intuition, their past building experience, and their knowledge of other wheels operating under similar conditions, they made the necessary decisions about design.

Based on over a millenium of accumulated design and construction experience this approach produced sound water wheels and good water-power systems. As we have seen, European millwrights and engineers by 1750 had evolved a variety of methods for constructing vertical water wheels and their auxiliary works. They had developed sufficient knowledge to tap the water power of large rivers and sometimes (as in the

Harz) of entire regions. And they were able to utilize the power collected by their dams and reservoirs and tapped by their water wheels in a large variety of manufactures. They even had the ability to transmit it mechanically for distances up to several miles.

The traditional millwright's reliance on past experience, intuition, and judgment by eye was sufficient so long as water was relatively plentiful and power needs relatively low. It began to fail only when it became important to utilize a limited and increasingly scarcer resource efficiently. Judgment by eye and intuition did not allow the traditional millwright to determine how much of a stream's potential power he was tapping and how much was going to waste. It did not allow him to reliably compare water wheels operating at different sites or powering different machinery. Because his methods did not allow such judgements, they impeded the development, recognition, and diffusion of improvements, and they became increasingly inadequate as industrial growth began to push water power to its limits.

Traditional methods of designing and building vertical water wheels were challenged and eventually replaced between 1750 and 1850, in large part under the pressure of continued industrial growth. Iron water wheels, with larger power output and a longer work life began to supplant the traditional wooden wheel. At the same time the traditional craft rules, which for centuries had governed the design of wooden vertical wheels, were supplanted by scientific theory and quantitative design criteria. These developments will be taken up in the next two chapters—the development of a quantitative methodology for water-power technology in Chapter 4, and the emergence of the iron industrial water wheel in Chapter 5.

# 4

# Analysis:
# Quantification and the Vertical Water
# Wheel, c1550 to c1850

The traditional millwright's success in water wheel design was predicated on two factors: (1) his woodworking skills and (2) the centuries of accumulated craft experience on which he could draw in constructing vertical wheels of small power. Between 1750 and 1850, as a result of continued industrial growth and increased pressure on limited water power resources, the importance of these factors was severely diminished. Iron began to replace wood as the primary construction material in water wheels; the power demands of industry grew well beyond the 5 to 7 horsepower commonly required of water wheels before 1750; quantitative concepts and methods began to be important, influencing the way water wheels were designed and operated, and diminishing the importance of craft experience.

In Chapter 5 we will look at the emergence of the iron breast wheel, the most frequently used prime mover of heavy water-powered industry the early nineteenth century. In this chapter, however, we will look at the origin and development of quantitative concepts and quantitative methods in water-power technology, especially since these methods played an important role in the emergence and design of the iron breast wheel.[1]

The origins of quantification in water power, unfortunately, are difficult to establish. The traditional millwright left no written records. This makes it impossible to determine whether he used quantitative measurements or, if he did, just how great their role was. Presumably, as indicated at the close of the last chapter, their use was slight, and numerical reasoning was restricted to simple gearing calculations. The earliest surviving evidence of numerical reasoning in water-power technology comes only from c1500 in the notebooks of Leonardo da Vinci (1452–1519).

# LEONARDO AND THE ANALYSIS OF THE VERTICAL WHEEL

Leonardo da Vinci's notebooks pose considerable difficulties to anyone attempting to reach a fair appraisal of his merits in any field. His observations occur in no predictable order. They are frequently obscure and sometimes contradictory. His notes on water power are typical. While Leonardo indicated in several notes his intention of writing a comprehensive treatise on hydraulics with a chapter on hydraulic machines, the treatise was never compiled.[2] Leonardo's notes on water power are, therefore, as scattered, contradictory, obscure, and disorganized as his notes on other subjects. For example, a large number of Leonardo's observations deal with the problem of how to lead water onto the blades of an impulse wheel. Is it best to drop the water vertically onto the blades so that it strikes them at a three o'clock position? Or is it better to direct the water onto the wheel at an oblique angle, where it strikes at the four, five, or six o'clock positions? In one note Leonardo implies that it is best to direct the water vertically onto the wheel.[3] In another place Leonardo says that it does not matter: the effect is the same irrespective of which delivery (vertical or oblique) is used.[4] In yet another passage Leonardo suggests that the oblique blow is best.[5]

Most of Leonardo's notes on water power are undistinguished, even when they are not contradictory, confused, or obscure, but there are one or two exceptions. In the *Codex Atlanticus* we find what is apparently the earliest attempt to apply mathematics to a relatively sophisticated problem involving water power. The note is quoted in full below, and the diagram which accompanied it is sketched in Figure 4–1.

If the line $Kv$ is 27, the line $vf$ is 20; therefore we will say: the water $Kv$ that is $Ke$, weighs 27 pounds, and presses on the blade $ea$ with half of its power, and the water $Kf$ is 47 pounds, and presses on the blade $fb$ with half of its power, that is 23½ pounds; therefore the power of the water $Ke$ is greater than the power $Kf$ by a half pound of water [*sic*].

Here is demonstrated the true power of a quantity of water which falls above the wheel of a mill through varying angles, but from the same height to the same depth and on the same type and size of wheel. Therefore we will say that the previously mentioned water leaves from the height $K$ and descends to the depth $St$, and among its varying angles is the line $Ks$, even though this line is not oblique, because the water falls perpendicular. The first [true] oblique will be the line $Ki$. ... The water $KS$ therefore falls from the point $K$ and hits the blade of the wheel $ea$ in the middle of its descent and with the force of 4, and this water hits the blade with all its weight because it does not hit anything else. The descent of the water $Ki$ follows, which hits the blade $fb$ in the point $f$. This water falls 1½ times the fall of $Ke$ as the line $nf$ shows; but it loses half of its weight as is indicated in $mo$ as opposed to $Ko$. Now it is necessary to see and to calculate further ... , and we will say: If the force of 4 moves the blade $ea$ and falls 1 braccio of height, with what strength will the water move the blade $fb$ falling from 1½ braccia with a force of 2? I will say this: if the force was 4, as the first and

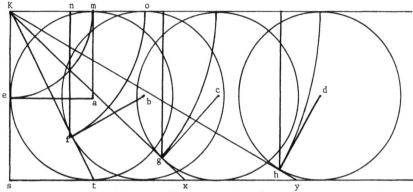

FIGURE 4–1. Geometrical diagram of Leonardo da Vinci for analyzing ways to apply water to water wheels. Leonardo used this sketch in his attempt to determine whether it was more advantageous to drop water perpendicularly on a water wheel (as *Ke*, the vertical line on the left), or at some angle (for example, *Kf*). This was perhaps the earliest attempt to solve a complicated problem involving the vertical water wheel with mathematics.

fell from 1½ times higher, it would be 1½ times more powerful than the first, i.e., 6; but because the force is nothing save 2, I have to divide six by ½, and say that the half of 6 is 3, from which we have lost a degree of force from the first to the second wheel.[6]

Many of Leonardo's notes are obscure and confusing because a precise, mathematical terminology is absent. Here, however, Leonardo's argument is clear. He believed that the velocity ($V$) of falling water, as well as its impacting force ($P$), varied directly as the height ($H$) fallen ($V \propto H$ and $P \propto H$). This force could be best applied, Leonardo contended, by dropping the water vertically onto the wheel, for then all of its weight would strike the blades. If the water impacted the wheel at an angle less than 90° (relative to the ground), only part of the weight of the water would strike the blade; the other part would press uselessly on the bottom of the channel. This loss would be partially compensated, since water, when it flowed at oblique angles, struck the wheel further down and had a greater velocity. But this gain, Leonardo calculated, would not offset the loss caused by the weight of the water on oblique chutes.

This analysis is deficient in several respects. The values Leonardo assigned the forces were arbitrary; the laws which he used ($V \propto H$, for example) were erroneous. But the note is significant as an early attempt to apply a mathematical methodology to water power.

There were several other notes where Leonardo stated problems quantitatively. For example:

Given the depth [volume?] of the fall of the water and its inclination, with the power of the wheel that is its object, we seek the height of the fall of the water in order to make it equal to the power of the wheel.[7]

In another passage Leonardo went even further:

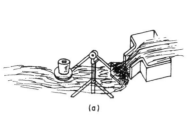 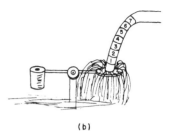

(a)                                     (b)

FIGURE 4–2. Experimental devices for measuring the impulses of a stream of water, suggested by Leonardo da Vinci. In *a* water flowing from a flume strikes against a flat board mounted opposite a weight on an inverted-V lever. In *b* the counterweight is on the opposite end of a conventional lever. Leonardo left no quantitative data derived from such experimental equipment.

> Given the resistance of the wheel and given the inclination and the descent of the fall of the water, we ask how great must its cross section be to be equal to the said resistance?
> Given the volume of the fall of the water and its length and inclination, we ask whether the power of the wheel is equal to the power of the water?
> Given the resistance of the wheel and the inclination of the water and its volume, we ask the length of the fall?[8]

Leonardo was apparently unable to answer these queries. His notes reveal no attempts at solutions. But the queries do indicate that he was groping towards a quantitative approach to the problems of the vertical water wheel.

Like mathematical analysis, experimental measurement was very rare in Leonardo's work. There are in his sketchbooks a few drawings of experimental apparatus which could have provided quantitative data to aid in vertical wheel design. For instance, on one page of the *Codex Atlanticus* is a sketch of a device for measuring the impulse of falling water. A flat plate is struck by spouting water on one end of an inverted-V lever; a counterweight to measure the impact is mounted on the other. A similar plan but with a straight lever is pictured elsewhere in the same codex (see Fig. 4–2),[9] and in *Codex Forster* Leonardo illustrated a weight suspended over an axle as a means of gauging the force of falling water and determining the power needed to operate a grain mill.[10] These drawings are suggestive. But nowhere is there any evidence that these devices were ever built or used. I have found no piece of quantitative data obtained from these devices anywhere in Leonardo's notes.[11]

At best, then, Leonardo can be considered only a precursor of quantification in water-power technology. In Leonardo there are a few rare but highly suggestive elements which link him to the future among many which link him to his contemporaries and to the past.[12] He seems to have sensed at times an alternative approach to the traditional millwright's intuition and judgment-by-eye methods, but he was unable to follow through on these insights.

To a large extent this was due to factors beyond Leonardo's control. The concepts by which machines were later to be measured, designed, and compared—power, work, energy, efficiency, impulse, force—had not yet been established. In mechanics, statics was still the dominant tradition, and statics could not easily be applied to moving machines like water wheels. Accurate quantitative laws in such vital areas as falling bodies, flowing water, and impact had not yet been developed. Direct and inverse proportions were well understood, but the more sophisticated form in which one quantity varied as some power other than unity of another quantity was not. Yet without these prerequisites the development of theoretical analysis in vertical wheel technology could not proceed very far.

## THE PREREQUISITES SATISFIED: GALILEO TO MARIOTTE (c1600–c1700)

The prerequisites for even the most simplistic mathematical analysis of the operation of the vertical water wheel were not available for over a century after Leonardo's death. They emerged only after the "scientific revolution" of the sixteenth and seventeenth centuries had completely altered the methodology of European science. Among the basic tenets of the new scientific methodology was the neo-Platonic belief that the underlying laws of nature were mathematical and that by applying mathematical and deductive methods and making quantitative measurements they could be discovered. The extension of this new methodology to mechanics and hydraulics in the early seventeenth century provided a number of the rules and concepts vital to the emergence of quantification in water-power technology.

Among the most able exponents of the new science was Galileo Galilei (1564–1642). Around 1600 he published a small treatise, *Le meccaniche*, designed for the instruction of his private pupils at Padua. It focused on an old subject, the five simple machines, but Galileo's treatment was novel.[13] He demonstrated that all of them could be reduced to a system of levers. Further, he extended the principle of the lever beyond the realm of statics into dynamics. This extension was based on Galileo's recognition that in an ideal machine, a machine free of all friction and other losses, the amount of force required to set it in uniform motion was greater only by an infinitely small increment than the force needed to maintain it in equilibrium. The application of this insight to the lever allowed Galileo to analyze it dynamically through the use of the principle of virtual velocities (Fig. 4–3). This analysis suggested that the product of a weight, force, or load, and its velocity could be used to measure the power of a body in motion or a machine. It also suggested that in a perfect machine the force put in and the force delivered should be identical.

The absence of well-defined and accepted quantitative concepts for

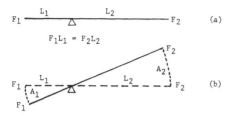

FIGURE 4–3. Galileo's dynamic analysis of the law of the lever. Figure *a* is a balanced lever. By the standard static law of the lever, $F_1L_1 = F_2L_2$. An infinitely small addition of weight to the left side of the lever will produce situation *b*. Since the weight added was infinitely small, it can, for all practical purposes, be neglected. Since the arcs $A_1$ and $A_2$ have the same ratio to each other as $L_1$ and $L_2$, the law of the lever can be rewritten $F_1A_1 = F_2A_2$, and since $F_1$ and $F_2$ travelled the distances $A_1$ and $A_2$ in the same time, the law can also be written as $F_1V_1 = F_2V_2$, where $V$ is the velocity of the weights. This suggests that the product of a weight, force, or load and its velocity can be used as a measure of the power of a moving body. This concept or measure was to be applied to water wheels by Parent almost a century after Galileo's lever analysis.

measuring power or work had been one of the factors behind Leonardo's gropings. Galileo's use of the product of force or weight and velocity contained the seeds of quantitative engineering concepts like power and work, while his notion of an ideal or perfect machine contained the germs of the concept of efficiency.[14] Galileo, however, did not apply his insights to the vertical water wheel or any other working machine. He was more interested in intellectual than practical pursuits.

Another factor behind Leonardo's inadequate analysis of water power was his use of erroneous laws. The basic law of hydraulics were established only near the middle of the seventeeth century as several of Galileo's associates extended the new deductive, quantitative, scientific methodology into that field. The first basic quantitative hydraulics law was deduced by Benedetti Castelli (1577–1644), a close friend and associate of Galileo. In 1630 Castelli published several elementary propositions which, in effect, established the relationship volume of flow ($Q$) = area of cross-section of a stream ($A$) × velocity of the stream ($V$). This is the so-called continuity equation or relationship. Before Castelli, Leonardo and others had apprehended the principle in special cases.[15] But the generalization of the relationship between stream cross-section, velocity, and volume became widely known only through Castelli's work.[16] Yet Castelli continued to maintain, as Leonardo had, that the velocity of spouting water or flowing water was directly proportional to its head or fall.[17]

A more accurate efflux law emerged only after further work by Galileo and Torricelli. In his mature work on mechanics, *Two New Sciences*, published in 1638, Galileo established the basic laws of falling bodies and demonstrated that they applied to bodies rolling down inclined planes and to the motion of the pendulum as well. In the 1640s Evangelista Torricelli (1608–47), one of Galileo's students, argued that the laws of free fall applied also to water spouting fron an orifice at the bottom of a reservoir, that a stream of water spouting under a head of, say, 6 feet (1.8 m) had the same velocity as a drop of water that had fallen 6 feet. By analogy to free-falling solids Torricelli established the efflux relationship: velocity of spouting water ($V$) is proportional to the square root of the head or fall of the water ($H$).[18] The efflux law, in the form it ultimately assumed, $V = \sqrt{2gH}$, was of vital importance to water wheel analysis.[19] Not only did it clear up old errors, but it permitted the computation of fall when only the velocity was known, or the velocity when only the fall was known.

The foundations laid by Galileo, Castelli, and Torricelli in the first half of the seventeenth century were built upon in the late seventeenth century, primarily at the Académie des sciences in Paris. Founded in 1666, the Académie, from its very beginnings, was closely associated with and supported by the French government. A sort of implicit agreement gained continued crown support for science in return for the academicians' willingness to work on projects of interest to the French monarchy. This close association between the Académie and the French government probably explains the very large role that members of the Académie were to have in the evolution of a mathematical and theoretical approach to water power. Hydraulics was one of the principal areas whose practical utility was recognized and promoted both by the government and its academicians. The standard work on the history of the Académie des sciences notes:

> By far the largest number of urban affairs brought to the Academy's attention were related to hydraulic phenomena. The problem of water—a resource essential for human life, a major source of power in pre-industrial society, and a dependable vehicle for transportation—took up much of the academy's time. Once-popular plans for elaborate fountain works for royal palaces were gradually replaced by schemes for water purification, storage, and distribution, by suggestions for... improved water mills.[20]

One of the very first projects undertaken by the newly formed academy was the experimental verification of Torricelli's efflux law. A committee of four was formed to undertake this task.[21] The two leading figures were Christian Huygens (1629–95) and Edme Mariotte (c1620–84). Huygens, the well-known Dutch savant, was one of the lights of the young academy. His contributions covered a wide variety of fields—astronomy, optics, mechanics—and are quite well known. Mariotte was another major figure, publishing works on the theory of heat and light, anatomy, meteorology, strength of materials, and hydraulics. He was best known for his experimental talents.

Between 1666 and 1670 Huygens and Mariotte confirmed the approximate validity of Torricelli's law through a variety of experiments. Not content with this, they extended their investigations, seeking to determine the impulse of spouting water against a plane surface at various velocities. Using experimental devices similar to those sketched by Leonardo, they found that the impulsive force ($P$) of water against a plane surface varied as the square of the water's velocity, and that its absolute value could be computed by the equation $P = dAH$ (which I will call the impulse formula), where $P$ = the force of impact, $d$ = the weight of water per unit volume, $A$ = the area of the orifice from which the water spouted or the surface area of the plane being struck, and $H$ = the head or height of the water above the orifice.[22]

Both Huygens and Mariotte saw that this formula had practical applications for water-power technology. Huygens in remarks before the academy in July 1669 declared: "The knowledge of the moving force of air and of water is useful, primarily for the construction of all types of water and wind mills, for with the first, the quantity and velocity of the water that we have at our disposal being given, which is easily measured, we can know in advance to which force of horses or men that of a mill would be equal."[23] Mariotte went further. He prepared a comprehensive treatise on hydraulics published posthumously in 1686. His *Traité du mouvement des eaux* contained the first full application of the new quantitative hydraulics to water wheels. In the *Traité* Mariotte derived and demonstrated the basic rules or laws of hydraulics which he and his predecessors had established—the continuity equation, the efflux law, the impact $\propto$ velocity squared ($V^2$) relationship, the impulse formula, and so on.[24] He then applied these to several problems, including the force of water acting on the mills of the Seine. Mariotte measured the dimensions of several of these mills and determined the velocity of the water which struck them by using small blades of grass and pieces of wood with a half-second pendulum. With this data he calculated that the blades of these watermills, which were 5 feet (1.62 m) long by 2 feet (0.65 m) deep, could sustain a force ($P$) of 233.33 pounds (114.3 kg) placed on a horizontal lever arm at the same distance from the axle as the center of their blades. He compared the force which these mills could sustain in equilibrium with that sustained by a typical windmill.[25] And recognizing that water wheels were used primarily in conjunction with dams, Mariotte demonstrated the theoretical advantages of placing the discharge chute of a dam at the maximum possible depth.[26]

Both Huygens and Mariotte seem to have expected (or at least hoped) that their findings would be usefully applied. They were to be disappointed. In great part this was because, as academic natural philosophers and theoreticians rather than engineers or craftsmen, they did not completely comprehend the needs of practical water-power technology. Their analyses were simplistic and misdirected. Their procedures permitted them to estimate the force striking the blades of a watermill only when those blades were held motionless—the static case. But work is performed only by motion, and neither Huygens nor Mariotte considered the case of a water wheel in motion.

# THEORETICAL ANALYSIS ESTABLISHED: PARENT AND HIS SUCCESSORS (c1700–c1750)

The analysis of the vertical wheel in dynamic terms first occurred in 1704, at the end of a five-year flurry of activity in the Academy of Sciences in the area of power. Interest in the problem was sparked by two papers in 1699. One was by Philippe de La Hire (1640–1718), who examined the force exerted by men carrying loads under various conditions and compared their exertions to those of horses. The other was by Guillaume Amontons (1663–1705), who had designed a steam wheel and tried to evaluate its performance in terms of the number of men or horses it could replace. Following upon their work, Antoine Parent (1666–1716) attempted to verify part of Amontons' work algebraically in 1700; in 1701 he considered the most advantageous inclination for windmill sails; and in 1702 he attempted to reduce the motion of animals to the laws of mechanics.[27]

One of the problems which faced all of these investigators was the absbnce of an accepted and objective manner of measuring and comparing the effects of different power sources in motion. This problem was most acute in the early papers of Amontons and La Hire. La Hire in 1699 had attempted to compare the force exerted by men and animals. As a basis for comparison he thought of them both as exerting force on a crank turning a horizontal winch drum which supported a suspended weight. But like Mariotte and Huygens some years earlier, he considered only the static case—the weight on the drum was held motionless.[28] Amontons in his paper needed to estimate the output of his steam wheel in terms of the manpower or horsepower it could replace. Unlike La Hire, he recognized that speed of motion was as important as static force. Galileo's *Le meccaniche* was widely known in France, and Amontons was a Galilean. Thus it did not require a radical break with the past for him to take the dynamic element which Galileo had introduced into the analysis of the lever and extend it to prime movers. He adopted as the basis for comparing men, horses, and steam wheels the exertions of plate-glass polishers, concluding that their effort was equivalent to continuously raising a weight of 25 pounds (12.25 kg) at 3 feet (0.97 m) per second. This standard was a dynamic one, in effect an application of Galileo's force × velocity.[29]

Amontons' paper moved the rationalization of power technology from statics into dynamics. His approach rather quickly acquired a permanent place in machine analysis, replacing the static approach of Huygens and Mariotte. In 1702, for example, in a paper on the force needed to draw boats upstream, La Hire adopted Amontons' method and considered not only the static force horses or men could exert, but the force they could exert while moving at a certain horizontal speed. This paper was important to the development of mathematical analysis in water power, for in that paper La Hire combined the hydraulics of Torricelli,

Huygens, and Mariotte with the dynamic concepts of Galileo and Amontons. The first problem he studied was the force needed to draw a boat up a stagnant body of water at a certain velocity. Applying Torricelli's efflux law and the Huygens-Mariotte rule ($P = dAH$) for calculating the absolute value of the impact of water against a plane (the bow of the boat), La Hire derived the equation $P = dA(v^2/2g)$, where $v$ is the velocity of the men or horses pulling the boat, $P$ is the force that they have to exert to move the boat at that velocity, and $A$ is the area of the bow of the boat. La Hire recognized that this equation would apply only to a boat being pulled through stagnant water. If the boat were pulled upstream against a current, the velocity at which the water impacted the boat and the force which horses or men had to exert to pull the boat at any given velocity would be greater. For this case La Hire modified his equation to $P = dA[(V + v)^2/2g]$, where $V$ is the velocity of the impacting water, and $v$, that of the barge. Thus, not only did La Hire combine the rules or laws established by Torricelli, Huygens, and Mariotte with the concepts of Galileo and Amontons, but he added to that combination the recognition that for moving bodies the velocity of impact was relative, depending on the speed of both the water and the impacted object.[30]

La Hire considered only the case of an object moving against a stream where the velocity of impulse was the sum of the velocity of the stream and the object struck ($V+v$). He did not consider the case of the water wheel. Here the impacted object (the blades) receded before the stream, and the relative velocity of impulse was determined by subtracting the velocity of the wheel from that of the stream ($V-v$). But La Hire's paper set the stage for Antoine Parent, who, in 1704, carried out the first sophisticated dynamic analysis of the vertical water wheel.

Antoine Parent is one of the central figures in the history of water wheel analysis. He was born in 1666 at Chartes and sent to Paris in his teens to study law. But he had acquired in his youth a taste for mathematics, and after completing his legal studies he returned to mathematics over the opposition of his parents, studying under La Hire and others. As he acquired competence in the field, he supported himself by tutoring other students, primarily in the area of fortification theory. In the 1690s, anxious to complement his mathematical and theoretical knowledge with practical experience, he attached himself for a period to an army unit. He was admitted to the Academy of Sciences in 1699.[31]

Although Parent contributed papers to the academy in a number of areas, the most significant was the one he read in 1704.[32] In that paper he attempted to establish the conditions under which a moving water wheel would produce the maximum possible effect (*effet*) or power.

This was no simple problem. Parent, on beginning his studies, saw that there were two limiting cases in which a water wheel would produce no effect. One was the state of equilibrium. If the blades of a wheel were held immobile in a stream by a counterweight (the case investigated by Mariotte), the impact of the water on the blades would be the greatest possible, but the water could have no effect since no velocity was imparted to the wheel or attached load. On the other hand, if the blades of the wheel moved at the same velocity as the impelling water, then the velocity of the wheel would be a maximum, but it would lift no load since the water would be exerting no pressure or force against the floats. Parent suspected that while the two quantities involved in the water wheel's effect—force, or load ($P$), and velocity ($V$)—could both vary widely between these two limiting extremes, there must be some ideal proportion between them where their product was the greatest possible.

Parent first reflected on this problem after visiting and calculating (apparently using Mariotte's methods) the force of hydraulic machines in and around Paris around 1700. He recognized after these visits that he needed principles for calculating the power of machines in motion. He also recognized that most previous work applied only to machines at rest (Huygens, Mariotte) or to the very special case of movement by an animate power (La Hire, Amontons).

After four years of work, interrupted, says Parent, by a number of other activities, he succeeded in solving the problem. In his 1704 memoir he asked his readers to imagine a vertical wheel loaded with a certain weight, $P$, of such magnitude that it would balance the impacting force of a stream moving at velocity $V$. The weight, $P$, he contended, would represent the greatest possible weight which a stream moving at a given velocity, $V$, could sustain. The greatest effect that such a stream could produce would then be the product of this weight, $P$, and the velocity of the stream, $V$. This product ($PV$) Parent defined as the natural effect (*effet naturale*) of the stream.

Parent then asked his readers to suppose that the weight, $P$, which held a water wheel motionless, was reduced to some smaller weight, $p$. When this occurred, he argued, the blades of the water wheel would begin to move and would accelerate up to a certain point, at which the velocity of impulse (the velocity of the water minus that of the wheel, or $V-v$) would be such that the reduced force of the water on the blades would equal the reduced load, and the motion of both load and blades would become uniform. The product ($pv$) of the reduced weight ($p$) and its velocity ($v$) Parent defined as the general effect (*effet generale*) of the water wheel. Parent's approach to the analysis of the water wheel was in the Galilean tradition, for it depended on the application of Galileo's principle that no more effort is necessary for uniform motion than is required to hold a load in equilibrium, and the measure that Parent adopted for the effect produced by the stream ($PV$) and the wheel ($pv$) were simply modifications of the Galilean force × velocity.

Accepting the law of Huygens and Mariotte which declared that the force exerted against a plane struck perpendicularly by the water was proportional to the square of the velocity of impact, Parent established the equation (ignoring gearing ratios):

$$\frac{P}{p} = \frac{V^2}{(V-v)^2}.$$

He solved it for $pv$, the power output of the water wheel, and then applied the newly invented differential calculus for the first time to an engineering problem to determine the maximum value of $pv$. Using this value he determined the velocity and load on the wheel which were necessary to attain this maximum.

Parent's analysis revealed that a water wheel had, at the very best, an efficiency of only around 15% ($\frac{4}{27}$), that is, it could deliver to its attached machinery no more than $\frac{4}{27}$ of the natural power (defined as $PV$) of the stream which moved it, neglecting all friction and other losses. He found that even this maximum could be produced only when the weight or load being lifted ($p$) was $\frac{4}{9}$ of the weight of equilibrium ($P$) and only when the center of the wheel's blades turned with $\frac{1}{3}$ the velocity of the stream. Parent was surprised at these findings, observing:

> There are, therefore, two very remarkable paradoxes; namely, first, that a power loaded to only $\frac{4}{9}$ of its capacity produces after a certain time a greater effect than if it carried very much more.
>
> The second is that no matter what changes we make to a machine, its product will never exceed $\frac{4}{27}$ of its natural effect, that is to say, of that which the motive force would produce without the machine.[33]

Despite these paradoxes, Parent's analysis of the water wheel soon became standard in European scientific literature, the basis of practically every attempt made in the early eighteenth century to analyze the water wheel theoretically. For example, Henri Pitot (1695–1771), inventor of the Pitot tube for measuring the velocity of flowing water, took up Parent's work in the 1720s. He simplified Parent's analysis and derived elementary algebraic formulas and arithmetical rules for calculating optimum load, optimum velocity, maximum effect, and blade area. He also used Parent's system to evaluate proposals to utilize the power of flowing water to move boats upstream (as in Fig. 2–21).[34] Bernard Forest de Bélidor (c1695–1761), one of the first great engineering handbook and textbook writers, gave Parent a prominent place in his magnum opus, *Architecture hydraulique*, in the 1730s. He declared that Parent had made one of the most important discoveries in the history of science and noted that when he first read Parent's 1704 paper, he considered it the most interesting thing he had ever learned in mechanics. Not only did Bélidor expound Parent's findings at great length, but he also utilized them to make calculations on a number of water-powered engines and to suggest modifications in their design.[35] Among other Continental authors who adopted Parent's findings were d'Alembert, Gravesande, and Leonhard Euler.[36]

In England, as on the Continent, Parent's analysis was widely accept-

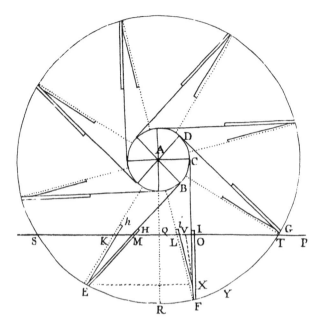

FIGURE 4-4. Pitot's comparison of radial and tangential blades on undershot wheels. Pitot theoretically compared the force of water acting on blades of equal length arranged radially (like *K* and *L*) and tangentially (like *M* and *O*). His analysis indicated that the radial situation was superior, since during most of the arc from *S* to *F* the radial paddle was in a more perpendicular position (compare *K* and *M*) and hence received the greater impact from the water. He also compared the two arrangements in the position of maximum impact—at *R* for radial blades and *F* for inclined blades—and again found that radial blades received more impulse from a stream of water.

ed. It was introduced to Britain in 1735 by John Muller (1699–1784), a German-born mathematician, teacher, and textbook writer, in a work on conic sections and the calculus.[37] It was then taken up by Colin Mac-Laurin (1698–1746), the ablest of Newton's English successors and the chief exponent of Newtonianism in Britain. MacLaurin recognized that Parent's method of measuring the force or power of motion (*PV*) was compatible with the orthodox Newtonian measure of momentum, *mv* (mass × velocity). He thus had little trouble incorporating Parent's work into the Newtonian system. It appeared in two of his works—in the *Treatise of Fluxions* of 1742 and, in a more simplified form, in the posthumously published *Account of Sir Isaac Newton's Philosophical Discoveries* in 1748.[38] Among other pre-1750 English authors to use Parent's findings were John T. Desaguliers and Benjamin Martin, both popular lecturers and textbook writers.[39]

Throughout the eighteenth century the main focuses of water wheel analysis were the problems established by Parent—maximum effect, optimum load, optimum velocity. But there were attempts after 1720 to rationalize other elements of vertical wheel design, notably floatboard or blade arrangement. This effort was initiated by Henri Pitot. In a paper presented to the Académie des sciences in 1729 he attempted to determine theoretically whether the blades of a water wheel should be prolongations of the radius (radial) or inclined (tangential) and how many blades a water wheel of a given size should have for optimum performance.[40]

Pitot's analysis of these problems was based on a law discovered in the

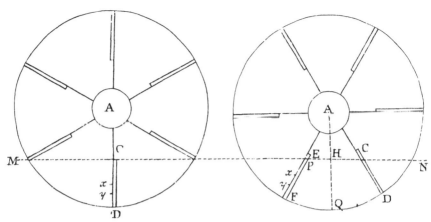

FIGURE 4–5. Pitot's system of blade distribution. Pitot's theory indicated that the blades on an undershot water wheel should be distributed according to the system indicated in the wheel on the left. That is, when one blade (D) was in a vertical situation and in position to receive maximum impulse, the blade which had proceeded it should be leaving the water and the trailing blade (M) just entering it. If other blades were inserted, Pitot argued, an oblique impulse would replace the perpendicular impulse over a portion of the blade D (Cy) and the amount of power delivered to the wheel would be decreased. Pitot's theory was later criticized on the grounds that when blades distributed as he specified were in the position shown in the wheel at the right, the water under a line from F to D would escape without acting on the wheel at all, while an intermediate blade at Q would have been able to use that water.

1670s by a Jesuit mathematician, Ignace J. Pardies. Pardies had found that the force of oblique impulse was to that of perpendicular impulse as the sine of the oblique angle (the angle of incidence) to 90°.[41] Using the water wheel shown in Figure 4–4, Pitot compared the force of a stream of water on both tangential and radial blades from the point at which they entered the water to the point where they were situated perpendicular to the flow of the stream. He found that the stream acted with much more force on radial blades both at the point of maximum impulse and over a much greater portion of their rotational arc.[42]

On this basis Pitot concluded that the number of blades a water wheel of given size should have was not arbitrary. The spacing between any two successive blades should be such that while the leading blade (D on the left in Fig. 4–5) was in a perpendicular position, the following one (M) was just entering the water. If a third blade, E, were placed between these two, Pitot argued, a portion of the leading blade, D, [Cy] would receive practically no impulse. The added impact on the submerged portion of the intervening blade would be less than the lost impulse on the leading blade, since a perpendicular impulse was replaced with an oblique one.[43] Pitot constructed the table reproduced here as Table 4–1 on the basis of these deductions, showing the optimum number of blades for water wheels of various sizes.

Pitot's analysis was extended by Bélidor in 1739. Bélidor accepted

TABLE 4–1. PITOT'S BLADE, OR FLOATBOARD, DISTRIBUTION TABLE, 1729

| Number of floats | Length of the floats |
|---|---|
| 4 | 1,000 ft |
| 5 | 691 |
| 6 | 500 |
| 7 | 377 |
| 8 | 293 |
| 9 | 234 |
| 10 | 191 |
| 11 | 159 |
| 12 | 134 |
| 13 | 114 |
| 14 | 99 |
| 15 | 86 |
| 16 | 76 |
| 17 | 67 |
| 18 | 60 |
| 19 | 54 |
| 20 | 49 |

How to use the table: If the radius of a water wheel is 8 feet and the length of its blades is to be 3 feet, you set up the proportion 8:3 :: 1000:$x$. This gives for the value of $x$, 375. Using the table this indicates that the wheel should have seven blades or floatboards.

SOURCE: Henri Pitot, "Remarques sur les aubes ou pallettes des moulins, & autres machines mûës par le courant des rivières," Académie des Sciences, Paris, *Mémoires* (1729).

Pitot's basic premises, but pushed the analysis further by determining the average force acting on the blades of a water wheel as they rotated from the position of minimum impact to the position of maximum impact. Geometrical analysis allowed Bélidor to estimate that the average impulse on a rotating wheel was $11/12$ of the impulse it received when its blades were in their most advantageous position (a single radial blade struck perpendicularly by the stream, the following blade just beginning to enter the water).[44]

Pitot's analysis of floatboard spacing complemented Parent's derivation of optimum velocity and load conditions and, like it, soon became standard in expositions of the application of mathematics to water wheels. Bélidor used it in 1739 to make recommendations for modifying the water wheel at the Samaratine water works in Paris.[45] D'Alembert presented it in his 1751 article "Aubes" (blades) in the *Encyclopédie*.[46] Desaguliers published an extended exposition of Pitot's arguments and reprinted his float distribution table in his *Course of Experimental Philosophy* in 1744.[47]

Parent's analysis of the optimum operating conditions for water wheels and Pitot's system of float distribution formed the heart of water wheel analysis in the first half of the eighteenth century. But Parent's and Pitot's studies of water wheels contained some very serious flaws and limitations. For instance, following Galileo, Parent had chosen the product $PV$ (force × velocity) as his measure of the natural effect of a stream, instead of $mgH$ (weight × height). This was, in itself, not a critical error. But he also accepted as valid the Huygens-Mariotte method of calculating the value of $P$ ($P = dAH$). This meant that the $P$ in Parent's equation for the maximum effect of a water wheel ($pv = \frac{4}{27} PV$) was expressed by the product $dAH$. The Huygens-Mariotte law was in error. Neither Huygens nor Mariotte had noticed the contraction of the vein in the spouting jets of water used in the experiments by which they derived this law (see below for Newton's work in this area). The correct theoretical value of $P$ was $2dAH$. This meant that the value used by Parent for the natural effect of a stream was high by a factor of two. Correcting Parent's calculations in this light, the maximum efficiency of a water wheel would be $\frac{8}{27}$ (30%), instead of $\frac{4}{27}$ (15%). The former figure is much closer to reality.

This was not the only flaw in Parent's analysis. Parent had also accepted as universally valid the experiments of Huygens and Mariotte on the relation of force of impact and velocity in hydraulics, $P \propto V^2$. Huygens and Mariotte had derived this relationship from experiments in which a broad stream of water impacted a flat plate always maintained perpendicularly to the flow. But in most water wheels, water was led against the wheel in a close-fitting channel, more than one blade was immersed, and the impact against some of the blades was oblique rather than perpendicular. If we imagine a wheel with an infinite number of blades in a confined channel, where none of the water can escape without acting on the wheel, the force on the blades is proportional simply to the velocity rather than to the velocity squared.

The third, and in many senses the most important, flaw in Parent's analysis was his implicit assumption, reflected in the title of his memoir ("On the greatest perfection possible in machines . . . having for a motive power any fluid body . . ."), that his analysis applied to all types of water wheels operating under any conceivable conditions. In reality it applied only to impulse wheels. The countercase of the overshot or gravity wheel was not even mentioned or, apparently, considered.[48]

It is easy to see why Parent missed the first two flaws in his analysis. He simply adopted laws regarded as firmly established in his time. The third flaw—his use of the undershot or impulse water wheel as the prototype for all water wheels—requires a little more comment. One of the reasons for Parent's failure to investigate the countercase of the gravity wheel was probably the prevalent conception in the scientific community that impulse and gravity were equally effective as motive powers. This idea had a long tradition, with seemingly strong support in the work of Galileo and his disciples. Torricelli, as noted above, had demonstrated in the 1640s, by analogy to Galileo's work on falling bodies, that the velocity of water spouting from an orifice under a head was exactly equivalent to the velocity a drop of water acquired in falling freely from the same height. This had been confirmed by the work of Mariotte and Huygens in the 1660s. If water attained the same velocity whether it fell or spouted, it seemed to follow that water wheels would realize the same power whether the water spouted against the base of an undershot wheel or fell in the buckets of an overshot wheel, so long as the fall was equal. Thus it did not matter which wheel—overshot or undershot—was chosen as the prototype for water wheel analysis.

The undershot was probably selected because of the location of the early scientific academies. The earliest permanent scientific institutions—the Royal Society of London and the Académie des sciences in Paris—were in regions of low falls and relatively high water volumes. As mentioned in previous chapters, undershot wheels usually were more prevalent in such topographical conditions. Certainly the most notable examples of water wheels in the Paris and London regions were undershot—the London Bridge water works, Paris's Pont Neuf and Notre Dame works, the Marly wheels near Versailles. Hence, the undershot-impulse wheel was more likely to be familiar to academicians and afford them opportunities for direct study.

Another factor which may also have influenced the choice of the undershot-impulse wheel as the prototype for water wheel analysis was the general scientific climate. Practically all of the major mechanical philosophies of the seventeenth century were based on the proposition that "all the phenomena of nature are caused by particles of matter in motion, acting on each other by contact alone."[49] This made the problem of impact a central issue to scientists of the era,[50] and thus, the impact or impulse wheel was more likely to attract their attention. While one might think that the gravity wheel, with its different mode of operation, would claim special consideration and a separate analysis, this was not the case. The tendency to see even gravitational force as due to the impact of tiny particles of matter in motion may well have tended to conceal the overshot-gravity wheel's distinctiveness.[51]

Pitot's work shared some of the same limitations as Parent's. Like Parent, Pitot completely neglected any mode of action save impulse. Thus he did not seriously consider the possibility that inclined blades might be able to make use of the weight or reaction of water and that this might compensate for their impulse deficiencies. Pitot also did not take into sufficient account the water which passed uselessly under the wheel when the blades distributed according to his recommendations were in the position shown on the right in Figure 4–5.

So subtle and difficult to detect were the faulty assumptions contained in the work of Parent and Pitot that they went long unrecognized. Dissent began to surface only around 1740. In 1738 Daniel Bernoulli suggested in an oblique manner that Parent's theory might need some revision, and in 1746 Pitot's blade-distribution system was attacked by DuPetit-Vandin.

The basis for Bernoulli's criticism lay in the work of Isaac Newton some 50 years earlier. In the portion of the *Principia* devoted to hydraulics Newton had noted in 1687 that his experiments revealed an apparent discrepancy between the Torricellian efflux law ($V = \sqrt{2gH}$) and the continuity equation ($Q = AV$). If the velocity of spouting water was what Torricelli said it was, the quantity of water delivered from an orifice according to the continuity equation was only about half of what it should have been. Newton initially thought Torricelli's law was at fault. Between 1687 and the second edition of the *Principia* in 1713 Newton found that the discrepancy was due to the contraction of the vein of water as it spouted from the orifice. He found that if he used the area of the cross-section of the contracted part of the efflux vein (the vena contracta) as the value of $A$, instead of the area of the orifice, the conflict between Torricelli's and Castelli's laws was resolved. He also found that the impulse of an effluent jet of water was then not $P = dAH$, but $P = 2dAH$, where $A$ is the area of the vena contracta's section.[52] The conflict over which value of $P$ ($dAH$ or $2dAH$) was correct occupied a number of theoretical hydraulicians, including Bernoulli, during the first third of the eighteenth century.

As part of his book on hydraulics, published in 1738, Daniel Bernoulli first analyzed machines driven by the impact or impulse of fluids in a more or less orthodox Parentian manner. He pointed out, however, that a change in the method of measuring the impulse of water against a plane would result in some revision of his results.[53] This was true. Newton's modified measure of impact implied that the maximum useful effect obtainable from water wheels was $\frac{8}{27}$ *mgH*, instead of the Parentian $\frac{4}{27}$ *PV*, and that the optimum load for a water wheel was $\frac{8}{9}$ *P*, instead of $\frac{4}{9}$ *P*. While Bernoulli intimated that the revised impact measure would alter his analysis of water-driven machines, he did not directly draw these conclusions. After presenting an analysis of fluid impact which supported Newton's findings later in the book, he did not re-analyze hydraulic machines in the light of the new theory. He simply pointed out obliquely that it might have application, and even warned against its general use: "I should wish that it be noted properly that I discuss here only solitary streams which the planes receive entirely, but not fluids surrounding bodies and making an impetus on the same, such as winds or rivers."[54]

Just as Parent's theory began to receive some criticism around 1740, so did Pitot's. In 1746 a French military officer and a corresponding member of the Academy of Sciences, Robert-Xavier Ansart DuPetit-Vandin (1713–90) took Pitot to task. Vandin's most telling argument was his observation that Pitot had not considered the water which passed beneath a wheel without ever impacting the blades if they were distributed in accordance with Pitot's recommendations, especially in the position shown on the right in Figure 4–5. These losses, Vandin argued, were avoidable if additional floats were inserted between $F$ and $D$. Theoretically, according to Vandin, one could not have too many blades, and he praised the practice of Dutch millwrights, who, he said, hardly ever built wheels with less than 32 blades and often used 48, while by Pitot's rules they should have used only 6 to 8.[55]

While some of the elements in the analytical edifice built up by Parent and Pitot had come under limited attack before 1750, its most serious deficiency, neglect of the overshot-gravity wheel, passed largely unnoticed, at least in southern and western Europe and in America. The general tendency of writers like Pitot, MacLaurin, and the others who followed Parent was to assume that Parent's analysis applied to "all types of machines moved by a current, or a fall of water." The section which Bélidor devoted to water power in *Architecture hydraulique* is particularly instructive here, for it illustrates how difficult it was in the early eighteenth century to break away from the impulse prototype. Bélidor attempted to demonstrate the inferiority of the overshot to the undershot wheel by comparing the forces acting on wheels of each type, both 9 feet (2.9 m) in diameter, operating under identical heads of 100 inches (2.7 m) with identical volumes of water. Bélidor calculated the impulse of the water against each of these wheels and, not surprisingly, found that it was about six times greater on the undershot wheel. He concluded, on the basis of this calculation, that the undershot wheel was approximately six times more efficient than the overshot. The work performed by the water in descending in the buckets of the overshot wheel was completely ignored.[56]

Even when it was recognized that the water in descending in the overshot wheel's buckets contributed to a wheel's power, this contribution was considered of less importance than impulse. For example, Henry Beighton (1686–1743), a prominent British engineer, studied the overshot wheel at Nun-Eaton in Warwickshire (Fig. 3–27). He calculated the static moment of the water in the buckets of this wheel and the static force the water would exert if it impacted stationary blades. He found that the moment of impulse was around 50 pounds (22.2 kg), while the moment of the water in the buckets was around 400 pounds (180 kg). Even though Beighton's analysis was static, it suggested the greater importance of weight to impulse in the operation of an overshot water wheel. Yet he concluded: "This Water-Mill is by most People accounted as good a one, as any the Country affords, for dispatching as much Business in the time, and doing it well . . . . It is a very agreeable Height 16 Feet [2.3 m], for the Wheel being made 20 feet [6.1 m] high in the Place it stands, it would not have been capable of doing so much business: *Of so much more service is the Impulse, Stroke, or Momentum of the Water, than is its bare Statical Weight*" (my italics).[57]

Even the perceptive theoretician Leonhard Euler believed that impulse was of more value than weight as a motive power and recommended the use of undershot over overshot wheels. Around 1750 he wrote:

When the fall of water is rather high, they make it fall on blades . . . that are hollow, in order that the water remain pushing there and contribute by its proper weight to the movement of the wheel. But although by this means the force acquires some augmentation, it will be often doubtful whether the augmentation would not be greater if they conducted the water to the bottom of the wheel, in order to make it strike the blades there with more velocity, without its weight contributing anything to make the wheel turn. This will be at least more advantageous if the wheel is very high, seeing that we can procure then for the wheel a very much greater velocity in conducting it down to the lowest blades.[58]

Thus, even when it was recognized that weight made some contribution to the operation of water wheels, this contribution was underrated vis-à-vis impulse.

While theoretical analysis in the early eighteenth century failed to indicate the shortcomings of Parent's analysis, suspicions that the undershot wheel was not as economical with water as the overshot were growing in practical circles. In 1729, for instance, Stephen Switzer (1682–1745), a builder of fountains, fancy gardens, and water works for the British aristocracy, declared that the undershot wheel used six times more water than the overshot.[59] Just how much insight Switzer had into the comparative efficiencies of impulse and gravity wheels, however, is not certain, for he did not specify whether the falls the wheels operated under were identical. John T. Desaguliers (1683–1744), however, concluded in 1744, after examining "many Under-shot and Over-shot Mills," that overshot ones ground as much as undershot with ten times less water, supposing, he added, that the fall of water on the overshot wheel was 20 feet (6.1 m) and that the fall on the undershot was around 6 or 7 feet (1.8 to 2.1 m). Thus Desaguliers clearly appreciated the superior efficiency of overshot wheels, rating them from 3.5 to 4.0 times more efficient than undershot wheels.[60] In 1747 Benjamin Martin declared that undershot wheels used more water than any other type of mill when all were operating under equivalent falls.[61]

But even those who noticed the superior water economy of the overshot wheel before 1750 were not certain whether its better performance was due to practical factors such as its ability to retain water longer in its buckets than the undershot wheel could retain water against its blades, or to the manner in which the water acted on the wheel (weight instead of impact). Apparently the former was considered the more likely explanation, and Parent's neglect of weight as a motive power continued to pass unnoticed.

As water power became increasingly more scarce and costly (see the opening pages of the next chapter), the need for experiments to determine the influence of such factors as impact, weight, velocity, and load on efficiency became steadily more desirable. By the 1740s this need was being voiced in both England and France.

In England, Desaguliers observed in 1744 that it was a "difficult thing" to determine whether considerable impulse should be given to an overshot wheel or whether it should receive the water without percussion. He declared: "For my part, I can determine nothing certain in this for want of sufficient Experiments; but I think there might be some Fall allow'd. . . . The determining this, to know what part of the Height of the Fall must be taken for the Diameter of an Over-shot Wheel, would be a useful Maximum [maxim?]."[62] Desaguliers added that the one-third velocity which Parent had prescribed as the optimum velocity for water wheels might apply to the overshot wheel, but confessed that without experiments he could not be certain.

Desaguliers also saw the need for an experimental confirmation of Parent's theory, and he took the first tentative steps towards this. In the second volume of his *Experimental Philosopy* (1744) he informed his readers that the "learned and ingenious Dr. Barker" had recently communicated to him the design of a machine which Barker thought would provide an experimental proof of Parent's proposition on the optimum velocity of a water wheel. This device—known variously as Barker's mill, Parent's mill, or Segner's mill—is well known to most students of hydraulics. Its trunk (see Fig. 4–6) was a vertical cylinder with a funnel at the top for the admission of water. Two boxlike arms, separated 180°, projected from the bottom of the cylinder. Water flowed down the vertical cylinder into the boxlike arms and then spouted out lateral holes, one on each arm, facing opposite directions. The reaction of the arms to the spouting water rotated the whole contraption. Desaguliers built a model of Barker's engine and demonstrated it to the Royal Society,[63] but unfortunately there are no surviving records of this demonstration.[64] This may be for good reason. Barker's device was a primitive reaction turbine. It would not have provided experimental proof of Parent's theory.

Barker and Desaguliers were not the only ones who saw and voiced the need for experiments in the 1740s. In France, d'Alembert, after noting many of the factors which tended to make theoretical calculations on water wheels inaccurate, noted: "All these circumstances so disturb calculations such as these, already thorny even without them, that I believe that it is only experiment which is capable of exactly resolving the problems in question."[65]

This general recognition of the need for experiment in the analysis of water wheel problems had been long in coming. Parent had not tested his work by experiment; neither had Pitot. In general, in the late seventeenth and early eighteenth centuries attempts to apply quantitative concepts and methods to water-power analysis had been largely the work of academic scientists and theoreticians, men more interested in general mathematical principles and in providing illustrations of the potential utility of science than in actual practical results. Practical engineers and practicing millwrights largely ignored the nascent work in theoretical analysis being carried out by academicians and continued to design,

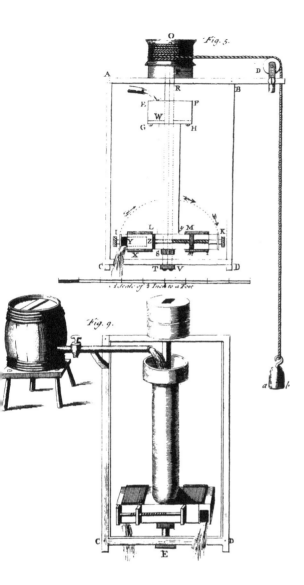

FIGURE 4–6. Barker's mill. This was the device proposed by Barker and Desaguliers for testing the validity of Parent's theory that water wheels operated most efficiently when their velocity was one-third that of the water. On a vertically situated cylinder, two boxlike arms were attached 180° apart. Water flowed down through the cylinder into the arms and out through two lateral holes, one on each arm. These holes were placed on opposite sides, so that the reaction from the spouting water turned the entire apparatus. The weight at *ab* in the upper figure was to be used to measure the mill's output.

build, and operate water wheels as they always had—relying on intuition, craft rules, and past experience. They lost little by ignoring the work of the theoreticians. The basic analyses, Parent's and Pitot's, were based entirely on assumptions too simplistic for actual practice and were not verified by practice or experiment. Moreover, as long as water-power sites were relatively plentiful and industrial growth slow, there was no pressure on those who built or owned watermills to consider the alternative methods for approaching the problems of water-power technology being developed in the scientific academies of Europe.

Towards the middle of the eighteenth century, however, the continued expansion of industry had begun to push water-power resources to the point of saturation in some areas of Europe and particularly in Britain (as we shall see at the beginning of the next chapter), making it increasingly necessary for mill owners and mill builders to devise means of using water in the most economical possible manner and to tap every possible unit of power they could. These circumstances provided an incentive for practicing engineers to begin to look at the quantitative methodology which had been slowly evolving in the scientific academies. This need for new methods to utilize water more economically, in conjunction with the rising recognition of the need for some experimental data to test the validity of existing water wheel theory and to solve some of the vexing problems of water-power technology forms the background for the decade of the 1750s, when practical engineers first began to apply quantitative techniques to water-power problems.

## THEORY CHALLENGED BY EXPERIMENT

Antoine de Parcieux (1703–68), a French mathematician and technician, was probably the first experimentalist to undertake a comprehensive reevaluation of the early eighteenth century's accepted water-power principles and challenge Parent's impulse prototype. De Parcieux was the son of a farmer. Orphaned while young, he was educated by the Jesuits, distinguishing himself in mathematics. In 1730 he moved from Lyons to Paris, where he worked for some years as a sundial maker. The technical skill and accuracy he brought to this vocation first attracted the attention of the French scientific community. Between 1730 and 1745 de Parcieux also designed an improved form of piston pump and published collections of astronomical and trigonometric tables. These accomplishments secured him admission to the Académie des sciences in 1746. Within the academy de Parcieux's primary concerns were with practical problems. He brought his mathematical skills to bear on subjects like water supply and life expectancy. He invented a new leveling instrument, studied the movement of rocking beams, investigated fountains and the traction of horses.[66]

A practical engineering problem provided the immediate stimulus which led de Parcieux to suspect the universality of Parent's theory. Louis XV's mistress, Mme. de Pompadour wanted a water supply for her chateau at Crécy to prevent accidents (what kind are not specified) which might occur during the king's visits. Several early proposals were rejected because they either could not furnish a sufficient supply of water or would cost too much. At this point Buffon of the Academy of Sciences was asked to review the problem, and he, in turn, called on de Parcieux for help. De Parcieux and Buffon concluded that the best way to handle the problem was to make use of the power of a small local river, the Blaise, to raise water to a reservoir some 163 feet (52.9 m) above river level. But during dry months the Blaise furnished only 4 to 5 cubic feet (0.14–0.17 m³) per second. De Parcieux realized that by Parent's theory only $\frac{1}{27}$ of that quantity could even be lifted to a height equal to the small fall available on the river and therefore only a very minute amount to the needed 163 feet, especially when friction and other losses were considered.[67]

The poor prospects which the project had if Parent's theory were valid led de Parcieux to seek an alternative to Parent's impulse prototype. He began to take a new look at the gravity or bucket wheel. De Parcieux says he "saw quickly" that he "could draw a much better portion from the weight of the water in considering it as weights which, in descending, lift up others."[68] He sensed an analogy between the action of water descending in the buckets of an overshot wheel, lifting water to a reservoir, and one weight lifting another by a cord over a pulley. He reasoned out the implications of this insight. He saw that the closer the magnitude of the descending and ascending weights, the slower the motion of the descending weight would be, but the greater the effect (defined by de Parcieux as the product of weight and height lifted) it would have in falling any given distance. He carried out several simple experiments with pulleys and weights to confirm this reasoning.[69]

De Parcieux then applied these findings to water wheels, arguing that the water-filled buckets of overshot wheels were like an infinite series of falling weights. The slower they descended, the closer the weight of the water they could raise would approach the weight of the water descending in the buckets. He asked his readers to imagine two water wheels of the same diameter, with the same number of buckets, carried on the same axle, but with the buckets of one wheel inclined in a direction opposite those of the other. This arrangement would permit the buckets of one wheel to lift water from a sump back to an elevated reservoir, while the other wheel utilized the water from the reservoir to power the lifting wheel. De Parcieux believed that this reversible cycle clarified his intuited model. If each bucket on the lifting wheel ascended with half of the water in the descending buckets, he reasoned, only half the amount of descending water would be lifted to the same height. But if each bucket ascended with $\frac{2}{3}$, $\frac{3}{4}$, or $\frac{4}{5}$ of the water in the descending buckets, the rotation would be proportionately slower, but a proportionately greater quantity of water would be elevated, with complete reversibility (100% efficiency) approached as the velocity neared zero. Thus, de Parcieux concluded, water acting by weight could theoretically have much more effect than water acting by impact. In the latter case the $\frac{1}{27}$ derived by Parent represented the maximum efficiency. But for gravity wheels it could be $\frac{2}{3}$, $\frac{3}{4}$, $\frac{4}{5}$, or even higher, depending on how slowly the wheel turned. Parent's analysis did not apply to wheels of this type.[70]

While de Parcieux relied on a combination of logic and intuition to deduce the superior efficiency of gravity to impulse wheels, he was familiar enough with practical engineering problems to recognize that reality did not always correspond to reason. He also believed that experiment was more convincing to practical technicians than abstract reason-

FIGURE 4–7. De Parcieux's apparatus for testing the optimum velocity of gravity wheels, 1754. The model wheel was 20 French inches (54 cm) in diameter, made of metal, and contained 48 buckets. The right side of its axle was divided into four drums of different diameter (a second axle with drums of other dimensions could also be used). The wheel's output was measured by means of a cord connected to one of the four drums, led over a pulley (top center), and connected to a weight. Water was supplied by the jar shown to the right of the wheel. De Parcieux was able to manipulate the velocity of the wheel by two means: utilizing different load weights and using axles of different diameters.

ing, which, in his words, "often only left the mind clouded." Thus, shortly after he had deduced the superior efficiency of the overshot wheel at slow velocities, he tested his conclusions by experiment. He utilized the apparatus pictured in Figure 4–7. A metal water wheel 20 inches (0.54 m) in diameter, with 48 buckets, was mounted on an axle on which drums of different diameters could be placed. The variable-diameter axle was then used to speed up or slow down the rotation of the wheel in order to determine whether the overshot wheel's efficiency was, as de Parcieux had postulated, dependent on speed. Output was measured by the product of a weight and the height to which it was lifted. De Parcieux's experiments were rather crude, designed only to confirm a general rule rather than to yield empirical coefficients to aid in practice. But de Parcieux was satisfied that they confirmed his hypothesis: the slower a gravity wheel turned, the greater its effect was for the same amount of water.[71]

De Parcieux's discovery that gravity was considerably more efficient than impulse led him also to question Pitot's float distribution theory. In the 1754 paper in which he announced his discovery of the advantages of gravity, de Parcieux briefly mentioned his suspicion that, contrary to Pitot's doctrines, inclined blades might be superior to radial blades. Inclined blades, he reasoned, might be able to utilize the weight of water as well as its impulse, and since weight-powered wheels had a superior efficiency, inclined blades might be advantageous. He noted also that he had begun experiments to test this idea.[72]

De Parcieux released the results of this work in 1759. Again he resorted to model experiments to prove his point. He built a wheel 32 inches (86.6 cm) in diameter with a dozen blades which were 7.5 inches (20.3 cm) wide by 8 inches (21.6 cm) high. These were hinged at the circumference so that their inclination could be varied without changing the wheel's diameter (see Fig. 4–8). De Parcieux again measured the output of the wheel by means of a weight lifted by a cord attached to the axle. The experiments he reported were carried out on May 5, 1758, on the Bièvre River, where the width was 5 feet (1.62 m), the depth 3 feet (0.97 m), and the velocity 3 feet (0.97 m) per second. They indicated that inclined blades were indeed superior to radial blades and that with 12 blades an inclination of 30° gave optimum results. No matter how many blades were used, de Parcieux found that inclined blades always yielded better results than radial blades. But in varying the number of blades on the wheel, he found that the best angle of inclination varied with the number of blades immersed in the water and with the velocity of the stream. Related experiments with the same model indicated that Pitot's system for determining the number of blades a wheel should have was also in error. De Parcieux found that the more blades his model wheel had, the greater its effect or output, confirming DuPetit-Vandin's 1746 critique of Pitot.[73]

De Parcieux's work with experimental models completely undermined

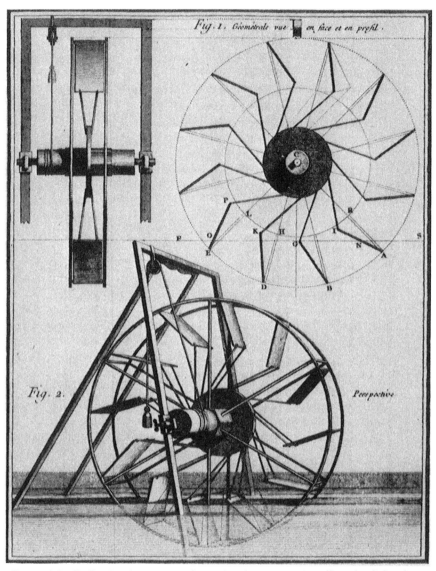

FIGURE 4–8. De Parcieux's apparatus for testing the comparative effect of radial and tangential blades, 1759. By means of hinges de Parcieux was able to set the blades so that they were radial or inclined to test Pitot's contention that radial blades were always more efficient than inclined blades. Output was measured by the weight being lifted by the axle. The wheel was 32 French inches (86.6 cm) in diameter with blades 7.5 French inches (20.3 cm) wide. With this device de Parcieux demonstrated the superior performance of inclined blades, as well as the advantage of using the maximum practical number of blades on a water wheel.

the theoretical edifice built up by Parent and Pitot in the first half of the eighteenth century. It was rightly regarded by later French theoreticians and engineers as a fundamental and important contribution to the rationalization of water-power technology. But his work was deficient in several respects. As de Parcieux recognized, his experiments established only general rules. More precise quantitative data that could be applied to detailed design required more extensive quantitative experiments, made under widely varying head and flow conditions, based upon observations on full-size wheels as well as models. Unknown to de Parcieux, at almost the same time he was carrying out his model experiments, John Smeaton was fulfilling these criteria.

John Smeaton (1724–92) was one of the most influential figures in the history of water-power technology. The son of a country lawyer, he demonstrated considerable mechanical talents while still a child. But his father intended him for the law and in the 1740s sent him to London to be trained in that field. Smeaton, however, had little taste for a legal career and eventually persuaded his father to allow him to take an apprenticeship as a scientific-instrument maker. On finishing this apprenticeship, he set up shop in London. But in the early 1750s, as the result of several commissions to design mills, he abandoned instrument-making for engineering. He worked on a variety of projects over the next 30 years. Besides designing both water and windmills, he was consulted on numerous canal and navigation improvement projects; he built bridges and lighthouses; and he improved harbors.[74]

Smeaton's decision to undertake model experiments to establish the optimum operating conditions for water wheels was probably the result of one of his early commissions to design a watermill.[75] After accepting the commission he investigated the literature on the subject and found only confusion. Bélidor indicated that undershot wheels were far superior to overshot; Desaguliers and Switzer said the opposite. "Finding that these conclusions were far from the truth and feeling from other circumstances, that the practical theory of making water and windmills was but very imperfectly delivered by any author I had then an opportunity of consulting; in the year 1751 I began a course of experiments upon this subject."[76] The bulk of Smeaton's experiments were carried out in 1752 and 1753. He announced his results to the Royal Society in 1759. The six- to seven-year delay in publication, he explained, was caused by his desire to test the deductions made from the model experiments "in real practice, in a variety of cases, and for various purposes."[77]

Smeaton built a model water wheel 2 feet (0.61 m) in diameter and supplied it with water from an adjacent reservoir (Figs. 4–9 and 4–10). Water from the reservoir, after passing through an adjustable sluice gate, struck the wheel and was recirculated to the reservoir by means of a discharge sump near the base of the wheel and a hand pump. A cord was attached to the wheel's axle and led around several pulleys to a balance pan. The load and velocity of the water wheel were varied by weights

*Philos. Trans. Vol. LI TAB. IV. p. 101.*

FIGURE 4–9. View of Smeaton's model for testing the comparative efficiency of undershot and overshot water wheels. The reservoir is on the right. Water from the reservoir rotated the water wheel, which by means of pulley and cord lifted a weight at R. The weight lifted by the wheel was used to determine output. Water was recirculated through the sump at BC back into the reservoir by the hand pump M.

placed in the pan, with the output measured by the product of weight and distance lifted. The operating head was calculated from the velocity of the spouting water. With this apparatus Smeaton systematically modified the key variables one at a time—load, wheel velocity, head, quantity of water, type of wheel.[78]

Smeaton undertook his experiments because he suspected that Parent's theory was incorrect. His experiments on undershot wheels confirmed this. His model undershot wheel yielded efficiencies (Smeaton called them ratios of "power" to "effect") roughly twice the maximum of

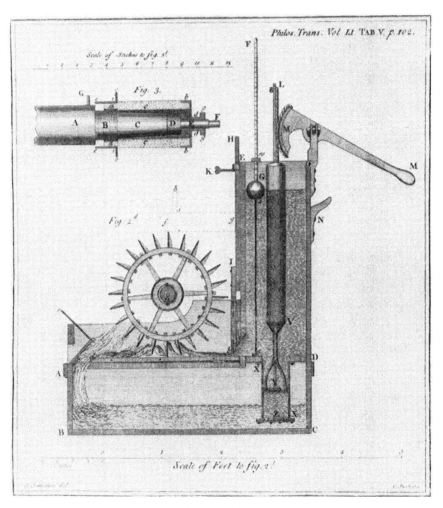

Philos. Trans. Vol. LI TAB V. p. 102.

FIGURE 4–10.  Cross-section of Smeaton's model for testing water wheels.

Parent's theory, or around 30%, with indications from certain tests that the theoretical maximum efficiency would be 50%, or more than three times Parent's $\frac{4}{27}$. When Smeaton measured the velocity of his wheel at the point at which it yielded the greatest effect, he found that the velocity of the wheel averaged around $\frac{2}{5}$ that of the water and approached $\frac{1}{2}$ when the sluice gate was opened wide. Parent's theory had indicated that it should be only $\frac{1}{3}$. Finally, Smeaton compared the load which would hold his wheel in equilibrium to the load it lifted when it produced the greatest effect. The proportion between the two by Parent's analysis was 9 to 4. Smeaton found that the closest he could possibly come to Parent's value was an experiment which yielded the ratio of 4 to 3.[79]

Smeaton's apparatus was easily adapted for an overshot wheel. After the undershot tests he closed off the lower orifice, added a channel at the top of the wheel, and replaced the undershot wheel with an overshot wheel. His experiments with the overshot wheel yielded even more drastic discrepancies from theory. Smeaton found, to his surprise, that his model overshot-gravity wheel had an efficiency roughly twice that of his undershot wheel and four to five times the maximum predicted by Parent! Depending on head and volume the maximum efficiency of his overshot wheel varied from 52% to 76%, with a medium of about 67%. Smeaton's experiments also indicated what de Parcieux had deduced—that the slower overshot wheels moved, within the limits of practicality, the better their efficiency. Smeaton recommended a peripheral velocity of about 3 feet (0.91 m) per second for good efficiency and performance. And, like de Parcieux, he postulated that the upper limit of the overshot wheel's efficiency was 100%.[80]

Smeaton was not sure how to account for the unexpected twofold superiority of gravity wheels to impulse wheels. He suggested that most of the difference was probably due to the power wasted by the water in turbulent impact against the blades of the undershot wheel. Half of the water's potential, he argued, was consumed simply in altering the figure or form of the stream on impact. Overshot wheels, particularly those where water filled the buckets slowly, would not suffer this loss; hence, their efficiency was higher.[81]

From a practical point of view the most significant discovery to emerge from the work of de Parcieux and Smeaton was, of course, the superior efficiency of weight or gravity over impulse as a motive force in water wheels. This discovery had a number of potentially significant repercussions on practical technology. It suggested, for example, that wherever water economy was a critical issue and wherever it was important to extract the maximum possible power from a stream, wheels operating on impulse should be replaced with wheels operating by gravity. For high falls this implied the use of overshot water wheels. For the more frequent medium to low falls, it implied the use of another type of wheel—the breast wheel. In a breast wheel the water enters at an intermediate point between the top and bottom of the wheel and is retained in buckets or between blades by a close-fitting casing so that it can turn the wheel by its weight. The emergence and diffusion of this type of wheel, along with associated devices to increase the gravity head under which such wheels operated, will be reviewed in more detail in the next chapter. Let it suffice for now to point out that under the influence of de Parcieux's and Smeaton's discovery, the breast wheel rapidly began to replace the traditional undershot wheel in industrial districts and in districts where economy of power was important.

The model experiments of Smeaton and de Parcieux were confirmed and supplemented between 1760 and 1775 by additional work carried out in France, Italy, and Britain by Bossut, Papacino d'Antoni, and Banks.

Charles Bossut (1730–1814), a prominent French hydraulician, used three different model wheels in tests carried out around 1770:

1. an undershot wheel 3 feet 2 inches (1.03 m) in diameter, 5 inches (14 cm) wide, operating in a closely fitting channel;
2. an undershot wheel 3 feet (0.97 m) in diameter, 5 inches (14 cm) wide, operating in a very wide (12–13 ft, 3.9–4.2 m) channel; and
3. an overshot wheel 3 feet (0.97 m) in diameter, 5 inches (14 cm) wide.

On the undershot wheels Bossut tested the effects produced by varying the number of blades, varying the inclination of the blades, varying wheel velocity, and encasing the wheel in a tightly fitting race. With the overshot wheel he studied the effect of wheel velocity on efficiency.

Bossut's experiments on blade number and inclination were more extensive than de Parcieux's and completed the work began by DuPetit-Vandin and de Parcieux in destroying Pitot's blade-distribution theory. Bossut found that the greater the number of blades, so long as the wheel was not overloaded or weakened by an overly large number, the greater the wheel's efficiency. On blade inclination Bossut expanded de Parcieux's work. De Parcieux had tested his model only in open stream. Bossut experimented on the effects of blade inclination in both confined channel and open stream. For open stream he found that inclining the blades slightly increased the wheel's power up to a certain angle, which varied depending on the blade's immersion and the water's velocity. Beyond that angle the power began to decline. In a confined channel Bossut found that traditional radial blades were slightly better than inclined blades.

Bossut's experiments on the optimum velocity of undershot and overshot wheels complemented the work of his predecessors. For undershot wheels he used both close-fitting and wide channels, finding that in both cases the optimum velocity was very close to 0.40 $V$, confirming Smeaton's work. His overshot wheel tests determined that within the bounds of practicality de Parcieux had been correct in concluding that the slower an overshot wheel moved, the greater its effect for a given quantity of water and fall.[82]

Additional model experiments were carried out by Papacino d'Antoni (1714–86), an Italian military engineer, around 1760, and by John Banks (fl. 1770–95), an itinerant British scientific lecturer, around 1770. Papacino d'Antoni constructed two wheels, each around 12.5 inches (c32 cm) in diameter. One was equipped with flat blades approximately 1.25 inches (c3.17 cm) wide and high; the other, with inclined bucket boards. Water was led onto the impulse wheel down a steep chute with a fall of approximately 12.5 inches (32 cm), and the wheel's output was measured by the product of weight lifted and the number of turns the wheel made per unit of time. Papacino d'Antoni's tests on undershot wheels supported Parent's theory, unlike those of Smeaton, de Parcieux, and Bossut. His experiments on the bucket or gravity wheel, however, supported the experimental work of his predecessors. They yielded an output at least twice that of experiments with the impulse wheel, and indicated that the higher the water was laid on the wheel, for a given fall, the greater the output.[83]

Banks in England used two wheels, one 20 inches (51 cm), the other 10 inches (25.5 cm) in diameter. With these wheels Banks studied the effects of diameter, slow motion, and point of delivery on the operation of bucket wheels, measuring output by the number of revolutions a flywheel geared into the test wheel made per unit of time. He also tested the overall efficiency of both undershot and overshot water wheels. Banks' investigations yielded little new data, and his experimental technique was clearly inferior to Smeaton's. But his studies did confirm the findings of his predecessors.[84] By the third quarter of the eighteenth century the deficiencies of Parent's theory were, therefore, glaringly obvious.

One other set of eighteenth-century model experiments should be mentioned, even though they exerted little influence on the mainstream of European water-power technology. Smeaton and de Parcieux were anticipated in their use of models for testing water wheels and in their discovery of the superior efficiency of the overshot wheel by the Swedish engineer Christopher Polhem (1661–1751). Polhem, who studied the natural sciences at the University of Uppsala in the late seventeenth century, was a well-known Swedish mining and manufacturing engineer, one of the earliest engineers to combine practicality with an appreciation for the value of theory.[85] Placed in charge of a mechanical laboratory sponsored by the Swedish monarchy, Polhem had one of his associates, Samuel Buschenfelt, construct in 1701–02 an experimental apparatus for testing water wheels.

Polhem's experimental device (Fig. 4–11) was quite sophisticated. Water was lifted by hand pumps to an elevated reservoir and led from this reservoir by means of adjustable races or troughs to one of two water wheels, each 18 Swedish inches (17.5 in, or 44.5 cm) in diameter. On one side of the model stand a water wheel was mounted on a pivoted, graduated arc which allowed Polhem to vary the angle at which water was directed on the water wheel. On the opposite side of the stand was a water wheel whose height with respect to the reservoir could be altered. A pivoting chute allowed water to be directed on this wheel overshot, breast, or undershot. Wheels with three different floatboard arrangements—radial blades, inclined blades, and elbow buckets—were available for tests. The velocity of the water striking the wheel was measured by the speed of the wheel when empty; the power produced was measured by the product of revolutions per minute and the weight elevated by a cord wrapped around one of several different-size cylinders mounted on the water wheel's axle.[86]

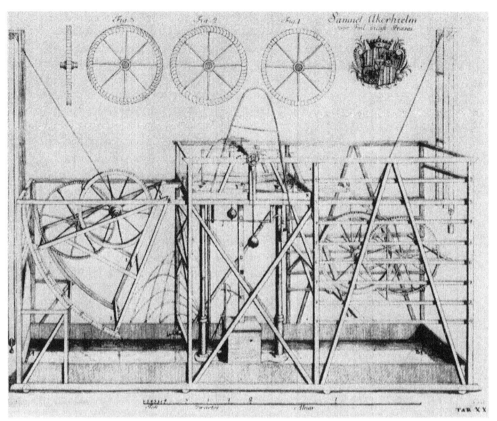

FIGURE 4–11. Polhem's apparatus for testing water wheels, c1701–1703. The reservoir which fed water to the wheels is at N (upper center). At the left a wheel is mounted on a pivoted arc to vary the angle at which water was delivered. The adjacent wheel is a gear wheel used to count the revolutions of the water wheel and determine its speed. At the right water wheels could be located at any of ten different levels. The types of water wheels tested are indicated at the top of the drawing.

With this apparatus Polhem attempted to determine the most advantageous manner for adapting a wheel to a waterfall—best slope for the head race, best wheel type, best height of the wheel with respect to the available fall, best wheel velocity, and so on. It was apparently from these experiments that Polhem was able to detect, long before Smeaton and de Parcieux, the superior efficiency of the gravity wheel.[87] Unfortunately, the results obtained from Polhem's experiments were not published. According to one of Polhem's biographers, the only data available are some tables compiled by his assistant, Vallerius. These are incomplete and deal only with experiments on undershot flumes.[88]

While Polhem's experimental model was far in advance of anything available before 1750 and apparently yielded some quite useful results, Polhem was critical of his handiwork within a few years. In 1710 he informed Vallerius that, even though the method was worthwhile, the experiments had been performed wrongly. They were, he declared, as "useless as the fifth wheel on a wagon."[89] In a letter written to Emanuel Swedenborg in 1716 Polhem noted how practical considerations seemed to make general rules (such as those gained from his water wheel experiments) difficult in application.[90] And in 1742 he pointed out several of the practical considerations that affected the performance of water wheels and of which he had been unaware at the time of his experimental investigations.[91]

Polhem's work was unknown outside of Scandinavia and had no lasting influence. The language barrier was undoubtedly the primary reason for this, but Polhem's failure to publish the results of his experiments (perhaps because of governmental restrictions) and his later scepticism about the value of his experimental findings probably contributed.

While the earliest attempts to rationalize the operation of the vertical water wheel emerged from the work of theoretical scientists like Parent, Pitot, d'Alembert, and Bernoulli in the first half of the eighteenth century, the earliest empirical tests of their theories were almost entirely the work of practical engineers and technicians like Polhem, Smeaton, and de Parcieux, instead of theoretical scientists. On reflection this is not surprising. To engineers, reliable data on the performance and operation of water wheels was a central concern, especially as the demands on limited water-power resources increased under the pressure of industrial growth. To academic scientists, on the other hand, the theory of the water wheel was a somewhat interesting but nonetheless narrow and unimportant problem, worthy of only a limited amount of time and expense. Academic scientists were more interested in broader questions which could lead to the discovery of general or universal natural laws.

There are other factors which help explain why practical technicians provided the first experimental data against which Parent's theoretical work could be tested. Owing to the expense of carrying out extensive manipulative, quantitative experiments on full-size water wheels, most of the experimental work in the eighteenth century was performed on models. The use of models to test designs had a long tradition in technology. In the 1500s, for example, the engineer Fontana had built a lead replica of the Vatican obelisk about 2 feet (0.61 m) high and experimented with various devices to lift and move it.[92] In the seventeenth century Riquet built a model canal on his estate to test the feasibility of his plans for linking the Atlantic to the Mediterranean, and the illiterate artisan Sualem, who built the Marly pumping station for Louis XIV in the late seventeenth century, apparently carried out tests on a model of the works before undertaking construction.[93] Moreover, according to Smiles, many of the great British engineers of the eighteenth century built in their youth model windmills or watermills, often with tools borrowed from a local millwright.[94] In a sense, then, the model water wheels which in the eighteenth century demonstrated the gravity wheel's superior efficiency and challenged the theories of academic scientists like Parent and Pitot were merely a continuation of a long technological tradition.

How dependent, then, were Polhem, Smeaton, and de Parcieux on the prior theoretical work of the academic theoreticians? Several authorities have considered Smeaton's experiments as an early example of the successful application of science to technology,[95] and presumably, de Parcieux's and Polhem's experiments would fit into the same category. But to term the work of these engineers an "application" of science conveys an erroneous impression of the relation between science and technology in the mid-eighteenth century.

First, it is clear that the discovery of the superior efficiency of weight was not directly dependent upon the conceptual structure of science. The prevailing theoretical idea was that weight and impulse were equally effective as motive powers, since water spouting under a given head and water falling freely from an identical height had the same velocity. De Parcieux observed that because of this, everyone had been led to think "that in whatever manner we employed a fall of water, whether by its weight, or by its impact ... we should expect only the same effect."[96] And Smeaton, in similar terms, declared: "In reasoning without experiment, one might be led to imagine, that however different the mode of application is; yet that whenever the same quantity of water descends thro' the same perpendicular space, that the natural effective power would be equal.... and consequently, that whatever was the ratio between the power and effect in undershot wheels, the same would obtain in overshot, and indeed in all others."[97] Smeaton later added that prior to his model experiments he "never had doubted" the truth of this doctrine.[98] It would seem, then, that the discovery of the superior efficiency of gravity was in no way a logical deduction from accepted mechanical laws. It was not, in other words, a product of applied science. It was, in de Parcieux's case, the product of mechanical intuition and thought experiment, confirmed by model testing, and in Smeaton's and Polhem's cases the surprise outcome of model testing alone.

Moreover, the fact that Polhem, Smeaton, and de Parcieux experimented cannot be used to argue that their work was an application of science, for experimentation is not the exclusive domain of the scientist. Experiment is something almost inherent in the nature of technological advance. Technicians experimented with methods and processes long before science emerged, and even systematic model experimentation of the kind carried out by Polhem, Smeaton, and de Parcieux was, as we have noted, not foreign to the technological tradition. It had roots which probably predated systematic experimentation as an accepted part of scientific inquiry.

This is not to say that science or theoretical systems had no influence on the work of Smeaton, de Parcieux, and Polhem. Even though their experiments were not applied science, they were clearly influenced by science. For instance, de Parcieux's imaginary double water wheel with its reversible cycle was largely a product of his mechanical intuition. But it was also an idealized model, a perfect machine which did not suffer losses from friction or spillage. The concept of a perfect machine, as Cardwell has pointed out, did not emerge from the technological tradition. It developed from Galileo's early work in mechanics.[99]

Likewise, the model experiments of Polhem, de Parcieux, and Smeaton were influenced by the scientific community and differed in at least one respect from the earlier model work of Sualem, Riquet, and Fontana. The formers' work had three essential components: (1) systematic methods of experimentation, (2) the use of working models, and (3) the application of quantitative measurements to key variables. The first two of these components the work of Polhem, de Parcieux, and Smeaton shared with the earlier experimental work of Sualem, Riquet, and Fontana. But in making quantitative measurements on the key variables involved in the operation of the vertical water wheel, Polhem, de Parcieux, and Smeaton were initiating a new approach in water-power technology. In place of asking traditional technological questions like, Will it work if I build it this way? or, If I change this element, will it work any better? their concerns had become more exacting. They began to ask, instead, questions like, What effect does increasing the speed of this water wheel by 5% have on its power output (measured by the product of a weight and the height to which it is lifted)? In a loose sense, what Polhem, de Parcieux, and Smeaton did was apply to the problems of practical water-power technology certain aspects of the quantitative methodology that the scientific community had found to be fruitful and which it had begun to use on a regular basis only a century or so earlier. The use of quantitative measurements allowed Polhem, de Parcieux, and Smeaton to give more precision to and extract more information from such traditional technological tools as mechanical intuition, physical analogy, and model experiment.

The particular method of experimentation used by Smeaton in particular, but also to a lesser extent by Polhem and de Parcieux, has been termed "parameter variation" by Walter Vincenti. What this means is that Smeaton's, Polhem's and de Parcieux's experiments were designed to measure the performance of water wheels while systematically varying the parameters—head, quantity of water acting on the water wheel, speed of the water wheel, load on the wheel, blade inclination, number of blades—which defined the primary operating conditions. Their experiments were among the earliest and most prominent examples of this particular engineering methodology. And the success they (and particularly Smeaton) achieved using parameter variation no doubt contributed to its later widespread adoption and use by the engineering community.[100]

Finally, it is probably significant that Polhem, de Parcieux, and Smeaton were all practicing engineers, as well as members of their coun-

TABLE 4–2. COMPARISON OF PARENT'S THEORETICAL PREDICTIONS WITH THE RESULTS OF SMEATON'S EXPERIMENTS

| | Parent (all water wheels) | Smeaton | |
| --- | --- | --- | --- |
| | | Undershot | Overshot |
| Optimum velocity of water wheel ($V$ = velocity of the water) | 0.33 $V$ | 0.40$V$ with indications that it would be higher for full-size wheels | |
| Optimum load on a water wheel ($P$ = load sustained when wheel is motionless) | 0.44 $P$ | 0.80 to 1.25 $P$ | |
| Maximum efficiency of water wheel | 15% | 33% with evidence that it would be higher for full-size wheels; 50% possibly the theoretical maximum | 67% with indications that 100% would be the theoretical maximum efficiency |

tries' leading scientific institution. This combination put them in a position to draw together elements from practical technology and theoretical science in considering problems like the performance of vertical water wheels and to go beyond their contemporaries in both the scientific and technological communities in solving these problems.

## THE RECONCILIATION OF THEORY AND EXPERIMENT: EULER AND BORDA

The quantitative data generated by the experimental work of the 1750s demonstrated very clearly the deficiencies of the commonly accepted theory of water wheels (Parent's), as Table 4–2 indicates. This discrepancy between Parent's predictions and the experimental findings of Smeaton, de Parcieux, and, later, Bossut and others was the central dilemma which faced those attempting to transform the technology of the vertical water wheel from a craft to an engineering science in the second half of the eighteenth century. Between 1750 and 1800 at least a half dozen theoreticians from both the scientific and engineering communities tried to bridge the gap between theoretical predictions and experiment. The two earliest attempts—by Johann Euler in 1754 and by Jean Charles Borda in 1767—came closest to success.

Johann Albrecht Euler (1734–1800) was the son and close associate of Leonhard Euler, Europe's preeminent mathematician. His career has been almost completely obscured by his father's, but Johann Euler was a competent mathematician and scientist in his own right. He served as perpetual secretary to the Academy of Sciences in St. Petersburg from 1766 to 1800 and won several prize competitions sponsored by European scientific societies in the late eighteenth century.[101] Among his victories was a competition sponsored in 1754 by the Royal Society of Göttingen on the question, "What is the most advantageous manner of employing the force of water or of any other fluid either in order to turn mills, or in order to produce some other effect?"[102] His prize-winning essay dealt with the impulse wheel, the gravity wheel, and the horizontal reaction wheel.

Johann Euler's analysis of the impulse wheel drew on Daniel Bernoulli's earlier work. Like Bernoulli, Euler suggested that under certain conditions the Parentian standard for measuring the impulse against a plane surface ($dAH$) was incorrect. If a jet of water struck a water wheel whose blades were much larger than the jet, then the correct measure was double that usually adopted, or $2dAH$. With this factor the figures for optimum velocity, optimum load, and maximum efficiency for water wheels would be $\frac{1}{3} V$, $\frac{8}{9} P$, and $\frac{8}{27} PV$, instead of Parent's $\frac{1}{3} V$, $\frac{4}{9} P$, and $\frac{4}{27} PV$. Euler's results in the area of optimum load and maximum efficiency corresponded much more closely to Smeaton's experimental results for undershot wheels (published five years later) than Parent's did. But, Euler declared, in most water wheels the stream was either much larger than the blades or the blades were only slightly larger than the stream. Thus for most wheels Euler erroneously conceded the validity of Parent's theory.[103]

Because of his reluctance to apply the doubled impulse measure to impulse wheels generally, Johann Euler's analysis of the undershot wheel did little to reconcile theory with practice. But, in dealing with overshot or gravity wheels he did move clearly beyond previous theoreticians. He saw that these wheels operated on a different principle than impulse wheels and detected their superior efficiency. In contrast to the experimentally oriented styles of de Parcieux and Smeaton, Euler's discovery emerged directly out of the conceptual structure of eighteenth-century rational mechanics. It was dependent, not on practical intuition, physical analogies, or model experiments, but on Bernoulli's application of the principle of vis viva to theoretical hydrodynamics and Leonhard Euler's mathematical treatment of the Segner, or Barker, wheel.

The principle of vis viva had emerged in the late seventeenth century. The dominate method of measuring the force of bodies in motion in the scientific community had been by the product of their mass and velocity, or $mv$. In 1686, however, Leibnitz suggested an alternative measure—the product of a body's weight (later mass) and the square of its velocity, $mv^2$—believing that this quantity (called vis viva, or living force) was more universally conserved. Since $v^2$ is proportional to $H$, however, $mv^2$ was often altered to $mgH$, or the product of a weight and a height.[104]

In 1738 Daniel Bernoulli adopted vis viva instead of the Newtonian mass × velocity as the basic measure of the force of moving water and made the conservation of vis viva fundamental in his hydraulic investigations. In outlining basic principles he declared: "The primary one is the conservation of live forces [vis viva], or, as I say, *the equality between actual descent and potential ascent.*"[105] Bernoulli's application of vis viva to hydro-dynamics contained elements which were potentially capable of challenging certain of Parent's assumptions. For example, the velocity factor in the Parentian measure of the natural effect of a stream ($PV$) tended to center attention on the impacting force of water and the undershot-impulse prototype. Bernoulli's use of the product of weight and height (which he called "absolute potential"), on the other hand, disassociated velocity from direct consideration and had the potential of directing attention away from the impulse prototype and more towards the gravity wheel, where one had a weight slowly descending from a height. Bernoulli claimed that the application of vis viva to hydrodynamics implied the equality of actual descent and potential ascent. This, in turn, suggested that some hydraulic engines should have a theoretical efficiency of 100% and posed the question of why, by Parent's analysis, only 15% was possible.

Daniel Bernoulli's use of vis viva in theoretical hydrodynamics was one of the elements which may have led Johann Euler towards a recognition of the superior efficiency of the overshot-gravity wheel. Another was his father's theoretical work with the Segner, or Barker, wheel. Leonhard Euler developed an interest in the horizontal reaction water wheel in 1750, shortly after a German savant, Andreas Segner, reinvented it.[106] It had, of course, been built earlier by Barker and tested by Desaguliers in England, apparently unknown to Segner. Euler was fascinated by the device, and in a series of papers written between 1750 and 1754 he analyzed it, redesigned it, and compared it to the vertical water wheel.[107]

Leonhard Euler's analyses of the vertical water wheel in the papers comparing it to the Segner wheel contained little that was novel. They varied little from Parent's.[108] But in even the earliest of his papers, Leonhard established that the reaction wheel's theoretical efficiency was far above the 1/27 traditionally accorded to all water wheels, approaching 100% under ideal conditions.[109] These findings could only have thrown serious doubts on the universal applicability of Parent's analysis and probably provided another of the elements which prepared Johann Euler for his recognition of the overshot wheel's distinctiveness.

Another possible source of Johann Euler's recognition of the distinctiveness of the gravity wheel was the work of the Swedish engineers Christopher Polhem and Pehr Elvius. As already noted, some Swedish engineers had recognized earlier than technicians in southern and central Europe the greater efficiency of the gravity water wheel. Polhem, for instance, declared in 1742 in an article published in the transactions of the Royal Swedish Academy of Sciences that gravity wheels theoretically developed twice the effect of impact wheels and that in practice the figure was more in the neighborhood of three to four times more.[110] These statements were probably based in part on the experiments he had carried out between 1700 and 1705, which have already been described. Also in 1742, Elvius, a close friend and associate of

Polhem, published an extensive book on water wheel theory which reached the same general conclusions.[111]

Euler may have known of the work of Elvius or Polhem in spite of the language barrier, since Abraham Kästner, who was translating into German the transactions of the Swedish Academy of Sciences, published the 1742 volume in Hamburg in 1750. This volume contained, in addition to Polhem's article, a long summary of the contents of Elvius's book.[112] But while it is possible that Johann Euler was aware of the Swedes, it is unlikely. Kästner admitted in 1797 that Elvius's work was not widely known in Germany and belatedly attempted to remedy the situation.[113] Kästner's translation of the Swedish transactions likewise did not bring much attention to the Swedish work on water wheel analysis, at least outside Germany. No one in England or France prior to 1790, as far as I have been able to determine, knew of either Elvius's or Polhem's contributions to water wheel analysis.[114] It is therefore likely that Euler's recognition of the superior efficiency of the gravity wheel was a deduction from Bernoulli's and his father's work rather than the result of Swedish work. His demonstration of the theory of the gravity wheel was certainly different. Polhem's recognition of the superior efficiency of the overshot wheel was probably based on experiment; Elvius's demonstration was lengthy, largely synthetic, and, according to Kästner, extremely difficult to follow.[115] Johann Euler's demonstration, in the tradition of theoretical mechanics established by Bernoulli and his father, was analytical and more directly to the point.

In analyzing the gravity wheel in his Göttingen Society prize essay, Johann Euler measured the natural effect (he called it the *momentum impulsionis*) of a stream by the product of the weight of the water ($K$) and its fall ($H$). Considering the moment of the water contained in the buckets on the loaded arc of an overshot wheel, he recognized from Galilean mechanics that its effective force would be identical regardless of whether the wheel was at rest or in uniform motion. He saw, then, that if the buckets were large enough to accommodate all the water ($K$) impinging on them, and that if the height of water fell while contained in the buckets was equal to the total fall ($H$), the overshot wheel's effect or output would be $KH$, the same as the effect or output of the stream which powered it. Its efficiency would be 100%. Since $KH$ was the theoretical maximum attainable from overshot wheels, while $\frac{8}{27} KH$ was the top impulse wheels could achieve, even under the most ideal conditions, the advantage of gravity was clear.

Euler's analysis also cleared up several of the problems that had plagued de Parcieux and Smeaton. Neither had been certain exactly why slow overshot water wheels were more efficient than fast ones. Both suggested that perhaps fast wheels moved away from the weight of the water in their buckets, preventing it from pressing as well. Euler correctly pointed out that water on leaving the buckets of an overshot wheel retained the velocity of the wheel and that this represented a loss in head, or fall, which could be calculated by the equation $h_w = v^2/2g$, where $h_w$ was the loss of head due to the wheel's velocity, $v$. Thus the useful gravity head of an overshot wheel would be the diameter of the water wheel ($H_g$) less the height ($h_w$) lost due to the wheel's velocity. The theoretical effect of an overshot wheel was, then, $K(H_g - h_w)$. Obviously the smaller the value of $h_w$, the head due to the velocity of the wheel (and the exiting water), the greater the output of the wheel.

For overshot wheels, where some portion of the available head was used to give impact to the water wheel, Euler combined his equations for the maximum effect of pure impact and pure gravity wheels.[116] For the maximum effect of an overshot water wheel his equation was $K(H_g - h_w)$ + $\frac{8}{27} Kh$. Here, $K$ = the weight of water acting on the wheel per unit time; $H_g$ = the diameter of the water wheel or the portion of the fall where the wheel was powered by weight; $h_w$ = the portion of the gravity head lost due to the velocity of the wheel; and $h$ = the distance the water fell before it struck the wheel.

Although Euler's analysis answered some questions left open by de Parcieux's and Smeaton's experiments, it seems, at first, to have received little attention. Smeaton apparently did not know of it when he published his experimental results in 1759, or even in the 1770s and 1780s when he attempted to explain the discrepancies between his experiments and theory. De Parcieux, who was aware of Euler's prize essay and mentioned it in 1759, did not notice his overshot wheel analysis, commenting only on Euler's advocacy of the reaction wheel.[117] Perhaps because Euler's prize essay was written in Latin and published by a small central European scientific society, it did not attract the same notice as the publications of the larger scientific bodies. Perhaps also the mathematical language that Euler used and the absence of any experimental data to back up his theoretical findings contributed to the neglect. Perhaps, finally, Euler's enthusiasm for the horizontal reaction wheel obscured his contribution to vertical wheel theory. Around 1770, however, Charles Bossut used Euler's analysis of the overshot wheel in his hydrodynamics textbook,[118] and it was noted toward the end of the century by several German hydraulicians.[119]

While significantly more advanced than any previous theoretical analysis of the vertical water wheel, Johann Euler's work did have several shortcomings. His correction was only a partial one and did not completely bridge the gap between experiment and theory, since he did not completely break with Parent's undershot wheel analysis. His treatment of the overshot wheel was a major breakthrough. It produced a partial explanation of Smeaton's and de Parcieux's experimental discoveries, and it did indicate that the overshot wheel was much more efficient than the undershot. But it did not clearly indicate why this was so. The 1767 mathematical analysis of Borda did.

Jean Charles Borda (1733–99) was a French military engineer. His early theoretical work dealt with projectile motion, and this work secured him admission in 1756 to the Academy of Sciences in Paris.[120] Around 1760 Borda developed an interest in hydraulics and began a series of experiments to test the validity of the accepted laws of fluid resistance. While pursuing these studies in 1766 he experimented on the immersion of perforated vessels. The experiments convinced him that, contrary to Bernoulli, vis viva ($mv^2$, or living force) was not always conserved in hydraulics. When water flowed through either abruptly constricted or abruptly enlarged tubes or was brought sharply into contact with other water or some other body, he found, vis viva was lost, and Borda established this loss as equivalent to $K[(V-v)^2/2g]$, where $K$ is the weight of the water involved, $V$ its velocity before obstruction, and $v$ its velocity after.[121]

Since the blades of water wheels abruptly impeded the motion or flow of water, they were an obvious application of this discovery. In 1767, the year after Borda first postulated a loss of energy in certain hydraulic phenomena, he presented a short but very important paper applying this hypothesis to water wheels (both horizontal and vertical).[122] In order to soften the impact of his new approach, however, he first analyzed in the paper water wheels in traditional terms, using the product of mass or weight and velocity to measure the potential power of the water and the power output or effect of the water wheels. Only in the second half of his paper did he attempt to show that identical results could be obtained by calculating the amount of vis viva lost by water in acting on water wheels.

In his more conservative analysis of the undershot wheel Borda corrected several of Parent's basic assumptions. For example, he contended that the force of impact of water against the blades of an undershot wheel in a confined channel was proportional to $V-v$, instead of Parent's $(V-v)^2$. He explained:

> The action of the water is not exercised against an isolated blade, but against several blades at the same time, and . . . these blades close off all the passages of the small channel, and take from the fluid the velocity it has above them. Thus the quantity of movement lost by this fluid, and consequently the impact that the blades experience, is no longer proportional to the square of the difference of the velocities of the fluid and blades, but only to the difference of these velocities.[123]

Further, Borda measured the moving power of the water by the product of the quantity of water acting on the wheel ($K$) and the height ($H$) it fell, rather than by the product of force of impact ($P$) and velocity ($V$). Borda's $KH$ had a value equivalent to one half of Parent's $PV$ and was a more valid measure of the "natural effect" of a stream. These modifications yielded theoretical predictions much closer to experiment than Parent's:

$v = 0.5\ V$, for the optimum velocity of undershot wheels;
$p = P$, for the optimum load for undershot wheels;

$0.50$ = the maximum efficiency of undershot wheels ($kh$ = $0.50$ $KH$).[124]

Borda's initial analysis of the overshot wheel in his 1767 paper was similar to that of Euler. He too found that the theoretical maximum efficiency of a pure gravity wheel was 100% and that the slower these wheels moved the closer they could approach that figure. Borda differed with Euler only in dealing with overshot wheels subjected to some impact. Euler used his modification of Parent's theory to account for the power transmitted to the wheel by an impacting stream, while Borda applied his new impulse wheel analysis. Thus, for an overshot wheel moving at a velocity due to the height ($h_w$) and struck by water moving at a velocity due to height ($h$), where $H_g$ is the distance that water falls in the water wheel, their equations for the maximum effect of the overshot wheel were

$$\text{Euler: } K(H_g - h_w) + \tfrac{8}{27} Kh$$
$$\text{Borda: } K(H_g - h_w) + \tfrac{1}{2} Kh.[125]$$

Borda's work in the first half of his 1767 paper, while in some ways conflicting with the analyses of Parent and his successors, was compatible in other ways. Like them Borda measured the potential of a stream by its momentum ($mv$) and assumed that momentum was conserved as water passed through a water wheel (that the momentum imparted to the wheel plus that remaining to the water after it had left the wheel were equal to the momentum of the incoming water). There was nothing wrong with this approach. As Borda demonstrated, it could be made to yield results which conformed moderately well with experiment. But, as Cardwell has pointed out, the momentum method yielded no insights; it did not indicate why, for instance, undershot wheels were less efficient than breast and overshot wheels.[126]

Borda apparently recognized this deficiency and in the second and more important portion of his 1767 paper remedied the matter by analyzing water wheels in terms of vis viva ($mv^2$) instead of momentum. As we have seen, Borda had discovered in 1766 that vis viva was not always conserved (as Bernoulli had believed) in hydrodynamic systems, but was destroyed or apparently destroyed whenever water was subjected to sudden changes. Borda used this discovery now to investigate what was lost when water acted on a water wheel instead of what was conserved (the traditional method). In looking at water wheels in this manner, he recognized that there were two major points at which vis viva was lost. When water moving at a certain velocity, $V$, entered a water wheel moving at a slower speed, $v$, it was abruptly stopped by the buckets or floats. This produced a loss which could be measured by the formula he had established a year earlier as $K[(V-v)^2/2g]$, or $\tfrac{1}{2} m(V-v)^2$. When the water left the wheel, it retained a velocity, $v$, equal to that of the water wheel. This,

Borda saw, also represented a loss of power which could be measured by the factor $Kv^2/2g$ or $\frac{1}{2}\,mv^2$. Borda's basic equation for the output of a water wheel was therefore (in a somewhat modernized form):

$$pv = mgH_g + \tfrac{1}{2}\,mV^2 - \tfrac{1}{2}\,m(V-v)^2 - \tfrac{1}{2}\,mv^2.$$

In this equation $pv$ is the wheel's output measured as the product of a weight and its velocity (assuming that no gearing is involved and that the weight is being lifted at the same velocity as the wheel's center of impact). $H_g$ is the gravity head, or the height that water falls in a water wheel acting by its weight; $V$ is the velocity of the water as it impacts the wheel; and $v$ is the velocity of the water wheel. Thus the first term in the above equation represented the vis viva imparted to the wheel by the action of gravity; the second term is the vis viva imparted to the wheel by the action of impulse; the third term is the vis viva lost because of the sudden impact of the water against the wheel; and the fourth term is the vis viva lost as a result of the velocity remaining to the water on leaving the wheel. This equation can be reduced to the form

$$pv = KH_g + Kh - Kh_i - Kh_w.$$

Here, $K$ is the weight of the water involved. $H_g$ and $h$ represent, respectively, those portions of a stream's fall where the water acts on the water wheel by gravity (or weight) and by impact, so that the total head or fall ($H$) is equal to $H_g + h$. In an undershot wheel, of course, the gravity head would be 0, and the first term could be eliminated. The term $h_i$ is the head due to the velocity of impact of the water against the wheel $(V-v)$; $h_w$ is the head due to the velocity of the wheel $(v)$ and thus represents the head lost because of the velocity retained by the water on leaving the wheel. Borda demonstrated that equations like these could be reduced by normal algebraic means to forms identical to those derived using the momentum method.[127]

Borda's analysis of the vertical water wheel in terms of vis viva was a major contribution to the rationalization of water power. In conjunction with his more conventional analysis, it substantially closed the gap between theory and experiment opened by Smeaton and de Parcieux in the 1750s. In addition, it answered some critical questions. Smeaton and de Parcieux had demonstrated that gravity wheels (i.e., overshot wheels) were substantially more efficient than impulse wheels and raised the question of why this was so. Traditional means of analyzing the water wheel provided neither conclusive answers nor reliable suggestions for design improvement. Borda's analysis pointed to the impact of the water against the blades of the undershot wheel, as well as to the high speed of the water leaving the wheel, as the causes for its poorer performance. It indicated that to approach 100% efficiency water wheels had to be designed so that water entered without impact and left without velocity. And Borda's equations provided technologists with a superior method for calculating the output of water wheels when impact and some exit velocity were unavoidable, as was invariably the case in practice.

Borda's work was extended by another French engineer, Lazare Carnot (1753–1823). In 1782 Carnot took the principle of vis viva loss which Borda had applied only to certain special cases, like water wheels, and generalized it to all machines. He also made explicit the intuitions of Borda and Smeaton on the necessity of avoiding impact in water wheels:

> We can conclude . . . that the means of producing the greatest effect in a hydraulic machine moved by a current of water, is not to apply a wheel whose blades receive the impact of the fluid. In fact, two good reasons prevent us from producing the greatest effects in this way: the first is . . . because it is essential to avoid all percussion whatever; the second is, because after the impact of the fluid there is still a velocity which it retains and it is a pure loss, since we should be able to employ this remainder to produce a new effect to be added to the first. In order to make the most perfect hydraulic machine—that is to say, one capable of producing the greatest possible effect—the true knot of difficulty lies therefore, first, in making it so that the fluid loses absolutely all its movement by its action upon the machine or at least that it retains only precisely the quantity necessary for escaping after its action; second, in making it lose all of its movement by insensible degrees, and without percussion, either on the part of the fluid, or on the part of the solid parts among themselves: the form of the machine would be of little consequence; for a hydraulic machine which will fulfill these two conditions will always produce the greatest effect.[128]

The criteria for maximum efficiency implicit in Borda's work (and in the work of Smeaton and de Parcieux as well) and made explicit here by Carnot eventually became the cardinal principle guiding the development of all water-powered prime movers: "For maximum effect the water must act on the wheel without impact and leave it without velocity."

Although the work of Borda, supplemented by Carnot, provided reliable principles to guide in water wheel design, it was neither widely known nor widely accepted for several decades. Much of this was due to a reluctance on the part of European theoretical physicists to depart from conservation principles and accept the idea that something (vis viva) could be lost, without compensation, in impact. For example, Joseph Louis LaGrange (1736–1813), one of Europe's foremost mathematicians, wrote to d'Alembert in 1771, responding to Borda's 1766 paper on fluid flow which first postulated the loss of vis viva: "I find [it] little worthy of him. . . . Do you not find very pitiful the reasons by which he pretends to prove that there is always a loss of living forces, etc.?"[129] So put out was LaGrange with Borda's theory that when he reviewed the literature on the problem of flow from orifices in his *Méchanique analytique* in 1788 he not only refused to admit the loss of vis viva in fluid masses; he did not even mention Borda's 1766 paper.[130]

Responding more directly to Borda was Charles-Louis DuCrest (1747–1824), who published a work on watermills in the 1780s. He complained that Borda had violated the "fundamental principle of all the sciences"—"the effect is equal to the cause." On this basis alone, he declared, Borda's work could be judged erroneous. No matter how logical a demonstration appeared, if it led to conclusions which indicated the inequality of cause and effect, one could assume that it was "very certainly false."[131] Conservation laws had been a foundation of theoretical physics in Europe for several centuries. It was apparently very difficult to concede that they were not always valid.[132]

Another factor behind the failure of Borda's work to gain acceptance was Borda himself. According to his biographer, Muscart, Borda was modest and did not worry about popularity. He was also lazy about writing and used the *Mémoires* of the Academy of Sciences as the exclusive outlet for his work. Finally, other activities diverted his attention from water power. In 1767, the very year he read his classic paper on water wheels to the Academy, he joined the French navy and devoted most of the remainder of his career to attempts to reform that service.[133]

When Borda died in 1799, his method of analyzing water wheels had yet to gain wide acceptance. His eulogist, Lefèvre-Gineau, scarcely mentioned the key 1767 paper on water wheels in his summary of Borda's accomplishments.[134] And in 1802 when Montucla reviewed the available literature on the theoretical analysis of water wheels, he did not refer to Borda's paper at all.[135]

Carnot's development of Borda's ideas did not significantly relieve this neglect, for Carnot's work suffered much the same fate. After initial publication in 1782, Carnot's *Essai sur les machines en général* faded into obscurity. The *Essai* began to gain recognition only after Carnot achieved political prominence as minister of defense in the French revolutionary government in the 1790s.[136] LaGrange, for example, cited it in his *Théorie des fonctions analytiques* in 1797, where he admitted that the hypothesis of vis viva loss was of importance to the theory of machines.[137] But as late as the second edition of the *Théorie* in 1811 he refused to concede the applicability of the concept to fluid bodies.

A new, enlarged edition of the 1782 *Essai* was published by Carnot in 1803 and fared little better. According to C. C. Gillispie, Carnot's most recent biographer, the new edition, titled *Principes fondamentaux le l'équilibre et motion* "fell into the same obscurity, lasting another fifteen years." "The book was occasionally mentioned prior to 1818, but rather by way of noticing its existence than because its point of view effected the treatment of problems."[138]

## THE ERA OF THEORETICAL CONFUSION (c1770–c1810)

Because Borda's definitive 1767 analysis of the vertical water wheel was neither widely known nor widely accepted during the eighteenth century, the reconciliation of theory with practice remained an important preoccupation of hydraulicians long after 1767. Between 1770 and 1810 a number of other theoretical analyses of undershot and overshot wheels circulated in Europe and America, none enjoying widespread acceptance, but all contributing to a general theoretical confusion.

In the absence of a widely accepted alternative, even Parent's outdated analysis continued to enjoy widespread acceptance, particularly in Britain. It was used in Petrus van Musschenbroek's physics text of 1769 and by Johann Lambert in a paper on water wheels delivered in 1775 to the Berlin Academy of Sciences.[139] Parent's one-third velocity maxim was used by James Ferguson in several of his books and in his popular "Mill-Wright's Table," which was reprinted in Chambers's *Cyclopaedia* (1781) and in the third edition of the *Encyclopaedia Britannica* (1797).[140] So widely accepted was Parent's theory in Britain that James Watt used it in 1784 to analyze Kempelen's proposal for an impulse steam turbine,[141] and as late as the 1820s and 1830s it was still appearing in British textbooks.[142]

Although Parent's analysis lingered on between 1750 and 1830, there were attempts other than Euler's and Borda's to derive a new theory. These attempts usually involved a denial of one of Parent's basic postulates or an attempt to avoid the assumptions Parent had made to simplify his calculations. The two basic postulates of Parent's theory most suspect were:

1. the impulse against a flat surface of area $A$ struck perpendicularly can be measured by the formula $P = dAH$; and
2. the impulse of water on a moving float or blade is proportional to the square of the velocity of impact, $(V - v)^2$.

Among the assumptions which Parent had made to simplify his analysis were:

1. the velocity of a stream of water is the same at all depths;
2. the velocity of all points on a submerged blade (and hence the velocity of impact on all of these points) is the same;
3. the impact against the blades on a water wheel can be approximated by assuming that only one blade is immersed at any given time and that the blade always receives the impact of water perpendicularly.

Even before Smeaton's and de Parcieux's experiments had demonstrated the deficiencies of Parent's theory, d'Alembert had pointed to the assumptions used by theoreticians to simplify water wheel calculations as a possible problem, and several attempts had been made to avoid some of Parent's simplifications and assumptions.[143] In 1746 DuPetit-Vandin and in 1767 Jacques André Mallet, for instance, tried to analyze water wheels either by considering the different velocities of the different points on the blades or by assuming that the velocity of flowing water varied with depth.[144] But their work yielded little clear direction for those seeking to apply theory to practice.

The culmination of attempts to retain Parent's basic postulates but eliminate the assumptions he had made to simplify calculations came with Charles Bossut. In a paper published in 1769, as well as in the various editions of his *Hydrodynamique* (1st ed., 1771; 3rd ed., 1796), Bossut attempted to derive a general equation which would take into account many of the major variables affecting the performance of impulse wheels and ignored by Parent, among them the radius of the wheel, the size of the blades, the proportion of the radius immersed, the different velocities of the impacting stream at different depths, the different velocities of various points on the wheel's blades, the number of blades, the inclination of the blades, and the varying angles at which the water impacted the blades.[145] The inclusion of all these factors made Bossut's calculations long and tortuous, without yielding much information of practical value. For example, when attempting to establish an equation for the moment of the water acting on a water wheel, Bossut confessed: "The equation that gives the maximum of the general moment is extremely complicated and almost beyond treatment."[146] Bossut's general equation, which included all of the factors involved in the operation of a water wheel mentioned above, was more than a page long, and again Bossut admitted: "It is not easy to find directly, by our general formula, the most advantageous number of blades for a wheel that turns with a finite velocity relative to that of a fluid, since the equation for the maximum is extremely complicated and almost untreatable."[147]

Thus Bossut's attempt to reconcile theory with practice by eliminating Parent's simplifications was a failure. It yielded no practical design data. The attempt choked on the complexity of the equations it yielded. In the end DuPetit-Vandin, Mallet, and Bossut were all to turn to methods of approximation to simplify their equations and thus found themselves using the same arbitrary methods they had set out to avoid.[148]

Somewhat more successful were several attempts to reconcile theory with practice by rejecting one of Parent's basic postulates. For instance, the Parentian measure for the impulse of water against a plane ($P = dAH$) was under suspicion after Daniel Bernoulli's work in 1738, if not before. Both Bernoulli and Johann Euler had suggested that a double measure, $P = 2dAH$, might be applicable in some cases, particularly where the blades of a water wheel were much larger than the impacting stream. Bossut in 1769, in a brief precursory analysis of a horizontal impulse wheel, applied this doubled measure.[149] In 1771, in the first edition of his *Hydrodynamique,* he extended it to undershot impulse wheels generally.[150] Experimental work carried out in the mid-1770s by a committee of the Academy of Sciences for the French government which involved towing objects across a large pool and in a narrow canal confirmed Bossut's insight. While the resistance of a plane to motion in the broad pool could be measured by $P = dAH$, in the narrow channel it was equal to $2dAH$.[151] For undershot wheels this modification yielded figures for optimum load ($\frac{8}{9}$ P) and maximum effect ($\frac{8}{27}$ PV) much closer to Smeaton's experiments.[152]

Not only did the basic measure Parent used for determining impulse come under challenge between 1770 and 1810; so did his assumption that the force of impact against a blade varied as the square of the difference between the velocity of water and that of the wheel, $(V-v)^2$.

At about the same time that Bossut incorporated the revised impulse measure into his analysis, a Spanish scientist, Jorge Juan y Santacilia (1712–74) suggested that the force of impact varied, not as the square of the velocity of impact, $(V-v)^2$, but simply as the velocity $V-v$. Juan was concerned primarily with the application of this new view of impulse to ship construction, but peripherally noted that its application to water wheels indicated that the optimum velocity of rotation would be 0.50 V, instead of the traditional Parentian 0.33 V. Juan did not pursue his analysis to the related questions of optimum load and maximum efficiency.[153]

But the same approach was adopted in 1792 by a little-known teacher at the Friends' Academy in Philadelphia, William Waring, apparently in ignorance of both Juan's and Borda's work. Waring pushed the analysis further than Juan, finding not only the optimum velocity ($v = 0.50$ V), but also the optimum load ($p = 0.50$ P). Although Waring did not compute the maximum effect of a water wheel operating under the conditions assumed in his analysis, his figures indicated that it would be 0.25 PV, or an efficiency of 0.25.[154]

Also rejecting Parent's assumption that the force of impact transmitted to a wheel was proportional to $(V-v)^2$ was Oliver Evans (1755–1819), an American millwright and inventor. In 1795 Evans attempted an analysis based on the assumption that the force transmitted to a water wheel was proportional to $V^2-v^2$. This yielded an optimum velocity of 0.58 V and a maximum efficiency of 38% for undershot wheels. The maximum efficiency of overshot wheels he calculated by an obscure method to be 76%. Evans believed that the results of his theory corresponded better to actual practice than any previous attempt.[155]

Finally, there was a rather confused analysis of the overshot wheel made around 1800 by the Scottish engineer Robertson Buchanan (1770–1816). Buchanan did not distinguish between dynamic concepts like "effect" and "power" and static concepts like "force" and "moment," and he focused on the distance water traveled on the circular arc of the overshot wheel, instead of on the vertical distance it fell. Thus his analysis yielded results in sharp contrast to the theoretical findings of Johann Euler and Borda, as well as to the experimental findings of Smeaton. Buchanan found, for example, that a water wheel produced its greatest effect only when water was laid on the wheel at a point 52.75° from its summit. When water was applied at the summit, Buchanan contended,

TABLE 4–3. COMPARISON OF THE PREDICTIONS OF VARIOUS THEORETICAL ANALYSES OF THE VERTICAL WATER WHEEL, c1700–c1800

| Investigator, date, and wheel type | Wheel's optimum velocity (V) | Wheel's optimum load (P) | Maximum efficiency (%) |
|---|---|---|---|
| Parent, 1704 | | | |
|   All water wheels | 0.33 | 0.44 | 15 |
| Euler, 1754 | | | |
|   Undershot wheels | 0.33 | 0.44–0.89 | 15–30 |
|   Overshot wheels | a.s.a.p. | | 100 |
| Borda, 1767 | | | |
|   Undershot wheels | 0.50 | 1.00 | 50 |
|   Overshot wheels | a.s.a.p. | | 100 |
| Bossut, 1769, 1771 | | | |
|   Undershot wheels | 0.33 | 0.89 | 30* |
| Juan, 1771; Waring, 1792 | | | |
|   Undershot wheels | 0.50 | 0.50 | 25 |
| Evans, 1795 | | | |
|   Undershot wheels | 0.58 | 0.67 | 38 |
|   Overshot wheels | 0.58 | | 67 |
| Buchanan, 1801 | | | |
|   Overshot wheels | | | 59 |

NOTE: $V$ = velocity of the water; $P$ = load which would hold the wheel stationary; a.s.a.p. = as slow as practically possible.
*Undershot wheels in a confined channel; figure lower by 50% if in open channel.

its theoretical efficiency was only 0.50, compared to 0.59 when applied at the optimum point.[156]

As Table 4–3 indicates, there were more than half a dozen analyses of water wheels in circulation by the late eighteenth or early nineteenth centuries. None was universally accepted, and the confusion generated by this situation created a mood of pessimism and, in some circles, a total rejection of the theoretical, mathematical approach to the problems of water-power technology. This attitude was particularly prevalent in Britain.

British science from the time of Newton had been predominantly experimental and generally nontheoretical, if not openly antitheoretical. In this atmosphere it is not surprising that some British savants were openly sceptical of the value of mathematical analysis in water-power technology. For example, Isaac Milner (1750–1820) of Cambridge argued that writers who published water wheel analyses really had no intention of making any useful improvements in practice. They were simply illustrating the use of algebra or the calculus. Too many arbitrary assumptions were used in such calculations for them ever to correspond with experimental tests.[157] Another British scientist, George Atwood (1746–1807), noted: "It is vain to attempt the application of the theory of mechanics to the motion of bodies, except every cause which can sensibly influence the moving power and the resistance to motion be taken into account: if any of these be omitted, error and inconsistency in the calculations deduced must be the consequence."[158] Atwood argued that theory had, therefore, never been of much value to the practical arts generally and to water power in particular.[159] In effect, Atwood and Milner conceded that the difficulties involved in reconciling theory with practice were too great to be overcome, and they all but conceded that theory would never be of use in water-power technology. The theoretical confusion which was in part responsible for this attitude, however, was to be ended in the early nineteenth century as Borda's work became widely known and accepted.

## THE CONFUSION ENDS: ACCEPTANCE OF THE BORDAN ANALYSIS (c1810–c1850)

Although a number of theories accounting for the operation of overshot and undershot vertical wheels were still in circulation around 1800, Borda's 1767 analysis began to emerge very early in the nineteenth century as the generally accepted basis for the rationalization of water-power technology. After almost 30 years of neglect, Borda's theory was mentioned in 1808 in Lanz and Bétancourt's popular textbook *Essai sur les composition des machines*.[160] Two years later, in 1810, Pierre-Simon Girard, a French engineer, translated Smeaton's "Experimental Enquiry" into French. In a long introduction Girard discussed the development of theoretical and experimental doctrines on vertical water wheels, giving Borda's work a prominent place and noting that Borda's 1767 paper seemed to contain "a complete solution to the question."[161] Also in 1810 Antoine Guenyveau, in his *Essai sur la science des machines*, adopted Borda's revisions to Parent's theory of undershot wheels and presented an analysis of overshot wheels similar to Borda's, though he continued to insist that momentum or quantity of movement ($mv$) was the "most natural" measure for the force of bodies in motion and used it instead of Borda's "living force" ($mv^2$ or $mgH$).[162] In 1818 a promising young physicist, Alexis-Thérèse Petit, published a demonstration of how the principle of living forces could be applied to the analysis of various prime movers, including the vertical water wheel.[163] A few months after Petit's paper the French engineer C. L. M. H. Navier reviewed the history of the principle of living forces in machine analysis, spotlighting Borda's contributions; and in 1819 he used Bordan theory in this notes to the new edition of Bélidor's *Architecture hydraulique*.[164]

In Britain diffusion of Borda's ideas was slower than on the Continent. Borda's general conclusions, but not his entire analysis, were brought to the attention of British readers by Thomas Young, David Brewster, and others around 1800.[165] But they encountered serious opposition. For example, Peter Ewart, a practical engineer of considerable eminence, rejected Borda's analysis.[166] And the practical success of Britain's engineers, reviewed in the next chapter, no doubt inhibited interest in things theoretical. As late as the 1830s the conflicting theories of Parent, Waring, Bossut, Evans, and Buchanan still had wide circulation in Britain. It was only after 1840 that Borda's analysis began to assume clear dominance and appear regularly in British engineering papers on water power.[167]

## THEORETICAL ANALYSIS IN THE NINETEENTH CENTURY

As Borda's work gained general acceptance, first on the Continent and then in Britain and America, the emphasis of analytical work shifted from attempts to derive a new theory to attempts to give more precision to Borda's basic equations. Navier around 1820, for example, added an expression to Borda's equations to take into account losses due to the immersion of a breast wheel in water.[168] Around 1830 Coriolis considered several other factors, including the case of a water wheel where the velocity of the water at exit was faster than the velocity of the wheel.[169] Poncelet around 1830 analyzed the losses due to impact when one or two walls of the buckets of an overshot wheel were struck obliquely. He also derived equations for overshot wheels operating at high velocities which included the effects of centrifugal force and spillage.[170] In the 1830s the German engineer Weisbach presented a detailed analysis of spillage losses in overshot wheels.[171] And d'Aubuisson around 1840 attempted to add to Borda's basic relations the losses due to water seeping between the blades and the breast or channel it operated in, losses due to spillage, losses due to centrifugal force, and the reduced gudgeon friction due to the wheel's loss of weight when its lower portion was immersed in a tail race, among other things.[172] The analysis of the French engineer and theoretician F. M. G. Pambour was typical of the type to which vertical water wheels were subject around 1850. To Borda's basic equation Pambour added factors to account for the loss of power due to the play of the wheel in its race, the loss of power due to the additional friction which the load put on a wheel, the loss of power due to air resistance, the loss of power due to the friction of the water wheel's axle, the loss of effect due to the elevation of the center of gravity of the water at the moment it traversed the blades, and the loss of power due to spillage.[173] Similar analyses may be found in the treatises of late-nineteenth-century German engineers like Bach and Müller.[174]

But even while theoreticians were deriving ever more complicated equations and pushing the mathematical analysis of the vertical water wheel to new limits, the engineers who actually designed and erected water wheels were often using a much simpler and more direct approach—the method of coefficients. Essentially, this method involved the use of Borda's basic equations in conjunction with an experimentally derived coefficient designed to bring them into close agreement with practice.

The use of a coefficient in water-power technology to make a theoretical equation accord with experimental results seems to date back to the

1780s. In the second edition of his *Principes d'hydraulique* the French hydraulician DuBuat adopted as the basic equation for the effect of an impulse wheel $pv = AV (V-v)^n$, where $n$ was a coefficient whose value was to be established by experiment. Using Bossut's model experiments, DuBuat found that for undershot wheels operating in a confined channel $n = \frac{5}{4}$, while for wheels operating in open channel it was $\frac{4}{3}$.[175] At about the same time Bossut also used experimental coefficients to modify Parent's basic equation. Bossut's experiments with d'Alembert and Condorcet in the mid-1770s on the resistance of water to a moving plane had indicated that resistance was almost twice as great in a confined channel as in broad stream. He therefore gave as the equation for the maximum effect of an undershot wheel the relation $pv = nA (V-v)^2 v$, where $n$ was an experimental coefficient whose value was either $\frac{1}{2}$ or 1, depending on whether the wheel operated in open stream or confined channel.[176]

The use of coefficients to modify theoretical equations became increasingly popular in engineering treatises in the early nineteenth century, though sometimes their application was implicit rather than explicit. For instance, Navier in 1819 noted that the theoretical maximum effect of an overshot wheel was $pv = 1,000K (H-0.5h)$ kil.-met., but later added that experiment revealed that the real effect was only $\frac{4}{5}$ of this, or $pv = 800K (H-0.5h)$ kil.-met.[177] Hence Navier's second formula could have been expressed as $pv = 1,000nK (H-0.5h)$ kil.-met., where $n = \frac{4}{5}$. Later works, particularly those written by practicing engineers or for practicing engineers, used experimental coefficients explicitly and frequently. For instance, one of the most practically oriented works on hydraulics was Jean d'Aubuisson de Voisins' *Treatise on Hydraulics for the Use of Engineers* (1840), which made regular use of coefficients to bring theoretical equations into agreement with practice.[178]

By the second half of that century the use of experimental coefficients had become so widespread that some theoreticians, like Pambour, were seriously concerned. In an 1872 paper he attacked the "method of coefficients."[179] Pambour complained that inexactness was inherent in coefficient equations, since resistance, friction, and all other losses were taken as a block and expressed by a constant coefficient. Every loss, he argued, depended on different circumstances; losses could not be experssed by a single relationship that was constant and made merely a function of the force of the wheel. Since all losses were included in one figure, it was impossible to study the influence of each on the wheel's performance. Finally, he pointed out, the coefficient method did not allow one to determine if a theory was really valid. Enormous reductions were required to make the predictions of existing theory match experiment: 50% for undershot wheels, 35% for breast wheels, 30% for overshot wheels. Pambour concluded: "It is not possible to admit, without proof, that the construction of these wheels is so imperfect as to cause such losses. The fact should at least need proof; and if the losses are so great, why do we make the calculation without introducing them into it?"[180]

The frequent application of experimental coefficients to adjust theoretical equations, however, was one of the things which distinguished the engineer's analytical methodology from the theoretical scientist's in the nineteenth century. To most engineers mathematical analysis was not an end in itself, but a means to an end. As Morin was to demonstrate (see below), the coefficient method yielded results that were in close agreement with practice, in many cases closer than the lengthy equations of Pambour, Weisbach, and other theoreticians. Since vertical wheel engineers worked in the real world, where no mathematical formulation was likely to yield exact results and where technicians for years had been working with "fudge factors," "safety factors," and the like, an error of a few percentage points in power or efficiency calculations usually made little difference. The exactness demanded by Pambour was unnecessary. The economy in time and effort afforded by coefficient equations and other approximate methods such as graphical solutions[181] was simply too attractive. It is thus no surprise that many of the most widely used engineering handbooks of the nineteenth century placed heavy reliance on coefficient equations, despite the strictures of Pambour.[182]

## EXPERIMENTAL WORK IN THE NINETEENTH CENTURY

Since the coefficients used to correct theoretical formulas were empirical in origin, their increasing popularity placed a premium on reliable experimental data. While the experiments of Smeaton, Bossut, and de Parcieux in the eighteenth century had been moderately extensive, they left significant lacuna. None of these experimenters tested breast wheels, the type that was rapidly displacing the traditional overshot and undershot in popularity by 1800. Also, all, or almost all, eighteenth-century experimentation was carried out on models,[183] and results based on models were regarded with suspicion in many circles. For example, John Sutcliffe, a British millwright, declared in 1816:

> For the last twenty years, gentlemen have gone through the nation, giving lectures upon the application of water . . . who have exhibited small models of water wheels and steam engines, by which they have pretended to shew the best mode of applying the powers of those two great agents of nature. By these paltry models and representations given of them, the public have been greatly misled; and large sums of money been expended to no purpose in making water wheels after such imperfect models. . . .
>
> These are the sad effects of representing mechanical powers by small models; for, as I have before observed, there is no judging of the merit of any design unless the model is as large as the machine it represents is intended to be.[184]

Nonetheless, some work with models was continued in the nineteenth century. For example, Gérard Joseph Christian (b. 1776), director of the

Conservatoire royale des arts et métiers in Paris carried out some tests on an impulse (undershot) wheel 2.09 feet (63.66 cm) in diameter, with blades 1.97 inches (5.0 cm) high by 3.94 in. (10.0 cm) wide. Christian placed this wheel in a canal 7.87 inches (20.0 cm) wide. He found the wheel's efficiency was 22%, its optimum velocity 0.47 that of the impelling water.[185]

In Britain, Smeaton's model-testing tradition was carried on by the Irish engineer Robert Mallet. In the 1840s Mallet tested two small wheels, one 25.5 inches (0.65 m) in diameter, the other 33 inches (0.84 m), focusing on two questions:

1. Was there any advantage to be gained by constructing a gravity wheel higher than the available head and feeding the water onto it at some point below the wheel's summit, or, in other words, did the high breast or pitchback configuration coming into use in this period have any theoretical advantages?
2. Did the use of a closely fitting casing to impede spillage from buckets enhance the performance of an overshot wheel?

Both Smeaton's speculations and subsequent theory (though not practice) had indicated that there was no advantage to be gained by building a water wheel higher than the available head. Mallet found that when the depth of water in the reservoir supplying a wheel fluctuated widely, but was lower than the mean height most of the time, the high breast or pitchback configuration would, on the average, deliver more power. The close-fitting conduit, Mallet discovered, increased the efficiency of an overshot wheel by as much as 8 to 11%, especially when the wheel's peripheral velocity was high (approaching 6 fps [1.83 mps]).[186]

Experiments on two model undershot wheels, one around 3 feet (0.91 m), and the other around 6 feet (1.82 m), in diameter were made around 1820 by Lagerhjelm, Forselles, and Kallstenius in Sweden. These tests on wheels of 72 and 144 blades, respectively, indicated that the efficiency of undershot wheels in a confined race could reach 30% to 35% and confirmed the advantage of using a large number of blades on such wheels.[187]

While model experiments continued to contribute some new information to vertical water wheel engineers in the nineteenth century, the focus of the experimental tradition increasingly shifted to full-size wheels. Model experimentation had tended to dominate experimentation on vertical wheels between 1750 and 1820 primarily because of convenience and economy. On a model one could easily and cheaply devise methods for systematically altering the critical factors—head, wheel size, point of delivery, velocity, etc. But with a full-size wheel, itself a major capital investment, even without supplementary frills, this was difficult. Moreover, the standard means of measuring output was by lifting weights with a wheel's axle. This method worked well with small engines, but the difficulties multiplied when one attempted to apply it to industrial wheels. The weights and heights needed to measure their output were too great to be convenient. And when a wheel was already installed and operating machinery, there was often not enough room for this method even to be considered. Two ways around this quandry were possible. One could either absorb the expense of building a special test apparatus for large water wheels, or one could devise a more compact and convenient way of measuring a wheel's output. The two most important experimental investigations of full-size vertical water wheels in the nineteenth century were due to the Franklin Institute in America (1829–31) and to Arthur Morin in France (1828–35). The Franklin Institute followed the first of these options; Morin, the second.

The Franklin Institute was one of a large number of mechanics' institutes that sprang up in Britain and America in the early nineteenth century with the goal of diffusing scientific knowledge to workingmen. But Philadelphia, the home of the institute, was the intellectual center of America in the period, so the institute attracted a membership with broader goals than most other mechanics' organizations. In the late 1820s the institute's leaders decided to create useful knowledge, as well as diffuse it. In 1829 they appealed to millwrights, millers, and other interested manufacturers for contributions to support a set of experiments on full-size water wheels. Eventually about $2,000 was raised, and the institute's Committee on the Power of Water built the apparatus illustrated in Figures 4–12 and 4–13.[188] With it the institute tested wheels of 20, 15, 10, and 6 feet (6.10, 4.57, 3.05, and 1.83 m) in diameter, and all three major vertical wheel types (overshot, undershot, breast). They determined the wheels' output by the product of the weights lifted by the wheels' axles and the vertical distances they were lifted.

The experiments of the Franklin Institute were the most extensive yet undertaken on vertical water wheels. In addition to the usual variables (velocity, flow, impact head, gravity head) and the usual goals (the determination of optimum velocity and maximum efficiency), the institute tested several different forms of penstock, studied the effects of varying wheel diameter, tested the effects of placing ventilation holes in the soaling, and used different bucket shapes and numbers. With the breast wheel the institute varied the proportion of the total head devoted to gravity and to impact between wide limits.

Most of the institute's tests yielding nothing revolutionary. Altering the form of penstock gate, for instance, proved to have little effect on wheel efficiency. Vent holes in the soal also had little effect; likewise, variations in wheel diameter. The institute's committee found, as Mallet did a decade later, that the use of breasts on overshot wheels increased efficiency by about 6 to 8%. Elbow buckets were found to be much more economical with water than the alternative forms, but this had already been recognized empirically, even though the exact magnitude of the advantage had not been known.

FIGURE 4–12. Side view of the Franklin Institute's apparatus for testing water wheels, 1829–1831. The 20-foot (6.1-m) water wheel is in place. Water was stored above the wheel in a reservoir and released on the wheel by levers which regulated its flow. After acting on the wheel, the water was collected in an afterbay (lower left) where it was accurately measured.

Many of the institute's experiments, of course, overlapped the work on models carried out by Smeaton, de Parcieux, Bossut, and others. In this area, also, there were no radically new findings, though there were some corrections of minor import. For instance, the institute found that overshot wheels could attain an efficiency of around 80%, somewhat higher than the 67% Smeaton's work had indicated. For undershot

FIGURE 4–13. End view of the Franklin Institute's apparatus for testing water wheels, 1829–1831. The output of the wheel was measured by a chain linked to the axle of the water wheel and led over a drum (top of drawing) to a weight-filled basket. The product of the weight and the distance it was lifted determined the wheel's effect. The experiments were carried out by a team of three. One set the reservoir gauge to mark the desired velocity head for the wheel. While maintaining this with a valve linked to the city's water mains, he would release water onto the wheel. When a mark on the chain suspending the basket of weights passed a certain point, it indicated to a second operator that the wheel was in uniform motion. He would then close off the tail race and the valve that allowed water to exit from the afterbay. This caused a bell to ring, which alerted a third operative, who noted the time. When the mark on the chain passed a second point, the second operator would reopen the tail race, causing the bell to ring again. This told the first operator to close the valve leading water to the wheel and the third operator to mark the time.

wheels the institute's work only confirmed Smeaton's. The institute determined the mean efficiency of the undershot wheel to be 28.5%, very close to Smeaton's results. For optimum velocity of both overshot and undershot wheels, the institute's experiments indicated $v$ to be 0.50–0.55 $V$, somewhat higher than the tests of Smeaton and Bossut, but very close to Bordan theory. Finally, the institute noted that velocities higher than the 3 feet per second (0.91 mps) recommended for overshot wheels by Smeaton, even up to around 7.5 feet per second (2.3 mps), caused only small efficiency losses.

For calculating the power output of a water wheel the Franklin Institute's committee recommended the use of two coefficients derived from their experiments—0.28 for the impact portion of a water wheel's fall; 0.90 for the gravity portion. Their equation for calculating the output of a water wheel operating at favorable speeds was, in effect:

$$\text{Power output} = 0.90K \ (H-h) + 0.28 \ Kh.$$

Here $K$ is the weight of water falling per unit time; $h$ is the impact head; and $H$ is the total head or fall available (so $H - h$ is the gravity portion of the fall). This formula, according to the Franklin Institute's committee on water wheels, yielded, in most cases, results within 4% of their experimental tests and usually much closer.[189]

Unlike the Franklin Institute's tests, Arthur Morin's tests were conducted on industrial water wheels in place. Morin was able to do this by applying the Prony brake, an invention of the 1820s that moved experimentation on large water wheels into the realm of possibility for individuals or groups with limited resources. The Prony brake, or dynamometer, consisted, in its original form, of two symmetrically shaped timber beams, hollowed out to fit an axle or drive shaft. These beams were clamped to opposite sides of an axle by means of nuts and bolts, which also served to adjust frictional resistance. A weight was then applied to one of the beams (or to a beam connected to the main part of the brake) and adjusted until, at some point, it held the brake steady midway between stops designed to hold it if the weight failed to do so (see Fig. 4–14). The moment arm of the weight served as a measure of the torque exerted by the mover, and from this the horsepower of the engine could be measured and computed. While not as accurate as lifted weights, the Prony brake did allow extensive experiments to be conveniently conducted on operating industrial water wheels.[190]

Arthur Morin (1795–1880), one of the most prolific French experimental engineers, was a former polytechnician. While an instructor of applied mechanics at the artillery and military engineering school at Metz he was charged by the French government with testing and comparing the engines used in government arms and munitions plants with similar engines used in private works. His initial experiments, carried out in 1828 and 1829, were on three large water wheels—one overshot and two breast. Between 1830 and 1835 Morin tested six additional

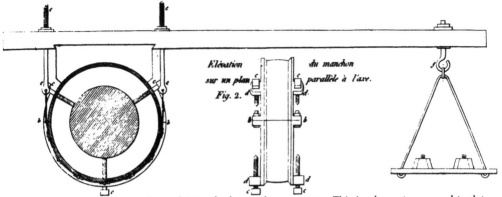

FIGURE 4-14. Prony friction brake, or dynamometer. This implement was used to determine the mechanical effect produced by the revolving shaft of a water wheel. In its simplest form it consisted of a beam with a scale pan (far right) and a concave wooden block with a metal band. By means of screw bolts the block and band were made to compress a revolving shaft. When one desired to know the power of the shaft, the bolts were screwed up and weights added to the pan until the beam was maintained in a horizontal position. In this situation the entire mechanical effect of the shaft or axle was expended in producing friction with the wood block and metal band, and the mechanical effect could be calculated from the speed of the shaft, the weight it sustained, and the length of the lever arm.

wheels—three breast and three overshot. He published the data on all nine wheels (several were iron wheels) in 1836.[191] On each of the nine wheels Morin carried out literally hundreds of tests, varying the loading on the wheel, the conditions of water discharge, the velocity, and the head (as far as possible). Once he had accumulated the data in tabular form, Morin compared them with the results yielded by calculations using Bordan theory, since one of the central aims of his experiments was to establish coefficients which, when applied to Bordan equations, would yield results compatible with practice.

The five breast wheels had diameters varying from around 13 feet (3.96 m) to 21.3 feet (6.5 m), under heads as low as 2.6 feet (0.80 m) and as high as 6.9 feet (2.10 m). Morin found that the theoretical formulas evolved by the Borda-Carnot school required in breast wheels a correctional coefficient of 0.755 when the wheel was operated from a traditional underflow sluice, 0.799 when operated from an overflow or sliding-hatch sluice. With these corrections, Morin declared, the equations would yield results within 5% of experiment in most cases. The overall efficiency of breast wheels he found to vary considerably. When load, velocity, and water delivery conditions were favorable, they attained at least 40 to 50% efficiency, and even as high as 60% for very well executed wheels.[192]

The overshot wheels Morin tested varied in diameter from 9.99 to 29.85 feet (2.74–9.0 m). So long as the velocity of their circumferences did not exceed around 6.6 feet per second (2 mps), Morin found, 0.78

was an adequate correctional coefficient, yielding results within 5% of experiment. Within reasonable velocity limits, Morin's tests revealed that the efficiency of overshot wheels was between 65 and 75%.[193]

While Morin's findings, like those of the Franklin Institute, resulted in no radical new breakthroughs, they were one of the landmarks of the emerging engineering science and rendered obsolete the model tests of the previous century. Poncelet, in commenting on Morin's work, declared:

> In spite of this conformity between Borda's theory and Smeaton's experiments and even though Smeaton announced, in his work, that he had taken care to verify the rules in question in their application to practice, it can not be denied ... that his experiments left much to be desired, both under the head of a more rigorous verification of the formulas of theory, whose coefficients were suitably established only around the maximum effect, and because these experiments were, in themselves, incomplete with respect to wheels ... which received their water between the axle and the lower point; and, finally, because of the notable differences that the arrangement of the model employed by this celebrated engineer present to those of the powerful wheels put in use by industry today.[194]

The experiments of Morin and the Franklin Institute were the most extensive ever carried out on full-size vertical water wheels, but their work was supplemented by the more limited investigations of other engineers. A few examples will suffice to illustrate the point. Around 1800 the Swedish engineer Nordwall, with financial support from the Swedish government, carried out 270 experiments on water wheels of around 15 to around 30 feet (4.6–9.1 m) in diameter, operating under falls of around 15 to around 37.5 feet (4.6–11.4 m). He varied wheel diameter, total fall, and impact head on overshot, undershot, and breast water wheels. Nordwall's results were published in Swedish and seem to have attracted little attention outside of Scandinavia.[195] In 1804 Louis-Charles Boistard studied the impact of water against the blades of floating mills.[196] In 1807 the mining engineer and hydraulician Jean d'Aubuisson de Voisins reported on an extensive series of tests on a very large overshot wheel (37.3 ft, or 11.4 m, high by 3.55 ft, or 1.08 m, wide) at the mine of Poullaouen in Brittany.[197] Around 1820 Christian in Paris constructed an apparatus in which he tested undershot, overshot, and breast wheels with diameters of 10.81 to 11.2 feet (3.3–3.4 m).[198] Poncelet in 1825 experimented on three boat mills on the Rhone near Lyons, and later in the same year he tested the effect of installing rims 1.35 inches (8 cm) thick on the blades of a water wheel at a gunpowder mill at Metz and measured the efficiency of flour mills operating under low water conditions.[199] At about the same time in Germany the engineer Egen measured the output of watermills in Westphalia at the request of the Prussian government.[200] Morin, Dieu, and Caligny, all French engineers, experimented around 1840 with partially flooded breast wheels.[201] Shortly after, Marozeau carried out additional experiments on a water wheel whose interior was divided into three compartments by two intermediate rims or rings to determine if applying water to only one of the compartments might be advantageous when only low water volumes were available.[202] A number of other experimental investigations on individual operating vertical water wheels in the mid to late nineteenth century could be cited as well.[203]

But after 1850 experimental, as well as theoretical, investigations relating to vertical water wheels became steadily less frequent, and after 1900 they were all but nonexistent. Weidner, for example, commented in 1913, after reviewing the available literature on overshot water wheels: "Most of the experimental work in relation to water wheels was done prior to 1840."[204] He found no major investigations after that date on the overshot wheel. A review of available literature indicates that similar statements could be made about most of the other major types and subtypes of vertical water wheel.

The reasons behind the demise of interest in the vertical water wheel are clear. Its importance as an industrial prime mover was being steadily eroded in the mid-nineteenth century by new power sources such as the water turbine and the steam engine. The vertical wheel's unsuccessful competition with these engines will be the focus of the concluding chapter of this work.

## THE EFFECTS OF QUANTIFICATION ON VERTICAL WHEEL TECHNOLOGY

Between 1750 and 1850, however, a science of the vertical water wheel had emerged and matured, a science characterized by both a widely accepted theoretical structure (Borda's analysis) and a large body of generally accepted experimental data. As early as the late eighteenth century this body of quantified engineering data had begun to have an influence on practical water wheel technology.

One of the early major contributions of the quantitative approach to the problems of vertical wheel technology was the general recognition, as a result of the experiments of de Parcieux and Smeaton, of the superior efficiency of the gravity wheel. This discovery led, as we shall see in the next chapter, to the general replacement of impulse wheels by gravity wheels, whether breast or overshot. It also played a role in the development of the overflow sluice gate, an important device designed to provide the highest possible gravity head to water wheels.

In the early nineteenth century additional improvements to the vertical wheel emerged as a result of attempts to design a water wheel which could meet the theoretical requirements laid down by Borda and Carnot—that water should enter a wheel without impact and leave it without velocity. For instance, the substitution of curved iron buckets for angular wooden buckets on overshot and breast wheels in the early 1800s was in part an attempt to reduce the impact of water on entering the wheel. Two new water wheels, the Poncelet and Sagebien wheels, moreover, were specifically designed to meet Borda's and Carnot's theoretical conditions.

The Poncelet wheel was designed by Jean Victor Poncelet (1788–1867), a graduate of the École polytechnique, a veteran of Napoleon's Russian campaign, an experienced military engineer, and a geometer of note. Poncelet's wheel was a modification of the old vertical undershot water wheel.

At the opening of the nineteenth century the undershot wheel with its flat radial blades, though much diminished in importance by the emergence of breast wheels, still held on in some areas. Although it was much less efficient than either form of gravity wheel, breast or overshot, it still had some advantages. The chief of these was simplicity and consequent economy of construction and maintenance. Where water was plentiful and efficiency a secondary concern, these factors made it viable. Moreover, the undershot wheel had a higher rotational velocity than the slow-moving gravity wheel. This meant that less gearing was needed to drive millstones or other implements from an undershot wheel than from a gravity wheel. Where very high speeds were required, as in sawmills, undershot wheels almost universally prevailed over the overshot.[205] Finally, the undershot wheel was the only form of vertical wheel which could work on rivers and in areas of very low falls.

Poncelet was familiar with Borda's work and the necessity, for an efficient water wheel, of having water enter without velocity and leave without impact. The basic problem he faced in redesigning the undershot wheel was how to accomplish this while retaining the practical advantages of the traditional construction—simplicity, low construction costs, high rotational velocity. Poncelet declared: "After having reflected on this, it seemed to me that we could fulfill this double condition by replacing the straight blades on ordinary wheels with curved or cylindrical blades, presenting their concavity to the current."[206] Thus, in 1823, Poncelet took the old undershot wheel and replaced its flat, radial blades with curved blades and angled its sluice gate to bring the water as close to the lower blades as possible (see Fig. 4–15).[207] These changes produced a wheel with all of the advantages of the undershot wheel plus a relatively high efficiency.

Poncelet's modifications, like many of the most fruitful scientific and technological concepts, were simple enough to have been carried out years earlier. They had, in fact, been foreshadowed by the inclined flat blades of de Parcieux and Bossut. That curved blades did not emerge earlier may have been due not only to the failure to recognize their advantage but also to the difficulty and expense involved in fabricating curved blades in wood. But sheet metal, which, as we shall see in the next chapter, was coming into use on the buckets of overshot and breast wheels in the early nineteenth century, was easily given a curved form

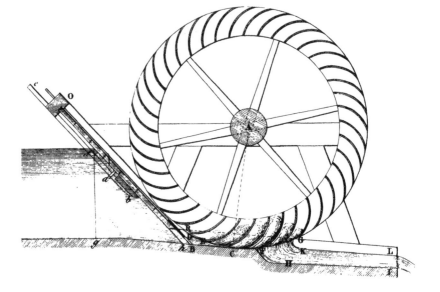

FIGURE 4–15. The Poncelet wheel. By using curved blades with an inclined sliding hatch sluice gate, Poncelet was able to lead water into the water wheel with a minimum of impact and to exit water from the wheel with little or no velocity.

and was common enough by the 1820s to make the idea more practical than it would have been some decades earlier.

The chain of logic which led Poncelet to believe in the superiority of curved to flat blades was more or less as follows. Two criteria had to be met for a wheel to have good efficiency—entrance without impact, exit without velocity. The first of these could be satisfied by curving the blades so that their edges would be tangential to the stream of inflowing water. Then the water, instead of suddenly impacting the blade, would enter without impact, as prescribed by Borda, and would slowly give up its momentum as it ascended the curve of the blade. The water would gradually slow down, stop, and then descend the blade, picking up its initial velocity, but in a direction opposite that of the wheel. If the wheel's velocity and the length of its blades were proportioned correctly, the water as it left the wheel would have a velocity equal to that of the wheel, but in an opposite direction, so that their relative velocity would be zero. Thus Borda's requirement that the water leave the wheel without velocity could be met as well.

Using Bordan theory Poncelet computed the necessary length for the blades of his wheels, finding that they had to have a height equal to $(V-v)^2/2g$, where $V$ was the velocity of the water, and $v$ that of the water wheel. He found that the optimum theoretical velocity for the wheel was the same as with an undershot wheel: half that of the impacting water. But its theoretical maximum efficiency was 100%. In theory, then, the Poncelet wheel combined the high rotational velocity, ability to handle low falls, high reliability, and simple construction of the traditional undershot wheel with the high efficiency of the overshot.[208]

Since he came from an engineering background, Poncelet was not satisfied with theory alone. He put his idea to the tests of experiment and practice. In August of 1824 he built a small wooden model of his invention, 1.64 feet (50.0 cm) in diameter, 2.6 inches (65.0 mm) wide, with curved blades spaced about 0.10 inches (2–3 mm) apart. His tests with this model indicated that his theory was correct. The velocity of maximum effect on the model was $v = 0.52\ V$, very close to the predictions of theory. He secured efficiencies varying from 65 to 72%, more than twice what Smeaton had found for ordinary undershot wheels with a model of comparable size. Poncelet estimated that the efficiency of full-size wheels might go as high as 80% for well-constructed engines at low velocities, and 70% at higher speeds.[209]

By 1825 several manufacturing establishments had erected full-size Poncelet wheels with curved blades, and Poncelet was able to supplement his model experiments with tests on full-size wheels. One of these wheels, in service near Metz, delivered 33% more power than the conventional undershot it had replaced despite neglecting some of Poncelet's design precepts. Experiments on another wheel 13.28 feet (4.05 m) in diameter indicated an efficiency of 73%.[210] By 1827 other Poncelet wheels were in service, and Poncelet carried out new experiments using the Prony brake. His results suggested that well-constructed wheels operating at design capacity would have an efficiency that would rarely fall below 60%, even when the wheels were under higher than normal heads (up to 6.56 ft, or 2.0 m). Their efficiency might reach 66% for normal heads (around 3.28 ft, or 1.0 m). When discharges through the wheel were too small or too large, however, the useful effect dropped to around 50 to 55%, still twice that of the traditional undershot wheel.[211]

Poncelet's wheel was one of the best-publicized developments in vertical wheel technology in the nineteenth century. His initial paper discussing the wheel, its theory, and his model experiments appeared in several widely read French engineering periodicals and was reprinted as a separate pamphlet.[212] In 1827 a second and much enlarged edition of Poncelet's paper was published as a monograph. In addition, the French Academy of Sciences awarded Poncelet its Prix de Mécanique.[213] The widespread publicity given Poncelet's wheel insured its rapid diffusion. According to John Reynolds, "Large numbers of Poncelet wheels were installed during the nineteenth century."[214] He is supported by a contemporary observer, the British engineer Lewis Gordon, who commented in 1842 that although the wheel had "not been much employed in Great Britain," it was "frequently used in France and Germany."[215]

The Poncelet wheel was important not merely as an instance of a vertical wheel which emerged more from theoretical than practical considerations. It was to play a major role in the emergence of the modern water turbine, a topic covered in Chapter 6.

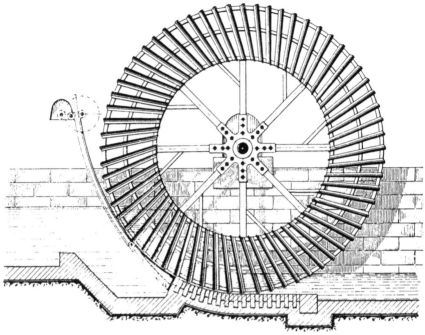

FIGURE 4–16. The Sagebien wheel. The blades of this wheel were inclined so that they would enter the water on the upstream side of the wheel at an angle as close to the vertical as possible. The enclosed area between the blades then acted as a tube, and water entered the wheel under atmospheric pressure, with minimum impact.

A second type of vertical wheel which emerged in the nineteenth century primarily as a result of theoretical considerations was the Sagebien wheel, named after its inventor, the French engineer Alphonse Sagebien. The Sagebien wheel was a modification of the low breast wheel. The basic change was the backward inclination of the blades (Fig. 4–16). Where de Parcieux and Bossut had inclined the blades of their undershot wheels upstream, Sagebien inclined the blades of his breast wheel in the downstream direction. The reason for this modification was Borda's injunction that water should enter a water wheel without impact. In a normal vertical breast wheel water flowed over a masonry crest or through a sliding hatch and fell with unavoidable impact against the blades or buckets of the wheel. Sagebien's wheel was designed to avoid this impact by relying on atmospheric pressure rather than the flow of water to fill the voids (buckets) between the blades. The blades of the Sagebien wheel entered the water on the upstream side of the wheel at as nearly a vertical angle as practically possible (which was usually a 40–45° angle to the surface of the water) and at a very slow speed. Water rose into the spaces between the blades under atmospheric pressure as it would rise into an open tube pushed into the water. This method of bringing water onto the wheel reduced the impact of the water at its

TABLE 4–4. EFFICIENCY OF THE SAGEBIEN WATER WHEEL FROM EARLY EXPERIMENTAL TESTS

| Water wheel of | Efficiency | Experimenter(s) |
|---|---|---|
| M. Pécourt, Amiens | 90% | De Marsilly |
| Hydraulic Establishment of Amiens | 94 | Lienard |
| Traill and Lawson, Beaurain | 88 | |
| Queste, Bonquerolles | 85 | |
| | | |
| Small wheel, 10 hp, at Chalons | 93 | |
| Raupp, Houlme | 88 | Slaweski |
| Duboc, Cany | 90 | |
| Coeurderoy, Brionne | 85 | |
| | | |
| Depoisses, Brionne | 85 | |
| Sement, Serquigny | 93 | Tresca, Faure, Alcan |
| De Croix, Serquigny | 82 | De Bernay |
| Greslé and Toury, Ivry | 86 | Hennezel, Leblanc |
| | | |
| Saint Mars, Labruyere | 85 | Julien, Ponion |
| Trilbardou | 70 | Belgrand, Hult |
| Forge-Thiry | 84 | |
| Maestricht | 93 and 87.7 | |

SOURCE: "Sagebien's Water Wheel," *The Practical Magazine* 6 (1876): 89.

entrance to a minimum, while the slow motion of the wheel reduced the velocity of the water at exit to a minimum. The result was a highly efficient water wheel.[216]

Sagebien developed the wheel around 1850. His first full-size wheel was a 6 to 7 horsepower model at a flour mill at Rouquerolles, in the department of Oise, in 1851. Its tested efficiency was around 85%. Sagebien's wheel at first attracted little notice, although it was shown at the Paris industrial exhibition in 1855. By 1858 Sagebien had constructed at least 17 wheels, and belatedly his design began to attract attention.[217] Experimental tests conducted by a number of engineers in the 1860s and 1870s confirmed that the wheel's efficiency was very high, perhaps higher than any other known type of vertical water wheel. In fact, one set of tests in December 1861 indicated that the wheel's efficiency was an impossible 103.5%, before improved methods of gauging water reduced this figure to 93%.[218] Subsequent tests did not usually yield figures quite that high, but did give results not much below, as Table 4–4 indicates.

Most of the figures in Table 4–4 are probably a little high, since the gauging of water in any but a specifically prepared test station is notoriously difficult. But the wheel's efficiency was certainly very high, probably at least 75 to 85%.[219] By 1868 Sagebien had installed more than 60 of his wheels in northwestern France,[220] and the French Academy of Sciences in 1875 recognized his accomplishment by awarding him its Fourneyron Prix.[221]

The form of vertical wheel against which the Sagebien wheel had to compete was the low breast wheel. Table 4–5 gives some indication of

| Category | Sagebien Wheel | Typical Low Breast Wheel |
|---|---|---|
| Efficiency | 75–90% | 50–70% |
| Quantity of water wheel can utilize in favorable operating situations | 35–55 ft³/sec (c 1–1.5 m³/sec) | usu. under 20 ft³/sec (usu. under 0.5 m³/sec) |
| Falls under which wheel can operate with reasonable ease and efficiency | 1–6.5 ft (c0.3–2 m) | 3–6 ft (0.9–1.8 m) |
| Typical velocity of rotation (peripheral speed) | 2–2.5 ft/sec (c0.6–0.8 m/sec) | usu. 3–8 ft/sec (usu. 1–2.5 m/sec) |
| Typical blade length | 5–6.5 ft (c1.5–2.0 m) | usu. less than 1.5 ft (usu. less than 0.5 m) |

the comparative advantages and disadvantages of the Sagebien wheel against this competitor. In brief, the Sagebien wheel could deliver higher efficiencies than the low breast wheel and deliver them even under exceptionally low falls. The Sagebien wheel could also handle larger quantities of water than the breast wheel under some circumstances. But the low breast construction had a higher rotational velocity and was often cheaper to construct because of the shorter and more conventional blades it used.

Poncelet and Sagebien were both representatives of the French school of water-power engineers. This school, which had its roots in the work of Mariotte, Parent, and Bélidor, was traditionally mathematical and theoretical in orientation. It is thus not surprising that the designs of the Poncelet and Sagebien wheels were heavily dependent on the engineering science that French technicians and scientists had played such a large role in creating.

There is evidence, however, that even the largely practical millwrights of Britain and America were also becoming increasingly dependent on quantitative theory and quantitative experiment as a supplement to their traditional craft rules and methods by the late eighteenth and early nineteenth centuries. For example, almost half of Oliver Evans' *Young Mill-Wright and Miller's Guide,* first published in 1795 (ultimately it went through 15 editions), was devoted to an elementary survey of the basic principles of mechanics and hydraulics. Evans, moreover, included the full text of Smeaton's classic experiments in the work.[222] Evans' millwright guide was not the only one to include quantitative theoretical and experimental material. The manuals of Gray, Nicholson, Overman, and Brunton, among others, also did.[223] Thus by the early nineteenth century, quantification had begun to play a role not only in the emergence of new forms and types of vertical water wheels, but also in the design, dimensioning, and operation of the more usual forms as well.

Traditional vertical water wheel technology had been a craft tradition, depending on centuries of accumulated experience to design low-to-medium-output power plants made largely of wood. The development of theoretical analysis and systematic, quantitative experimental methods for dealing with the problems of water power between 1700 and 1850 significantly reduced the importance of craft knowledge to water power technology. The importance of traditional craft methods was being further undermined at the same time by other developments, notably the rise of the iron industrial breast wheel. The general substitution of iron for wood, outlined in the following chapter, reduced the importance of the traditional millwright's wood-working skills. Machine shops and centralized engineering firms capable of fabricating metal parts significantly reduced the role and importance of the itinerant craftsman in the manufacture of water wheels.[224]

# 5

# Change:
# The Emergence of the Iron Industrial
# Water Wheel, c1750 to c1850

The vertical water wheels in use in industrialized areas of Europe and America by the mid-nineteenth century differed from the traditional wheels described in Chapter 3 in several respects. They were typically breast wheels, instead of pure overshot or pure undershot wheels; they were fabricated largely of iron, rather than wood; they were often much wider than the typical wooden wheel; they used a sliding hatch or overflow sluice for water delivery, instead of the traditional underflow gate. Some of these changes were in large part the outgrowth of mathematical analysis and quantitative experimentation. As indicated in Chapter 4, the widespread adoption of the breast wheel and the use of the sliding hatch or overflow sluice gate emerged from this tradition. Other changes grew out of trends present in water-power technology for centuries. For example, iron had already begun to replace wood for certain parts in the medieval period.

But even if the emergence of the wide iron breast wheel was primarily due to a combination of theoretical analysis, quantitative experimentation, and the internal dynamics of water-power technology, the location and timing of its birth—Britain between 1750 and 1850—were determined by external factors. The primary one was the surge of economic growth and technological innovation that occurred in Britain in the late eighteenth and early nineteenth centuries, commonly called the "industrial revolution." As we have seen, this surge was not entirely unprecedented and had roots stretching as far back as the medieval period. But the scale and rate of growth which began in Britain in the late eighteenth century were greater than that of earlier centuries.

A number of the technological and economic changes which occurred during Britain's industrial revolution played a role in the demise of the traditional wooden water wheel. The substitution of coke for charcoal as a fuel in the iron industry, for instance, freed that industry from the limitations that had long restricted iron production and kept iron prices high. Between 1750 and 1850 iron production jumped from around

25,000 tons annually to over 2,000,000 tons.[1] The greater availability and declining price of iron undoubtedly encouraged the replacement of wood with iron parts on water wheels. Steam emerged as an alternative to water power in the late eighteenth century. The competition of the steam engine stimulated the designers, manufacturers, and users of vertical water wheels to develop wheels of higher output and to seek more efficient ways of utilizing water power.[2]

The pressure of Britain's rapidly expanding industrial economy on available water-power resources, combined with that nation's geography, was also an important stimulus for the development of the iron industrial wheel. These pressures provided an incentive to British mill owners to seek more economical means of using the water they had available, and these means often required using a water engine more efficient than the traditional wooden water wheel. There is a natural limit set by climate and geography to the number of mills that a stream can carry. In Britain, unfortunately, this limit is often rather low, especially compared to the potential in better watered areas like the northeastern United States and large parts of France and Germany.[3] The limited water-power resources of Britain, coupled with the fact that British industrial utilization of water power had been steadily growing from the medieval period onward and had accelerated in the late sixteenth century and again in the late eighteenth century, meant that the ceiling for water power began to be approached earlier in Britain than elsewhere.

There are no data for Britain as a whole, but a number of local studies clearly indicate the expanded use of water power in Britain between 1750 and 1850 and the saturation of British streams with watermills. There was no more room for additional mills on some streams and in some areas of Britain by 1700, and on many streams and in many areas of Britain by the nineteenth century. The number of watermills on the River Kent and its tributaries in relatively unindustrialized south Westmorland, for example, almost tripled between 1750 and 1850.[4] Ian Donnachie, studying southwestern Scotland (Galloway), discovered that during this same period the number of mills there doubled.[5] By 1700 nearly all the available water-power sites within a 5-mile (8-km) radius of Birmingham had already been occupied, and by mid-century all except one of the possible mill sites were gone. Shortly after 1760 there were none left.[6] In Lancashire, on Cheesden and Naden brooks, a total of 6 miles (9.7 km) of stream, there were 23 watermills by 1848.[7] The Spodden, a small stream near Manchester, had 18 mills on 9 miles (14.5 km) of stream.[8] Of the 900 feet (c275 m) of fall on the River Irwell between Baucup and Bolton, 800 feet (c245 m) were occupied by more than 300 mills in the 1830s.[9] Sheffield streams (Fig. 3–1) were crowded with 161 water-driven works in 1770,[10] and there were 46 watermills within a 4-mile (6.4-km) radius of Perth in Scotland in 1797.[11]

There are many other examples of the saturation of British streams with watermills. The River Leen, a sluggish stream only a dozen miles (c19 km) long and with a fall of only 156 feet (47.6 m), near Nottingham, had 17 mills by 1784.[12] In southern England the River Frome and its tributaries were overloaded with mills. According to a local tradition, undoubtedly exaggerated, a stretch of 6 miles (9.7 km) between Garson and Beckington once had 200 mills.[13] In Kent, the River Len, only around 10 miles (16 km) long, had almost 30 watermills, and the River Loose had 13 on only around 2 miles (3.2 km) of stream.[14] The River Wandle, a small tributary of the Thames, near London, had 38 mills with 52 water wheels on 9 miles (14.5 km) of its course in 1853.[15] And Ulster had over 1,700 water wheels in operation by the mid-1830s.[16]

The overcrowding of British streams with watermills towards the early nineteenth century is also evidenced by complaints about the magnitude of legal problems involving watermills on congested streams. The mill-wright-engineer Sutcliffe complained early in the 1800s: "In consequence of so many water mills, the country is never free from litigations and vexatious law suits, respecting erecting, repairing, or raising mill-weirs, by which the peace and harmony of neighbors and friends are often destroyed."[17] He added: "Some years ago, the annual expense of water cases and arbitrations, in consequence of them, in the counties of York and Lancaster, were estimated at £10,000, and should trade revive, it is probable in a few years they will exceed this sum."[18] A contemporary of Sutcliffe made similar observations:

> Upon most rivers in this country all the falls of water are fully occupied, and at every mill there is a weir, which pens up the water as high as the mill above can suffer it to stand without inconvenience. Each miller is anxious to obtain the greatest possible fall, and he can at any time augment the fall, by raising the surface of his weir; but as this may produce an inconvenience to the mill above, ... it is a constant source of dispute and litigation.[19]

By the late eighteenth century, it would seem, most of the adequate water-power sites around Manchester, Nottingham, Birmingham, Sheffield, and other major British industrial towns had been occupied.[20]

The saturation of British streams with watermills was in part due to the continued expansion of already existing water-powered industries—fulling, flour milling, paper manufacturing, and so on. But the pressure on Britain's limited water-power resources was considerably intensified after 1770 by the rapid growth of a new water-powered industry—cotton textiles. As noted in Chapter 3, before 1750 mechanization had already begun to penetrate textile production. By that date, one stage of wool and silk production and several stages of linen cloth production could use water-powered equipment. But the most time-consuming steps in textile manufacturing—the processes required to clean and pre-pare cotton, wool, and most other fibers for spinning, as well as the processes of spinning and weaving—were done largely by hand. In the last half of the eighteenth century, however, the series of inventions associated with the names of Hargreaves, Arkwright, Crompton, and Cartwright radically transformed the cotton textile industry.[21] By 1800 or 1810 the production of cotton cloth had been mechanized almost from beginning to end, and modern factories with hundreds of power machines were turning out cotton thread and cloth in unprecedented quantities. The demand for cheap cotton textiles skyrocketed. British raw cotton imports rose from 2,359,000 pounds in 1760 to 664,000,000 pounds in 1850.[22] Further, by the early nineteenth century a start had been made in mechanizing, in a similarly comprehensive manner, the production of other textiles, especially wool.

The cotton textile industry exerted increased pressure on Britain's limited water-power resources in two ways. First, a cotton textile mill required more power than a typical water-powered industrial plant of the pre-1770 era. The latter required, usually, only around 5 horsepower; in contrast, even the early cotton textile mills, owing to the vastly expanded scope of mechanized equipment, required 10 to 20 horsepower, more than most British water-power sites could provide (Fig. 5–1).[23] Second, the rapid rate of growth of the new industry quickly exhausted whatever adequate water-power sites were available near industrial areas. Arkwright built the first water-powered cotton mill in 1771. As early as 1782, because of a lack of good water-power sites, Joseph Thackeray, an early Manchester spinner, was forced to move from the Manchester area, the center of Britain's cotton textile industry, to the more abundant water-power resources of Cark-in-Cartmel, 80 milles (129 km) away.[24] At about the same time, for the same reason, another Manchester spinner, William Douglas, moved the focus of his operations to Holywell, on the north coast of Wales.[25] By 1788 there were around 150 cotton textile mills in Britain, with 40 water-powered cotton mills in the southern part of Lancashire alone.[26] By 1790 in Oldham parish, near Manchester, most small mills were being driven by horses because all of the good water-power sites were already taken.[27] By 1811 the lateral streams feeding 5 miles (8 km) of the River Aire in the remote upper regions of Lancashire were occupied by 44 cotton mills,[28] and in thc Shuttlcworth valley, north of Bury, there were 10 mills on only 1 mile (1.6 km) of stream where a total of 500 feet (152 m) of fall was available.[29] By the mid-1830s vertical water wheels were providing 10,000 to 12,000 horsepower to the cotton textile industry, and more than that to the combined wool, worsted, flax, and silk industries.[30]

The early history of the cotton textile industry furnishes a good example of the influence a prime mover may have on the way an industry develops and the limits it can place on its growth. The new textile machinery introduced by Arkwright, Crompton, and others created a demand for power on an unprecedented scale and made it difficult to secure sufficient power from man or animal. Water power quickly be-

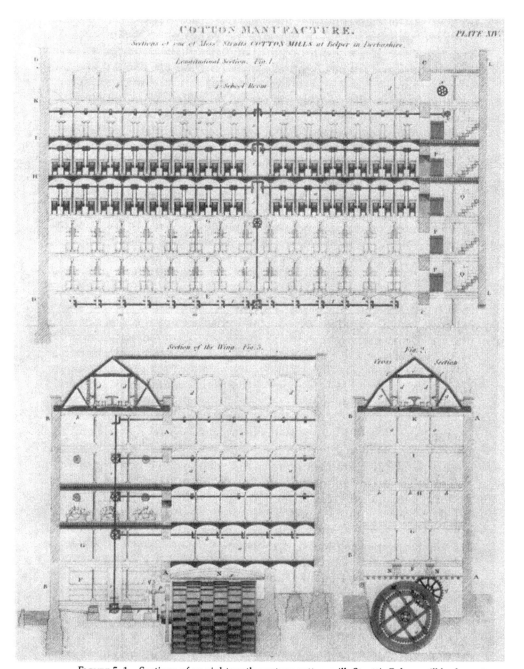

COTTON MANUFACTURE.

PLATE XIV.

Sections of one of Mess.ʳ Strutts COTTON MILLS at Belper in Derbyshire.

FIGURE 5–1. Sections of an eighteenth-century cotton mill. Strutt's Belper mill had a water wheel 18 feet (5.5 m) in diameter by 23 feet (7 m) wide. It provided power via mechanical linkages to more than 4,000 spindles on about 70 spinning machines on the lower floors. It also powered 64 carding machines, 72 finishing machines, and an assortment of drawing frames, roving machines, reeling, doubling, and twisting engines on the upper three floors.

came vital to the growth of the cotton textile industry. But the best water-power sites near existing industrial centers had already been occupied by earlier users of water power, or were quickly occupied by the pioneer cotton spinners. Thus, the rapid growth of cotton textile production soon compelled entrepreneurs to seek water-power sites far out in the country, where few workers and skilled technicians were available. To attract and hold a labor force in these isolated areas, the cotton mill owners often had to expend considerable capital to provide complete, attractive communities to meet the needs of their mills.[31] Or, if a mill owner settled for an inferior site near existing urban centers, he often had to expend enormous sums of money improving his site with dams, reservoirs, and canals. Both of these options probably prolonged the period during which the domestic producer of cotton thread or cotton cloth could compete with the new mechanized industry.

Moreover, the shortage of water-power sites adequate for the newly mechanized cotton industry and the crowded conditions of British streams did more than influence the location of the industry and the facilities provided for its early work force. It imposed a relatively low ceiling on the expansion of the textile industry. Attempts to raise this ceiling took two paths. One involved a search for a new prime mover without the limitations of the water wheel, and this path eventually led to the general adoption of steam power by industry (see the next chapter). The other path involved attempts to use water power more efficiently, to insure that no inch of fall or drop of water was wasted.[32] This path sometimes required the use of quantitative methods (see Chap. 4) to improve the performance of water wheels. But it also led, among other things to

1. the wider use of systems to efficiently harness and regulate the flow of streams;
2. the substitution of breast wheels for traditional undershot and overshot wheels;
3. the development of more efficient means of feeding water onto vertical wheels;
4. the replacement of wooden parts with iron parts; and
5. the construction of more powerful, more efficient, and longer-lasting vertical water wheels.

These developments form the focus of this chapter and will be taken up, in turn, below.

## SYSTEMS TO REGULATE THE FLOW OF STREAMS

Dams, reservoirs, and canals, as we have seen, were used in Europe well before 1750 to regulate the flow of streams and to create better operating conditions for vertical water wheels. The larger complexes, however, had generally been restricted to mining areas like the Harz or

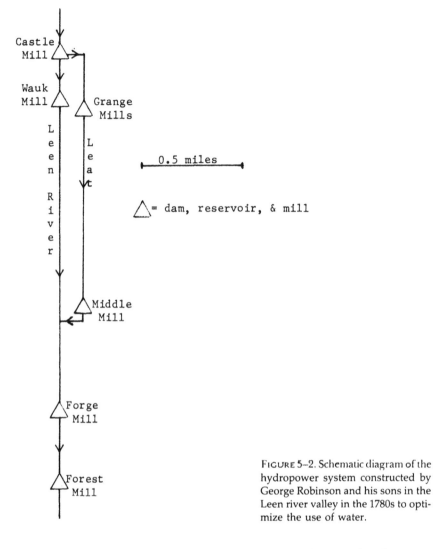

Castle Mill

Wauk Mill

Grange Mills

Leen River

Leat

0.5 miles

△ = dam, reservoir, & mill

Middle Mill

Forge Mill

Forest Mill

FIGURE 5–2. Schematic diagram of the hydropower system constructed by George Robinson and his sons in the Leen river valley in the 1780s to optimize the use of water.

Potosi. In the late eighteenth and early nineteenth centuries the pressures placed on Britain's limited water-power resources by industrial expansion led to the expanded use of networks of dams, reservoirs, and canals outside of mining areas.

One of the early systems designed to draw maximum power from a small stream was constructed in the 1780s on the River Leen in the East Midlands near Papplewick. George Robinson and his sons, John and James, first dammed the Leen to form a reservoir for their highest mill, Castle Mill (Fig. 5–2). From this reservoir part of the water was taken by extensive embankments and a large aqueduct which paralleled the Leen to a nearby reservoir which supplied the Grange Mills. From there the aqueduct led the water to yet another reservoir, this one supplying Middle Mill. The Leen itself, after driving Castle Mill, fed the reservoir of Wauk Mill downstream. About 1 mile (1.6 km) below Wauk Mill the Leen was reunited with the water taken from it at Castle Mill, and the two streams together fed the pool above Forge Mill. After providing power to Forge Mill, the water flowed into the reservoir of Forest Mill. Altogether, the Robinsons' reservoir system covered 30 acres (12.1 ha). By careful conservation and direction, they were able to obtain a considerable amount of power from a rather small flow of water.[33]

Other textile manufacturers also built dams and reservoirs to tap water-power resources more effectively. In the 1780s, for example, Arkwright dammed a small stream falling into the Irwell near Manchester to create a 30-foot (9.2-m) fall for a textile mill.[34] And on the River Wye, the millwright-engineer Hewes built for Arkwright a dam with a reservoir which covered nearly 24 acres (9.7 ha).[35] At Belper, Strutt had a semicircular dam built across the Derwent, creating a lake of 14 acres (5.7 ha) for one of his mills.[36] Near Stockport, in Lancashire, a system of dams, reservoirs, and tunnels was constructed around 1790 to supply a series of textile mills.[37] Finally, the dam built for Longfords Mill in Gloucestershire was about 30 feet (9.2 m) high and 450 feet (137.2 m) long. It created a reservoir of 15 acres (6.1 ha).[38]

Many streams in the British Isles are subject to flooding in spring and drought in summer. As the pressure on available water resources grew acute, extensive systems of dams and reservoirs were erected to regulate the flow of some of these streams. For example, in 1836 the mill owners on the Upper Bann River in northeastern Ireland asked J. F. Bateman and William Fairbairn to regulate the Bann's flow. Fairbairn and Bateman adapted a natural lake, Lough Island Reavy (see Fig. 5–3), into a reservoir for flood waters by constructing an embankment which increased the depth of the lake 35 feet (10.7 m). Into this reservoir they led surplus water from the River Muddock. This enlarged Lough Island Reavy from 92.5 acres (37.5 ha) to 253 acres (101.2 ha) and provided a storage capacity of 287 million cubic feet (8.13 million m³). In addition, a second lake, Corbet Lough, was modified. Embankments deepened it by 18 feet (5.5 m), creating a reservoir of 74.5 acres (30.2 ha) with a capacity of 47 million cubic feet (1.33 million m³). A third storage reservoir site was selected, but the reservoir was never constructed.[39]

In 1837 mill owners on the River Eagley, near Bolton, erected the first British dam to exceed 100 feet (30 m) in height for reasons similar to those which prompted the Lough Island Reavy works. The Entwistle dam was 108 feet (32.9 m) high when completed, although 128 feet (39 m) had been originally contemplated. The 360-foot-long (109.8 m) structure impounded 3,000 acre-feet (3.7 million m³) of water to regulate the Eagley's flow.[40]

The best-known and largest scheme to develop additional water power in Britain was Shaw's Water Works at Greenock, Scotland. Greenock had long suffered from a general lack of water, so in the early 1820s

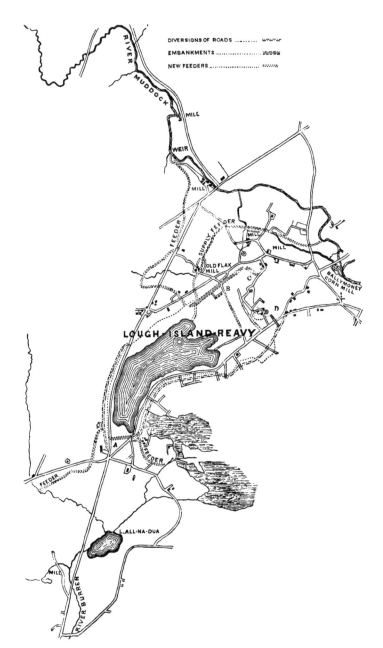

Figure 5–3. Map of Lough Island Reavy. This shows the original disposition of the lake and the modifications made by Bateman and Fairbairn to develop it as a storage reservoir for the mills on the River Bann. Embankments were constructed at A, B, C, D. The dotted lines connecting these embankments indicate the new boundaries of the lake. The feeders from nearby rivers and streams used to supply the reservoir with water are also shown.

the city hired Robert Thom to survey the area for water to satisfy the drinking and sanitary needs of the city. Thom reported in 1824 that a system of reservoirs and aqueducts could secure an adequate supply for the city and that this supply could be used to develop power as well. A company was quickly formed to implement Thom's proposals. By 1827 work was completed. The principal reservoir in the system was formed by an embankment 66 feet (20.1 m) high and 1,400 feet (427 m) long; it covered 295 acres (119.5 ha) and stored 285 million cubic feet (8.07 million m³) of water. A compensation reservoir covered 40 acres (16.2 ha) and stored 14.5 million cubic feet (0.41 million m³) of water; an auxiliary reservoir of 10 acres (4.0 ha) contained an additional 4.7 million cubic feet (0.13 million m³). There were, in addition, five other smaller reservoirs with a capacity of more than 6 million cubic feet (0.17 million m³), bringing the total for the system to 310 million cubic feet (8.8 million m³). Around 1,200 cubic feet (33.6 m³) per minute were led from the main reservoir along a 6-mile-long (9.7-km) aqueduct to the top of a tall hill overlooking Greenock. Here the channel was split in two and dropped 512 feet (156 m) in a series of artificial waterfalls. Thirty-three mills were located adjacent to these falls. The total power developed by these mills was between 1,500 and 2,000 horsepower; the ultimate capacity of the system was estimated at 3,000 horses.[41]

The Greenock, Lough Island Reavy, and Entwistle systems were designed mainly to aid manufacturing industries. In the mining industry, where the construction of extensive systems for water regulation already had a long history, additional systems were erected. The Wheal Friendship Mine near Tavistock in southwestern England, for instance, had a complex of 17 overshot wheels worked by a 526-foot (160.4-m) fall.[42] At the Devon Great Consols Mine, one of Europe's richest copper producers for several decades, three long leats, or power canals, provided the power needed for mine dewatering and other tasks. The "Great Leat" at Great Consols was 18 feet (5.5 m) wide. It took water from a dam, or weir, on the Tamar River and then followed a path that roughly paralleled that stream for 2 miles (3.2 km) until it gained enough head to drive a wheel 40 feet (12.2 m) in diameter and 12 feet (3.7 m) wide which developed approximately 140 horsepower. Great Consol's other two leats totaled 6.5 miles (10.5 km) in length and brought water to wheels of comparable size.[43] The leat which provided power to Cornwall's Wheal Emma in the 1850s was 10 miles (16.1 km) long.[44] The parish of St. Just in Cornwall was 12 miles (19.3 km) from the nearest river, but by 1842 mine owners there were utilizing diversion dams and long canals to provide power to 50 water wheels.[45]

Mining ventures further north in Britain were also using dam and canal systems to increase the amount of water power available. On Swinehope Burn, near Allenheads, mine owners erected a dam 15 feet (4.6 m) high, 50 feet (15.2 m) thick at base, and 500 feet (152.4 m) long to develop water power.[46] Near Stanhope other mine owners built the

Burnhead Dam, also 500 feet (152.4 m) long, but 100 feet (30.5 m) thick at base, and 40 feet (12.2 m) high.[47] Finally, on Blea Beck three dams were built to supply water power to mines on Grassington Moor. The largest of the dams was 25 feet (7.6 m) high by 450 feet (137.2 m) long. This system, through 6 miles (9.7 km) of canal, supplied water to about a dozen water wheels, ranging from 15 to 50 feet (4.6–15.2 m) in diameter.[48]

Some rather extensive systems for regulating the flow of water in British rivers to alleviate Britain's growing shortage of water power were planned but never completed because of the displacement of water power by steam towards the middle of the nineteenth century. For instance, by the early 1800s the natural power available from the River Irwell was being heavily used. As already noted, of the 900 feet (274.4 m) of fall between Baucup and Bolton, 800 feet (244 m) were occupied by mills. These mills, like those on the Bann in Ireland, suffered from an overabundance of water in winter and a lack of it in summer. Around 1835 plans were drawn up and construction begun on a system of remedial dams and reservoirs. Eighteen reservoirs were contemplated in the hills which formed the catchment basin of the Irwell. They were to cover 270 acres (109.4 ha) and impound 241 million cubic feet (6.8 million m³) of water. The wheels utilizing this system were expected to develop around 6,600 horsepower. One of the reservoirs was completed and another begun, but the system was never put into operation.[49]

An ambitious scheme produced by Bateman in 1844 for tapping the River Kent and its tributaries, the Mint and Sprint, for power purposes suffered a similar fate. Bateman's design called for six reservoirs covering 468.5 acres (189.7 ha), with a storage capacity of 304 million cubic feet (8.6 million m³). These were to regulate water flow on approximately 21 miles (33.8 km) of stream. The head reservoir was built in the late 1840s. But its cost considerably exceeded Bateman's estimate. Moreover, the coming of the railroad to the region made the transportation of coal into the area convenient, and cheap coal allowed manufacturers to turn to steam power instead of water. The River Kent project was terminated.[50]

Britain, of course, was not the only country to construct extensive systems for the development of water power between 1750 and 1850. A number of non-British systems were, in fact, much larger than even the largest British complexes because of the larger streams found in continental situations. The largest and the most extensive non-British hydropower developments in the nineteenth century were American.

Perhaps the best-known American plant was at Lowell, Massachusetts, adjacent to the Pawtucket Falls on the Merrimack River. Late in the eighteenth century a shipping canal had been constructed around these falls. This canal and the adjacent properties were purchased in the early 1820s by a group of Boston capitalists interested in power development rather than transportation. They enlarged the canal and built a dam 950 feet (289.8 m) long at the head of the falls. This created a storage reservoir of 1,120 acres (454 ha) and increased the available head at the site to around 35 feet (10.7 m). A system of branch canals was dug to better utilize the power available. Lands adjacent to these canals and water rights were then sold to manufacturers. By mid-century almost 9,000 horsepower were available to mills adjacent to the Lowell works, and this figure eventually grew to 12,000 horsepower.[51]

At Lawrence, Massachusetts, and Manchester, New Hampshire, also on the Merrimack River, works modelled after Lowell were erected. Extensive development at Lawrence began around 1845. A dam about 40 feet (12.2 m) high and 900 feet (274.5 m) long was erected, giving the site a fall of 26 to 30 feet (7.9–9.1 m) and forming a reservoir of around 640 acres (259 ha). As at Lowell, branch canals were constructed and lots were leased to manufacturers. By 1880 the power available to manufacturers at Lawrence was 11,000 horses.[52] Manchester did not lag far behind. Extensive development began in the period 1837–45. By 1880 the water wheels of Manchester manufacturers were drawing some 11,500 horsepower from the river.[53]

To augment the water supply available from the Merrimack, in the mid-1840s the owners of the Lawrence and Lowell complexes began to acquire the natural lakes on the upper reaches of the river, some 80 to 90 miles (129–45 km) distant. By constructing dams with sluice gates at the outlets of these lakes and arranging for regulation of outflow, the Lawrence and Lowell textile interests were able to moderate the floods of the wet season and relieve the droughts of the dry. The reservoir system constructed on the headwaters of the Merrimack in New Hampshire ultimately covered an area of 100 square miles (259 km²).[54]

Works similar to those at Lowell, Lawrence, and Manchester were erected on other American streams. Holyoke, Massachusetts, for example, constructed a very large rubble and timber dam on the Connecticut River in 1849. This structure was 1,017 feet (310.2 m) long by 35 feet (10.7 m) high and was anchored to bedrock with 3,000 iron bolts. Branch canals were dug here also, and by 1880 over 12,000 horses could be tapped.[55]

While Lowell, Holyoke, Manchester, and Lawrence were among the very largest developed water-power systems in America in the mid to late nineteenth century, by 1880 there were probably 50 others which delivered as much or more power than Britain's celebrated works at Greenock.[56]

Hydropower development projects such as these required heavy capital investments which raised the price of water power. The rising cost of water power—together with the growing scarcity of easily tapped streams, mill crowding, the need of industry for steadily larger amounts of power, the competition of the steam engine, and the greater availability of iron—provided powerful incentives towards the development and application of more efficient and powerful vertical water engines. The stimuli provided by these developments led to the adoption of the breast wheel and, eventually, the iron breast wheel.

## THE RISE OF THE BREAST WHEEL

### The Emergence of the Low Breast Wheel

Perhaps the most significant of the changes in vertical wheel technology between 1750 and 1850 was the replacement of first undershot and then overshot wheels with breast wheels in heavy industry. The importance of this development was clearly recognized by contemporary observers. A writer in 1775, for instance, called the breast wheel "one of the principal modern improvements" in water wheels.[57]

Although the breast, or breastshot, wheel could have either blades or buckets, it differed from overshot and undershot wheels in the point at which it received water. The overshot wheel took water over its summit; the undershot wheel took water from beneath. The breast wheel received its water at an intermediate point, between one and five o'clock if the water was led to the wheel from the right, or between the seven and eleven o'clock positions if water was brought in from the left. Breast wheels were often subdivided into high and low varieties. High breast wheels (sometimes called pitchback wheels) received water above the level of the axle; low breast, below.

The breast wheel was probably initially conceived as a compromise between the undershot and the overshot wheel, an attempt to combine the actions of impulse and weight. It was applied to falls between approximately 4 and 10 feet (1.2–3.0 m) where there was some doubt over whether an overshot or an undershot wheel should be installed.

Just when the breast wheel was invented is not certain. Gille would make it medieval in origin, for he believes that the watermill sketches in the *Rentier* of Aundenarde, c1270, depict breast wheels (see Fig. 2–35).[58] Reti says that Leonardo (c1500) sketched a breast wheel in one of his manuscripts, but his interpretation of the sketch is open to challenge.[59] Only in the sixteenth century is it certain that breast designs were in use. Around 1550 Juanelo Turriano illustrated several vertical water wheels without casing impacted by water at approximately axle level (in a manner similar to Fig. 5–5),[60] and the British agricultural writer Fitzherbert mentioned the breast wheel by name in 1539.[61] Breast wheels were nonetheless neither popular nor well known before 1700. The machine books published between 1550 and 1700 by Agricola, Ramelli, Böckler, Zeising, and Zonca show only the traditional overshot and undershot wheels.

In 1724 Jacob Leupold pictured and described several breast wheels (see Fig. 5–4, e.g.) in his *Theatrum machinarum generale,* calling them "halb ober schlächtiges rads" (half-overshot wheels).[62] Leupold's wheels not only received the water at an intermediate position, like Turriano's,

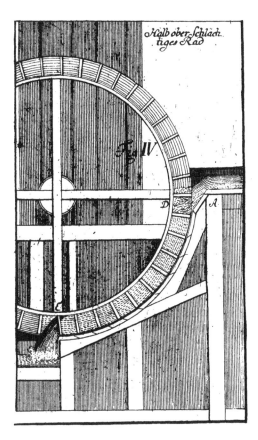

FIGURE 5–4. A "halb ober schlächtiges rad," or breast wheel with close-fitting casing from Leupold, 1724.

but they also contained the very important close-fitting casing (or breast) which held the water on the wheel until the bottom of its revolution. This construction allowed the wheel to utilize the weight of falling water and significantly raised its efficiency. Bélidor in his *Architecture hydraulique* of 1737 also devoted some little space to the breast design. He noted that some builders led water into the buckets of a water wheel at approximately axle level. The wheel he used to illustrate this practice (Fig. 5–5) was like an overshot wheel, differing only in the point at which water entered. But it was built without the close-fitting casing, or breast, of Leupold's wheels. Belidor condemned this design, noting that these wheels spilled much of their water. He recommended, instead, a design where the water was led down a steeply inclined race curved at its bottom to fit the arc of the wheel. This revised design was similar in principle to Leupold's encased "half-overshot wheels," though it had a much larger portion of its fall devoted to impact.[63]

Despite the growing awareness of breast wheels, exemplified by Leupold and Bélidor, they remained, in the early eighteenth century, a deviant design of little importance. Bélidor devoted only a paragraph or two to them. Stephen Switzer in his *Hydraulicks and Hydrostaticks* of 1729

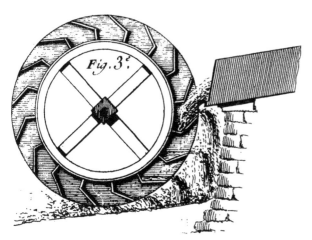

FIGURE 5–5. Usual design of the breast wheel in the 1730s according to Bélidor. Bélidor condemned this arrangement because of spillage from the buckets.

and Beyer in his *Theatrum machinarum molarium* of 1735 completely ignored them.

There are two possible reasons why breast wheels remained unpopular. One is technological: the critical innovation which made breast wheels an effective prime mover—the close-fitting casing—may not have been invented or widely known to millwrights before the 1730s. This interpretation is supported by evidence from Desaguliers. In 1744 Desaguliers observed that breast wheels were sometimes used when there was not enough water to turn an undershot wheel or enough fall to use an overshot. But, he added, "few of these Wheels are good for any thing, because they seldom receive their Water well, and generally part with it too soon; that is, before it comes to the Bottom."[64] Desaguliers noted that this common defect could be remedied by equipping the wheel with a circular channel or trough that came very close to, but did not quite touch, the edges of the blades of the water wheel. This close-fitting circular channel, Desaguliers declared, would make the breast wheel equal to the overshot. Apparently unfamiliar with Leupold's treatise, and not having noticed Bélidor's design, Desaguliers attributed this improvement to a friend and associate, Dr. Barker.[65]

Another possibility is that the technology to make the breast wheel efficient was available some time before the eighteenth century but was not adopted for economic reasons. It is quite conceivable, for example, that the close-fitting casing, which Desaguliers attributed to Barker, had already been developed by Leonardo or some engineer some centuries earlier, but had been either rejected as impractical or neglected since the additional expense required to erect it was not worth the gain in efficiency or power in an era when water power was still relatively plentiful.

Whatever the case, the breast wheel came into extensive use only after 1750. The key figure in its rise in popularity was John Smeaton. As noted in Chapter 4, Smeaton in the 1750s carried out extensive quantitative model experiments on water wheels. These experiments indicated

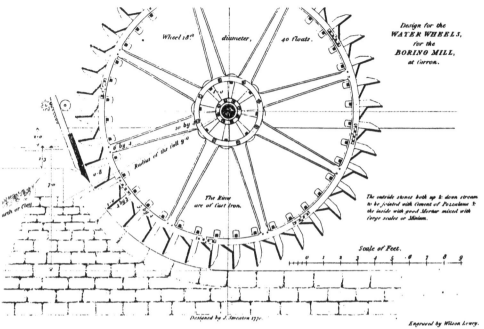

Design for the
**WATER WHEELS,**
for the
**BORING MILL,**
at Carron.

Wheel 18.ᵗ diameter, 40 floats.

The Rims
are of Cast Iron.

The outside stones both up & down stream
to be jointed with Cement of Pozzulana &
the inside with good Mortar mixed with
forge scales or Minium.

Scale of Feet.

Designed by J. Smeaton 1770.

Engraved by Wilson Lowry.

FIGURE 5–6. Low breast wheel designed by John Smeaton for the Carron Ironworks, 1770. This was typical of Smeaton's low breast designs. The masonry breast, or casing, is at the left.

that wheels which utilized the weight of water were at least twice as efficient as wheels which depended on impulse. Smeaton recognized that the breast wheel, when equipped with the close-fitting casing, was, in effect, a weight-activated wheel. Smeaton was widely used as a consultant on mill projects all over England between 1750 and 1780. He applied his findings by substituting, wherever possible, low breast wheels (for example, Fig. 5–6) for undershot wheels.[66] Undershot wheels, in practice, had an efficiency only on the order of 20–30%. Smeaton's low breast wheels could deliver at least 40–50%. With Britain's growing water-power shortage this twofold gain in efficiency assumed an importance it would not have enjoyed a century earlier and made the extra expense of the casing a profitable investment.

The Kilnhurst Forge hammer mill was a typical Smeaton-designed installation. The fall available at the mill site was rather low, less than 6 feet (1.8 m). In order to make optimum use of this fall Smeaton installed a masonry casing, or breast, almost 4 feet (1.2 m) high, which embraced the lower part of the mill's water wheel. The breast increased the height at which the water began acting on the wheel and confined it to the spaces between the floatboards so that it would act during its fall by weight instead of by inefficient impulse.[67]

The effects of Smeaton's discovery of the superior efficiency of weight-driven wheels and his extensive application of the breast design to utilize this discovery on British water wheel practice were noted by a number of later observers. For example, in 1808 the millwright-en-gineer Peter Ewart (1767–1842) declared that Smeaton's use of gravity-powered low breast wheels wherever possible was largely responsible for the disappearance of undershot wheels from England: Smeaton's meth-od of designing water wheels, he said, was "afterwards so generally adopted and improved upon by himself and by other engineers in this country, that although undershot water-wheels were, about fifty years ago, the most prevalent, they are now rarely to be met with, and wherev-er economy of power is an object, no new ones are made."[68] Rees in 1819 said much the same thing: "It was one of the continual occupations of Mr. Smeaton, during forty years to improve the old water-mills, by substituting breast-wheels for undershot; and the advantages were uni-formly so great, that these mills were copied, until scarcely any of the original construction remained."[69] And John Farey (1790–1851), an en-gineer and writer, noted a few years later: "The numerous specimens [of watermills] which he left in all parts of the kingdom, were copied by others in their respective neighbourhoods, till the improved methods of constructing water-mills became general and was attended with great benefit to trade."[70]

In the eighteenth century the low breast design was also applied on occasion on the European continent, perhaps due to the influence of de Parcieux, Johann Euler, or Smeaton, for it appears in several of the plates of Diderot and d'Alembert's *Encyclopédie*.[71]

*The Sliding Hatch and the Emergence of the High Breast Wheel*
The most serious defect in most early low breast wheels was inadap-tability to varying water levels. The portion of the available head of water devoted to action by weight, which Smeaton had found should be as high as practically possible, was fixed by the height of the breast. If the water level rose several feet over the breast there was no way of utilizing the extra head, save by inefficient impact. On the other hand, if the water level fell below the crown of the breast, the wheel would not operate at all. Smeaton attempted to alleviate this problem by using a false, or movable, crown, a piece of wood which was added to the top of a masonry breast, raising its height when the water level was high. When the water sank too low to run over the movable crown, the crown was removed to admit water at a lower level.[72]

A device which more conveniently and efficiently performed the same service—the sliding hatch, or overflow sluice gate—was introduced in the 1780s by John Rennie (1761–1821), another of Britain's great en-gineers.[73] In the sliding hatch, water flowed *over* the gate which admit-ted water to the water wheel, instead of under as in the traditional control gate. This allowed the operator to control the height at which the water flowed over the gate and adjust the gate to different stream conditions.

FIGURE 5–7. Sliding hatch in operation at the Royal Armoury Mills, Enfield Lock. The upper part of the breast, or sweep, was made of cast-iron plate curved to the form of the wheel. The shuttle (B) was applied to the back of this plate and was raised or lowered by rack and pinion so that a thin stream of water flowed over the shuttle and onto the water wheel from the maximum possible height.

Rennie first conceived of the possibility of an overflow gate while experimenting in his youth on a water wheel at an oat and barley mill on his brother's estate in East Lothian. The mill was flooded on one occasion, and Rennie saw that water running over the closed sluice gate was causing the wheel, even when partially submerged, to turn. In 1783, with this incident in mind, he began to experiment with overflow gates, and in 1784 put the idea into practice at the water-powered rolling mill which he erected at Soho for Matthew Boulton, James Watt's partner.[74]

In its simplest form Rennie's sliding hatch (Fig. 5–7) consisted of a flat sluice gate or shuttle *over* which water flowed. Its operation was the reverse of the traditional control gate. The traditional sluice gate was raised when a wheel needed water, and the water flowed under it to impact the water wheel. Rennie's hatch was lowered to initiate operation, and the water flowed over it onto the wheel. Rennie's gate was usually fitted to the flat upstream portion of a masonry breast and was worked by rack and pinion. It was adjusted as the water level rose or fell, so that just a thin sheet of fluid flowed over the gate and onto the wheel and the highest available head was always utilized to power the wheel by weight, as Smeaton had suggested.

To gain additional control over the way water was fed onto the wheel, some sliding hatches were equipped with horizontally situated metal slats. These were placed (as in Fig. 5–12) just beyond the overflow gate and were inclined at angles designed to assist the smooth flow of water onto the water wheel.[75] Rennie later developed a double hatch (Fig. 5–12 also) which had some additional features. Just upstream of a curved iron grating which fit the sweep of the wheel, an overflow shuttle was installed. Above this was installed a second shuttle or gate. By adjusting the two gates, not only could one regulate the height at which water was admitted to a water wheel, but (by varying the opening between the two gates) one could also control the amount of water and its velocity. Usually the additional or upper hatch dipped slightly below the surface of the water in order to give the water jetting between the two gates the proper velocity.[76] This type of gate was often used in textile mills, where fine regulation of water speed was necessary.

Britain's water-power shortage insured the domination of efficient low breast wheels in industrial areas with low falls where the flow of water was at a fairly constant level. But it was Rennie's sliding hatch which insured its general domination over the undershot wheel *wherever* heads were low and *wherever* there was a need to economize water, for it was only after the introduction of Rennie's hatch that the low breast wheel was able to operate under variable heads with the same ease as undershot wheels. By 1800 the low breast wheel had replaced the undershot wherever water economy was a problem.

It was not until well after 1800, however, that the breast wheel began to replace the traditional overshot wheel. Although Smeaton played the key role in the adoption of the low breast design in Britain, he played little role here. He did construct a few high breast wheels, but for high heads he mostly used the traditional overshot wheel, although he sometimes installed a casing similar to that used on breast wheels to inhibit spillage on the lower portion of the wheel's loaded arc.[77]

While superior efficiency was the key technical factor behind the replacement of undershot by low breast wheels, efficiency played little role in the transition from overshot to high breast, since overshot and high breast wheels were roughly equivalent in that category. Ability to handle variable heads and backwater were more important. Backwater (or partial flooding) was always a serious problem with vertical water wheels. Few streams remain at even close to a constant level year around, and when high water came it usually forced watermills to operate with their wheels partially submerged. Long experience had taught millwrights that a wheel which rotated in the same direction as the flow of water turned much better in backwater than one which rotated in the opposite

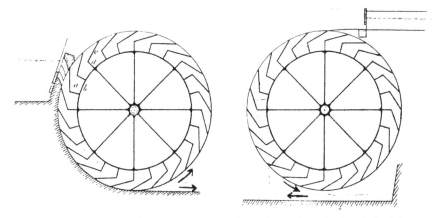

FIGURE 5–8. Comparison of high breast and traditional overshot wheels. The high breast wheel is at the left; the traditional overshot at the right. As the diagram indicates, the direction of rotation of the high breast wheel was identical with the flow of the water in the tail race. This enabled the high breast wheel to operate effectively even when partially flooded. The traditional overshot wheel, on the other hand, rotated in a direction opposite the flow of the tail water, making operation very difficult in backwater.

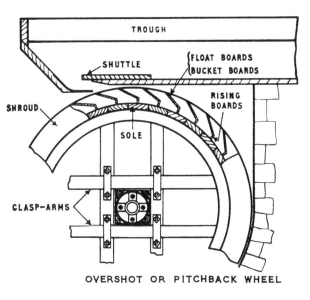

OVERSHOT OR PITCHBACK WHEEL

FIGURE 5–9. Smeaton's shuttle, or pentrough, for reversing the rotation of traditional overshot wheels. This enabled them to operate better in flood waters.

direction. The traditional overshot wheel, as Figure 5–8 indicates, was at a disadvantage vis-à-vis the high breast wheel in this regard. Smeaton, recognizing this, had reversed the direction of some of his overshot wheels by using a penstock with a backward slanted shuttle (Fig. 5–9).[78] Rennie's sliding hatch, combined with the high breast (or pitchback) configuration, achieved the same effect in a more efficient and convenient manner.

Another factor behind the popularity of high breast wheels was their adaptability to varying falls. With the traditional overshot wheel the available fall had, at all times, to be greater than the diameter of the wheel. If it fell below that level, the wheel could not operate. Moreover, increases in head could be utilized only by inefficient impact. The high breast wheel, which could be equipped with a sliding overflow hatch, was more adaptable. Since water was fed onto its side, the water could be directed with an arc-shaped hatch into buckets at any of a wide range of levels, utilizing at all times the maximum possible gravity head.[79] Rennie's invention thus played an important role in the emergence of both high and low breast wheels as alternatives to the traditional overshot and undershot designs.

It is not possible to date precisely when the breast wheel became more popular than the traditional forms of the vertical wheel in heavy industry. But it is fairly certain that this had occurred, in Britain and America at least, by the period 1825–50. In 1823, for example, the British textbook writer John Millington declared that the breast wheel was "by far the most common" type of water wheel.[80] Writing around mid-century, the American hydraulic engineer James B. Francis observed that breast wheels had come to dominate in the principal manufacturing establishments in New England some years previously.[81] William Fairbairn, one of the nineteenth century's best-known British engineers, declared in the 1840s that the breast wheel had "taken precedence of [sic] the overshot wheel," and, finally, David Scott, another British technician, commented at about the same time that while the pure overshot configuration was not uncommon, it was "generally abandoned by millwrights who make pretensions to a superior knowledge of the principles which ought to govern the transmission of hydraulic power."[82]

## THE IRON WATER WHEEL

### The Introduction of Iron Parts

Rivaling in importance the substitution of the breast wheel for the traditional overshot and undershot wheels was the substitution of iron for wood in the construction of vertical water wheels, a step encouraged by the growing availability and declining price of iron in late-eighteenth- and early-nineteenth-century Britain.

As a material for fabricating water wheels, wood was deficient in a number of respects.[83] Because wood is a relatively weak material, wooden members must be relatively large to have reasonable strength. This limited both the efficiency and the size of wooden wheels. For example, to admit water with minimum impact (recognized as critical to high efficiency after the work of Smeaton, de Parcieux, Borda, and Carnot) water wheels needed buckets with thin partitions. But the minimum thickness of a wooden bucket partition, consistent with adequate strength, was in the neighborhood of 1 inch (2.54 cm). When wooden bucket boards of this thickness were placed close together to inhibit spillage from buckets in the lower portion of an overshot wheel's loaded arc, the relatively thick partitions obstructed the entry of the water into the buckets, caused considerable impact and splashing losses, and therefore significantly reduced the wheel's efficiency. Because of the weakness of wood, moreover, wheels built larger than around 40 feet (12.2 m) in diameter or 20 to 25 feet (6.1 to 7.6 m) in width tended to warp under their own weight. This made it difficult to design wooden vertical wheels which could generate more than around 20 to 25 horsepower.

The alternate wetting and drying of the wooden wheel was also a serious problem. As its planks expanded and shrunk, the joints, pins, and bolts worked loose. Wooden wheels often got out of balance as one side absorbed more water than the other. In winter water thrown from the buckets accumulated as ice on the arms, shrouds, and shafts, seriously impeding and sometimes stopping operations. As a result of these factors and other problems (like rot) a wooden water wheel required frequent maintenance and had a relatively short life span. George Woodbury, relying in part on experience in rehabilitating and operating a wooden wheel, estimated that repairs to a wooden vertical water wheel required the complete replacement of every part of the wheel in five years. The director of the Springfield Armory reported that wooden wheels lasted not more than eight years. And the millwright Hobart estimated that wooden construction lasted seven years before repair, ten before replacement with new material.[84]

The substitution of metal for wood in water wheels offered a solution to many of these problems. This had long been recognized, and some replacement of wood, as noted in Chapter 3, had occurred long before the eighteenth century. The traditional vertical wheel ran on iron gudgeons and had axles reinforced with iron hoops. Probably only the high price of iron and the absence of a heavy demand for power in units above 5 to 10 horsepower had retarded further development in this direction.

The first major step towards the iron industrial water wheel was the introduction of the iron axle. The shaft of a wooden wheel was one of its weakest links, especially since good oak timbers of the size necessary for large wheels were scarce by 1800 in Britain.[85] On the traditional water wheel the axle had to both support the weight of the wheel and transmit its power. It had, in other words, to resist torque stresses as well as tensile and compressive stresses. Due to natural irregularities in wood, or to overloading, oak axles frequently failed and fractured. In many cases the natural strength of a wooden axle was weakened by the conditions under which it operated (alternately wet and dry and at a variety of temperatures) and by the mortises made for the wheel's spokes and gudgeons. Iron axles did not suffer from these defects. There was no shortage of iron in Britain by the late eighteenth century. Iron axles could be forged or cast to any desired dimension or shape; they were less likely to have natural flaws; they were not susceptible to rot; and because of their high heat conductivity, ice did not accumulate as readily on them. Iron axles could be made either stronger (for the same size) or smaller (for the same strength) than comparable wood axles.

John Smeaton, seeing these advantages, attempted early in his career, amid considerable scepticism, to introduce iron shafts. In 1755 he designed a cast-iron axle for a windmill. Looking back on this event he commented: "I applied them as totally new subjects, and the cry then was, that if the strongest timbers are not able for any great length of time to resist the action of the powers, what must happen from the brittleness of cast iron? It is sufficient to say that not only those very pieces of cast work are still in work, but that the good effect has in the north of *England,* where first applied, drawn them into common use, and I never heard of any one failing."[86] Smeaton did not apply iron to the axles of watermills until almost 15 years later. In the 1760s he was made a consulting engineer for the Carron Iron Works, Falkirk, Scotland, an association which lasted for over a decade and which probably encouraged the extension of his experiments with iron. Thus, when the oak shaft of the water wheel of Carron Blowing Engine no. 1 fractured in 1769, Smeaton recommended replacing it with a cast-iron axle.[87] It was not until 1770 or 1771 that his recommendation was followed, but success here led him to extend the innovation to other mills.[88]

The introduction of iron axles was apparently not without problems, for both Rees and Farey note that Smeaton's wheels suffered a number of axle failures.[89] Cast iron was brittle. Even though a cast-iron shaft would usually take torque strain quite well, a sudden, sharp application of force would sometimes break it. This danger was particularly acute on cold, frosty days. Especially prone to breakage were the flanges cast on the axle to receive the spokes or arms (see Fig. 5–10). These flanges tended to cool faster than the main shaft after casting, leaving a heavily porous area where they joined the axle proper. According to Rees (1819), this problem eventually forced Smeaton to adopt tubular cast-iron axles in place of his early solid cast-iron ones, but Farey a few years later (1827) claimed that Smeaton met the problem in another way—by casting flanges separately and fastening them to the axle with wedges afterwards.[90] Paul N. Wilson, who studied Smeaton's water wheel designs, was unable to find any evidence that Smeaton ever used tubular iron axles, so Farey's analysis is probably the correct one.[91] The tubular design was most likely a post-Smeatonian development. In any case, despite early problems with solid cast-iron axles, Smeaton's innovation was eventually successful, for, according to Farey, the modified cast-iron axles were "very permanent" and "came into very general use in the course of twenty years."[92]

After the success of his early experiments with cast-iron axles, Smeaton expanded the use of iron parts. In 1770 he used cast iron for

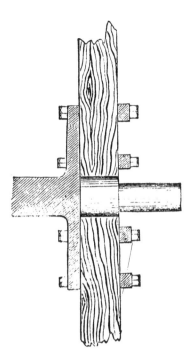

FIGURE 5–10. View of the end of one of Smeaton's solid cast-iron axles. This view shows the flanges to which the arms of the water wheel were bolted.

the rims of the water wheel that powered the Carron boring mill to provide a flywheel effect. The iron was cast in segments and assembled much as wooden rim pieces had been on traditional wheels.[93] In 1776 Smeaton used a similar design for the wheel of the River Coquette Rolling Mill and Wright's Oil Mill at Hull in Yorkshire.[94] In 1771 Smeaton proposed using strips of wrought iron approximately 1.25 to 1.50 inches (3.2–3.8 cm) wide to strengthen the edges of the bucket boards on overshot wheels.[95] And in 1780 he suggested the use of wrought-iron plates for the bucket boards themselves.[96] Finally, Smeaton pioneered in the introduction of iron gearing. He used iron gears as early as 1765 at Kilnhurst Forge in Yorkshire.[97] Carron Blowing Engine no. 1 had a cast-iron spur wheel in 1768.[98] And in 1778 Smeaton used cast-iron gearing at Brooks Hill, Deptford, London.[99]

The foundations laid by Smeaton were built upon by several other prominent British engineers. Among them was John Rennie, who perpetuated Smeaton's practice of using iron parts in the construction of watermills. In 1784, for example, at Boulton and Watt's at Soho, Rennie made extensive use of cast iron in the gearing and shafting of the power transmission system leading from the water wheel. And at the steam-powered Albion Mills, constructed between 1784 and 1788, he replaced wooden gears and shafts entirely with cast iron.[100]

The all-iron water wheel slowly emerged between 1770 and 1800 or 1810. Exactly who built the first water wheel made totally or almost totally of iron and where it was erected are not certain. Hills in his study of power in the industrial revolution suggested that the millwright-engineer Peter Ewart may have built the first all-iron or almost all-iron wheel at Styal in either 1800 or 1807.[101] But it might also have been constructed by T. C. Hewes of Manchester, for there was a Hewes-designed wheel in the Strutt Mill at Belper before 1811 which used wood only in the buckets.[102]

### The Suspension Wheel: The Iron Wheel Refined

Most of the early industrial wheels which incorporated iron in their construction were in overall design and operation very similar to the traditional wooden water wheels, with power developed and transmitted as it had been for almost two millenia. Only around 1810 was it recognized that the use of iron made a different system of construction and power transmission possible.

In the traditional watermill, torque was transmitted from the rim through the arms and the axle to a gear wheel set in a pit inside the mill. From this pit wheel, power was taken and transmitted to the mill's machinery. This arrangement required that the arms and axle of the water wheel be of sufficient size and strength to both support the weight of the wheel *and* to transmit the torque forces developed on the wheel's rim. The torque strains were already considerable in water wheels prior to the eighteenth century. But they were greatly increased after the efficiency advantages of slow motion were recognized by de Parcieux and Smeaton around 1750 (see Chapter 4) and as mill owners attempted to draw greater power from their engines under the pressure of water-power shortages.

One method of relieving the arms and axle of their power transmission chores had been suggested as early as the 1600s, for Jan van der Straet around 1580 illustrated several water-powered mills where power was taken from pinions mortised into the rims of the water wheel (see, for example, Fig. 2–12).[103] The suspended wheel shown in Figure 5–11, from the *Encyclopédie* of Diderot and d'Alembert indicates that the idea was still alive in the mid-eighteenth century.

Rim, or peripheral, gearing offered several advantages over the conventional arrangement. The dimensions of axle and arms could be reduced to a size just sufficient to support the weight of the wheel and water, since they no longer had a role in power transmission. This would mean a lighter wheel and less power lost through gudgeon friction. In addition, taking power from the rim of the wheel gave the gearing maximum velocity, obviating the need for some step-up gearing in the mill and saving both power and money.

Despite these advantages, rim gearing was never popular in wooden water wheels. A number of factors were probably responsible. Wooden gear teeth mortised into the rim of a wooden wheel would have seriously weakened it. Maintenance was a problem. Gear teeth worked loose rapidly in rims because of the expansion and contraction of the wood as

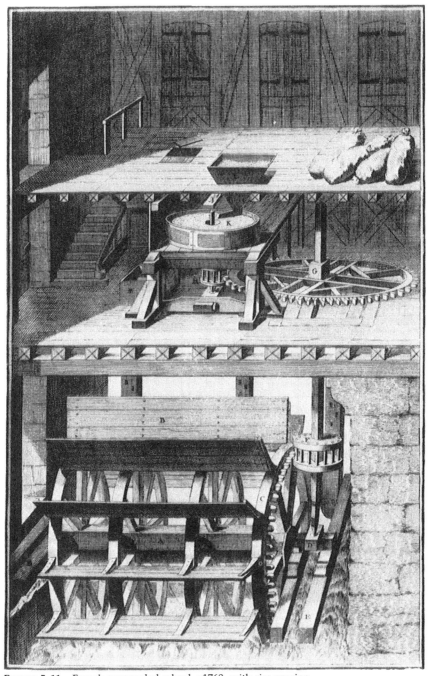

FIGURE 5–11. French suspended wheel, c1760, with rim gearing.

it was alternately wetted and dried. Similar teeth on a conventional gear wheel mounted in a pit inside the mill and protected from the elements would not have been troubled as much. Finally, it may have been difficult to take advantage of the reduced dimensions of arm and axle that rim gearing made possible because of the non-uniform nature of wood and its weakness in torsion.

Iron, with its greater strength, more uniform texture, and ease of forming and attachment, provided a way to avoid these difficulties. Thin wrought-iron rods could be used in place of wooden arms; light tubular axles could replace solid oak beams; cast-iron gear segments could be securely bolted into cast-iron shrouds without seriously weakening them, or the gear teeth could be cast as one piece with the shroud. The need of textile mills for high shaft speeds and for power in large units probably provided the incentive for a reconsideration of rim gearing within the context of an increasingly iron-oriented technology.

Exactly when the peripheral cast-iron gear was introduced is not certain. It was probably around 1800, for by 1806 there were indications that it was becoming standard practice in textile mill construction.[104] Thomas C. Hewes (1768–1832) is usually credited with being the first to recognize the full implications of the new power transmission system on the dimensioning of iron axles and shafts.[105] Hewes was a millwright and engineer who settled in Manchester, the heart of Britain's cotton textile industry, in 1792. Little is known of his background, though he is said to have had a good knowledge of mechanics in general and hydrostatics in particular. Hewes had a general interest in the substitution of iron for wood, using it to frame buildings, shaft mills, and support bridges. But his greatest contribution was the introduction of the iron suspension (or spider) wheel.[106]

Basically, the suspension wheel was an iron water wheel with peripheral (or rim) gearing where the function of the arms and axle was reduced to supporting the weight of the wheel and holding it to shape. This involved substituting thin wrought-iron tie rods, stretched so tight that they supported the wheel by tension alone, for the large square or rectangular wooden arms or thick cast-iron bars previously used. The suspension wheel was in principle similar to the modern bicycle wheel, and it shared the bicycle wheel's characteristic light weight and airy appearance.

Little is known about how Hewes developed the wheel. William Fairbairn, who adopted the Hewes design, refined it, and used it extensively from around 1820 on, provides the bulk of the available information:

> It was reserved for Mr. T. C. Hewes, of Manchester, to introduce an entirely new system in the construction of water wheels, in which the wheels, attached to the axis by light wrought-iron rods, are supported simply by suspension. I am informed that a wheel on this principle in Ireland was actually constructed with chains, with which, however, from the pliancy of the links, there was some difficulty. But the principle on which

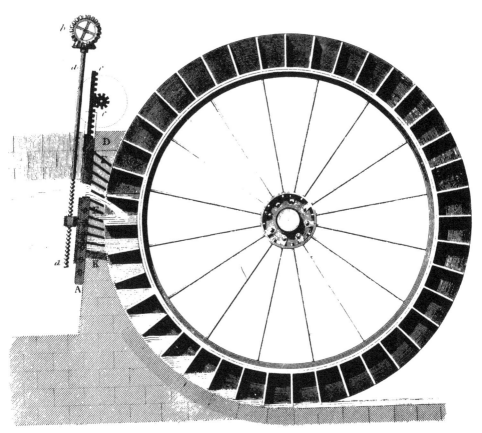

FIGURE 5–12. Hewes suspension wheel at Belper, c1810. This wheel was 21.5 feet (6.55 m) in diameter and 15 feet (4.6 m) wide, and was the first iron suspension wheel. It was also one of the first wheels to use a double sliding hatch and the flyball governor. By means of the double hatch (*A* and *B*) water could be let onto the wheel at any point between *D* and *K*, and the quantity and velocity of the water very closely controlled.

this wheel was constructed was as sound in theory as economical in practice, and is due originally, it is said, to the suggestion of Mr. William Strutt, and was carried out some fifty years ago [c1810] by Mr. Hewes, whilst at the same time Mr. Henry Strutt applied the principle to cart wheels.... Mr. Hewes employed round bars of malleable iron in place of the chains, and this arrangement has kept its ground to the present time [c1860], as the most effective and perfect that has yet been introduced.[107]

The first suspension wheels were probably the two designed and con-structed by Hewes for William Strutt's cotton mills at Belper, sometime before 1811 (Fig. 5–12). These were breast wheels 21.5 feet (6.55 m) in diameter, 15 feet (4.6 m) wide, operated by a 14-foot (4.3-m) fall. The two wheels were placed in line, separated only by a wall which supported the bearings of their common axle. They functioned as one wheel, for while it would have been possible to build a single iron wheel 30 feet

(9.15 m) wide, Hewes felt his design was more convenient and practical. Power was taken from these wheels off a ring of cogs screwed to the circular rim of the wheel on the inside edge. A pinion mounted adjacent to the loaded side of the wheel and worked by the ring of cogs conveyed power to the mill. For the arms, whose power transmission function was eliminated, Hewes used light wrought-iron rods only 2.0 to 2.25 inches (5.1–5.7 cm) in diameter. To sustain the wheel sideways, diagonal bars were run from the hubs on the axle to a circular shroud in the middle of the wheel.[108]

The Hewes wheels at Belper were distinguished by other refinements beyond the introduction of the suspension system. They were among the first to be regulated by a double hatch. They may also have been the first water wheels whose motion was regulated by a fly ball governor,[109] although there is evidence that governors, perhaps in a more primitive form, were in use in the textile mills owned by the Robinson near Papplewick as early as the mid-1780s.[110]

Particularly in textile mills, like those at Belper, a smooth, constant motion was necessary so that the threads being mechanically spun were not broken by jerks from the prime mover. Early forms of the water wheel governor were simple. For example, one early water-powered textile mill used smith bellows. The bellows were driven by a cam off the water wheel shaft which acted on the lower board. The air compressed by the motion of the cam was directed from the bellows through a pipe fitted with a stopcock. If the lower board worked faster than the air could escape, the bellow's upper board would rise; if it worked slower, the upper board would sink. The motion of the upper board was used to regulate the sliding hatch, or overflow sluice, of the water wheel by throwing into action the wheel work of the mill to raise or lower it.[111]

This primitive governor gave some speed control, but it was not capable of the fine regulation which textile mills needed. The centrifugal (or fly) ball governor was. It had been applied in primitive form to windmills in the 1780s, may have been used on watermills at about the same time, and was adapted by James Watt to steam engines around 1790. The operation of the fly ball governor is relatively easy to understand. The vertical shaft to which the balls were linked and about which they rotated was driven off the millwork. As the velocity of the millwork increased, the balls were thrown up and outward, lifting a loose collar on the shaft; as the velocity decreased, the balls fell toward the shaft and lowered the collar. On the steam engine the motion of the collar actuated the steam admission control valve.[112]

Water wheels were not as easy to regulate with the fly ball governor because moving the sliding hatch or shuttle of a large water wheel required more force than the motion of the governor could provide. Hewes and Strutt overcame this problem at the Belper mills in a rather ingenious way. They constructed a large water cistern adjacent to the water wheel. Pipes controlled by stopcocks led water from the mill pond to the cistern and from the cistern to the tail race. Almost filling the cistern was a large floating chest. The chest rose and fell with the water level in the cistern and was linked by rack and gearing to the machinery which controlled the water wheel's sliding hatch sluice gate. Linkages attached to the centrifugal ball governor controlled the cocks which regulated the entrance and exit of water into the cistern. When the mill moved at its intended speed, both cocks were shut. If the mill began to move too slowly, the governor activated the cock in the supply pipe; the water level in the cistern rose, lifting the floating chest; the motion of the chest lowered the shuttle and increased the flow of water on the wheel. If the mill began to turn too fast, the governor activated the relief pipe; this lowered the water level in the cistern, and the falling of the chest raised the shuttle and diminished the water flow.[113] By 1820 this cumbersome and indirect system of speed control was replaced with more direct linkages between fly ball governor and sliding hatch.[114] In these the motion of the governor's collar engaged the mill's wheel work via a sliding clutch mechanism to raise or lower the sluice. Figure 5–13 illustrates a system of this type.

Hewes initiated the iron suspension wheel and developed an effective governor to regulate its motion. The iron breast wheel was further refined by William Fairbairn (1789–1874). Fairbairn was born in Scotland of farming stock and in 1804 was apprenticed to a millwright in a Northumberland colliery. While learning the millwright trade, he imposed on himself an intensive program of self-study in science and mathematics. In 1811, at the end of his apprenticeship and following some limited millwrighting activity, he moved to London. His stay there was short. He toured Britain for several years and then in 1813 settled permanently in Manchester. He found a job in 1816 and 1817 working as a draftsman for Hewes, but soon split with Hewes and formed with James Lillie, also a former Hewes employee, a millwrighting partnership. The firm of Fairbairn and Lillie soon became well known, achieving between 1817 and 1832 (when the partnership broke up) a worldwide reputation as builders of suspension wheels. Even after the split both Lillie and Fairbairn continued to manufacture water wheels, although the focus of Fairbairn's interests increasingly shifted to other applications of iron, notably in ships, and to the manufacture of locomotives, steam engines, and boilers.[115]

In designing water wheels Fairbairn basically followed Hewes. He built all-iron wheels on the suspension principle. But he made several improvements on his mentor's designs. One of these was in the connection between the wrought-iron spokes and the axle. In Hewes's suspension wheels the wrought-iron rods used for the arms and diagonal braces were usually threaded and secured to a flange on the hub of the wheel by nuts (Fig. 5–14). These nuts could be adjusted to center the wheel and keep it running true. But in practice there were problems. The nuts tended to work loose, and the wheels frequently got out of balance. In

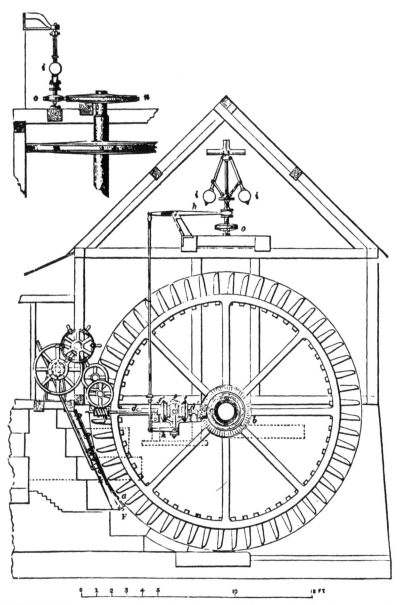

FIGURE 5–13. Flyball governor used to regulate the speed of water wheels. The motion of the wheel was transmitted to the shaft of the governor at *o* by millwork. This motion caused the metal balls (*i*) to spin at a distance from the vertical shaft of the governor dependent on the speed of the shaft. The position of the revolving balls, through linkages, raised or lowered a metal collar which, through other linkages (*h*), activated a sliding clutch. This clutch through gearwork raised or lowered the sliding hatch, and the increased or decreased amount of water flowing over the hatch raised or lowered the speed of the water wheel.

FIGURE 5–14. Center, or hub, of a Hewes cast-iron axle. The drawing shows Hewes's method for attaching the arms (*c*, *c*) and braces (*e*, *e*) to the axle. Hewes threaded the arms and braces and used bolts to adjust and tighten them down.

1824 Fairbairn modified the Hewes system. The ends of the arms to be attached to the axle were forged square and fixed in sockets in a cast-iron hub. They were secured there by means of gibs and cotters instead of nuts (Fig. 5–15). The gibs and cotter pins, Fairbairn found, did not work loose like nuts and bolts. Wheels built by Fairbairn could run for years with minimum maintenance.[116]

A more important improvement was Fairbairn's ventilated bucket. The efficient operation of a gravity wheel involved several factors. Two of the more important were (1) the free entrance of water into and exit of water from the buckets, and (2) minimum spillage from the buckets on the loaded arc of the wheel. One solution for the spillage problem was the use of very constricted bucket openings. But when this solution was adopted, the entrance and exit of water was seriously impeded. John Robison in 1797 commented on this problem:

> There frequently occurs a difficulty in the making of bucket wheels, when the half-taught mill-wright attempts to retain the water a long time in the buckets. The water gets into them with a difficulty which he cannot account for, and spills all about, even when the buckets are not moving away from the spout. This arises from the air, which must find its way out to admit the water, but is obstructed by the entering water, and occasions a very

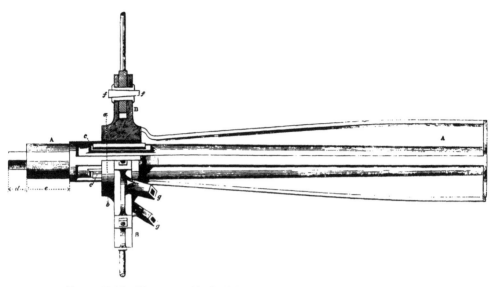

FIGURE 5–15. Elevation of half of the main axle and one hub, or center, of a Fairbairn water wheel. The center, or hub, of this wheel is keyed to the axle rather than cast integrally with it (the keys are *c, c*). Fairbairn used gibs and cotters (*f, f*) instead of Hewes's nuts and bolts to link arms to hubs. The arms and braces were secured in slots cast into the center or hub (*B, B*, for the arms; *g, g*, for the diagonal braces of the wheel).

> great sputtering at the entry.... This obstruction is vastly greater than one would imagine.[117]

As the wheel rose from the bottom of its cycle, there was the related problem of allowing the water to leave the buckets freely so that it did not have to be lifted. Here, too, constricted openings caused problems. Quoting Robison again:

> There is another and very serious obstruction to the motion of an overshot or bucketed wheel. When it moves in back water, it is not only resisted by the water, when it moves more slowly than the wheel,... but it lifts a great deal in the rising buckets. In some particular states of back water, the descending bucket fills itself completely with water; and, in other cases, it contains a very considerable quantity, and air of common density; while in some rarer cases, it contains less water, with air in a condensed state. In the first case, the rising bucket must come up filled with water, which it cannot drop till its mouth get out of the water. In the second case, part of the water goes out before this; but the air rarefies, and therefore there is still some water dragged or lifted up by the wheel, by suction.[118]

To ease the entrance and exit of water the traditional millwright had several options. He could make the mouths of the buckets very large, so that the air would have plenty of space in which to escape. He could make the delivery spout considerably narrower than the buckets, allowing the air room to escape on the side. Or he could bore air holes in the sides of the buckets or in the soal. Theoretically these holes would facilitate both the inflow and outflow of air and water, and efficient buckets with small openings could be used. These modifications, however, caused other problems. Very large buckets with wide mouths spilled water early. When holes were bored in the buckets or soal, water poured uselessly out after the air had escaped. Moreover, if the holes were in the soal, the water fell into the interior of the wheel, shortening the life of wheel and axle. More complicated remedies, such as tubes or boxes attached to the soal plates and extending upwards from the buckets to allow air, but not water, to escape from the holes in the buckets, worked no better. In brief, attempts to reduce spillage in water wheels created obstructions to the free entrance and exit of water, while attempts to eliminate the entrance and exit problem resulted in increased spillage. Thus, Fairbairn noted, "it has been found more satisfactory to submit to acknowledged defects [spillage], than to incur the trouble and inconvenience of partial and imperfect remedies."[119]

In 1825 Fairbairn and Lillie were contracted to deliver several very large iron water wheels to works owned by James Finlay and Company. It was while involved in the construction of these wheels that Fairbairn first seriously directed his attention to the problem of ingress and egress of water. The wheels made for Finlay were completed around 1827, with the problem still unsolved. Fairbairn next built a breast wheel for Andrew Brown of Linwood. Shortly after this wheel was installed, complaints came in. When the wheel was loaded and in flood waters, the air in its buckets could not escape fast enough to admit the incoming water and created a water blast. The compressed air in the buckets, according to Fairbairn, forced water and spray to a height of 6 to 8 feet (1.8–2.4 m) above the shuttle that admitted water to the wheel.[120]

Pressured to correct this fault, Fairbairn cut small openings in the soal plates and attached small interior buckets to the inner soal (b, b in Fig. 5–16A). The use of iron sheets instead of wood for the soal of the wheel made this modification possible with a minimum of expense and inconvenience. Fairbairn believed that this modification would allow air to escape upwards into the interior of the wheel with little or no water spillage. It was successful. Fairbairn reported that the buckets were effectively cleared of air while they were filling and that when the wheel operated in backwater the same modification facilitated the readmission of air and the discharge of water. According to Fairbairn the effect produced by his modification "could scarcely be credited." The "wheel not only received and parted with the water freely, but an increase of nearly one-fourth of the power was obtained."[121]

Sensing that he had hit on something important, Fairbairn improved on his makeshift design by introducing in 1828 the ventilated bucket and the ventilated bucket with closed soal plates (Figs. 5–16B and 5–16C). These worked even better and either reduced or completely eliminated spillage into the interior of the water wheel.[122]

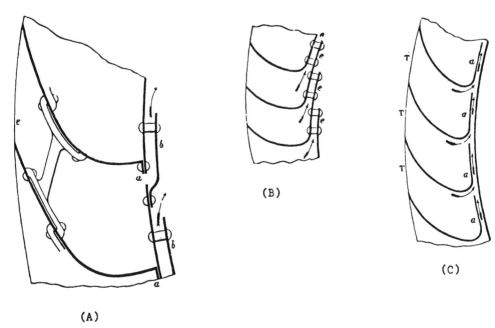

(B)

(C)

(A)

Figure 5–16. Fairbairn's ventilated buckets. (A) Modification made by Fairbairn to the Brown water wheel at Linwood to permit easier entrance of water. Openings were cut across the soal plates (a, a) and small internal buckets (b, b) were attached. The openings allowed air to escape freely from the main buckets as water entered, while the smaller internal buckets inhibited spillage. (B) Method used by Fairbairn to ventilate buckets on wheels operating under low falls. The wheel had no soal, but the buckets were bent and extended to overlap each other, leaving openings for the escape of air (e, e). (C) Ventilated buckets with closed soal plates were used by Fairbairn for very high falls and low water volumes. The closed soal and the continuous passage left for air between the buckets and the soal plates completely eliminated spillage into the interior of the wheel while facilitating the admission of air.

### The Iron Water Wheel: Construction Details

The iron industrial water wheel, which emerged with the work of Smeaton, Rennie, Hewes, Strutt, and Fairbairn, was in overall appearance somewhat similar to the traditional wooden water wheel of the eighteenth century. The parts, for example, were essentially the same—axle, arms, shrouding, soal, and buckets or floats. But the way these parts were fabricated and the options available in assembling them differed considerably from previous practice.

The axle of the traditional wooden water wheel was almost invariably a solid oak timber. The axle of the iron water wheel could be cast either solid or tubular and be made of either wrought iron or cast iron. Smeaton's axles, as noted, were cast solid (Fig. 5–10). And some of the large wheels of the nineteenth century used forged wrought-iron axles. But by the early 1800s the tubular cast-iron axle was generally preferred. The tubular axle combined strength with lightness. It was some-

FIGURE 5-17. Iron water wheel axles.
The top figure is a solid ribbed cast-iron axle with gudgeons ($A$, $A_1$) cast
with the axle and turned down on a
lathe. The bottom figure is a tubular
cast-iron axle.

times cast with ribs in an attempt to further increase its resistance to
tensile, compressive, and torque stresses (Fig. 5–17). For very wide water
wheels, axles were sometimes cast in two or three parts and then put
together at the construction site.

The gudgeons of solid iron water wheels were normally cast as part of
the shaft itself, though turned on a lathe to ensure good roundness. The
gudgeons of tubular shafts could be fabricated in the same manner, but
were more often separate pieces of iron work, as in the traditional wheel,
and had to be secured to the axle. Often holes were bored into the butt
of such a shaft and the gudgeons turned to fit tightly into these recesses,
as in Figure 5–20. Sometimes annular flanges were cast onto the ends of
tubular axles and the gudgeons bolted to these.

In many iron wheels the simple gudgeon support bearings used on
water wheels in previous centuries were replaced with capped sockets.
These were frequently equipped with removable brasses and oil holes
where cups could be set for automatic lubrication (Fig. 5–18).[123]

The radial arms of an iron wheel could be either cast-iron bars or
wrought-iron rods (or wood in a hybrid wheel). Cast-iron arms were
frequently used with wheels whose diameter was less than 20 feet (6.1
m); wrought-iron arms were widely used on iron wheels of all sizes. The
connection between the arms and the axle varied. Most commonly, the
axle of the iron wheel was cast with several flat flanges or hubs to which
the arms were bolted (Figs. 5–10, 5–12, 5–20). But frequently, and especially when cast-iron arms were used, the axle was cast with bosses,
hubs, or centers which contained recesses into which the arms were
rigidly secured (Fig. 5–15). Sometimes the rims were not cast with the
wheel, but separately, and then secured by pins or wedges to the axle.

The outer ends of the arms on an iron wheel were normally secured to
the shrouds. In many cases they were given a T-shape and bolted to

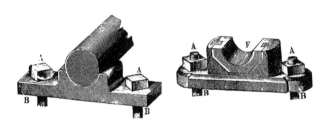

FIGURE 5–18. Bearings for iron water wheels. The bearing at top left is a simple open bearing of the type used for centuries in various forms on water wheels. The bearing at the top right is a more sophisticated form featuring a removable and replacable brass, *F*. The bearing at the bottom is a covered bearing which has removable brasses and an oil hole at *L* to which an oil cup could be fixed for automatic lubrication.

recesses on the shroud's inner periphery (Fig. 5–19); in other cases the arms were secured in recesses cast into the side of the shrouds (Fig. 5–20).

In addition to radial arms, most of the larger iron wheels had diagonal arms or braces, which were installed to prevent lateral motion. These, too, were small wrought-iron rods, about 1.5 to 2.5 inches (3.8–6.4 cm) in diameter. The braces usually ran from the shroud on one side of an iron wheel to the hub on the opposite side. Where a middle shroud or rim was built on a wheel, the usual configuration was for the braces to run from the central rim or shroud out to the two external hubs (Fig. 5–20).

The shrouds or sides of an iron water wheel were built up from cast-iron or wrought-iron plates fabricated in 10- to 15-inch (25.4- to 38.1-cm) segments. These plates were usually overlapped and joined by riveting, but in some cases the edges of the plates were simply butted and bolted tightly or secured by having the arm of the wheel overlap the joint (Fig. 5–20). The shroud segments were usually made with a perpendicular flange along their lower edge (their internal circumference). It was to this flange that the soal plates, around ⅛ inch-thick (3.2 mm) sheet iron, were riveted or bolted with lap joints. Unlike traditional water wheels, which usually had rims as well as shrouds, iron wheels required only shrouds. Iron shrouds had sufficient strength to support the wheel without a separate rim.[124]

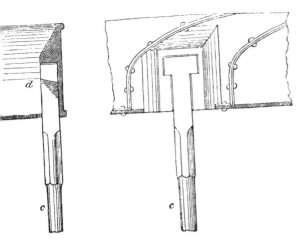

FIGURE 5–19. Linkage between arms and shrouds of an iron water wheel. The outer end of the arm or spoke was shaped like the letter $T$ and bolted into a recess cast into the shroud. Fig. 5–20 shows another method, where the arms were bolted to the outside of the shroud at the point where the shroud segments met.

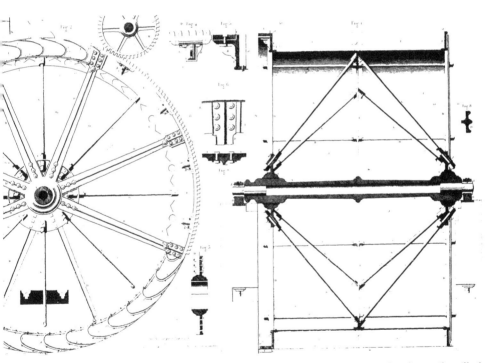

FIGURE 5–20. French iron water wheel of 1841. This wheel was erected at the textile mill of Charles Bellot at Angecourt, near Sedan. It was 13.4 feet (4.1 m) in diameter and 9.6 feet (2.92 m) wide. The wheel had peripheral gearing, curved sheet-iron buckets riveted directly to the soal plates, and an intermediate rim or shroud to which the cross braces ($G$, $G$, $G$) were secured. The drawing also illustrates the use of a tubular iron axle with inserted gudgeons. This wheel delivered around 17 hp at 78% efficiency.

When the buckets of an iron wheel were made of wood, and sometimes even when iron buckets were used, grooves or recesses were cast into the shroud plates to receive the bucket boards. But as wrought-iron sheets replaced wooden bucket boards in the early nineteenth century, other methods were more widely used for securing buckets to the wheel. In many iron wheels, wrought-iron sheets were bent into either a J- or a V-shape and riveted directly to the soal plates. In other wheels angle iron was attached to the soal, and the buckets were attached to the angle iron (see Fig. 5–21).

The substitution of iron for wood in buckets offered opportunities for extensive changes in form, shape, and arrangement. The bucket boards of the traditional wooden wheel were relatively thick. The volume occupied by these boards restricted capacity and caused splashing. Further, it was difficult to shape wood to the best form for avoiding impact and retaining water on the downward arc. Rolled wrought-iron plates were thin (about ⅛ inch, or 3.2 mm). They provided less of a barrier to the entrance of water and occupied a much smaller volume. The sheets were easily molded and could be given the curvilinear form which reduced impact and splash losses at entry and retarded spillage later. With iron buckets water began to work earlier than with wooden buckets and stayed on the wheel longer. The iron wheel also turned better in backwater, since there was less friction between iron and water than between wood and water.[125] As a result of these advantages, by the 1840s wood was fast disappearing in the construction of buckets in industrial water wheels.[126]

Despite the advantages of iron, it nonetheless replaced wood slowly. Iron had begun to supplant wood on an extensive scale with Smeaton's work at Carron around 1770, but it did not completely supersede wood until very late in the nineteenth century. There were many large industrial water wheels, as well as small mill wheels, built well after 1850 that were hybrid—i.e., part wood, part iron. The higher cost of iron in many parts of the world was largely responsible for this situation. In France, for example, the high price of iron meant that wheels of mixed construction were still more frequently employed there in the 1850s than all-iron wheels.[127]

Wood-iron hybrids (largely wood, but with some iron parts) were often used in small mills, where a cheap yet reliable method for securing the water wheel to its axle was needed. As previously noted, the compass arm technique required mortising which weakened the axle, while clasp arms were not particularly strong where the spokes were wedged to the axle tree. A cast-iron center or hub provided the solution for the mill owner who wanted some of the advantages of iron construction without heavy expense. An oak axle was fitted with two cast-iron centers or hubs, each with six to eight sockets or recesses cast into its periphery. The two hubs were tightly wedged to the wheel's axle, and the arms of the water wheel were inserted into the sockets.

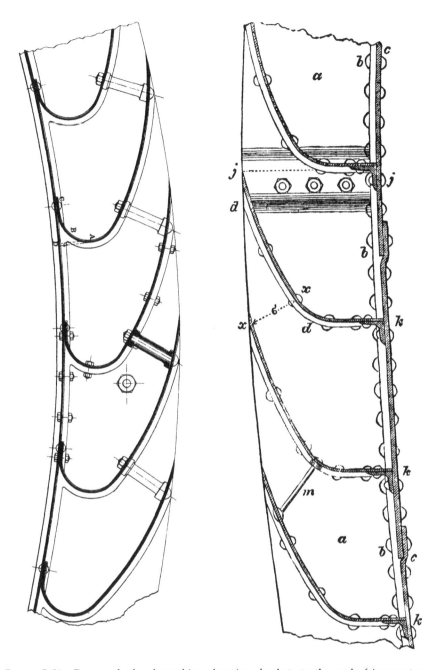

FIGURE 5–21. Two methods of attaching sheet-iron buckets to the soal of iron water wheels. *Left*, Buckets directly riveted to the soal of the wheel without the use of angle iron. Reinforcing spacers were also frequently used with this design. *Right*, Buckets secured to the soal and shrouds (*a, a,*) by pieces of angle iron (*k, k, k*) and reinforced by spacers (*m*).

Iron was sometimes used at other points on a predominantly wooden wheel, for example at the ends of the axle. The mortising required to fit a gudgeon to the butt of a wooden axle weakened it. And even after the gudgeon was fitted, keeping it tight was a perpetual problem. These difficulties could be avoided by fitting the ends of an axle with an iron butt or cap. This cap was tightened down with screws or bolts. The top of the cap was cast with notches to receive a cross-arm gudgeon, held rigidly in place by bolting. In other wood-iron hybrid wheels, sheet-iron buckets were used in place of wooden ones.[128]

Just as there were hybrid wheels made largely of wood, but with a substantial number of iron parts, there were other wheels made mostly of iron but retaining some wooden parts (as iron-wood hybrid). For instance, the very large and powerful iron water wheel built by Henry Burden in the 1850s and described in more detail later in this chapter retained wooden soal and buckets.[129]

The overall effect of the general substitution of iron for wood was, for wheels of comparable size and power, a much lighter wheel. Sutcliffe noted in 1816, for example, that a wheel 5 feet (1.5 m) wide and 15 feet (4.6 m) in diameter weighed nearly 2 tons less when made of iron instead of oak.[130] Redtenbacher in 1848 also found that many (but not all) iron wheels weighed less than wooden wheels per unit of horsepower.[131] But lighter weight per unit of horsepower was not the major advantage of iron construction. More important was the fact that iron made possible the construction of larger, more durable, more efficient, and more powerful wheels.

### The Iron Industrial Water Wheel: Performance

The magnitude of the increases in the dimensions, efficiency, power, and durability of vertical water wheels between 1750 and 1850 is difficult to determine with exactness. Of these parameters the efficiency improvement is perhaps the most difficult to establish, for efficiency was not commonly measured before 1750. But that significant gains occurred is certain. For example, Smeaton's substitution of low breast and overshot wheels (efficiencies between 40 and 70% in practice) for undershot wheels (efficiencies of only 15 to 30% in practice) resulted in almost a 100% improvement wherever it was carried out. The use of the sliding hatch or overflow gate probably added from 5 to 10% to the overall efficiency of both low and high breast wheels. The traditional wooden overshot wheel was the least susceptible to improvement, since its efficiency was already relatively high, in the neighborhood of 55 to 65%. But by 1850 to 1900 this figure had probably been increased to 70 to 80% by modification to high breast operation, by the application of the sliding hatch, by the use of the sheet-iron buckets and ventilated buckets, by slower operating speeds, and by lighter construction. Some iron overshot and high breast wheels even attained efficiencies as high as 85 to 90%. Table 5–1 compares the efficiencies of traditional wooden water

TABLE 5–1.  AVERAGE EFFICIENCY OF WATER WHEELS, 1750–1900

| Type of Wheel | Fall | Volume* | Efficiency |
|---|---|---|---|
| *Traditional wooden water wheels* | | | |
| Undershot | | | 15–35% |
| Overshot | | | 50–70 |
| *Wooden breast wheels* | | | 40–65 |
| *Iron water wheels* | | | |
| Low breast | | | |
| small falls | 1.0–3.8 ft | 7.0–180 ft$^3$ | 50–75 |
| | (0.3–1.15 m) | (0.2–5.0 m$^3$) | |
| medium falls | 3.9–11.5 ft | 21.0–180 ft$^3$ | 65–75 |
| | (1.2–3.5 m) | (0.6–5.0 m$^3$) | |
| high falls | 9.8–14.8 ft | 14.0–140ft$^3$ | 70–80 |
| | (3.0–4.5 m) | (0.4–4.0 m$^3$) | |
| High breast | 11.5–19.7 ft | 0.7–28.0 ft$^3$ | 60–80 |
| | (3.5–6.0 m) | (0.02–0.8 m$^3$) | |
| Overshot | | | |
| small falls | 9.8–16.4 ft | 2.6–31.8 ft$^3$ | 65–75 |
| | (3.0–5.0 m) | (0.075–0.9 m$^3$) | |
| medium falls | 16.4–26.2 ft | 2.1–24.7 ft$^3$ | 75–80 |
| | (5.0–8.0 m) | (0.06–0.7 m$^3$) | |
| high falls | 26.2–39.4 ft | 1.4–17.7 ft$^3$ | 80–88 |
| | (8.0–12.0 m) | (0.04–0.5 m$^3$) | |

SOURCES: Wilhelm Müller, *Die Eisernen Wasserräder*, 1 (Leipzig, 1899): 12; Paul N. Wilson, "British Industrial Waterwheels," International Symposium on Molinology, *Transactions* 3 (1973): 26; William Fairbairn, *Treatise on Mills and Millwork*, 3rd ed., 1 (London, 1871): 126; Fitz Water Wheel Co., *Fitz Steel Overshoot Water Wheels*, Bulletin no. 70 (Hanover, Pa., 1928), pp. 6, 10–12; and F. Redtenbacher, *Theorie und Bau der Wasserräder* (Mannheim, 1846), pp. 158, 308.
*Of water flowing on the wheel per second.

wheels with the efficiencies attained by iron water wheels in the late nineteenth century. The widespread adoption of the breast wheel, the use of iron construction, and the emergence of new forms of vertical water wheels due to the application of quantitative analysis (notably the Poncelet and Sagebien wheels) meant that by 1850 vertical water wheels could tap falls from between 1 foot (0.3 m) and 50 feet (15 m), under appropriate flow conditions, with an efficiency exceeding 50% (see Table 5–1 and Fig. 5–22).

Iron not only contributed to raising the efficiency of water wheels. Its superior strength permitted the construction of much larger and more powerful water wheels. Wooden wheels, as mentioned in Chapter 3, did not perform reliably when built more than 40 feet (12.2 m) in diameter, even though larger sizes were sometimes possible. The corresponding limit for iron wheels was in the neighborhood of 70 to 80 feet (21.3–24.2 m). James Smith of Deanston, for example, built an iron wheel (Fig. 5–27) 70 feet 2 inches (21.4 m) in diameter around 1830–35 for a cotton mill operating on Shaw's Water Works near Greenock.[132] The wheel used by the Great Laxey Mining Company on the Isle of Man beginning in 1854 was 72.5 feet (22.1 m) tall (Fig. 5–28).[133] Bryan Donkin, a British

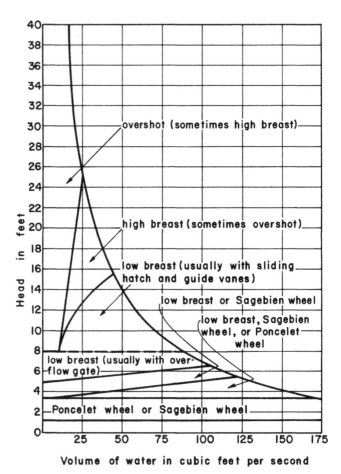

FIGURE 5–22. Vertical water wheels available for operation at various heads and volumes with efficiencies above 50%.

engineer, built a wheel 76.5 feet (23.3 m) diameter for an Italian firm,[134] and John and Robert Mallet of Dublin constructed perhaps the tallest iron vertical water wheel ever built, an engine 80 feet (24.4 m) in diameter for a paper mill near the Irish capital.[135]

While the upper limits attainable with iron construction were significantly greater than with wood, the average height of water wheels increased only slightly. For traditional wooden wheels the average height was probably around 14 to 15 feet (4.3–4.6 m). The average diameter of 69 water wheels erected or designed by Smeaton between 1750 and 1780 was slightly over 18 feet (5.5 m).[136] Fairbairn's wheels, in the first half of the nineteenth century, averaged a little over 23 feet (7.0 m) in diameter.[137] And tables with data on 132 German iron wheels published in 1898 indicate that their average diameter was only around 17 feet (5.2 m).[138]

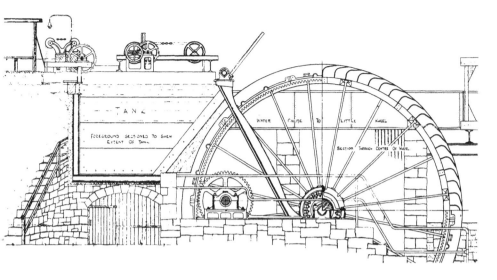

FIGURE 5–23. The all-metal high breast suspension wheel designed by Hewes and Wren for the Arkwright Mill at Bakewell in 1827. This wheel utilized both peripheral gearing and a sliding hatch. It developed around 140 hp and was replaced by a turbine only in 1955.

If the average height of water wheels increased only slightly during the transition from wood to iron, this was not the case with their width. The typical wooden water wheel, especially in the overshot configuration, was rather narrow in proportion to its height. While a few wooden wheels were 20 to 25 feet (6.1–7.6 m) wide, and a very exceptional one or two much wider,[139] most were around 2 to 3 feet (0.6–0.9 m) in breadth. The ratio of breadth to height on the traditional wheel was thus somewhere around 1 to 5 or 1 to 6.

The typical iron industrial wheel, on the other hand, was much broader in proportion to its height. For example, the wheels installed by Hewes for Strutt's Belper mills were 15 feet (4.6 m) wide, but only 18 feet (5.5 m) tall.[140] Another wheel built by Hewes and Wren in 1827 at Bakewell measured 18 feet (5.5 m) wide by 25 feet (7.6 m) tall (Fig. 5–23).[141] Most of Smeaton's wheels were built largely of wood, but the tendency towards wider wheels was already visible in his work. Smeaton's wheels averaged about 4 feet 3 inches (1.3 m) wide, and the ratio of breadth to height was 1 to 4.24.[142] As all-iron wheels replaced the wooden wheels of the eighteenth century, the trend towards wideness continued. The average width of 57 German iron overshot wheels listed in tables published by Wilhelm Müller around 1900 was 5.87 feet (1.79 m), and 75 breast wheels listed in similar tables had an average width of 6.66 feet (2.03 m).[143] The ratio of width to diameter on these 132 wheels was 1 to 2.7. Fairbairn's wheels were even wider. The average width of 15 water wheels constructed by Fairbairn whose dimensions were published in the *Engineer and Machinist's Assistant* was slightly under

TABLE 5–2. DIMENSIONS OF WATER WHEELS, c1750–c1900

| | No. | Ave. diameter | Ave. width | Ratio width:diameter |
|---|---|---|---|---|
| Traditional wooden water wheels, c1700–1750 (est.) | — | c15 ft (c4.6 m) | c2.5 ft (c0.75 m) | c1:6 |
| Wheels designed or erected by Smeaton, c1750–c1780 | 69 64 | 18.28 ft (5.57 m) | 4.31 ft (1.31 m) | 1:4.24 |
| Wheels constructed by William Fairbairn, c1817–c1870 | 62 15 | 23.23 ft (7.09 m) | 15.79 ft (4.81 m) | 1:1.5 |
| German overshot water wheels, 1898 | 57 | 14.01 ft (4.27 m) | 5.87 ft (1.79 m) | 1:2.4 |
| German breast wheels, 1898 | 75 | 19.22 ft (5.86 m) | 6.66 ft (2.03 m) | 1:2.9 |

16 feet (4.9 m), or three to five times greater than the average of wooden wheels, and the ratio of width to height on these wheels was approximately 1 to 1.5.[144] (See Table 5–2.)

Since the average width of a water wheel increased significantly between 1750 and 1850 and the average diameter slightly, and since efficiency improved, one might naturally expect significant increases in the average power output of vertical water wheels. This did occur. The average output of water wheels between 1700 and 1750 was probably about 4 to 7 horsepower, as I indicate in Chapter 3. This figure rose between 1750 and 1850 to around 12 to 18 horsepower—at minimum, double that of wheels a century earlier, and perhaps three to four times greater. My estimate of the average power of nineteenth-century iron water wheels is based on the data given in Table 5–3.

While these figures indicate the average power output of water wheels in the nineteenth century, they do not reflect the much higher output obtainable when the technology was pushed to its limits. Table 5–4 lists some of the large iron wheels built in the nineteenth century. Some of the giant vertical water wheels constructed between 1820 and 1870 developed more than ten times as much power as the average iron wheel.

William Fairbairn constructed several of these giant wheels. Among the best known were the two wheels he built for the Catrine Works, Aryshire, Scotland, between 1825 and 1827 (Fig. 5–24). Originally four wheels were planned for Catrine's 48 feet (14.6 m) of available fall. But the two wheels Fairbairn constructed were found sufficient, and the others were never built. These wheels were 50 feet (15.2 m) in diameter by 10.5 feet (3.2 m) wide, had an efficiency of around 75%, and developed 120 horsepower apiece. Some 35 years after their construction Fairbairn boasted: "Taking into consideration the height of the fall, these wheels, both as regards their power and the solidity of their con-

TABLE 5–3.   AVERAGE POWER OF VERTICAL WATER WHEELS, c1830–c1900

| | | |
|---|---|---|
| 1835 | Report of the Factory Inspectors on water power developed in British textile mills: 1,037 water wheels reportedly developed 12,138 hp, or an average power per wheel of.................................. | 11.70 hp |
| 1836 | In West Riding there were 102 water wheels which generated 873 hp, an average for each wheel of.................................. | 8.56 hp |
| 1837 | In the Bolton district 60 water wheels developed 1,171 hp, or an average per wheel of.......................................... | 19.52 hp |
| 1838– 39 | Report of the Factory Inspectors on the power developed in British textile mills: 2,230 water wheels produced 27,983 hp, or an average per wheel of..................................................... | 12.55 hp |
| 1842 | Power delivered by 8 overshot pumping wheels operating in the mines of southwestern Britain averaged........................... | 28.90 hp |
| 1853 | Survey of watermills on the River Wandle: a total of 38 mills with 52 water wheels, approximately half flour mills, half industrial mills (snuff, oil, paper, copper, felt, print): the 52 wheels generated 781 hp, or an average per wheel of.................................. | 15.02 hp |
| 1868 | Average power of 193 water wheels and 3 water-pressure engines operating in the Harz area of Prussia with 2,896 hp total............ | 14.78 hp |
| 1870 | United States Census of 1870: 51,018 water wheels generating 1,130,431 hp. Many of these were undoubtedly water turbines, but the average power per wheel was............................... | 22.16 hp |
| 1880 | United States Census of 1880: 55,404 water wheels generating 1,225,279 hp. Many of these were probably water turbines, but the average was.................................... | 22.12 hp |
| 1899 | Average power of 57 iron overshot water wheels in a table published by Wilhelm Müller........................................... | 13.24 hp |
| 1899 | Average power of 75 low and middle breast water wheels in tables published by Müller........................................... | 19.72 hp |

SOURCES: Andrew Ure, *Philosophy of Manufactures*, 3rd ed. (London, 1861), p. 482; J. James, *History of the Worsted Manufacture* (London, 1857), p. 450; A. E. Musson, "Industrial Motive Power in the United Kingdom, 1800–70," *Economic History Review*, 2nd ser. 29 (1976): 433; Anthony Rouse, "Overshot Water Wheels employed in pumping water at Wheal Friendship Lead and Copper Mines, near Tavistock, in July, 1841," Institution of Civil Engineers, *Minutes of Proceedings* 2 (1842): 97–102; Frederick Braithwaite, "On the Rise and Fall of the River Wandle . . . ," Institution of Civil Engineers, *Minutes of Proceedings* 20 (1860–61): 191–210; Denys Barton, *Essays in Cornish Mining History*, 1 (Truro, Cornwall, 1968): 174–75; Wilhelm Müller, *Die Eisernen Wasserräder*, 1 (Leipzig, 1899): 72–77, and 2 (Leipzig, 1899): 110–13, 156–57; G. F. Swain, "Statistics of Water Power Employed in the United States," American Statistical Association, *Publications* 1 (1888): 35.

struction, are even at the present day among the best and most effective structures of the kind in existence."[145] They were, indeed, excellent wheels. They were dismantled only in 1947, after 120 years of continuous operation, and were found to be true within less than ⅛ inch (3.2 mm). Every few years the spokes had had to be adjusted, but otherwise the wheels had given little trouble.[146]

A Fairbairn-built wheel at the flax mill of Thomas Ainsworth of Cleator (Fig. 5–25) was 20 feet (6.1 m) in diameter by 22 feet (6.7 m) wide. This wheel was equipped with Fairbairn's closed-soal ventilated buckets. Operating under a fall of 17 feet (5.2 m), it was rated by Fairbairn at 100 horsepower.[147]

Perhaps the greatest concentration of waterpower within a single British mill was at Deanston, owned by the same proprietary as Catrine

| Location (designer) | Year built | Diam. | Width | Head | Est. h.p. |
|---|---|---|---|---|---|
| Wm. Strutt, North Mill, Belper (designer unknown) | 1804 | 18.0 ft (5.5 m) | 23.0 ft (7.0 m) | 13.5 ft (4.1 m) | 130 |
| J. J. & W. Wilson, Castle Mills, Kendal (firm's millwright and Baleman & Sherrat, Manchester) | 1806 | 14.0 ft (4.25 m) | 12.0 ft (3.65 m) | 7.2 ft (2.2 m) | 20 |
| Wm. Strutt, West Mill, Belper (Hewes) | before 1811 | 21.5 ft (6.55 m) | 15.0 ft (4.6 m) | 14.0 ft (4.3 m) | 80 |
| Crystal Factory, Baccarat, France (Aitken & Steel) | 1816 | 6.5 ft (2.0 m) | 12.8 ft (3.9 m) | 6.9 ft (2.1 m) | 20 |
| Royal Armoury Mills, Enfield Lock (Lloyd & Ostel) | before 1819 | 18.0 ft (5.5 m) | 14.0 ft (4.3 m) | 6.0 ft (1.8 m) | 30 |
| Samuel Greg, Quarry Bank Mill, Lancashire | 1820 | 32.0 ft (9.75 m) | 21.0 ft (6.4 m) | 30.0 ft (9.15 m) | 120 |
| Woodside Paper Mill, Aberdeen, Scotland (Hewes & Wren) | 1826 | 25.0 ft (7.6 m) | 21.0 ft (6.4 m) | 18.0 ft (5.5 m) | 150 |
| James Finlay, Catrine Mills, Aryshire (Fairbairn & Lillie) | 1827 | 50.0 ft (15.25 m) | 10.5 ft (3.2 m) | 48.0 ft (14.6 m) | 120 |
| Richard Arkwright, Bakewell, Derbysh. (Hewes & Wren) | 1827 | 25.0 ft (7.6 m) | 18.0 ft (5.5 m) | 22.0 ft (6.7 m) | 140 |
| Schlumberger & Cie., Spinning Mill, Guebwiller, France (unknown Eng. designer) | before 1830 | 29.9 ft (9.1 m) | 10.3 ft (3.15 m) | 25.4 ft (7.75 m) | 50 |
| James Finlay, Deanston Mill, Perthsh. (Fairbairn & Lillie and James Smith) | 1831 | 36.0 ft (11.0 m) | 12.0 ft (3.65 m) | 33.0 ft (10.1 m) | 100 |
| Shaw's Cotton Spinning Co., Greenock, Scotland (James Smith) | around 1830 | 70.2 ft (21.4 m) | 13.0 ft (4.0 m) | 64.3 ft (19.6 m) | 200 |
| Thomas Ainsworth, Cleator, Cumberland (Fairbairn & Lillie) | 1840 | 20.0 ft (6.1 m) | 22.0 ft (6.7 m) | 17.0 ft (5.2 m) | 100 |
| Wheal Friendship Mine, nr. Tavistock, Devon (unknown) | before 1841 | 51.0 ft (15.55 m) | 10.0 ft (3.05 m) | 51.0 ft (15.55 m) | 60 |
| Tuscan Felted Cloth Mill, St. Marello, Florence, Italy (Bryan Donkin) | 1843 | 76.5 ft (23.3 m) | 2.0 ft (0.6 m) | 76.0 ft (23.2 m) | 30 |
| Izmet Wool Mill, Izmet, Turkey (Fairbairn) | 1843 | 30.0 ft (9.15 m) | 13.0 ft (4.0 m) | 28.0 ft (8.5 m) | 35 |
| Iron low breast wheel with sliding hatch described by Redtenbacher | before 1846 | 19.7 ft (6.0 m) | 13.6 ft (4.15 m) | 9.8 ft (3.0 m) | 55 |

| Location (designer) | Year built | Diam. | Width | Head | Est. h.p. |
|---|---|---|---|---|---|
| Iron-wood hybrid high breast wheel described by Redtenbacher | before 1846 | 22.5 ft (6.9 m) | 12.9 ft (3.9 m) | 16.9 ft (5.15 m) | 55 |
| Large overshot wheel described by Redtenbacher | before 1846 | 39.4 ft (12.0 m) | 6.2 ft (1.9 m) | 41.3 ft (12.6 m) | 30 |
| Flour mill on Essonne R., Corbeil, France (Cartier & Armengaud) | before 1850 | 21.3 ft (6.5 m) | 21.25 ft (6.5 m) | 8.2 ft (2.5 m) | 30 |
| Great Consols Mine, nr. Tavistock, Devon (Smith, mine's engineer) | around 1850 | 40.0 ft (12.2 m) | 12.0 ft (3.7 m) | 40.0 ft (12.2 m) | 140 |
| Burden Iron Works, Troy, N.Y. (H. Burden) | 1851 | 62.0 ft (18.9 m) | 22.0 ft (6.7 m) | 71.7 ft (21.85 m) | 280 |
| Lady Isabella, Great Laxey Mining Co., Isle of Man (Rbt. Casement) | 1854 | 72.5 ft (22.1 m) | 6.0 ft (1.8 m) | 72.5 ft (22.1 m) | 230 |
| Marly Pumping Works, Marly, France (Dufrayer) | around 1855 | 39.4 ft (12.0 m) | 14.8 ft (4.5 m) | 9.0 ft (2.75 m) | 70 |
| Rishworth Mills, Halifax, Yorksh. (Taylors of Marsden) | 1864 | 57.5 ft (17.5 m) | 12.0 ft (3.65 m) | 57.5 ft (17.5 m) | 240 |
| Killhope Lead Mine, nr. Stanhope, Durham | around 1870 | 39.4 ft (12.0 m) | 6.2 ft (1.9 m) | 39.4 ft (12.0 m) | 90 |
| Dinorwic Slate Quarry, Llanberis, N. Wales (De-Winton & Co.) | 1870 | 50.5 ft (15.4 m) | 5.25 ft (1.6 m) | 50.5 ft (15.4 m) | 80 |
| Iron high breast wheel described by Bach | before 1886 | 23.0 ft (7.0 m) | 8.2 ft (2.5 m) | 15.3 ft (4.65 m) | 55 |
| Iron overshot water wheel, figures given by Müller | before 1899 | 21.3 ft (6.5 m) | 9.85 ft (3.0 m) | 24.9 ft (7.6 m) | 70 |
| Drape factory of M. Crozel at Vienne, France with Sagebien wheel (Gonnard) | before 1899 | — | — | 7.5 ft (2.3 m) | 55 |
| Sagebien wheel at cotton mill of Augustin Vy, Serquigny, France (Boudier) | before 1899 | — | — | 7.2 ft (2.2 m) | 55 |
| Iron breast wheel, figures given by Müller | before 1899 | 23.0 ft (7.0 m) | 14.8 ft (4.5 m) | 6.55 ft (2.0 m) | 70 |

SOURCE: I have used the table in Paul N. Wilson, "British Industrial Water Wheels," International Symposium on Molinology, *Transactions* 3 (1973): 22, as a model for this table and for some of the data contained in it. But I have corrected some of his figures and added information on a number of wheels not included in his table.

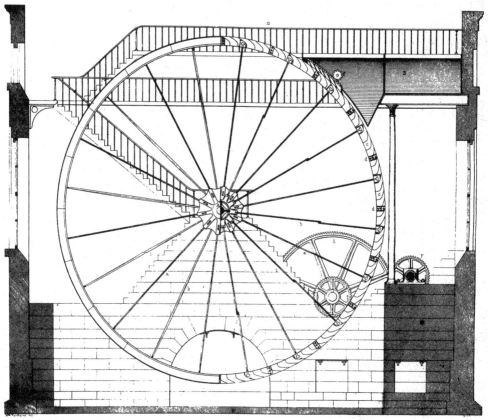

FIGURE 5–24. Fairbairn's Catrine water wheel. One of the 50-foot (15.25-m) diameter suspension wheels designed by William Fairbairn for the Catrine Mills of James Finlay in Ayrshire, this high breast wheel developed around 120 hp at 75% efficiency and operated in the mills until 1947.

(James Finlay). At Deanston two water wheels had been erected for a textile mill around 1780. Plans developed in the 1820s called for replacing these two wheels with eight iron high breast wheels. Each of the eight wheels was to develop around 100 horsepower. Fairbairn was hired to design the first two wheels, but their construction and erection were taken over by the Deanston firm's engineer, James Smith. Smith not only erected the two Fairbairn-designed wheels, but designed and erected two additional ones, so that the mill had 400 horsepower available from its water engines. The other four wheels were never built. At Deanston, water was led into the mill in a W-shaped wrought-iron pentrough (Fig. 5–26) supported on iron columns. Water was delivered to the water wheels, each 36 feet (11.0 m) in diameter, on each side of the pentrough through sliding hatches. These wheels powered the Deanston works for almost 125 years. Only in 1949 were they, like the Catrine wheels, dismantled and sold for scrap. They were replaced by water turbines and alternators.[148]

FIGURE 5–25. Fairbairn and Lillie wheel installed for Thomas Ainsworth of Cleator, 1840. This wheel was 20 feet (6.1 m) in diameter and 22 feet (6.7 m) wide. Unlike many Fairbairn and Lillie high breast wheels, this one was provided with a close-fitting casing. The sliding hatch is visible at A. The wheel was equipped with closed-soal ventilated buckets.

Smith later built a very large iron suspension wheel for Shaw's Cotton Spinning Company of Greenock, Scotland. This wheel (Fig. 5–27), as noted previously, was 70 feet 2 inches (20.8 m) in diameter and 13 feet (3.8 m) wide. Operating under a head of 64 feet 4 inches (19.0 m), with 166 buckets and 75% efficiency, it delivered around 200 horsepower.[149] This wheel was replaced by a turbine in 1881 after a half century of service.

Perhaps the best known of the giant iron water wheels is the Lady Isabella wheel (Fig. 5–28), designed by Robert Casement for the Great Laxey Mining Company on the Isle of Man. Lady Isabella was 72.5 feet (22.1 m) high and 6 feet (1.8 m) wide. Like most of the large iron wheels

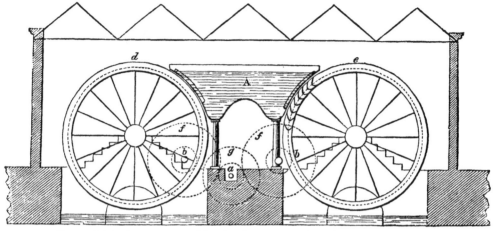

FIGURE 5–26. Arrangement of the shuttle, or penstock, and water wheels at the Deanston mills. Fairbairn and Lillie designed two 36-foot (11.0-m) diameter water wheels and this penstock arrangement for the Deanston mills. Two additional wheels were designed by the firm's millwright, James Smith. All four wheels were fed water through sliding hatches from the W-shaped pentrough.

of the nineteenth century it was pitchback, or high breast, in construction. It was equipped with a wooden casing to inhibit splashing and spillage. Water was delivered to Lady Isabella in a rather unusual manner. Just behind the wheel was a large cylindrical tower. Water was led to the wheel through the center of this tower by the inverted siphon principle. The Laxey wheel was not all iron, but an iron-wood hybrid. The shaft was built up of wrought iron and was 17 feet (5.2 m) long by 21 inches (53.3 cm) in diameter. The rims were cast-iron segments, and the braces were wrought-iron rods. But the arms of the wheel were made of wood, as were the buckets. The Laxey wheel was used to power mine pumps several hundred feet away by means of trussed wooden field rods. It delivered approximately 230 horsepower. Like Fairbairn's and Smith's wheels, the Laxey wheel was durable. It operated almost continuously from 1854 to 1929, when the mines it dewatered were closed down. For almost ten years the wheel stood idle. In 1938 the wheel was renovated, and it still operates today, though only as a tourist attraction.[150]

Even more powerful than the Lady Isabella wheel was an American suspension wheel designed and built by Henry Burden, a Scottish-born and -educated engineer. Around 1840 Burden became owner of an iron works near Troy, New York. These works were located on a small tributary of the Hudson River near a 50-foot (15.2 m) waterfall which had been utilized to power five small wheels. Burden increased the head available on the site to 72 feet (22.0 m) by damming up the stream and leading the water to the iron works in a leat. In 1838 he attempted to

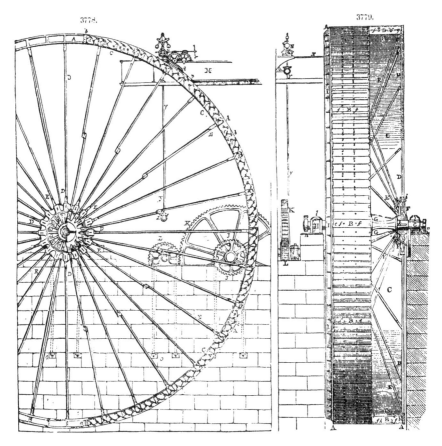

FIGURE 5–27. Suspension wheel of Shaw's Cotton Spinning Company, Greenock, Scotland. This wheel was designed by James Smith, c1830, and was one of the largest and most powerful vertical wheels ever constructed. Over 70 feet (20.3 m) tall and 13 feet (3.8 m) wide, it delivered around 200 hp.

construct a single large wheel to replace the five smaller ones. This wheel was only a moderate success, so in 1851 he rebuilt it.

Burden's new wheel was overshot, 62 feet (18.9 m) in diameter by 22 feet (6.7 m) wide (Fig. 5–29). Like the Lady Isabella it was an iron-wood hybrid. The axle was of cast iron, the spokes were wrought-iron rods 1.5 inches (3.8 cm) in diameter, and the shrouding was of ⅜-inch-thick (9.5 mm) cast-iron sections; but the buckets were 2-inch-thick (5.1 cm) Georgia pine planks, and the soal was built up from 10-by-10-inch (25.4 × 25.4 cm) Georgia pine timbers. Both buckets and soal, however, were iron-reinforced. A wrought-iron rod was run through each of the soal timbers, and several sets of cast-iron spacers circled the wheel between the shrouds to support the wooden buckets. The wheel, when loaded with water, weighed 250 tons. Working under normal loads it produced

FIGURE 5–28.  The Lady Isabella water wheel of the Great Laxey Mining Company, Isle of Man. Constructed in 1854, this giant is 72.5 feet (22.1 m) in diameter by 6 feet (1.8 m) wide and during its industrial use delivered around 230 hp. This power was transmitted over 600 feet (183 m) by mechanical linkages to pumps for dewatering mines. Water was delivered to the wheel via an inverted siphon which ran up through the cylindrical tower to the left of the wheel. The large crank in the foreground was intended for starting the wheel.

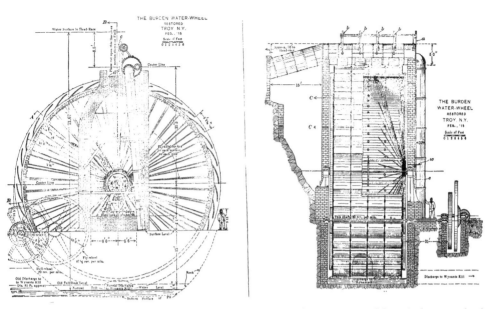

FIGURE 5–29. The Burden water wheel. Perhaps the most powerful vertical water wheel ever constructed, the Burden wheel was 62 feet (18.9 m) high by 22 feet (6.7 m) wide and under normal loads delivered 280 hp with almost 80% efficiency.

278 horsepower at 84.25% efficiency, but its maximum capacity was almost 500 horsepower. Like many iron wheels it was long-lived. It operated continuously for almost 45 years. The Burden works closed down in 1896, and the wheel was allowed to decay and rust to ruin.[151]

The wheels of Fairbairn, Smith, Casement, and Burden represented the pinnacle of vertical water wheel technology, and indicated the limits in power output, efficiency, and durability which could be attained. While traditional wooden water wheels might generate, at the very maximum, 40 to 60 horsepower, large iron industrial wheels could produce 200 to 300 horsepower, and their upper limit was possibly as high as 500 horsepower. Where wooden wheel efficiency was usually no higher than 60%, iron industrial wheels, under favorable conditions, developed around 80 to 85%. The life expectancy of the traditional wheel was ten years or less; iron suspension wheels operated for over a century with minimum maintenance (see Table 5–5).

## CONCLUSION

The improvements adopted by the builders of vertical water wheels between 1750 and 1850 significantly increased the power, efficiency, and durability of the vertical water wheel. Some of these improvements were in large part the result of the application of a quantitative methodolgy to vertical wheel technology, as noted in Chapter 4. Among them can be numbered the adoption of the breast wheel and the sliding hatch,

TABLE 5–5. LIFE SPANS OF SOME IRON AND IRON-WOOD HYBRID WATER
WHEELS OF THE NINETEENTH CENTURY

| Wheel (type) | Date erected | Date ceased operation | Life span | Fate |
|---|---|---|---|---|
| Castle Mills, Kendal, Westmorland (iron-wood hybrid, low breast) | 1806 | c1900 | 94 yrs. | replaced by turbine |
| Woodside Paper Mill, Aberdeen, Scotland (iron-wood hybrid, high breast) | 1826 | 1966 | 140 yrs. | preserved in museum* |
| Catrine Mills, Aryshire (iron, high breast) | 1827 | 1947 | 120 yrs. | replaced by turbine |
| Arkwright Mill, Bakewell, Derbyshire (iron, high breast) | 1827 | 1955 | 128 yrs. | replaced by turbine |
| Shaw's Cotton Spinning Co., Greenock, Scotland (iron, high breast) | c1830 | 1881 | c51 yrs. | replaced by turbine |
| Deanston Mills, Perthshire (iron, high breast) | 1831 | 1949 | 118 yrs. | replaced by turbine |
| Burden Iron Works, nr. Troy, N.Y. (iron-wood hybrid, overshot) | 1851 | 1896 | 45 yrs. | allowed to rust to ruin |
| Lady Isabella Wheel, Great Laxey Mining Co., Isle of Man (iron-wood hybrid, high breast) | 1854 | 1929 | 120 yrs.+ | preserved as museum† |
| Rishworth Mills, Halifax, Yorkshire (iron, high breast) | 1864 | 1949 | 85 yrs. | replaced by turbine |
| Donorwic Slate Quarry, Llanberis, N. Wales (iron, overshot) | 1870 | | 100 yrs.+ | preserved as museum‡ |

SOURCE: Drawn largely from information in Paul N. Wilson, "British Industrial Water Wheels," International Symposium on Molinology, *Transactions* 3(1973): 22–25.

*Presented to the Royal Highland Museum, Edinburgh, in 1966 and re-erected there; in operative condition in the 1970s.

†The works closed in 1929, and the wheel lay idle for ten years, but beginning around 1940 it was restored to operative condition and is now a major tourist attraction.

‡The quarry workshop and the wheel have been turned into a museum.

or overflow sluice gate. Other improvements resulted largely from the replacement of wood with iron—for instance, the ventilated bucket, successful rim gearing, the suspension principle, and larger wheel dimensions with greater power outputs. But it was the pressures being placed on Britain's limited water-power resources by the growth of the British economy in general and the cotton textile industry in particular that made increased power, better efficiency, and a longer life important and which led to the general adoption of these improvements by heavy industry in spite of the fact that they increased the capital cost of water-power development.

# 6

# Demise:
# The Decline of the Vertical Water Wheel

## THE STEAM ENGINE AND THE VERTICAL WATER WHEEL, c1700–c1850

Despite the improvements in vertical water wheel design which came from quantitative investigations and the introduction of iron, some of the most serious problems of water power—geographical inflexibility, natural power limitations, irregularity, and unreliability—remained almost irremedial. These deficiencies were felt in some degree by every manufacturer who used water power and became steadily more acute as the pace of economic expansion picked up in the late eighteenth and early nineteenth centuries.

They were, however, first keenly felt in the mining areas of Europe. Mines and water power were almost equally immovable. Some mines simply could not make use of water power because of unfavorable location. At others, the power requirements of the mine increased at a faster rate than nearby water power could be economically developed, as the demand for metals grew and deeper levels were mined. In this light it is not surprising that the mining industry pioneered in the use of elaborate networks of canals and reservoirs for the development of water power and in long-distance mechanical power transmission using field rods (stagenkunst).

But elaborate dam-reservoir-canal systems and forests of field rods were, at best, partial solutions to the power problems of European mines. Thus, an incentive was provided for developing and applying an alternative prime mover, one without the inherent problems of water. The steam engine, the prime mover that ultimately displaced the vertical water wheel, first emerged, therefore, in the mining areas of Britain. Very late in the seventeenth century and very early in the eighteenth century, Savery and Newcomen introduced steam-powered pumping engines as an alternative to water-powered mine pumps. While these early steam engines were very cumbersome, inefficient, and expensive,

321

they already possessed a number of important advantages over the vertical water wheel. They were geographically flexible and did not have to be located adjacent to a stream of water. They could be placed anywhere that coal could be hauled. While the number of water wheels operable at a given location was fixed, there was no such limit on the number of steam engines that could be moved to a site. As a result the use of Newcomen engines (and an occasional Savery engine) spread slowly but steadily in the mining areas of Britain in the eighteenth century.[1]

The Savery and Newcomen engines were only pumping engines, as was the early form of Watt's improved engine, with its separate condenser and direct use of steam pressure. This restricted the early application of steam power to tasks which involved pumping water. Steam engines could be used to drain mines and dry docks and in public water supply systems. But they could not be applied directly to other industrial tasks, particularly those requiring rotary motion, like textile production.[2] In all but a few areas the water wheel continued its unchallenged sway throughout most of the eighteenth century.

However, the availability of a steam-powered pumping engine, coupled with the growing power needs of Britain in the late eighteenth century, led some manufacturers and millwrights and engineers to attempt to use steam power in conjunction with water wheels. These efforts involved working Savery or Newcomen engines, in periods of short water, to lift water from a tail race back into a reservoir which supplied an overshot water wheel.

The idea of using a steam engine to recirculate water over a water wheel had emerged with the steam engine. It had been suggested, for example, by Thomas Savery in his 1698 patent and in his *Miner's Friend* of 1702, a work designed to publicize his form of steam engine.[3] Another of the early steam engine pioneers, the French savant Denis Papin, designed several systems in the early 1700s which involved using the pressure of steam to force water up to the reservoir of an overshot water wheel (Fig. 6–1).[4] The ideas of Savery and Papin were not quickly adopted, probably because of the high cost of the combination and because Britain's water-power shortage was not yet critical. According to John Farey, one of the early recorders of steam engine history, the first steam engine–water wheel combination was erected only in 1752 when Champion of Bristol began using a Newcomen steam engine to supply several overshot wheels at his brass works.[5] The combination, however, seems to have been used a decade or so earlier at Newcastle and elsewhere, unknown to Farey.[6]

Several prominent eighteenth-century British engineers were advocates of the combination of steam engines and water wheels. John Smeaton, for instance, installed a steam engine at the Carron Iron Works in the 1760s to supply water wheels with water during the dry summer months. In 1762 Oxley erected a steam engine–water wheel combination to raise coal in Northumberland. His plant was apparently

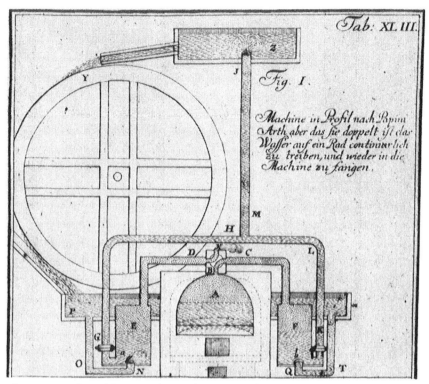

FIGURE 6–1. One of Papin's schemes for compounding a steam engine with an overshot water wheel to produce rotary motion. Steam from the boiler, *A*, is alternately admitted into containers *E* and *F*. The pressure of the steam forces water out of these containers through valves *G* and *K* into rising pipes and up to the reservoir at *Z*. The water from this reservoir is led onto the overshot wheel, *Y*, and recirculated into the system at *P*. Water is admitted to the pressure chambers through the valves at *N* and *O* as the rotating plug valve above the boiler alternately vents steam from one pressure cylinder and then the other.

the first to continuously recirculate water without benefit from natural stream flow. Smeaton adopted this closed-cycle idea for an engine he designed in 1777 for Long Benton Colliery. This plant had a Newcomen engine with a 26-inch (66-cm) cylinder worked by waste coal. It raised 140 cubic feet (3.96 m³) of water per minute to a reservoir which supplied water to an overshot wheel 30 feet (9.15 m) in diameter and 16 inches (40.6 cm) wide. Its output was 5.22 horsepower, and it replaced 16 horses and four men working a 12-hour shift. This installation worked so well that in 1785 a larger plant of the same type was erected. In 1782, at Walker Colliery, Newcastle, Smeaton built yet another combination steam engine–water wheel with an output of 8.44 horsepower.[7]

Where Smeaton specialized in the erection of units combining a Newcomen steam engine with an overshot water wheel, Joshua Wrigley, a Manchester engineer, specialized in erecting units which combined a

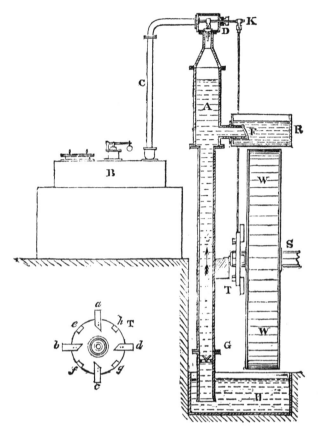

FigURE 6–2. Wrigley's combination of Savery steam pump and water wheel. Steam from the boiler at *B* was admitted into the working cylinder at *A*. When this steam was condensed through the injection of a stream of water, the vacuum produced at *A* caused water to be forced upwards under atmospheric pressure from the pump at *H*. This water was then fed into the reservoir at *R*, which directed it on the overshot water wheel at *W*. The cycle was controlled by the wooden wheel *T*, which was fixed on the axle of the water wheel. Projections from this wooden wheel pulled levers which controlled the admission of steam and condensing water into the working cylinder at *A*.

Savery steam engine with an overshot water wheel. The Savery engines Wrigley used were not of the original Savery design, which was grossly inefficient and dangerous because of the high pressures it employed. Wrigley's engines eliminated the high-pressure (force) portion of the original Savery cycle and used steam only to create suction to lift water 16 to 20 feet (4.9–6.1 m) to a reservoir, and Wrigley utilized Newcomen's internal water-injection system to condense the steam more efficiently. Perhaps even more important, Wrigley devised a method by which the rotation of the water wheel opened and shut the valves which admitted steam and condensing water into the work chamber of the engine so that the plant could work without an attendant (Fig. 6–2).[8]

Wrigley erected a large number of combination steam and water plants for early textile mills in the Lancashire area. The younger James Watt, for example, noted in a letter to his father in March 1791 that Wrigley had "orders for 13 engines for this town & neighbourhood, all of them intended for working Cotton Machinery of one kind or another by the Medium of a Water Wheel." As Musson and Robinson note, "This means that, apart from an unknown number of engines which Wrigley had previously erected, he had on his order books at this date one third of the number of engines erected by Boulton and Watt in the same area by 1800."[9]

It is clear from a number of recent studies that the use of steam engines in combination with water wheels enjoyed considerable vogue in certain areas of Britain in the late eighteenth century.[10] This system continued to be preferred for rotary motion by many despite the application of the crank and flywheel to the Newcomen engine around 1780 and despite the emergence of the Watt double-acting, rotative steam engine later in that decade. Smeaton summed up the advantages of the combination system over single-action, rotative steam engines in 1781:

> I apprehend that no motion communicated from the reciprocating beam of a fire engine [steam engine] can ever act perfectly equal and steady in producing circular motion, like the regular efflux of water in turning a water-wheel, and much of the good effect of a water-mill is well known to depend upon the motion communicated to the mill-stones being perfectly equal and smooth, as the least tremor or agitation takes off from the complete performance.
>
> Secondly, all the fire-engines that I have seen are liable to stoppages, and that so suddenly, that in making a single stroke the machine is capable of passing from almost the full power and motion to a total cessation. . . .
>
> By the intervention of water, these uncertainties and difficulties are avoided. . . .
>
> For these reasons, were I to establish a work of this kind at my own cost, I should certainly execute it by the intervention of water, and therefore must greatly prefer it.[11]

The improvements made by Watt to the steam engine between 1780 and 1800—double action, the direct use of steam pressure, expansive operation, sun-and-planet gearing, the fly-ball governor—eliminated many of Smeaton's objections. They made it possible for the steam engine to be applied to industries, like the cotton textile industry, which required rotary motion. This ultimately spelled the doom of the vertical water wheel as the most important prime mover of European heavy industry. By around 1815 to 1820 steam engines had displaced water wheels as the most important prime mover in the British textile industry, and by 1840 almost 75% of the power in use in the cotton textile industry was being steam generated (see Table 6–1). By 1900 only a very small portion of the British textile industry was still water powered (see Table 6–2).

Why was steam able to displace water as the leading power source of European and American industry in the nineteenth century? In large part it was due to three major deficiencies of water power: (1) irregularity and unreliability; (2) geographical inflexibility; and (3) natural power limitations.

Water power was irregular and unreliable. Drought, floods, and ice could either stop or seriously impede the operation of a vertical water

TABLE 6–1. POWER GENERATED BY WATER AND STEAM IN BRITISH COTTON
MILLS, c1835

| | | Horsepower | |
| District | No. of mills | Steam | Water |
| --- | --- | --- | --- |
| Lancashire | 657 | 21,387 | 2,831 |
| Yorkshire | 140 | 956 | 1,430 |
| N. Staffordshire, Denbighshire, and | | | |
| Flintshire | 10 | 284 | 65 |
| Cheshire | 71 | 3,210 | 847 |
| High Peak Hundred of Derbyshire | 56 | 676 | 921 |
| Cumberland | 12 | 98 | 78 |
| Nottinghamshire and Derbyshire (S.),with | | | |
| Staffordshire and Middlesex | 54 | 438 | 1,172 |
| Scotland | 125 | 3,200 | 2,480 |
| Ireland (N.) | 15 | 372 | 234 |
| Parts of central and western England not | | | |
| included above, the southern counties of | | | |
| Wales, and southern Ireland (est.) | 14 | 232 | 146 |
| All districts | 1,154 | 30,853 | 10,204 |

SOURCE: Edward Baines, *History of the Cotton Manufacture in Great Britain* (London, c1835), pp. 386–94.

wheel. In northeastern Scotland, for example, one company lost 8.6% of its working hours because of ice in 1826. Floods caused another British company in the same era to lose up to 20 days a year in working time. And drought, or low water, the most serious of the three, frequently shut down water-powered mills for a larger portion of the year than floods.[12] As long as all manufacturers used water power, the irregularity and unreliability of water power were troublesome but not crippling. Competitors were affected in at least some degree by the same elements.

But the emergence of the steam engine, a prime mover not halted by drought, flood, and ice, transformed troublesome problems into serious ones. Water power's irregularity and unreliability now became serious deficiencies. Manufacturers who continued to depend on water power were placed at an important competitive disadvantage vis-à-vis manufacturers using steam power.[13]

A second deficiency of water power made critical by the emergence of steam power was water power's geographical inflexibility. Water power was often not close to established transportation routes, markets, raw materials, and a labor supply, and usually it was not economically feasible to move water power to the proximity of these essentials. Instead, manufacturers had to locate where a sufficient quantity and fall of water were available. In an era of primitive transportation facilities and medium- or small-scale industry dominated by local or, at best, regional markets, power costs were more important than the cost of labor or transportation, and industry had to move to where power was available.[14]

The geographical inflexibility of water power, however, placed it at a very serious competitive disadvantage with the steam engine. Steam en-

TABLE 6–2.   POWER GENERATED BY WATER AND STEAM IN THE COTTON, WOOL, FLAX, JUTE, HEMP, AND CHINA GRASS FACTORIES OF BRITAIN, 1838–1870/71

| Date | Horsepower (×1,000) | | Percentage of total | |
|---|---|---|---|---|
| | Steam | Water | Steam | Water |
| 1838 | 72 | 27 | 72.7 | 27.3 |
| 1850 | 108 | 25 | 81 | 19 |
| 1856 | 136 | 23 | 85.5 | 14.5 |
| 1861 | 372 | 28 | 93 | 7 |
| 1867 | 323 | 30 | 91.5 | 8.5 |
| 1870/71 | 462 | 25 | 95 | 5 |

SOURCE: B. R. Mitchell, *Abstract of British Historical Statistics* (Cambridge, 1962), pp. 185, 198, 203, 210.

gines could be located anywhere fuel could be delivered. Hence, areas not blessed with sufficient water or an appropriate topography for water-power development, eagerly turned to the new prime mover. Further, as mass production, national markets, and a railroad network (made possible by the mobile steam engine) emerged, power became only one of many determinants in industrial location. Increasingly, proximity to a labor supply, to raw materials, to markets, and, especially, to established transportation lines became more important than power costs.[15] The steam engine gave industry the freedom to locate close to these. Water power, because it was geographically fixed, did not. Thus, the geographical inflexibility of water power was to provide a very powerful incentive for manufacturers in the nineteenth century to switch from water power to steam power.

Another serious problem with water power was its natural limitations. Only a certain number of watermills could be crammed into a given stretch of stream, and only a certain amount of power could be developed at a specific site. Further, conditions were rare, in Britain at least, where more than 20 to 30 horsepower could be developed economically and conveniently. This amount of power was adequate for the scale of the early textile industry, but it became increasingly inadequate after that. Because any number of steam engines could be concentrated at a specific site, steam did not have the natural limitations that plagued water power. And as mills grew in size, resort to steam became almost inevitable.

The growing scale of industry and the limits placed by nature on the amount of water power that could be developed in any specific location, moreover, steadily eroded one of water power's most important advantages over steam—cost. Particularly in the late eighteenth century, the very high first cost of steam engines, combined with their high operating and maintenance costs, gave water power an edge that usually was sufficient to offset its deficiencies in other areas. This situation, however, began to change slowly in favor of steam power early in the nineteenth century. The development of more efficient steam engines slowly cut the operating costs of steam power. Technical improvements reduced maintenance costs. The availability of second-hand engines and the extended manufacture of steam engines began to reduce their initial costs. At the same time, the cost of developing water power in units adequate for the growing power requirements of heavy industry were rising. In order to develop the 20 to 30 horsepower required for small mills or the 100 horsepower required for large mills toward mid-century, mill owners had to erect iron wheels and large dam and reservoir systems. These were expensive and increased the per unit cost of water power significantly. Thus, as the power requirements of heavy industry grew in the nineteenth century, the per unit cost of power rose for water power while it declined for steam power.[16]

There were, of course, other factors besides unreliability, geographical inflexibility, and natural limitations on power development that contributed to the displacement of water power by steam. Among these could be numbered increased competition for water use by other interests, especially urban areas in need of a water supply, and the larger size of manufacturing firms.[17]

In spite of the numerous disadvantages it faced in its competition with the steam engine, however, the vertical water wheel did not give way quickly or easily. Even in Britain, the birthplace of the steam engine, steam replaced water as the primary source of industrial power very gradually. As late as 1835 in cotton textiles, the industry that was the first to make massive use of steam, water power was still responsible for around 25% of all power produced (see Table 6–1). And if Lancashire is excluded, the continued importance of water power in Britain in the nineteenth century is even more obvious. Outside of Lancashire 40% of the power used in British cotton mills c1835 was water power. Water continued to play an important, though diminishing role, in the British textile industry throughout the remainder of the nineteenth century, as Table 6–2 suggests. And dependence on water power was much greater outside of the textile industry.[18]

Britain was not particularly well endowed with water power. In countries where water power was more abundant—in France and the United States, for example—the dominance of steam was much longer in coming.

In the United States the number of watermills increased from around 7,500 in 1790 to more than 55,000 in 1840.[19] There were still more water wheels than steam engines in manufacturing in the United States as late as 1870, and the advantage of steam in terms of total horsepower at that date was very slight (51.82% to 48.18%).[20] Early in the nineteenth century water was clearly the dominant form of power. Temin found, for instance, that around 1838 only 13% of the total power in American textile industries was steam-generated. For the metal industry the figure was 12%; for food products, 33%; for lumber and wood products, 23%;

and for all other industries, only 19%.[21] Even as late as 1869, Temin pointed out, less than one-third of the flour and grist mills in the United States used steam power.[22] He concluded:

> Although steam power was used widely in manufacturing by 1840, most of its use was concentrated in a few industries and it provided the main power supply for almost none. The direct costs of steam power were higher than the costs for waterpower, and industries used steam only when the freedom of location gained by using steam was large. In other words, in the years before it became important as a supplier of land transportation, the steam engine functioned as a *substitute* for such transportation, allowing power to be brought to the raw materials when it was expensive to bring the materials to waterpower sites.[23]

The picture in France was much the same as in the United States. Water power resisted the inroads of steam and remained the dominant inanimate prime mover far into the nineteenth century, retaining considerable importance even towards the end of that century. In 1826, for instance, Charles Dupin estimated that there were at least 66,000 water wheels in France and that they produced more than three times the power of all French steam engines.[24] If anything, Dupin's estimate for steam power output was high. Data from France at mid-century indicate the continuing dominance of water power. In iron manufacturing in 1844, for example, the proportion of water to steam horsepower was 3.6 to 1.[25] And in 1856 Alexandre Moreau de Jonnes found that of 734 cotton spinning mills in France, 478 used water power, only 244 steam, with a few using both.[26] The industrial census of 1861–65 indicated that in the textile, mining, metallurgical, and metal-working industries steam had displaced water power by a margin of 2 to 1. But in grain milling the margin for water power was 20 to 1, and for all French industries 2 to 1.[27] By around 1900 the water turbine had largely displaced the vertical water wheel, but water-powered prime movers of one sort or another were still producing around 30% of the inorganic horsepower in French industries.[28]

A similar picture could be painted for most of the other major European industrial countries—the water wheel still the dominant prime mover in terms of total power output well into the nineteenth century. In Belgium, to give just one other example, there were 2,600 water wheels in 1846 and only 1,500 steam engines.[29]

The reasons for the persistence of water power were many. The principle ones, however, were probably lower first cost and lower operating costs, especially in mills with relatively low power output where the auxiliary works (dams, reservoirs, canals) requirements were relatively small. Chapman, who compared power costs for water and steam in Britain, concluded that "water was capable of competing with steam power throughout the long period (1780–1850) of transition from the domestic to the factory system in cotton textiles where sites near large towns could be developed." He also found that as late as 1840 the total costs for water power were lower, even for a 100 to 110 horsepower cotton mill with relatively extensive auxiliary works, than the total costs for steam power.[30] Only towards mid-century, as the power requirements of industry exceeded what was available at most water-power sites, and as transportation and technical improvements brought the initial, operating, and maintenance costs of steam power down, did this situation change.

In addition to lower costs, there were other things which delayed the triumph of steam. Among them were improvements made to the vertical water wheel between 1750 and 1850. As the last two chapters have indicated, the cumbersome and only modestly efficient water wheels which saw the birth of steam power around 1700 were replaced between 1750 and 1850 in heavy industry by iron or iron-wood hybrid breast wheels which had a longer life and developed significantly more power at higher efficiencies. Quantitative investigations like those of Smeaton, Borda, and Poncelet and the improvements based on those investigations helped keep the vertical wheel competitive. John Farey, for instance, credited Smeaton's innovations in water-power technology with delaying the coming of steam: "The natural waterfall being thus employed to advantage [as a result of Smeaton's findings], the necessity of employing steam power for mills was not felt until some years afterwards, when manufactories had become greatly extended beyond their former scale; and having no other resource, fire-engines were then erected to return the water, and make up for the deficiency of the natural supplies."[31]

The vertical water wheels of the nineteenth century developed almost as much power as their steam rivals. The average power of stationary steam engines in France between 1840 and 1870 was only around 12 to 14 horsepower.[32] The average power of British steam engines in the first half of the nineteenth century was a little higher, around 20 horsepower.[33] Since vertical wheels could easily be built to develop this output and a fair number of sites were available in this range, the typical steam engine prior to 1850 was not that much more powerful than the typical vertical water wheel.

Another factor contributing to the persistence of the vertical water wheel was its simplicity. This simplicity meant less liability to breakdown and ease of repair if something did go wrong. The operation of the engine was easy to understand, and this, conceivably, may have influenced some mill owners to favor it over steam. Finally, the vertical water wheel was a safe energy source. It did not present the dangers of fire or explosion offered by steam.

In summary, lower total costs, competitive power output, improved design, superior mechanical reliability, and familiarity all contributed to the persistence of water power through the eighteenth century and far into the nineteenth century, even in some of the most industrialized areas of the Western world.

The steady displacement of the vertical water wheel by the steam engine did not mean that the vertical water wheel was obsolete. Where power requirements were relatively small (and, in a few areas, even where they were relatively large) it could have continued to play an important and continuing, if diminished, role as the major *water-powered* prime mover in industry. But this was not to be, for at the same time the vertical wheel was being successfully challenged by the steam engine, it also faced challenges from several new water-powered prime movers— the water pressure engine, the hydraulic ram, and the water turbine. It was all but completely displaced by the last of these.

## THE WATER-PRESSURE ENGINE

One of the lesser known but more powerful water-powered prime movers of the eighteenth and nineteenth centuries was the water-pressure engine. In engines of this type the pressure of a column of water acting on one or both sides of a piston in a cylinder provides power in a manner analogous to a steam engine.

The water-pressure engine has been cited as an example of a development within water-power technology inspired by the Newcomen steam engine.[34] While certain eighteenth- and nineteenth-century water-pressure engines, like those of Westgarth and Trevithick, mentioned below, were almost certainly inspired or influenced by contemporary developments in the rival power technology, the development of this form of engine was not necessarily dependent on steam power. Water-pressure engines were conceived long before Newcomen's engine. Robert Fludd, an English astrologer and alchemist, for example, described a single-acting water-pressure engine as early as 1618, although it apparently never passed beyond the idea stage.[35]

After Fludd the water-pressure engine largely disappeared for over a century. It reappeared in 1731, when two French inventors, Denisart and de la Deuille, described a crude water-pressure engine in the collection of machines and inventions approved by the Académie des sciences.[36] Bélidor in his *Architecture hydraulique* of 1739 reviewed this device and described a water-pressure engine of his own.[37] Both of these engines, however, were designed to act only under low heads and were never actually put into operation.

The first large-scale and practically successful water-pressure engine was constructed for the mines of Schemnitz in Slovakia around 1750 by the Austrian engineer Höll. This engine (see Fig. 6–3) utilized a column of water almost 300 feet (91.5 m) high to lift a large and heavy piston and piston rod. This action caused a pump rod on the opposite side of a rocking beam to descend. When the water under the piston was released, the heavy piston and piston rod overweighed the pump rod and, via the rocking beam, lifted it to its original position. Höll's engine had some similarities to the contemporary Newcomen engine. Like the New-

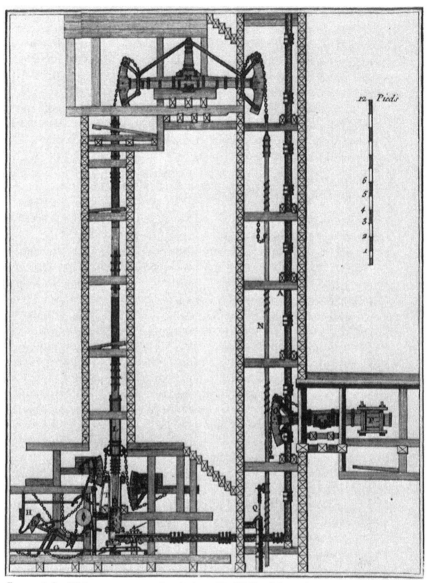

FIGURE 6–3. Water-pressure engine designed by Höll at Schemnitz, c1750. The pressure of a column of water was led downward through the pipe $A$ to the working cylinder $C$, where it lifted the very large piston and piston rod $D$ and $L$. Through the rocking beam $M$, this movement lowered the pump rod $N$, which operated a pump at $RQ$. When the water was released from the cylinder, $C$, the heavy piston and piston rod overweighed the pump rod and lifted it into position for another cycle of operations. The machinery shown to the left of the working cylinder was designed to make the engine self-acting.

comen engine, it depended on the differences in pressure on opposite sides of a piston for motion, it utilized a rocking beam, it was single-acting, and it was used almost exclusively for mine pumping. Its advantage over conventional vertical water wheels was in size (it would fit easier in a mine shaft) and in adaptability to high falls (vertical wheels could not conveniently be used with heads of more than c50 feet [c15 m]).[38]

Höll erected a number of water-pressure engines at Schemnitz in the mid-eighteenth century, but they spread slowly. In 1765 the British inventor and engineer William Westgarth reinvented the water-pressure engine without ever having heard, apparently, of Höll's machines, basing his design very closely on the analogy to the Newcomen engine.[39] John Smeaton and Westgarth both erected several water-pressure engines of similar design in the late eighteenth century.

Around 1800 the British engineer Trevithick also designed and erected several water-pressure engines, also apparently in ignorance of earlier work. Trevithick's engines were analogous to Watt's double-acting steam engines, not to Newcomen's single-acting engines, as were those of Höll, Westgarth, and Smeaton. In Trevithick's engines the pressure of a high column of water alternately acted on top of and beneath a piston. The Wheal Druid engine constructed by Trevithick in 1799 worked under a fall of 204 feet (62.2 m). Even more impressive was his engine at the Alport Mines, built in 1803. With a cylinder 25 inches (63.5 cm) in diameter, a stroke of 10 feet (3.0 m), and a head of 150 feet (45.7 m), it operated almost continuously until the 1840s.[40] The engine which replaced it was erected by John Taylor in 1842. This new engine had a cylinder 50 inches (127 cm) in diameter and operated under a head of 132 feet (40.2 m) with a 10-foot (3.0-m) stroke (see Fig. 6–4). Its estimated efficiency was 70%, and it produced 168 horsepower.[41]

By the early nineteenth century water-pressure engines were in operation in southern Germany and in France, as well as in Britain and Hungary. The German engineer Reichenbach, for example, erected nine water-pressure engines, some single-acting, some double-acting, in southern Germany. They pumped brine a distance of 60 miles (96.6 km) through a 7-inch (17.8-cm) pipeline, and produced around 9 to 10 horsepower each.[42] Water-pressure engines operating under very high heads were erected in the early 1800s at a number of mines. At the Alte Mordgrube mine near Freiberg one utilized a 366.5-foot (111.7-m) fall.[43] At Clausthal in the Harz one worked under a head of 630 feet (191.5 m).[44] And, finally, at Leupold Mine at Schemnitz a dual-cylinder engine with 11-inch (27.9-cm) pistons and an 8-foot (2.4-m) stroke operated under a 748-foot (227.4-m) head.[45]

Because water-pressure engines could operate efficiently under very high heads, some engineers preferred them to vertical wheels in certain circumstances. Guenyveau, a French engineer, for instance, recommended the use of water-pressure engines instead of vertical water

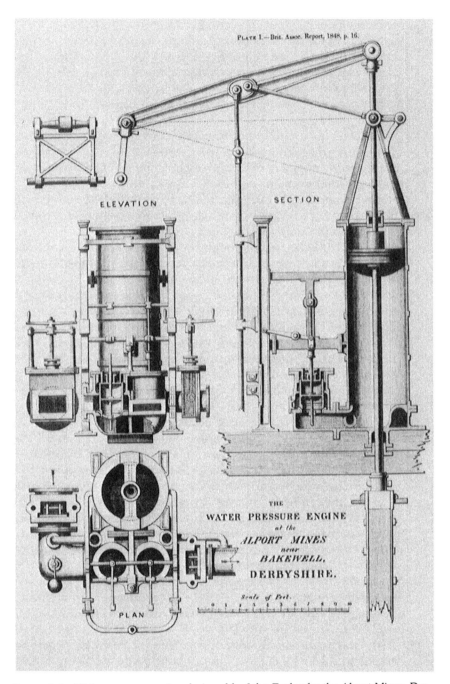

FIGURE 6–4. Water-pressure engine designed by John Taylor for the Alport Mines, Derbyshire, 1842. One of the largest water-pressure engines ever constructed, its cylinder was 50 inches (127 cm) in diameter, and it operated under a head of 132 feet (40.2 m), producing 168 hp at about 70% efficiency. This engine was single-acting and had the pump rod directly connected to the working piston.

wheels for heads above 43 to 46 feet (13–14 m).[46] Weisbach, the German engineer, felt that up to around 80 feet (24.3 m) the conventional vertical water wheel was still the best choice and that two water wheels, one placed over the other, were cheaper and more convenient than a single water-pressure engine. But, he concluded, when more than two vertical wheels were necessary to tap a fall, the water-pressure engine was the best choice.[47]

Since the water-pressure engine was economically competitive with the vertical water wheel only for falls above 50 to 80 feet (15–24.3 m), its challenge to the vertical wheel was a limited one. About 90% of all waterfalls are 10 feet (3 m) or lower,[48] and only a very small fraction are more than 50 feet (15 m). The effective challenge of the water-pressure engine was thus restricted largely to Europe's great mining areas.[49]

## THE HYDRAULIC RAM

Another water-powered prime mover which offered a limited challenge to the vertical water wheel in the nineteenth century was the hydraulic ram. The principle behind the operation of this device had long been recognized. If one had a fair volume of water and this water was made to fall through a constricted passage, the energy of that fall could be converted into work by suddenly stopping the flow. The momentum of the arrested water would create an increase in pressure which could be used to force a small portion of the water to a considerable height.

A crude device using this principle was built in 1775 by John Whitehurst of Derby (1713–88). Water with a fall of 16 feet (4.9 m) was led through a pipe nearly 600 feet (182.9 m) long and 1.5 inches (3.8 cm) in diameter. A manually operated stopcock was used to abruptly halt this flow. The pressure developed in the line was used to force some of the water above the level of its source. Whitehurst's ram (Fig. 6–5) was not successful. It had no self-acting mechanism, and the noise and vibrations occasioned by the ram action made it inappropriate for its intended function—domestic water supply.[50]

In 1797 Joseph de Montgolfier (1740–1810) reinvented the hydraulic ram and added a self-acting valve. Montgolfier's self-acting valve was a loaded, upward-seating type (see Fig. 6–6). To begin operation the loaded valve was pushed downward. Water would begin to flow through the system. As it attained full velocity the pressure of the water would lift the valve against its seat. This suddenly arrested the flow of the water and the increased pressure in the line caused the water to rise through another valve, into an air vessel, and up an ascent pipe. As the pressure decreased in the main line as a result of this action, both the valve leading to the air vessel and the self-acting valve dropped and the cycle began anew. This series of operations could be repeated between 40 and 200 times per minute to give almost continuous discharge.[51]

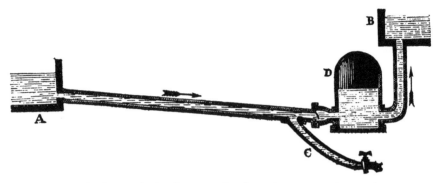

FIGURE 6–5. Whitehurst's hydraulic ram. The flow of a stream of water from *A* was abruptly stopped by manually closing the tap at the end of *C*. This action increased the pressure of the water in the line, forcing it to flow into air vessel *D* and up to reservoir *B*.

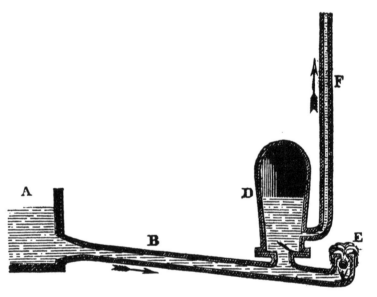

FIGURE 6–6. Montgolfier's hydraulic ram. Montgolfier's self-acting ram had a weighted, upward-seating valve at *E*. As water began to flow from *A* through *B* and out the valve at *E*, pressure built up and forced valve *E* up against its seat, closing off the flow of water through *E*. This increased the pressure in the line, forcing the water up through the valve leading to air vessel *D* and up through *F*. This decreased the pressure in the line, allowing valve *E* to fall from its seat and automatically beginning the cycle anew.

Montgolfier's first ram was put to work at his paper mill near Voiron, France, to raise water under a 10-foot (3-m) head. Much larger units were soon constructed. Montgolfier's son erected one at Mellor, near Clermont-sur-Oisne, with a body of cast iron pipe 0.354 feet (10.8 cm) in diameter and 108.2 feet (33 m) long. It made 60 blows per minute under a head of 37.3 feet (11.4 m), raising water to an elevation of 195.01 feet (59.5 m). Its efficiency was around 65%.[52]

The efficiency of hydraulic rams was high, comparable to or slightly better than a good vertical water wheel and pump combination. Experiments carried out by Montgolfier on hydraulic rams operated in the vicinity of Paris under falls from 3 to 35 feet (0.9–10.7 m) indicated efficiencies of from 55 to 65%. A much more extensive set of experiments conducted by the German hydraulician Eytelwein on two rams in 1804 produced efficiencies varying from less than 20% to above 90%, depending on the number of beats per minute, the height of the fall, and the volume of water raised. High efficiencies, Eytelwein found, were possible when the height to which water was raised was small compared to the fall. When a small fall was used to raise water to a large height, the efficiency was very low.[53]

Besides having a very high efficiency under certain conditions, the hydraulic ram, like the water-pressure engine, was durable. Dickinson cites a case where a ram had not been touched through 27 years of continual operation and another that worked for 104 years with occasional renewal of valves.[54]

Despite its low costs, reliability, and high efficiency, the hydraulic ram was able to challenge the vertical water wheel in only one area—pumping. And even here it was able to do so only under certain conditions—when the height water was to be lifted did not exceed the available fall by more than a ratio of 12 or 13 to 1. Moreover, the output of the hydraulic ram was small, usually less than a horsepower. D'Aubuisson reported that the largest rams constructed in France (c1840) delivered no more than 144.7 foot-pounds (20 kg-m) per second, or well under a half horsepower.[55] The hydraulic ram was usually inappropriate for raising large volumes of water owing to the violent shock on the valves, the noise, and the vibrations of the engine. To increase its power output required larger and thicker pipes and valves and extensive foundations to resist the shocks and vibrations and to hold the machine in place, and those drove its cost up considerably. Thus, d'Aubuisson felt that while the hydraulic ram was "otherwise so remarkable," it would probably always be "restricted in its use" and "inadequate" to furnish a water supply sufficient for manufacturing needs.[56] Another factor which prevented the hydraulic ram from challenging the vertical water wheel over a wide spectrum was the skill required to design and erect a ram. One manufacturer of the implement noted: "We know of no machine which requires more skill and experience to design and adapt properly to the conditions under which it has to work than a hydraulic ram."[57]

Thus both the water-pressure engine and the hydraulic ram were able to challenge and displace the vertical water wheel only in a few areas and under fairly limited conditions. They did not seriously threaten the overall dominance of the vertical water wheel within water-power technology. Between 1820 and 1850, however, another challenger emerged—the water turbine. Unlike the water-pressure engine and the hydraulic ram, it was able to challenge the vertical water wheel over its entire range of applications and operating conditions. It could be used on high falls as well as low falls, with high volumes of water as well as low volumes, for the production of rotary motion as well as for pumping water.

## THE WATER TURBINE

The water turbine deserves a monograph of its own, so I will not delve too deeply into its origins and history here. The water turbine was, in some sense, a descendent of the horizontal water wheels that the vertical wheel had relegated to the backwaters of Europe during the medieval period. The primitive horizontal water wheel, however, had been steadily improved during the late Middle Ages and the Renaissance. In southern France, in Spain, and elsewhere the flat blades of the primitive version were replaced with curved blades, and some of the wheels were placed in a tub or surrounded by a masonry casing to reduce the volume of water which escaped without acting on the wheel. Although most of these modified horizontal wheels developed efficiencies no higher than the vertical undershot wheel, their simplicity and low construction costs encouraged further development. During the eighteenth and early nineteenth centuries a number of engineers attempted to design horizontal water wheels which could challenge the dominant vertical wheel. Besides the curved-bladed tub wheels of southern France described by Bélidor early in the eighteenth century, there were the reaction wheels of Barker and the Eulers at mid-century. Early in the nineteenth century the French engineer Claude Burdin attempted to improve on both curved-bladed wheels (à la Bélidor) and reaction wheels (à la Euler); and in 1826, to encourage further work in this area, the French Société d'encouragement pour l'industrie nationale offered a prize of 6,000 francs to anyone who could successfully apply on an industrial scale horizontal wheels with curved blades of the type described by Bélidor.[58]

The search for an efficient and practical horizontal water wheel was one of the roots of the modern water turbine. But the vertical water wheel probably played an equal, or possibly even greater, role in the birth of that engine. The "missing link" between the conventional vertical water wheel and the early modern water turbine is the Poncelet wheel, described in Chapter 4.

The Poncelet wheel, as noted, was in one sense a modification of the traditional undershot vertical water wheel, and it shared the general appearance of that machine. But it was, in another sense, a radically new and different type of water engine. In a traditional vertical water wheel the motion of the water relative to the blades or buckets after it entered the wheel was only incidental and unimportant to its operation. In the Poncelet wheel, as in water turbines generally, the motion of the water relative to the blades after entrance was essential. The Poncelet wheel was, in essence, a pressureless turbine. It only required moving the

Poncelet wheel from the vertical to the horizontal plane to create an engine much more recognizable to modern eyes as a water turbine. Poncelet, in fact, suggested this arrangement in 1826 in order to remedy one of the deficiencies of his design—the reversal of direction of the water within the machine because the water entered and exited at the same point. Poncelet proposed dealing with this difficulty by setting the wheel horizontally on a vertical axle and introducing the water from the exterior circumference and exhausting it through the interior of the wheel. He also contemplated admitting the flow either at a single point or, alternately, all around the periphery. Poncelet suggested, in other words, the essential idea of an inward-flow radial turbine with either partial or complete admission.[59] Several decades later his partial admission design was actually constructed by several European turbine manufacturers (Fig. 6–7).[60]

Poncelet's wheel and the modifications which Poncelet proposed for it probably had a significant influence on Benoit Fourneyron (1802–67), the creator of the water turbine in its modern form.[61] At the opening of the paper in which he revealed the secrets of his turbine designs, Fourneyron acknowledged the prior work of Poncelet and correctly observed that Poncelet's wheel was "essentially different from all others."[62] It is also probably no accident that Fourneyron succeeded in perfecting a workable, efficient turbine only in 1827, three years after Poncelet had announced his invention and a year after he had suggested placing his wheel with curved blades on its side to gain superior operating characteristics.

Fourneyron's first successful water turbine was an experimental model developed and erected at Pont-sur-l'Ognon. The wheel was small. It had a diameter of 9.5 feet (2.9 m) and developed only around 6 horsepower under falls which varied from 3.9 to 4.6 feet (1.2–1.4 m). As in all of Fourneyron's turbines, water was led into a stationary inner core and directed by horizontally situated, fixed, curved guide vanes simultaneously against all of the curved blades which formed the mobile outer wheel of the turbine (Fig. 6–8). The water in flowing over the curved outer ring of blades exerted pressure and produced mechanical power at the wheel's shaft. On testing the Pont-sur-l'Ognon wheel, Fourneyron found that it developed an efficiency above 80% and maintained this high figure even when completely flooded by tail water.[63]

Shortly after this success Fourneyron moved to Besançon, where he devoted considerable time to the further development of the turbine with financial assistance from F. Caron, a forge master. His second successful engine, installed at Dampierre in the Jura region of France, had an exterior diameter of only 2.9 feet (0.9 m). Operating under heads which varied from 9.8 to 19.7 feet (3.0–6.0 m), this turbine produced 7 to 8 horsepower at close to 80% efficiency.[64] Fourneyron's third turbine, the Fraisans turbine, was completed in 1832. It had an exterior diameter of 9.5 feet (2.9 m) and was only 1.2 feet (0.36 m) high. With discharges

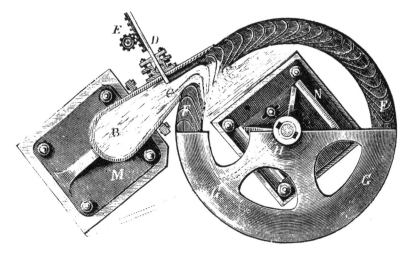

FIG. 464.

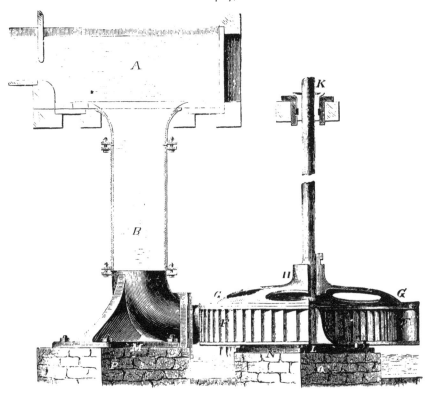

FIGURE 6–7. Inward-flow tangential turbine with partial admission, constructed by Zupinger for Escher, Wyss & Company, Zurich. This turbine was, in effect, a Poncelet wheel laid on its side. Poncelet suggested this configuration as early as 1826 in order to correct a fault in his vertical design—the reversal of flow within the wheel.

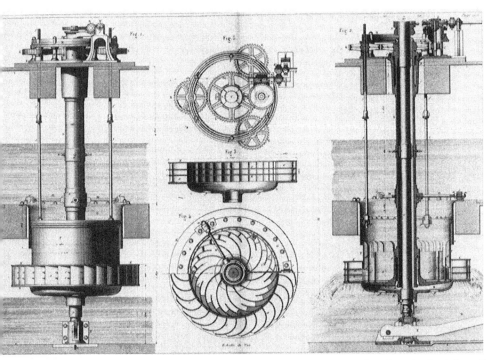

FIGURE 6–8. Typical Fourneyron turbine. Water was admitted to the wheel axially and directed against a mobile outer ring of curved blades by a fixed inner ring of curved blades.

varying from 53 to 117 cubic feet per second (1.5–5.0 m³/sec) under falls of from 9 inches to 4.6 feet (0.227–1.4 m), it, too, had an efficiency greater than 80%, and it developed around 50 horsepower.[65]

In 1832, with these successes behind him, Fourneyron took out a patent on his design. He also submitted a memoir on his work to the Société d'encouragement pour l'industrie nationale and was awarded the society's prize of 6,000 francs in 1834. The manuscript which Fourneyron produced for the society was published in its *Bulletin* that same year.[66]

In 1834 Fourneyron built a turbine with an exterior diameter of 9.51 feet (2.9 m) and a height of 0.98 feet (0.3 m) for a spinning mill at Inval, near Gisors, northwest of Paris. As a result of the publicity being given his work, this turbine was subjected to a large number of experimental tests by French engineers. The results were favorable. D'Aubuisson commented: "What machine, other than the turbine, under the small fall of 3.77 ft. [1.15 m], could acquire more than three quarters of the motive force and a force of thirty horsepowers? or, under the slight fall of 0.984 ft. [0.3 m], could take more than three fifths, and that, too, when entirely submerged in the water? Truly, the wheel of M. Fourneyron has an undoubted superiority in certain respects over all others; it is an admirable machine."[67]

The most notable of Fourneyron's early turbines was one erected at St. Blasien in 1837. One of the falls at St. Blasien, in the Black Forest of Germany, was 354 feet (107.9 m). Previous attempts to make use of this fall had taken the form of a series of overshot wheels, one situated above the other. Fourneyron replaced these with a single turbine which made use of the entire fall. Water was led to the turbine under pressure in a 1,640-foot-long (500 m) pipe. The results were astounding. The St. Blasien turbine was only 1.5 feet (0.46 m) in diameter and weighed less than 40 pounds (18.2 kg), yet it was able to utilize effectively a head of more than 350 feet (106.7 m) and develop around 60 horsepower at 2,300 rpm with an efficiency of around 80%. The German engineer Moritz Rühlmann considered this the most important of Fourneyron's engines. The impression that this machine made on him must have been experienced by others:

> I can best describe this turbine by detailing what I myself saw and learned upon the spot.... Already, half an hour before arriving at the remarkable locality of St. Blasien, situated in one of the most beautiful, but also one of the wildest and loneliest parts of the Schwartzwald of Baden, a curious noise announces the uncommon spectacle, which becomes more extraordinary as you approach.
>
> On entering into the wheel-room, one learns that what had been heard at a distance about this place was not merely mystification, but reality.
>
> One then feels seized with astonishment, and wonders, more than in any other place, at the greatness of human ingenuity, which knows how to render subject to it the most fearful powers of nature.
>
> At every moment the powerful pressure appears likely to burst in pieces the little wheel, and the spiral masses of water issuing from it threaten to destroy the surrounding walls and buildings. Often when I went out of the wheel room, and looked at the enormous height from which the conducting tubes brought down the water to the wheel, the idea forced itself upon me, 'that it was impossible,' but the idea passed away when I went back into the little room.[68]

During the 1830s and through the 1840s the new water engine spread rapidly in those areas of Europe where water power was still in extensive use. Fourneyron's biographer has noted, for instance, that by 1843 there were at least 129 factories with wheels built according to Fourneyron's specifications, and these plants were located not only in France and Germany, but in Austria, Italy, Russia, Poland, and even Mexico.[69] Fourneyron's turbine was introduced into the United States by Ellwood Morris in the early 1840s and temporarily found a home there as well, before being replaced by improved American-designed turbines.[70]

It was quickly recognized that the water turbine was a very versatile and adaptable engine. Even though all of Fourneyron's turbines were radial outward-flow reaction turbines, with complete admission, a legion of other arrangements were possible. Flow, for instance, could be inward as well as outward, and the direction of flow could be radial, axial, or mixed. Impulse as well as reaction could be used to power turbines, and admission of water could be complete or partial. Once Fourneyron had pointed out the advantages of the basic turbine principle and after certain shortcomings in his original design were recognized (it operated very efficiently at full gate, but rapidly lost efficiency at part gates), new turbine types were introduced at a rapid rate, especially in the United States.[71] The modifications and improvements to Fourneyron's invention made a good machine even better, so that as early as the 1840s the water turbine had become a very serious threat to the vertical water wheel.

The water turbine was able to offer a serious challenge to the centuries-old dominance of the vertical wheel at such an early date because, even in its early form, it had significant advantages over its rival. One of the most important of these advantages was the ability to continue to develop power effectively and efficiently even when flooded, for unlike vertical water wheels, water turbines could operate completely submerged. This meant that the water turbine could utilize every inch of head available; it did not have to be suspended a foot or two above the tail race to avoid minor flooding. This was very important. Vertical wheels utilizing a 10-foot (3-m) fall often were built as much as 2 feet (0.6 m) above the bottom of a fall to avoid flooding.[72] The ability to operate submerged meant that the water turbine could better deal with the irregularity of streams than the vertical wheel. It also meant that turbines could be more easily protected from icing over in winter and from the danger of floating objects.

Another major advantage of the water turbine was its much higher rotational velocity. Turbines, it was quickly discovered, operated most efficiently at velocities ranging from 0.7 to 1.0 of the velocity due to the available head. Thus they rotated much faster than the slow-moving gravity wheel and significantly faster than even the swifter-moving undershot wheel. Higher shaft speed meant less gearing was necessary to step axial speeds up to what was needed for attached machinery, a savings in both power and money.

Moreover, the turbine had an efficiency as high as or higher than that of the best vertical water wheels and could operate, thanks to its ability to make use of pressure pipe, under heads varying from less than a foot (0.3 m) to hundreds (and later even thousands) of feet, instead of only in the 2 to 50 foot (0.6–15.2 m) range of the traditional mover.[73] Certainly one of the most important advantages of the turbine was its smaller size. Because turbines operated at higher velocities than vertical water wheels and because water was applied over the entire surface of a turbine simultaneously (instead of only on one bucket or blade at a time), a turbine had more water moving through it than through a vertical wheel per unit of time. Thus the turbine could be much smaller than a vertical

FIGURE 6–9. Comparison between an overshot wheel and a turbine. Both water engines in this illustration are working under a 24-foot (7.3-m) head. The overshot wheel is 22 feet (6.7 m) in diameter; the turbine, 11.5 inches (0.29 m). The smaller size and number of shafts and gears required by the higher-velocity turbine installation are evident.

water wheel of equivalent power output. Similarly, for wheels of the same size, the turbine's power capacity was much greater. These factors gave the turbine a tremendous (and critically important) advantage in size, as Figure 6–9 indicates.

Smaller size, of course, usually meant smaller cost as well. For example, in the 1850s James Prince, an American manufacturer, planned to install at a cotton mill in Mexico an overshot wheel 100 feet (30 m) in diameter to produce 140 horsepower under a 160-foot (49-m) head. The estimated cost of this wheel, ordered from Manchester, was $25,000. Prince was persuaded by Alexis Du Pont to utilize turbines instead. The two turbines installed cost only $2,300. Thus the cost of turbines was only around $16.40 per horsepower, while the vertical wheel would have cost $178.60, or more than ten times as much.[74] In 1882, in a similar manner, the British engineer John Turnbull demonstrated the advantage of turbines. He compared one of the largest iron industrial wheels ever built, the 70-foot (21-m) diameter vertical wheel at Shaws in Greenock, with a turbine. The Greenock wheel, Turnbull estimated, weighed around 115 tons (128.8 U.S. tons, or 117 metric tons) and cost around £3,000. This giant, he suggested, could be replaced by a

| | High breast wheel (iron-wood hybrid) 1846 | High breast wheel (iron) 1886 | Overshot wheel (iron) 1899 | Turbine (Victor) 1882 |
|---|---|---|---|---|
| Head | 16.9 ft (5.15 m) | 15.3 ft (4.65 m) | 16.7 ft (5.1 m) | 18.2 ft (5.55 m) |
| Quantity of water used per second | 35.3 ft³ (1.0 m³) | 35.3 ft³ (1.0 m³) | 30.0 ft³ (0.85 m³) | 27.9 ft³ (0.8 m³) |
| Horsepower developed | 53.6 | 52.7 | 45.0 | 49.0 |
| Diameter of water wheel | 22.5 ft (6.9 m) | 23.0 ft (7.0 m) | 14.8 ft (4.5 m) | 2.5 ft* (0.76 m) |
| Width of water wheel | 12.9 ft (8.9 m) | 8.2 ft (2.5 m) | 12.5 ft (3.8 m) | 2.3 ft† (0.7 m) |
| Weight | 58,476 lb (26,500 kg) | 31,115 lb (14,143 kg) | 27,676 lb (12,580 kg) | 1175 lb (534 kg) |
| Weight per unit horsepower | 1091 lb (494 kg) | 590 lb (268 kg) | 615 lb (280 kg) | 24 lb (11 kg) |

SOURCE: F. Redtenbacher, *Theorie und Bau der Wasserräder* (Mannheim, 1846), pp. 267–76, 308; C. Bach, *Die Wasserräder* (Stuttgart, 1886), pt. 1, pp. 220–24; Wilhelm Müller, *Die Eisernen Wasserräder* . . . (Leipzig, 1899), pt. 1, pp. 76–77 (seventh water wheel in Table III); and Robert Grimshaw, *The Miller, Millwright and Millfurnisher: A Practical Treatise* (New York, 1882), pp. 86, 91–92.

*Diameter with control gate; without the gate the turbine's diameter was only 1.67 ft (0.51 m).

†Width estimated, including gate; estimated width without the gate was only around 1 ft (0.3 m).

"Hercules" turbine only 1.25 feet (0.38 m) in diameter, weighing (without its case) only a half ton (0.56 U.S. tons, or 0.51 metric tons), and costing only about £50. And Turnbull believed that the turbine would deliver greater efficiency.[75]

While indicative of the cost advantages that the turbine's smaller size gave it (for a given amount of power), these figures may overestimate the exact magnitude of the turbine's advantage. The price of vertical wheels 70 to 100 feet (21–30 m) in diameter may have been higher per unit of power than vertical wheels of a more common size. I have been unable to locate reliable cost figures for turbines and vertical water wheels operating under more usual conditions. One way to approach this problem, however, is to compare the weight per unit of horsepower of these engines when operating under comparable conditions and assume that the amount of material going into water engines bears some direct relationship to the price charged for the engine. Table 6–3 compares a water turbine with vertical water wheels operating under a head of around 16 to 18 feet (4.9–5.5 m) and producing about 50 horsepower. The turbine, as the table indicates, weighed only around 25 pounds (11.3 kg) per horsepower. The vertical wheels weighed from 20 to 40 times more. A vertical wheel built in the 1840s working under similar conditions with approximately the same output as the turbine weighed over 1,000 pounds (453.6 kg) per horsepower, while even late-nineteenth-century vertical wheels still weighed around 600 pounds (272.2 kg) per horsepower. Thus, where manufacturers had only a given amount of power available, the water turbine offered very clear economic advantages in terms of cost per unit of power.[76]

Because turbines can develop more power per unit of size, they can also be built to a much greater power capacity than conveniently possible with vertical wheels. The average power of vertical wheels, even by the late nineteenth century, was under 20 horsepower, and the upper range of their power was no greater than 300 to 400 horsepower. Turbines, on the other hand, were already developing significantly higher outputs by the mid-nineteenth century. The average power of Swiss turbines cast between 1844 and 1869 was already 48 horsepower, and by 1889 to 1899 the figure was around 180.[77] In the United States, the average power of turbines manufactured by Stilwell and Bierce of Dayton, Ohio, a major American manufacturer, in the 1886–91 period was 117 horsepower.[78] In other words, by the late nineteenth century the *average* power of water turbines was fast approaching the *upper limit* of power available from vertical water wheels. Single turbine wheels, of course, had far surpassed the upper limits of the vertical wheel. The Niagara Falls turbines of the 1890s, for instance, had a capacity of 5,000 horsepower, more than ten times that of the most powerful vertical wheel ever built. Since the cost per unit of power tends to decline as the power goes up,[79] the superior power capacity of the turbine gave it an important additional cost advantage in its competition with the vertical water wheel.

While the major cost advantages of the water turbine early emerged as an important factor in its success, at exactly what point the water turbine began to make serious inroads on the domain of the vertical water wheel cannot be exactly determined. Possibly it occurred as early as the 1840s in countries like France and the United States, where water power was important. That serious inroads had been made by the mid-1850s is indicated by James B. Francis, who reported in 1855 that the turbine had begun to replace the breast wheel at Lowell, Massachusetts, the most influential American water-power center.[80] An 1866 survey of 47 New England cotton mills indicated a ratio of turbines to breast wheels of 119 to 88, while an 1875 Massachusetts census reported that of 2,806 water-powered installations only around 500 were still using traditional vertical water wheels.[81] And Robert Thurston reported that by 1882 only turbines were in use at Holyoke, another of the large American water-power complexes.[82] Thurston also observed: "The forms of motor employed in the utilization of the water-power of the United States, although very numerous, are almost universally of the class known as turbines."[83] Unwin, at about the same time, was already calling the vertical wheel "cumbrous and antiquated."[84] The trend towards turbines and away from the vertical wheel is reflected in American milling manuals of the same period. William Hughes's *American Miller* (1873), Robert Grimshaw's *Miller, Millwright, and Millfurnisher* (1882), and James Abernathey's *Mill Building* (1880), for example, spent much more time discussing the turbine than the vertical water wheel,[85] and Abernathey observed: "Comparatively few water wheels other than turbine wheels are now used as motions for flour mills, or, in fact, for any other kinds of mills using water power." He added: "The turbine wheels have many defenders because there are many makers, while the overshot has no defenders, unless, perhaps now and then a conservative individual, who does not take up with new-fangled notions very readily."[86]

In Britain the transition from vertical wheel to turbine was not nearly so rapid, and in some places probably did not occur at all. This was primarily because by the time the turbine had emerged, Britain was strongly committed to steam power and because of the general British poverty in water power. As Paul Wilson has pointed out, "The water turbine had certain advantages over the waterwheel, but for the heads and flows of water generally used in Britain, and for the type of machinery to be driven, the advantages were not overwhelming."[87] Thus a 1962–63 survey of Dorset watermills indicated a continued prevalence of vertical water wheels. The surveyors, Addison and Wailes, found evidence of 99 vertical wheels, but only 26 water turbines.[88]

In France and Germany, where the water turbine was first applied on a major scale, I know of no data, but one suspects the transition to the turbine came with approximately the same rapidity as in the United States.

As the superiority of the turbine over the vertical water wheel manifested itself in the mid-nineteenth century, the vertical water wheel began a decline which was to render it all but obsolete by the twentieth century. The overshot wheel was the most tenacious of the vertical species. All-steel overshot wheels could be constructed with an efficiency approaching 90%, much superior to the small turbines they competed against. The efficiency of small turbines seldom exceeded 70 to 75%. Unlike turbines, the overshot wheel was able to run just as efficiently at quarter gate or half gate as at full gate. This gave it an advantage on small streams with highly variable volumes. The simple construction, ease of repair, reliability, and longevity of the iron or steel overshot wheel allowed it to remain competitive between heads of 10 and 40 feet (3–12 m) and discharges of between 2 and 30 cubic feet per second (56.6–850 l/sec).[89] Almost equally tenacious for medium to low falls and low power outputs were modifications of the old breast wheel, like the Sagebien wheel. For example, when the French government decided to build a water-powered plant at Marly around 1850, at approximately the site of the famous seventeenth-century works, vertical breast wheels of the Sagebien type were selected over turbines (Fig. 6–10). Their slow motion was considered more appropriate for the piston pumps planned

FIGURE 6–10. The breast wheels installed for the Marly Pumping Plant in the 1850s. Vertical water wheels were adopted for this plant because of the simplicity of their mechanism, the ease of power transmission to pumps, and the facility of repair. Each of the six wheels was composed of 64 blades of sheet iron and wood 9.8 feet (3.0 m) long. The wheels were 39.4 feet (12.0 m) in diameter and 14.8 feet (4.50 m) wide. They operated under a fall of approximately 10 feet (3 m) with a volume of 91.8 to 105.9 cubic feet per second (2,600–3,000 liters per second).

for the installation, and their simplicity and ease of repair were considered critical advantages over turbines.[90] As late as 1902 breast wheels were being recommended for manufacturers with low power demands by some engineers.[91]

While the cost and other advantages of the water turbine had made the vertical water wheel a second class prime mover by the late nineteenth century, what finally rendered even high efficiency overshot and breast wheels obsolete was the emergence of electrical power generation and transmission between 1880 and 1900. In electric generators high rotational velocity is critical to cost, efficiency, and output. The turbine was able to develop high shaft speeds with high efficiency. The vertical wheel could not without elaborate, costly, and inefficient gearing or belting. And as the use of electricity spread and electric motors became available, the last remaining footholds of the vertical water wheel— pumping plants, small industries, mines—were lost. All that remained by the 1920s and 1930s were a few very small-scale grist mills operated, often, by conservative millers whose affection for the old prime mover outweighed considerations of economy and efficiency. Thus, by the mid-twentieth century the vertical water wheel, once the mightest industrial prime mover available to the Western world, was a curiosity, operating more often for display than for production.

# Notes

INTRODUCTION

1. Lewis Mumford, *Technics and Civilisation* (New York, 1934), pp. 107–267.

2. A. R. Ubbelohde, *Man and Energy* (New York, 1955), p. 13.

3. R. J. Forbes, *Studies in Ancient Technology*, 2 (Leiden, 1955): 78–79; Forbes, "Power," in Charles Singer et al., eds., *A History of Technology*, 2 (Oxford, 1956): 589–90; and Forbes, *La technique et l'énergie au cours des siècles* (Conférence faite au Palais de la Découverte, Paris, 2 Juin 1956, série D, no. 42).

4. Charles H. Loring, "The Steam Engine in Modern Civilization," American Society of Mechanical Engineers, *Transactions* 14 (1892–93): 52–58, esp. 53–55.

5. Wilhelm Ostwald, "The Modern Theory of Energetics," *The Monist* 17 (1907): 510–11.

6. George G. MacCurdy, *Human Origins: A Manual of Prehistory*, 2 (New York, 1924): 134.

7. For example, Chauncey Starr, "Energy and Power," in *Energy and Power* (San Francisco, 1971) p. 5; Martin Ruhemann, *Power* (London, 1946), p. 110; and Earl Cook, *Man, Energy, Society* (San Francisco, 1976), p. 166.

8. Leslie White, *The Science of Culture* (New York, 1949), pp. 362–93, and idem, "The Energy Theory of Cultural Development," in Morton H. Fried, ed., *Readings in Anthropology*, 2 (New York, 1959): 139–46. See Carlo M. Cipolla, *The Economic History of World Population* (Sussex and New York, 1978), for a similar approach.

9. Lynn White, Jr., "Cultural Climates and Technological Advances in the Middle Ages," *Viator* 2 (1971): 175–76, and elsewhere. This essay was reprinted in White, *Medieval Religion and Technology: Collected Essays* (Berkeley and Los Angeles, 1978), pp. 217–54.

10. Fred Cottrell, *Energy and Society: The Relation between Energy, Social Change, and Economic Development* (Westport, Conn., 1970), p. 2.

11. Forbes, *Studies*, 2:90; Luigi Jacono, "La ruota idraulica di Venafro," *L'ingegnere* 12 (1938): 853.

12. R. E. Palmer, "Notes on Some Ancient Mine Equipments and Systems," Institution of Mining and Metallurgy, *Transactions* 36 (1926–27): 299–336; J. G. Landels, *Engineering in the Ancient World* (Berkeley and Los Angeles, 1978), pp. 69–70.

13. Vannoccio Biringuccio, *The "Pirotechnia" of Vannoccio Biringuccio*, transl. with an introduction and notes by Cyril Stanley Smith and Martha Teach Gnudi (Cambridge, Mass., 1966), p. 22.

14. Jacob Leupold, *Theatrum machinarum generale: Schau-Platz des Grundes mechanischer Wissenschaften...* (Leipzig, 1774), pp. 126–27. The work was originally printed in 1724.

15. E.-M. Carus-Wilson, "An Industrial Revolution of the Thirteenth Century," *Eco-*

nomic History Review 11 (1941): 39–60. For the importance of water power on industrial location see also W. H. K. Turner, "The Significance of Water Power in Industrial Location: Some Perthshire Examples," *Scottish Geographical Magazine* 75 (1958): 98–115, and Leslie Aitchison, *A History of Metals*, 2 (London, 1960): 310, 311, 343 (for the metal industries).

16. Lewis C. Hunter, "Waterpower in the Century of the Steam Engine," in Brooke Hindle, ed., *America's Wooden Age: Aspects of Its Early Technology* (Tarrytown, N.Y., 1975), pp. 179–84, and Hunter, *A History of Industrial Power in the United States, 1780–1930*, vol. 1, *Waterpower in the Century of the Steam Engine* (Charlottesville, Va., 1979), pp. 118–39, 181–203.

CHAPTER 1: ORIGINS

1. For an elementary overview of the technical aspects of water-power development see, for example, Hans Thirring, *Energy for Man: Windmills to Nuclear Power* (Bloomington, Ind., 1958), pp. 230–39.

2. I define a vertical water wheel as a wheel whose plane of rotation is in the vertical plane and a vertical watermill as a mill which is powered by a vertical water wheel. Similarly, I call a wheel whose plane of rotation is horizontal a horizontal water wheel, and I call a mill driven by such a wheel a horizontal watermill. In using this approach I believe I am following the lead of the majority of historians. But E. Cecil Curwen, "The Problem of Early Water-mills," *Antiquity* 18 (1944): 130, preferred the term *vertical mill* for what I call a horizontal mill because the shaft or axle of such a mill was situated vertically.

3. On the history of flour milling in general and the rotary quern in particular see L. A. Moritz, *Grain-Mills and Flour in Classical Antiquity* (Oxford, 1958); L. Lindet, "Les origines du moulin à grains," *Revue archéologique*, 3rd ser. 35 (1899): 413–27; ibid. 36 (1900): 17–44; John Stock and Walter D. Teague, *Flour for Man's Bread: A History of Milling* (Minneapolis, Minn., 1952); Richard Bennett and John Elton, *History of Corn Milling*, 4 vols. (London, 1898–1904); R. J. Forbes, *Studies in Ancient Technology*, 3 (Leiden, 1955): 138–48; and Forbes, "Food and Drink," in Charles Singer et al., eds., *A History of Technology*, 2 (Oxford, 1956): 106–12.

4. On ancient water-raising devices see Forbes, *Studies in Ancient Technology*, 2 (Leiden, 1955): 1–79; Thorkild Schiøler, *Roman and Islamic Water-Lifting Wheels* (Odense, 1973); and Thomas Ewbank, *A Descriptive and Historical Account of Hydraulic and Other Machines for Raising Water . . .* , 12th ed. (New York, 1851).

5. The mechanism I call the water lever has been given a variety of names. Joseph Needham, *Science and Civilisation in China*, 4, pt. 2 (Cambridge, 1965): 363, called it a "spoon tilt-hammer." Bradford Blaine, "The Application of Water Power to Industry During the Middle Ages" (Ph.D. diss., University of California, Los Angeles, 1966), pp. 8–10, termed it a "fill-and-spill device," and Martha Zimiles and Murray Zimiles, *Early American Mills* (New York, 1973) p. 26, used the phrase "plumping mill" to describe it.

6. Needham, *Science and Civilisation*, 4, pt. 2: 361.

7. Schiøler, *Water-Lifting Wheels*, pp. 88–89.

8. Philo of Byzantium *Pneumatica*, chs. 54, 61–62, 65. On Philo see also Theodor Beck, "Philon von Byzanz," *Beiträge zur Geschichte der Technik und Industrie* 2 (1910): 64–77.

9. A. G. Drachmann, *Ktesibios, Philon and Heron: A Study in Ancient Pneumatics* (Copenhagen, 1948), pp. 64–66; Schiøler, *Water-Lifting Wheels*, pp. 65–66, 163.

10. Strabo *Geographica* 12. 556, ed. and transl. Horace Leonard Jones, 5 (London, 1928): 428–29.

11. Curwen, "Problem of Early Water-mills," p. 135; Bennett and Elton, *History of Corn Milling*, 2:9; and Paul N. Wilson, *Watermills: An Introduction*, a Society for the Protection of Ancient Buildings booklet, rev. ed. (London, 1973), pp. 7–8, among others, claim that Strabo's mill must have been of the horizontal type. Arthur W. Parsons, "A Roman Water-Mill in the Athenian Agora," *Hesperia* 5 (1936): 89, and C. L. Sagui, "La meunerie de Barbegal (France) et les roues hydrauliques chez les anciens et au moyen âge," *Isis* 38

(1947): 230, believe that the Strabo wheel was of the vertical type. The evidence is not sufficient to support either of these positions. Further, I think that it is at least conceivable that Strabo's *hydraleta* was a water-raising wheel of some type, rather than a watermill.

12. *Greek Anthology*, bk. 9, no. 418, ed. and transl. W. R. Paton, 3 (London, 1917): 232–33. The poem is attributed to Antipater of Thessalonica by most authorities, but H. Stadtmüller, *Anthologia Graeca*, vol. 2 (Leipzig, 1906), gives its author as Antipater of Byzantium, and Shemuel Avitsur, "Watermills in Eretz Israel, and Their Contribution to Water Power Technology," International Symposium on Molinology, *Transactions* 2 (1969): 389, claims that the writer of the poem was Antipater of Sidon and that the mill was in upper Galilee, not Greece.

13. The most extended argument for the Antipater mill's being of the horizontal variety was made by Curwen, "Problem of Early Water-mills," pp. 134–35. His position, however, was supported by Forbes in *Studies in Ancient Technology*, 2:86–87, and also in "Power," in Singer, *History of Technology*, 2:593; by Bennett and Elton, *History of Corn Milling*, 2:9; and by Lynn White, *Medieval Technology and Social Change* (Oxford, 1962), p. 80, among others.

14. Moritz, *Grain-Mills*, pp. 132–34; Parsons, "Roman Water-Mill," p. 81n; and J. G. Landels, *Engineering in the Ancient World* (London, 1978), p. 17, have supported the idea that the Antipater mill was an overshot mill. The possibility of an undershot wheel with steeply inclined chute has, so far as I know, never been suggested before. Wheels of this type, sometimes called flutter wheels, were frequently used much later where rapid motion was needed, as in sawmills.

15. Lucretius *De rerum natura* 5. 515–16 (. . . *ut fluvios versare rotas atque huastra videmus*).

16. White, *Medieval Technology*, p. 80, n. 5; Moritz, *Grain-Mills*, p. 131, n. 4; Blaine, *Application*, p. 160; and Schiøler, *Water-Lifting Wheels*, p. 163, agree that Lucretius could only be referring to a noria. There are some who disagree. Hugo Blümner, *Technologie und Terminologie der Gewerbe und Künste bie Griechen und Römern*, 2nd ed., 1 (Leipzig, 1912): 47, n. 4, stated that the Lucretius wheel was undershot. Forbes, *Studies in Ancient Technology*, 2: 87, suggested that Lucretius was speaking of a horizontal water wheel, but in his article "Power," p. 596, he made the Lucretius wheel vertical undershot.

17. Vitruvius *De architectura* 10. 5. 1–2, ed. and transl. Frank Granger, 2 (London, 1934): 305, 307.

18. L. A. Moritz, "Vitruvius' Water Mill," *Classical Review* 70 (1956): 193–96, has argued convincingly that the usual translation of the portion of Vitruvius dealing with mill gearing is wrong. In the eighteenth century it was the practice to establish gearing ratios so that the millstones turned faster than the wheel by making the vertical gear wheel larger than the horizontal gear wheel. Most translators have assumed that this is what Vitruvius must have meant and have amended the manuscripts to make them read in that manner. Moritz, referring to various archeological finds (discussed later in this chapter), noted that this may not have been the Roman practice and that the amendations to Vitruvius' text may be incorrect. I have altered Granger's translation of Vitruvius' passage on watermills to take into account Moritz's findings.

19. Needham, *Science and Civilisation*, 4, pt. 2: 370.

20. Ibid., p. 392.

21. Luigi Jacono, "La ruota idraulica di Venafro," *L'ingegnere* 12 (1938): 850–53.

22. F. Mayence, "La troisième campagne de fouilles à Apamée," *Bulletin des Musées Royaux d'Art et d'Histoire* (Brussels), 3rd ser. 5 (1933): 5, fig. 5, reproduced in Needham, *Science and Civilisation*, 4, pt. 2: facing 359.

23. For the watermill complex in France see Ferdinand Benoit, "L'usine de meunerie hydraulique de Barbegal (Arles)," *Revue archéologique*, 6th ser. 15 (1940): 19–80; for the Athenian watermill see Parsons, "Roman Water-Mill," pp. 70–90. These will be discussed in more detail in the text later.

24. Axel Steensberg, *Bondehuse og Vandmøller i Danmark gennem 2,000 År (Farms and Water Mills in Denmark during 2,000 Years)*, in Danish with an English resumé (Copenhagen, 1952), pp. 52–64 (in Danish), 294–97 (in English). Needham, *Science and Civilisation*, 4, pt.

2:366n, comments that the horizontal watermills reported by Steensberg are "hard to visualise since nothing was left but the stones of the presumed mill-races," and Edward M. Fahy, "A Horizontal Mill at Mashanaglass, County Cork," *Cork Historical and Archaeological Society, Journal* 61 (1956): 47, says that "in the absence of structural evidence for a mill on that site, the excavator's claim must be treated with reserve." Blaine, "Application," pp. 7–8n, accuses Fahy of failing to see the structural evidence implicit in the patterns of silt deposits carefully described by the excavators. Steensberg's interpretation has been accepted by most authorities—e.g., Blaine, "Application," p. 7; White, *Medieval Technology*, pp. 81, 160; Forbes, "Power," p. 594; John Reynolds, *Windmills and Watermills* (New York, 1970), p. 11.

25. Bennett and Elton, *History of Corn Milling*, 2:6–11, 31–36; Curwen, "Problem of Early Water-mills," pp. 130–32, 145; Forbes, *Studies in Ancient Technology*, 2:36, 38–39, 88.

26. Abbott Payson Usher, *A History of Mechanical Inventions* (1929; rev. ed., Cambridge, Mass., 1954), p. 161.

27. Moritz, *Grain-Mills*, p. 134.

28. J. Troup, "On a Possible Origin of the Waterwheel," *Asiatic Society of Japan, Transactions*, 1st ser. 22 (1894): 109–14.

29. Needham, *Science and Civilisation*, 4, pt. 2:405.

30. Ibid., pp. 361–62, 405.

31. Ibid., pp. 405–6.

32. Ibid., p. 406.

33. Bennett and Elton, *History of Corn Milling*, 2:10–11, assume that it was a Greek invention; Needham, *Science and Civilisation*, 4, pt. 2:369, and Curwen, "Problem of Early Water-mills," p. 144, attribute it to the Chinese, as does C. Reindl, "Die Entwicklung der Wasserkraftnutzung und der Wasserkraftmaschinen," *Wasserkraft Jahrbuch*, 1924, p. 3. A northern European, or "barbarian," origin has been suggested by White, *Medieval Technology*, p. 81, and Blaine, "Application," p. 12.

34. There is an extensive literature on the horizontal watermill. Some examples are Paul N. Wilson, *Watermills with Horizontal Wheels* (Kendal, 1960); Louis C. Hunter, "The Living Past in the Appalachias of Europe: Water-Mills in Southern Europe," *Technology and Culture* 8 (1967): 446–66; Curwen, "Problem of Early Water-mills," pp. 130–46; H. W. Dickinson, "The Shetland Watermill," *Newcomen Society, Transactions* 13 (1932–33): 89–94; Fahy, "Horizontal Mill," pp. 13–57; Daniel Gade, "Grist Milling with the Horizontal Waterwheel in the Central Andes," *Technology and Culture* 12 (1971): 43–51; Gilbert Goudie, "On the Horizontal Water Mills of Shetland," *Society of Antiquaries of Scotland, Proceedings* 20 (1886): 257–97; A. T. Lucas, "The Horizontal Mill in Ireland," *Royal Society of Antiquaries of Ireland, Journal* 83 (1953): 1–36; Robert MacAdam, "Ancient Water-Mills," *Ulster Journal of Archaeology* 4 (1856): 6–15; Joseph P. O'Reilly, "Some Further Notes on Ancient Horizontal Water-Mills, Native and Foreign," *Royal Irish Academy, Proceedings* 24, sec. C (1902–4): 55–84; Ladislao Reti, "On the Efficiency of Early Horizontal Waterwheels," *Technology and Culture* 8 (1967): 388–94; Kenneth Williamson, "Horizontal Water-Mills of the Faeroe Islands," *Antiquity* 20 (1946): 83–91; as well as Usher, *Mechanical Inventions*, p. 166 f, and Needham, *Science and Civilisation*, 4, pt. 2: 366–69.

35. For the history of the noria see B. Laufer, "The Noria or Persian Wheel," *Oriental Studies in Honour of C. E. Pavry* (Oxford, 1933), pp. 238–50. See also Needham, *Science and Civilisation*, 4, pt. 2: 360–62, who points out that Laufer sometimes confuses the noria with other ancient water-raising devices. Other extensive studies of the noria include Julio Caro Baroja, "Norias, azudas, aceñas," *Revista de dialectologia y tradiciones populares* 10 (1954): 29–160, and G. S. Colin, "La noria morocaine et les machines hydraulique dans le monde arabe," *Hesperis* 14 (1932): 22–60, and "L'origine des norias de Fes," ibid. 16 (1933): 156–57.

36. William L. Westermann and Casper J. Kraemer, *Greek Papyri in the Library of Cornell University* (Ithaca, N.Y., 1926), pp. 42–43. Schiøler, *Water-Lifting Wheels*, pp. 88–89, believes the passage refers to tread-operated wheels.

37. Moritz, *Grain-Mills*, pp. 103–5.

38. Derek J. de Solla Price, "Gears from the Greeks: The Antikythera Mechanism—A Calendar Computer from ca. 80 B.C.," American Philosophical Society, *Transactions*, n.s. 64, pt. 7 (1974). See pp. 53–62 for a history of early gearing.

39. The supposition of an eastern Mediterranean origin for the watermill gains some incidental support from the fact that Vitruvius used its Greek name, according to Bertrand Gille, "Le moulin à eau: Une révolution technique médiévale," *Techniques et civilisations* 3 (1954): 1.

40. Needham, *Science and Civilisation*, 4, pt. 2: 369–70, 396, 405–8. See also Joseph Needham, *The Grand Titration: Science and Society in East and West* (London, 1969), pp. 94–97, and "The Pre-Natal History of the Steam Engine," Newcomen Society, *Transactions* 35 (1962–63): 29–33.

41. Blaine, "Application," pp. 8–10, and Robert Brittain, *Rivers, Man, and Myths: From Fish Spears to Water Mills* (Garden City, N.Y., 1958), pp. 253–56, also suggest that the early Chinese passages on water power may refer to the water lever.

42. Needham, *Science and Civilisation*, 4, pt. 2: 378, n. *a*.

43. Ibid., pl. 225, facing p. 364. The engine was designed by the Swedish engineer Triewald.

44. Ibid., p. 370.

45. Ibid., p. 396.

46. Ibid., p. 393.

47. Ibid., pp. 363–65, 393.

48. Ibid., pp. 406–7, 435–46. See also Joseph Needham, Wang Ling, and Derek J. de Solla Price, *Heavenly Clockwork: The Great Astronomical Clocks of Medieval China* (Cambridge, 1960), pp. 21–22, 77, 100–101, 104–5, 110, and elsewhere.

49. Hans E. Wulff, *The Traditional Crafts of Persia* (Cambridge, Mass., 1966), p. 291, illustrates a traditional Persian rice-husking mill. The blades on this mill were shaped in a manner similar to Chinese water levers. Vertical undershot wheels with spoon-shaped blades are also depicted in, e.g., Donald R. Hill, ed., *"The Book of Knowledge of Ingenious Mechanical Devices..." by Ibn al-Razzaz al-Jazari* (Dordrecht, Holland, and Boston, Mass., 1974), fig. 126, p. 183.

50. Vitruvius *De architectura* 10. 1. 6.

51. Pliny *Historia naturalis* 18. 23 (*mairo pars Italiae nudo utibur pilo, rotis etiam quas aqua verset obiter et mola*); Saul Liberman, ed., *Tosefta Sabbath* 1. 23 (New York, 1962), p. 6, cited by Blaine, "Application," p. 39. The Pliny passage is too short and ambiguous to furnish grounds for positive identification of the implement involved, but Forbes, *Studies in Ancient Technology*, 2:87, claimed that it was a horizontal mill, as did Bennett and Elton, *History of Corn Milling*, 2:8–9. H. P. Vowles, "Early Evolution of Power Engineering," *Isis* 18 (1932): 415–16, suggested that the Pliny reference might be to a vertical water wheel operating pestels by means of cams or pegs projecting from its axle. (See also Vowles, "Pliny's Watermill," *Nature* 127 [1931]: 889.) And H. Glendinning and D. W. T. Glendinning, "Pliny's Water-mill," *Nature* 127 (1931): 974, postulated the water lever.

52. See note 74, below.

53. Suetonius *De vita Caesarum* (Caligula) 39.

54. *Edictum Diocletiani* 15. 54.

55. W. H. Buckler and D. M. Robinson, "Greek and Latin Inscriptions," *Sardis* 7, pt. 1 (1932): 139, no. 169. The term used here for "water wheel" or "watermill" is the same term used by Strabo: *hydraleta*.

56. Georgius Cedrenus *Compendium Historiarum* 516, in J. P. Migne, ed., *Patrologiae Graeca* 121. 562.

57. Ausonius *Mosella* 361–64. White, *Medieval Technology*, pp. 82–83, argues that Ausonius's authorship of the *Mosella* is suspect on several grounds. There is no other evidence of water-driven saws until the thirteenth century; there is no marble in the vicinity of the Moselle; and the poem occurs only in late and marginal manuscripts of Ausonius (tenth century and later). Also the *Mosella*, according to White, is a poem "notably and suspiciously above the level of Ausonius' certainly authentic works both in literary style and in

sensibility towards nature." White thus believes that the *Mosella* was a medieval addition to Ausonius' works.

58. Procopius *De bello Gothico* 5. 19. 8, ed. and transl. H. B. Dewing, 3 (London and New York, 1919): 187.

59. Prudentius *Contra orationem symmachi* 2. 950, ed. and transl. H. J. Thompson, 2 (Cambridge, Mass., and London, 1949): 83.

60. *Codex Theodosianus* 14. 15. 4 (for a translation see Clyde Pharr, transl., *The Theodosian Code and Novels and the Sirmondian Constitutions* [Princeton, N.J., 1952], p. 416); Palladius *De re rustica* 1. 41.

61. L. Allodi and G. Levi, eds., *Il registo Sublacense de secolo XI* (Rome, 1885), no. 28, p. 68, cited by Blaine, "Application," p. 40.

62. See note 23, above, for the Athenian mill.

63. *Codex Justinianus* 11. 42. 10 (see Bennett and Elton, *History of Corn Milling*, 2:41, for a translation).

64. *Digestorum Justinianus* 39. 2. 24, and 42. 12. 12 (see Bennett and Elton, *History of Corn Milling*, 2:41–42, for a translation).

65. Procopius *De bello Gothico* 5. 19. 19–27. The text does not make it clear whether Belisarius' engineers invented the boat mill or merely applied a technique previously developed elsewhere. An anonymous author more than a century earlier had used a paddle wheel mounted on a boat for propulsion instead of power (see note 91, below), but there is no evidence to indicate that this idea was followed up. Nor is there any evidence of boat mills before Belisarius'.

66. Herodotus *Historiae* 1. 174, transl. A. D. Godley, vol. 1 (Cambridge, Mass., 1960), p. 219. See J. Donald Hughes, *Ecology in Ancient Civilizations* (Albuquerque, N.M., 1975), pp. 48–55, for an overview of the Greek attitude towards nature.

67. Plutarchus *Vitae parallalae: Marcellus* 17, transl. Bernadotte Perrin, 5 (New York and London, 1917): 479. For additional discussion on how the attitude of the upper classes in antiquity could have retarded technology generally and the adoption of the water wheel in particular see H. W. Pleket, "Technology and Society in the Graeco-Roman World," *Acta Historiae Neerlandica* 2 (1967): 1–24, esp. 17–23, and M. I. Finley, "Technical Innovation and Economic Progress in the Ancient World," *Economic History Review*, 2nd ser. 18 (1965): 29–45.

68. Columella *De re rustica* 3. 3; Pliny *Historia naturalis* 18. 36, 38, 43.

69. Finley, "Technical Innovation," pp. 38–39.

70. David W. Reece, "The Technological Weakness of the Ancient World," *Greece and Rome*, 2nd ser. 16 (1969): 46–47, stresses the geographical explanation of the failure of classical antiquity to adopt the water wheel. Blaine, "Application," p. 43, also inclines in that direction.

71. Suetonius *De vita Caesarum* (Vespasian) 18.

72. Pliny *Historia naturalis* 36. 195; Petronius *Satyricon* 51; Cassius Dio Cocceianus *Historia* 57. 21. 7.

73. For additional discussion on how the labor surplus affected the course of ancient technology see Pleket, "Technology and Society," pp. 13–16, and Finley, "Technical Innovation," pp. 43–44.

74. L. Jacono, "La ruota idraulica di Venafro," *L'ingegnere* 12 (1938): 850–53, and *Annali dei lavori pubblici* (Rome) 77 (1939): 217–20; C. Reindl, "Ein Römisches Wasserrad," *Wasserkraft und Wasserwirtschaft* 34 (1939): 142–43; F. M. Feldhaus, "Ahnen des Wasserrades," *Die Umschau* 40 (1936): 472–73, 476.

75. F. Gerald Simpson, *Watermills and Military Works on Hadrian's Wall: Excavations in Northumberland, 1907–1913*, ed. by Grace Simpson, with a contribution on watermills by Lord Wilson of High Wray (Kendal, 1976), pp. 26–42.

76. Arthur W. Parsons, "A Roman Water-Mill in the Athenian Agora," *Hesperia* 5 (1936): 70–90.

77. Ibid., p. 80.

78. Ferdinand Benoit, "L'usine de meunerie hydraulique de Barbegal (Arles)," *Revue*

*archéologique*, 6th ser. 15 (1940): 19–80; Benoit, "'Usine' de meunerie hydraulique à l'époque Romaine," *Annales d'histoire sociale* 1 (1939): 183–84; and Sagui, "Meunerie de Barbegal," pp. 225–31.

79. Benoit's assumption that the wheels were overshot has been accepted by most authorities, including Forbes, "Power," pp. 298–99; Sagui, "Meunerie de Barbegal," pp. 225–31; and Reynolds, *Windmills and Watermills*, p. 12. But Blaine, "Application," p. 25, also questions whether the wheels were overshot.

80. Sagui, "Meunerie de Barbegal," pp. 226–27.

81. Albert William Van Buren and Gorham Phillips Stevens, "The Aqua Trainana and the Mills on the Janiculum," American Academy in Rome, *Memoirs* 1 (1915–16): 59–62; ibid. 6 (1927): 137–46; and idem, "The Antiquities of the Janiculum," ibid. 11 (1933): 69–79.

82. J. Collingwood Bruce, *Handbook to the Roman Wall*, ed. Ian A. Richmond, 11th ed. (Newcastle-upon-Tyne, 1957), p. 85; Arthur Stowers, "Observations on the History of Water Power," Newcomen Society, *Transactions* 30 (1955–57): 244–45; Moritz, *Grain-Mills*, p. 136; Simpson, *Watermills*, pp. 43–49. A barrel-shaped stone with slots found at this site was interpreted by Richmond, Stowers, and Moritz as the core of a composite water wheel hub built of wood, stone, and iron (for an illustration of the hub see Stowers, "Observations," p. 244). Simpson discussed the Chesters Bridge site at length and concluded that the barrel-shaped stone was not part of a water wheel but, more likely, was part of a capstan or a balance weight for lifting and lowering a portcullis below the bridge.

83. R. C. Shaw, "Excavations at Willowford," Cumberland and Westmorland Antiquarian and Archaeological Society, *Transactions*, 2nd ser. 26 (1926): 429–506; Simpson, *Watermills*, pp. 49–50.

84. Vitruvius *De architectura* 10. 4. 1–3.

85. For archeological material on Roman water-lifting wheels see R. E. Palmer, "Notes on Some Ancient Mine Equipments and Systems," Institution of Mining and Metallurgy, *Transactions* 36 (1926–27): 299–336; George C. Boon and Colin Williams, "The Dolaucothi Drainage Wheel," *Journal of Roman Studies* 56 (1966): 122–27; G. Gosse, "Las minas y el arte minero de España en la antigüedad," *Ampurias* 4 (1942): 43–68; and Baroja, "Norias, azudas, anceñas," pp. 38–50.

86. This is noted also by Storck and Teague, *Flour for Man's Bread*, p. 108.

87. Norman Smith, *A History of Dams* (London, 1971), pp. 25–49.

88. E. A. Thompson, ed. and transl., *A Roman Reformer and Inventor, Being a New Text of the Treatise "De rebus bellicis,"* with commentary (Oxford, 1952), pp. 47–49, 121 (ch. xviii, 9), for example.

89. Palladius *De re rustica* 1. 41.

90. See note 57, above.

91. Thompson, *Roman Inventor*, pp. 50–54, 119–20 (ch. xvii).

CHAPTER 2: DIFFUSION AND DIVERSIFICATION

1. Lynn White, "Technology and Invention in the Middle Ages," *Speculum* 15 (1940): 141–59, esp. 155–56.

2. In addition to the work cited in note 1, above, see also White's *Medieval Technology and Social Change* (Oxford, 1962), esp. pp. 79–134, and his *Medieval Religion and Technology: Collected Essays* (Berkeley and Los Angeles, 1978), pp. 18–22, 75–80, 143–44.

3. White, *Medieval Technology*, p. 87.

4. The sections of this chapter on the geographical diffusion of the vertical water wheel and its industrial diversification owe much to the Ph.D. dissertation of Bradford Blaine, "The Application of Water-Power to Industry during the Middle Ages" (University of California, Los Angeles, 1966). For additional information on these aspects of medieval water power, readers should consult this well-written and excellently documented work. A spin-off from this dissertation is Blaine's "The Enigmatic Water-Mill," in Bert S. Hall and Delno C. West, *On Pre-Modern Technology and Science* (Malibu, Calif., 1976), pp. 163–76.

5. Cassiodorus *Variarum libri duodecim* 3. 31, in Migne, *Patrologia Latina* (henceforth *PL*) 69. 593 (Bennett and Elton, *History of Corn Milling*, 2:42, translate the passage); Cassiodorus *De institutione divinarum litterarum* 29, in Migne, *PL* 70. 1143 (translated in Bennett and Elton, *History of Corn Milling*, 2:42).

6. Blaine, "Application," pp. 47–48.

7. Paul Aebischer, "Les dénominations du 'moulin' dans les chartes italiennes du Moyen Age," *Archivum Latinitatis Medii Aevi (Bulletin Du Cange)* 7 (1932): 49–109.

8. Blaine, "Application," pp. 45–48; Aebischer, "Dénominations," pp. 54–57 et passim.

9. Gregory of Tours *Historiae Ecclesiasticae Francorum* 3. 19, in Migne, *PL* 71. 250; *Vitae patrum* 18. 2, ibid., col. 1085. These passages are translated in Bennett and Elton, *History of Corn Milling*, 2:71.

10. Venantius Fortunatus *Operum Omnium Miscellanea* 3. 12. 37–38, in Migne, *PL* 88. 137.

11. *Leges Visigothorum* 7. 2. 12, and 8. 4. 30, in *Monumenta Germaniae Historica* (henceforth *MGH*), *Leges*, sec. 1, vol. 1 (Hanover, 1892), pp. 293, 344; *Lex Salica* 24, in Frederick Lindebrog, ed., *Codex legum antiquarum*, (Frankfurt, 1613), p. 324. Bennett and Elton, *History of Corn Milling*, 2:42, 71–73, translate a number of the relevant passages from these law codes. See also Marc Bloch, *Land and Work in Mediaeval Europe: Selected Papers by Marc Bloch*, transl. J. E. Anderson (Berkeley and Los Angeles, 1967), pp. 137–39, 144, and Blaine, "Application," pp. 48–49.

12. Blaine, "Application," pp. 50–52.

13. Ibid., p. 52. For the spread of watermilling in France see also Bloch, *Land and Work*, pp. 137–38, 143–44, and Germain Sicard, *Les moulins de Toulouse au Moyen Age* (Paris, 1953), pp. 29–31.

14. Marius d'Avenches, *Marii Episcopi Aventicensis Chronica, CCCCLV–DLXXXI*, in *MGH*, *Auctores Antiquissimi*, 11, pt. 2 (Berlin, 1894): 237.

15. Blaine, "Application," pp. 56–57.

16. Heinrich Beyer, ed., *Urkundenbuch zur Geschichte der jetzt die Preussischen Regierungsbezirke Coblenz und Trier bildenden mittelrheinischen Territorien*, 1 (Coblenz, 1860): 9.

17. *Lex Alamannorum* 80, 83, in Lindebrog, *Codex*, p. 385.

18. Blaine, "Application," pp. 53–56.

19. Ibid., p. 54; "Ludwig der Deutsche," in *MGH*, *Diplomata regum Germaniae ex stirpe Karolinorum*, 1 (Berlin, 1934): 161 *(molinam unam)*.

20. Lubor Niederle, *Manuel de l'antiquité slave*, 2 (Paris, 1926): 198–99. There is no documentation in the French translation, but Slavic sources are cited in Niederle's *Zivot starych Slovanu*, 3 (Prague, 1921): 119–20.

21. This is noted also by Blaine, "Application," p. 69.

22. *Helmholdi Presbyteri Chronica Slavorum a. 800–1172*, 1. 12, in *MGH*, *Scriptores*, 21 (Hanover, 1869): 19, a twelfth-century report of remnants of watermills among the debris of tenth-century German settlements destroyed by a Slav offensive.

23. Blaine, "Application," pp. 63–64.

24. Niederle, *Manuel*, 2:198–99; Bertrand Gille, "Le moulin à eau: Une révolution technique médiévale," *Techniques et civilisations* 3 (1954): 3, citing Wasivtynski, *La regale du moulin dans le droit polonais du Moyen Age* (Warsaw, 1936), p. 10. Maria Dembińska, *Przetwórstwo Zbożwe w Polsce Średniowiecznej (X–XIV Wiek)* (Warsaw, 1973), p. 76, places the earliest Polish mills in 1178 and 1198.

25. Cornel Irimie and Coneliu Bucur, "Typology, Distribution, and Frequency of Water Mills in România in the First Half of the Twentieth Century," International Symposium on Moninology, *Transactions* 2 (1969): 421.

26. Albertus Aquensis *Incipit Historia Hierosolymitanae expeditionis* 1. 10, in Migne, *PL* 166. 396; John Beckmann, *A History of Inventions and Discoveries*, 3rd ed., 1 (London, 1817): 244–45.

27. Dembińska, *Przetwórstwo*, p. 263.

28. Irimie and Bucur, "Water Mills in România," p. 423.

29. V. V. Danilevskii, *History of Hydroengineering in Russia before the Nineteenth Century*,

transl. from the Russian by the Israel Program for Scientific Translations (Jerusalem, 1968), p. 5.

30. *Codex diplomaticus aevi Saxonici,* no. 108, ed. John M. Kemble, 1 (London, 1839): 132–33; Bennett and Elton, *History of Corn Milling,* 2: 97, translate the passage.

31. Blaine, "Application," pp. 60–61. For the report on the ninth-century watermill near Windsor see David M. Wilson, "Medieval Britain in 1957, 1: Pre-Conquest," *Medieval Archaeology* 2 (1958): 184–85.

32. *Ancient Laws and Institutes of Wales,* ed. and transl. Aneurin Owen (London, 1841), pp. 147, 189, 332, 522, 563, 740.

33. There is an extensive literature on watermills in Ireland. See, for example, Blaine, "Application," pp. 57–59, and the items mentioned in note 34 of the previous chapter by Goudie, Fahy, Lucas, MacAdam, and O'Reilly.

34. Blaine, "Application," pp. 61–63. S. M. Imamuddin, *Some Aspects of the Socio-Economic and Cultural History of Muslim Spain, 711–1492 A.D.* (Leiden, 1965), pp. 104–5, says that both Christian and Muslim records make it clear that many watermills were in operation in Spain as early as the eighth and ninth centuries.

35. Margaret T. Hodgen, "Domesday Water Mills," *Antiquity* 13 (1939): 261–79. Hodgen found the greatest concentration of mills in the parts of England facing the Continent, indicating that the watermill probably spread to England from there. Bennett and Elton, *History of Corn Milling,* 2:131–80, give a long list of the Domesday watermills with their valuations. Reginald V. Lennard, *Rural England, 1086–1135* (Oxford, 1959), p. 278, suggests that Hodgen's figure of 5,624 mills is "almost certainly too low" and that, for the 15 counties surveyed more closely since 1939, the evidence seems "to indicate that the number in this area was appreciably larger than Miss Hodgen's total of 2,628 mills at 1,600 places." However, Jennifer Tann, "Multiple Mills," *Medieval Archaeology* 11 (1967): 253–55, points out that the word "mill" frequently meant "the machinery in a building and not the building itself." This consideration has led Tann to question whether Domesday "mills" should always be regarded as separate buildings and hence separate mills. Thus, in Tann's opinion, Hodgen's count may be too high.

36. Sicard, *Moulins de Toulouse,* p. 37.

37. Ibid., p. 37, n. 78.

38. Prosper Boissonnarde, *Essai sur l'organization du travail en Poitou,* 1 (Paris, 1900): 114.

39. Dembińska, *Przetwórstwo,* p. 266.

40. Forbes, *Studies in Ancient Technology,* 2:105.

41. Bertrand Gille, "The Medieval Age of the West," in Maurice Daumas, ed., *A History of Technology and Invention,* 1 (New York, 1969): 550.

42. Gille, "Moulin à eau," p. 3.

43. Georges Duby, *The Early Growth of the European Economy,* transl. Howard B. Clarke (London, 1974), p. 187.

44. Forbes, *Studies in Ancient Technology,* 2:105–6.

45. Gille, "Medieval Age," p. 550.

46. Theodore Sclafert, *Le Haut-Dauphiné au Moyen Age* (Paris, 1926), pp. 1–161 passim; Sicard, *Moulins de Toulouse,* pp. 30–31; Gille, "Moulin à eau," p. 3.

47. Jean Gimpel, *The Medieval Machine: The Industrial Revolution of the Middle Ages* (New York, c1976), p. 57.

48. See Louis Hunter, "Waterpower in the Century of the Steam Engine," in Brooke Hindle, ed., *America's Wooden Age: Aspects of Its Early Technology* (Tarrytown, N.Y., 1975), pp. 179–84, and Hunter, *A History of Industrial Power in the United States, 1780–1930,* vol. 1, *Waterpower in the Century of the Steam Engine* (Charlottesville, 1979), pp. 118–39, 181–203.

49. Procopius *De bello Gothico* 5. 19. 19–27.

50. See note 14, above.

51. Martin of Trier, "De calamitati abbatiae sancti Martini Treverensis," in *MGH, Scriptores,* 15, pt. 2 (Hanover, 1884): 740.

52. See note 26, above.

53. Sicard, *Moulins de Toulouse*, p. 37 and n. 78 on that page.

54. Antoine Fontanon, ed., *Les edicts et ordonnances des roys de France depuis S. Loys iusques à present*, 2nd ed., 2 (Paris, 1585): 422.

55. C. Rivals, "Floating-Mills in France: A Few Notes on History, Technology, and the Lives of Men," International Symposium on Molinology, *Transactions* 3 (1973): 149–58; Cornel Irimie, "Floating Mills on Boats in România," ibid. 2 (1969): 437–45; M. Ferrendier, "Les anciennes utilisations de l'eau," *La houille blanche* 3 (1948): 331–34; ibid., 5 (1950): 779–81; Blaine, "Application," pp. 27–30.

56. Abu 'Abd Alla Muhammad al-Idrisi, *Géographie*, transl. P. Amédée Jaubert from the Arabic, 2 (Paris, 1840): 63–64 (fol. 135r).

57. Blaine, "Application," p. 32.

58. Gille, "Moulin à eau," p. 6, does not indicate the source of this plan. Bennett and Elton, *History of Corn Milling*, vol. 2, frontispiece and p. 76, reproduce a fifteenth-century sketch of bridge mills under the guise of "The Mills of Babylon" (Roy 15 E vi, 5b).

59. For information on bridge mills (often considered a type of *moulin pendant*, or suspended or hanging mill) see Ferrendier, "Anciennes utilisations," 5:776–79; Blaine, "Application," pp. 31–33; and D. H. Jones, "The *Moulin Pendant*," International Symposium on Molinology, *Transactions* 3 (1973): 169–74.

60. For the association of bridges and mills see Marjorie Boyer, "Bridges and Mill Sites in Medieval France," XIIᵉ Congrés international d'histoire des sciences (Paris, 1966), *Actes* 1B: 13–17, and Boyer, *Medieval French Bridges: A History* (Cambridge, Mass., 1976), p. 167. Gimpel, *Medieval Machine*, p. 17, says there were 13 mills under the Grand Pont in 1323.

61. David Luckhurst, *Monastic Watermills: A Study of the Mills within English Monastic Precincts*, a Society for the Protection of Ancient Buildings booklet (London, n.d.), pp. 14–15.

62. Forbes, "Power," in Singer, *History of Technology*, 2:611.

63. Luckhurst, *Monastic Watermills*, pp. 9–10 and fig. 4; Frank Nixon, *The Industrial Archaeology of Derbyshire* (Newton Abbot, 1969), p. 100.

64. Norman Smith, *A History of Dams* (London, 1971), p. 158.

65. Ibid., p. 150.

66. Ibid., p. 158.

67. Ibid.

68. Constance H. Berman, "The Cistercians in the County of Toulouse, 1132–1249: The Order's Foundations and Land Acquisition" (Ph.D. diss., University of Wisconsin, 1978), pp. 286 and 308, n. 79.

69. Gille, "Medieval Age," p. 451, and Gille, "Moulin à eau," p. 4.

70. Wilson, "Medieval Britain," pp. 184–85.

71. Gimpel, *Medieval Machine*, p. 4.

72. Gille, "Medieval Age," p. 452.

73. See, for example, Luckhurst, *Monastic Watermills*.

74. Gille, "Medieval Age," pp. 550–51.

75. Fernand Braudel, *Capitalism and Material Life, 1400–1800*, transl. Mirian Kochan (London, 1973), p. 262.

76. Lynn White, "The Expansion of Technology, 500–1500," in Carlo M. Cipolla, ed., *The Fontana Economic History of Europe*, 1 (London, 1972): 157.

77. [Antonio di] Filarete *Treatise on Architecture*, transl. John R. Spencer (New Haven, 1965), p. 277 (fol. 161v).

78. Hodgen, "Domesday Water Mills," pp. 266–67.

79. Smith, *History of Dams*, p. 165.

80. Bennett and Elton, *History of Corn Milling*, 2:185.

81. Ibid.

82. Sicard, *Moulins de Toulouse*, pp. 38–43; Smith, *History of Dams*, pp. 158–59.

83. Sicard, *Moulins de Toulouse*, pp. 145–326 passim; Gimpel, *Medieval Machine*, pp. 17–23.

84. For additional information on medieval tide mills see C. Rivals, "Tide-Mills in

France: A Few Notes on History, Technology, and the Lives of Men," International Symposium on Molinology, *Transactions* 3 (1973): 159–68; Rex Wailes, "Tide Mills in England and Wales," Newcomen Society, *Transactions* 19 (1938–39): 1–33; Blaine, "Application," pp. 33–36; and Ferrendier, "Anciennes utilisations," 5:782–83. Referring to the "tide mills" at Venice, Rivals says: "It is hard to imagine these in a tideless sea" (p. 164). A. G. Keller in the discussion following Rivals's paper suggested that the Venetian mills might have been designed to exploit the currents between sandbars in the Venetian lagoons (p. 165). All the early evidence on tide mills is reviewed by W. E. Minchinton, "Early Tide Mills: Some Problems," *Technology and Culture* 20 (1979): 777–86. He considers the Venetian tide mills to be doubtful.

85. White, *Medieval Technology*, pp. 84–85.

86. Mariano Taccola *De ingeneis* [1433] 3. 33v–34r, 44v–45r, in Frank D. Prager and Gustina Scaglia, *Mariano Taccola and His Book "De Ingeneis"* (Cambridge, Mass., 1972), pp. 84–85, 106–7.

87. Bradford Blaine in his dissertation, "The Application of Water-Power to Industry during the Middle Ages" (see note 4, above), reviews the earliest indications of the use of water power in a wide variety of industries. The reader wishing additional information on the industrial diversification of water power during the Middle Ages should consult this work, on which I am heavily dependent in this portion of my monograph.

88. For example, John P. Arnold and Frank Penman, *History of Brewing Industry and Brewing Science in America* (Chicago, 1933), p. 88, describe malt-grinding mills as being basically identical to conventional flour mills.

89. Georges Tessier, *Recueil des Actes de Charles II Le Chauve, Roi de France*, vol. 2 [861–877] (Paris, 1952), p. 2, document 225.

90. Blaine, "Application," pp. 73–75; Gille, "Moulin à eau," p. 7; Anne-Marie Bautier, "Le plus anciennes mentions de moulins hydrauliques industriels et de moulins à vent," *Bulletin philologique et historique* 2 (1960): 601–3.

91. Georgius Agricola *De re metallica*, transl. from the first Latin edition of 1556 by Herbert C. Hoover and Lou Henry Hoover (New York, 1950), pp. 295–99.

92. Blaine, "Application," pp. 75–77; Forbes, "Power," p. 610; Gille, "Moulin à eau," p. 7; John Reynolds, *Windmills and Watermills* (New York, 1970), pp. 166–72.

93. Blaine, "Application," p. 77; Gille, "Moulin à eau," p. 7. Blaine, p. 80n, also mentions the possible existence of water-powered pepper mills in the thirteenth century. These could have used either edge runners or conventional millstones.

94. Blaine, "Application," pp. 78–80, and Noel Deerr, *The History of Sugar*, 1 (London, 1949): 77–78.

95. Blaine, "Application," pp. 97–98; Gille, "Moulin à eau," p. 7.

96. Blaine, "Application," p. 161.

97. Reynolds, *Windmills and Watermills*, p. 166.

98. Bautier, "Moulins hydrauliques," p. 595.

99. Ibid., pp. 594–601; Blaine, "Application," pp. 94–97; Gille, "Moulin à eau," p. 7; Ferrendier, "Anciennes utilisations," 4:132.

100. Bautier, "Moulins hydrauliques," pp. 603–6; Blaine, "Application," pp. 142–45; Ferrendier, "Anciennes utilisations," 4:130–31.

101. Robert S. Woodbury, *History of the Lathe to 1850* (Cleveland, 1961), p. 46; Blaine, "Application," pp. 158–60; Gille, "Moulin à eau," p. 7. Leonardo da Vinci *Codex Atlanticus* 289r mentions lathes for shaping vases of jasper and porphyry in a list of industrial activities that could be run by water power if the Arno River were modified.

102. Gille, "Machines," in Singer, *History of Technology*, 2:655, fig. 599, also believes that the wheel in the illustration may be a water wheel.

103. Blaine, "Application," pp. 150–52; Ferrendier, "Anciennes utilisations," 4:129–30.

104. Blaine, "Application," pp. 147–50.

105. Ibid., p. 147.

106. Ibid., p. 171.

107. Agricola *De re metallica,* Hoover ed., pp. 183, 185, 187, 189 (piston pumps), 199 (mine hoist), 206 (fan for ventilation).

108. Blaine, "Application," pp. 99–100; Ferrendier, "Anciennes utilisations," 4:131.

109. Leonardo da Vinci *Codex Atlanticus* 289r.

110. F. M. Feldhaus, *Die Technik der Antike und des Mittelalters* (New York, 1971), p. 286.

111. Filarete *Treatise on Architecture,* Spencer ed., p. 277 (fol. 161v).

112. Gille, "Moulin à eau," pp. 8–9; Usher, *Mechanical Inventions,* p. 140; White, *Medieval Technology,* p. 79, n. 3; and Needham, *Science and Civilisation,* 4, pt. 2:392–94, discuss the emergence of the cam.

113. According to Needham the usual form of the trip-hammer in the West prior to 1500 was the vertical mallet, while the Chinese, very early, were already using the more advanced recumbent hammer (Needham, *Science and Civilisation,* 4, pt. 2:390–96). But Blaine, "Enigmatic Water-Mill," pp. 167–69 and p. 176, n. 89, points out that the type of trip-hammer used in the medieval West cannot be determined from extant documents. The West could very well have used the recumbent hammer long before 1500.

114. Blaine, "Enigmatic Water-Mill," p. 169 and p. 176, n. 88. Drawings of both types of hammer from the so-called "Hussite engineer's manuscript" are reproduced by F. M. Feldhaus, *Die Technik der Vorzeit der geschichtlichen Zeit und der Naturvolker* (Munich, 1965), col. 915, fig. 600 (vertical) and col. 1077, fig. 717 (recumbent).

115. Agricola *De re metallica,* Hoover ed., pp. 287, 313, 314, 320, and elsewhere; Olaus Magnus *Historia de Gentibvs Septentrionalibvs*... (Rome, 1555) p. 207 (6. 6).

116. For reproductions of the plan of St. Gall see, for example, Walter Horn, *The Plan of St. Gall,* 3 vols. (Berkeley and Los Angeles, 1979); Walter Horn and Ernest Born, "The Dimensional Inconsistencies of the Plan of Saint Gall and the Problem of the Scale of the Plan," *Art Bulletin* 48 (1966): fig. 1; and John Arnold, *Origin and History of Beer and Brewing* (Chicago, 1911), p. 209.

117. Blaine, "Enigmatic Water-Mill," p. 168.

118. Blaine, "Application," pp. 83–88 passim.

119. Bautier, "Moulins hydrauliques," p. 583 et passim.

120. Eleanora Carus-Wilson, "An Industrial Revolution of the Thirteenth Century," *Economic History Review* 11 (1941): 39–60.

121. For the diffusion of the fulling mill see ibid.; Blaine, "Application," pp. 81–94; R. V. Lennard, "An Early Fulling-Mill," *Economic History Review* 17 (1947): 150, and Lennard, "Early English Fulling-Mills: Additional Examples," ibid., 2nd ser. 3 (1951): 342–43; Gille, "Moulin à eau," p. 10; Bautier, "Moulins hydrauliques," pp. 581–94, 621–26. On fulling by water power see also Ferrendier, "Anciennes utilisations," 3:503–8; R. A. Pelham, *Fulling Mills,* a Society for the Protection of Ancient Buildings booklet (London, n.d.); Reynolds, *Windmills and Watermills,* pp. 115–22; and Victor Geramb, "Ein Beitrag zur Geschichte der Walkerei," *Wörter und Sachen* 12 (1929): 37–46.

122. Bautier, "Moulins hydrauliques," pp. 569–94; Blaine, "Application," pp. 81–86; Gille, "Moulin à eau," pp. 10–11; Sclafert, *Haut-Dauphiné,* pp. 103–5.

123. Blaine, "Application," pp. 101–15; Gille, "Moulin à eau," pp. 12–13; Ferrendier, "Anciennes utilisations," 3:497–502; Andre Blum, *On the Origin of Paper,* transl. Harry M. Lydenberg (New York, 1934), pp. 28–33; A. F. Gasperinetti, "Paper, Papermakers, and Paper-Mills," in E. J. Labarre, ed., *Zonghi's Watermarks* (Hilversum, Holland, 1960), pp. 63–82.

124. Leslie Aitchison, *A History of Metals* (London, 1960), p. 311.

125. Blaine, "Application," pp. 117–18. In 1197 the archbishop of Lund gave to the Danish Cistercian monastery of Søro (Sjaelland) a village in Halland (Sweden). In 1224 this village had a mill for fabricating iron.

126. Blaine, "Application," pp. 117–32; Gille, "Le moulin à fer et le haut-fourneau," *Métalaux et civilisations* 1 (1946): 89–94, as well as his "Les origines du moulin à fer," *Revue d'histoire de la sidérurgie* 1 (1960–63): 23–32, and "Moulin à eau," pp. 11–12; Reynolds, *Windmills and Watermills,* pp. 159–65; and Ferrendier, "Anciennes utilisations," 4: 126–29.

127. Blaine, "Application," p. 131.

128. Blaine, "Application," pp. 132–40; Ferrendier, "Anciennes utilisations," 4: 126–29; and the articles by Gille cited in note 126, above.

129. Blaine, "Application," pp. 140–41; Ferrendier, "Anciennes utilisations," 4: 122–23.

130. Agricola *De re metallica,* Hoover ed., p. 284.

131. Villard de Honnecourt, *Sketchbook of Villard de Honnecourt,* ed. Theodore Bowie (Bloomington, Ind., 1959), p. 129 and pl. 59.

132. Blaine, "Application," pp. 153–58; Gille, "Moulin à eau," p. 12. The water-powered stone saw may also have been a medieval invention. As noted in Chapter 1, note 57, above, Lynn White has challenged the authenticity of the *Mosella* as Ausonius' work, suggesting that it was probably, instead, a tenth-century work. The earliest definite evidence Blaine could find of hydraulic stone saws was a document from 1584. See White, *Medieval Technology,* pp. 82–83, and Blaine, "Application," pp. 153–54.

133. Sclafert, *Haut-Dauphiné,* pp. 197–98, 202, 435.

134. White, *Medieval Technology,* pp. 103–19, 166–67. Examples of crank-activated systems can be found in the machine books of the Renaissance period, for example, Agostino Ramelli, *Le diverse et artificiose machine* (Paris, 1588), chs. 36, 97, 100, 116, 135, 136, 137.

135. Vannoccio Biringuccio, *Pirotechnia,* transl. Cyril S. Smith and Martha T. Gnudi from the 1st ed. of 1540 (Cambridge, Mass., 1966), pp. 377–81; Blaine, "Application," pp. 145–46.

136. Cipriano Piccolpasso, *The Three Books of the Potter's Art . . . ,* transl. Bernard Rackham and Albert Van de Put (London, 1934), pp. 41–43 (fols. 35v–37v).

137. John U. Nef, *The Conquest of the Material World* (Chicago, 1964), p. 141. Nef developed the idea of an early British industrial revolution and the continuity of British industrial development not only in this work but also in *The Rise of the British Coal Industry,* 2 vols. (London, 1932); in *War and Human Progress: An Essay on the Rise of Industrial Civilization* (Cambridge, Mass., 1950); in *Industry and Government in France and England, 1540–1640* (Ithaca, N.Y., 1964); and in numerous essays.

138. Melvin Kranzberg and Joseph Gies, *By the Sweat of Thy Brow: Work in the Western World* (New York, 1975), p. 84.

139. Usher, *Mechanical Inventions,* p. 170.

140. Dembińska, *Przetwórstwo,* p. 264.

141. Bennett and Elton, *History of Corn Milling,* 2:73 (British Library, Harl. MSS. 334, 71b).

142. There are a number of very similar sketches of water wheels in the *Veil rentier.* See Léo Verriest, ed., *Le polyptyque illustré dit "Veil rentier" de Messire Jehan de Pamele-Audenarde* (Gembloux, Belgium, 1950), fols. 13r, 15r, 24v, 47v, 98v, 116v.

143. For example, Helmuth T. Bassert and Willy F. Storck, eds., *Das Mittelalterlich Hausbuch* [c1475] (Leipzig, 1912), plates 18, 40; Mariano Taccola *De ingeneis* 3. 34v–35r, 56r, in Frank D. Prager and Gustina Scaglia, *Mariano Taccola and His Book "De Ingeneis"* (Cambridge, Mass., 1972), pp. 86, 120; Francesco di Giorgio Martini, *Trattati di Architettura, Ingegneria, e Arte Militare . . . ,* vol. 1 (Milan, 1967), fols. 34, 36, 36v, 48v.

144. Karl von Amira, ed., *Die Dresdener Bilderhandschrift des Sachsenspiegels* [c1350] (Leipzig, 1902), fols. 25, 33b, 35, 55b, 77b. Blaine says that the thirteenth-century manuscript *Liber Anselmi qui dicitur Apologeticum* (British Library, Cottonian MSS., Cleopatra C XI, fol. 10a) contains a sketch of an overshot wheel and that Bennett and Elton, *History of Corn Milling,* 2:74, are incorrect in calling it an undershot. See Blaine, "Enigmatic Water-Mill," pp. 166 and 173, n. 44.

145. Taccola *De ingeneis* 1. 30v, 40r, 3. 44v–45r, in Prager and Scaglia, *Taccola,* pp. 41, 42, and 106; Martini, *Architectura,* vol. 1, fols. 34, 35–35v, 36v–37v, et passim; vol. 2, fol. 95. Singer, *History of Technology,* 2:596, reproduces a 1423 woodcut of an overshot water wheel.

146. The dominance of the compass arm construction may be due merely to the artist's desire for a more aesthetically attractive wheel. The Science Museum in London has several models of "Saxon" water wheels which have clasp arm construction. A photograph of these wheels has been published by Leslie Syson, *British Water-Mills* (London, 1965), fig.

9, facing p. 27. I have been unable to find out what evidence these reconstructions were based on.

147. See Hunter, "Living Past"; Goudie, "Horizontal Water Mills of Shetland"; O'Reilly, "Ancient Horizontal Water-Mills"; Needham, *Science and Civilisation*, 4, pt. 2:368; Bennett and Elton, *History of Corn Milling*, 2:6–30; and Danilevskii, *Hydroengineering in Russia*, p. 173.

148. Felix F. Strauss, "'Mills without Wheels' in the Sixteenth-Century Alps," *Technology and Culture* 12 (1971): 23–42; John Muendel, "The Horizontal Mills of Medieval Pistoia," ibid. 15 (1974): 194–225.

149. Usher, *Mechanical Inventions*, p. 180; Hodgen, "Domesday Water Mills," pp. 261–62; and Sylvia Thrupp, "Medieval Industry, 1000–1500," in Cipolla, *Fontana Economic History*, 1:234, suspect that some of the Domesday mills were probably horizontal mills.

150. Cyril S. Smith, "Granulating Iron in Filarete's Smelter," *Technology and Culture* 5 (1964): 390 and 390n; Theodore Wertime, "Asian Influences on European Metallurgy," ibid., pp. 394–97; Joseph Needham, "Chinese Priorities in Cast Iron Metallurgy," ibid., pp. 398–404.

151. Ladislao Reti, "A Postscript to the Filarete Discussion: On Horizontal Waterwheels and Smelter Blowers in the Writings of Leonardo da Vinci and Juanelo Turriano," *Technology and Culture* 6 (1965): 428–41.

152. Hunter, "Living Past," p. 456n, citing a letter from Hilmar Stigum, curator, Norske Folkemuseum, Oslo.

153. In the Irish law code, *Senchus Mor*, which, in the form we have it, was probably written c900 A.D. In the *Senchus Mor* the principal parts of a watermill are listed. Since the parts given fit the horizontal mill and since the all-important gearing of the vertical mill is not listed among them, the assumption is that a horizontal mill was being described. See *Ancient Laws of Ireland*, 1 (Dublin, 1865): 124, 141; Usher, *Mechanical Inventions*, p. 178; Curwen, "Problem," p. 140; and Lucas, "Horizontal Mill," pp. 29–34.

154. One of the first extant drawings of a horizontal water wheel comes from the notebooks usually ascribed to an anonymous "Hussite engineer" and dated c1430. This drawing has been reproduced in a number of publications, including Strauss, "Mills without Wheels," p. 36, fig. 9, and Theodor Beck, *Beiträge zur Geschichte des Maschinenbaues* (Berlin, 1900), p. 279, fig. 324. A recent student of the manuscript of the Hussite engineer, however, has argued that the Hussite codex is in two sections by two authors and should be dated around 1480: see Bert S. Hall, "The So-Called 'Manuscript of the Hussite Wars Engineer' and Its Technological Milieu: A Study and Edition of *Codex latinus monacensis* 197, Part 1" (Ph.D. diss., University of California, Los Angeles, 1971). If Hall is correct, this would mean that the horizontal wheel pictured by Taccola, c1433, is the earliest extant drawing of a horizontal mill in the West (Taccola *De ingenesis* 3. 33v–34r, in Prager and Scaglia, *Taccola*, p. 84).

155. For example, A. G. Keller, *A Theatre of Machines* (London, 1964), p. 29, quoting Abraham Goelnitz of Danzig's *Ulysses Belgico-Gallicus* (1631), and Michael Montaigne, *The Diary of Montaigne's Journey to Italy in 1580 and 1581*, transl. E. J. Trechmann (New York, 1929), p. 50.

156. Jacob Leupold, *Theatrum machinarum generale* (Leipzig, 1724), pp. 208–9.

157. Anders Jespersen, *A Preliminary Analysis of the Development of the Gearing in Watermills in Western Europe* (Virum, Denmark, 1953), p. 8.

158. Usher, *Mechanical Inventions*, pp. 178–82.

159. *Ancient Laws of Ireland*, 3 (Dublin, 1873): 281, 283; also Bennett and Elton, *History of Corn Milling*, 2:90–91.

160. Usher, *Mechanical Inventions*, p. 179. See also the standard works on milling laws in the Middle Ages: Carl Koehne, "Die Mühle im Rechte der Völker," *Beiträge zur Geschichte der Technik und Industrie* 5 (1913): 27–53; Koehne, *Das Recht der Mühlen bis zum Ende der Karolingerzeit*, Untersuchungen zur Deutschen Staats- und Rechtsgeschichte, vol. 71 (Breslau, 1904); and Bennett and Elton, *History of Corn Milling*, 4 vols. (London,

1898–1904). In Koehne's *Recht der Mühlen* see esp. pp. 71 f. for provisions relating to the mill in the early Teutonic law codes.

161. L. Levillain, ed., "Texte des Statuts d'Adalhard," *Le moyen age,* 1st ser. 13 (1900): 351–86.

162. Koehne, "Die Mühle," pp. 36–37; Usher, *Mechanical Inventions,* p. 181.

163. Muendel, "Pistoia," p. 194.

164. Wilson, "Medieval Britain," pp. 184–85.

165. Strauss, "Mills without Wheels," p. 33.

166. Hunter, "Living Past," p. 455, for example, noted that even though horizontal mills predominated in many areas of the Balkans for flour milling, vertical wheels almost invariably drove the fulling mills.

167. Muendel, "Pistoia," pp. 210–14.

168. Thrupp, "Medieval Industry," p. 234. See also Bloch, *Land and Work,* pp. 149–60 passim; Usher, *Mechanical Inventions,* pp. 180–81; and Jespersen, *Preliminary Analysis,* p. 8.

169. See, for example, Bloch, *Land and Work,* pp. 157–59, and Gimpel, *Medieval Machine,* pp. 15–16.

170. Bloch, *Land and Work,* p. 159. In one case it may be possible to date the enforced transition from individually owned and operated horizontal mills to the manorial mill (presumably vertical). Goudie, "Horizontal Water Mills," p. 282, cites a complaint by the inhabitants of Orkney and Shetland against Lord Robert Stewart in 1575 to the effect that he was "taking away sucken fra the auld outhal mills of Orkney, whilk were observit of before inviolate." The suggestion seems to be that Lord Robert had erected mills of his own and was compelling his tenants to grind at them, relinquishing use of their old "outhal mills," presumably horizontal.

171. Hans E. Wulff, "A Postscript to Reti's Notes on Juanelo Turriano's Water Mills," *Technology and Culture* 7 (1966): 398–401, and Ladislao Reti, "On the Efficiency of Early Horizontal Waterwheels," ibid. 8 (1967): 388–94, have already pointed out that the efficiency and power output of horizontal mills was not necessarily low.

172. It is possible, despite the impressions gained from available written materials, that the vertical mill did not displace the horizontal mill in many rural hinterlands. The sparsity of published materials relating to the horizontal mill could be due largely to the fact that the majority of texts containing technical descriptions came from authors familiar only with urban centers and the densely populated areas of Europe, environments where vertical water wheels were most likely to have predominated. It may be significant that fifteenth-century Italian illustrated technical manuscripts, done by men in closer contact with the rural hinterland than many of the non-Italian technical writers, contain a higher percentage of horizontal wheels than do other works from the same period. This possibility was suggested to me by Bert S. Hall of the University of Toronto.

173. Lynn White, "Cultural Climates and Technological Advance in Middle Ages," *Viator* 2 (1971): 171–201; White, "What Accelerated Technological Progress in the Western Middle Ages?" in A. C. Crombie, ed., *Scientific Change* (London, 1963), pp. 272–91. The role of the monasteries in the diffusion of the vertical water wheel is also mentioned by Blaine, "Application," pp. 69–70; Gimpel, *Medieval Machine,* pp. 3–6; and Arnold Pacey, *The Maze of Ingenuity* (Cambridge, Mass., 1976), pp. 32–40.

174. See Saint Benedict, *The Rule of Saint Benedict,* ed. and transl. Abbot Justin McAnn (Westminster, Md., 1952), esp. ch. 48, pp. 111–13.

175. Ibid., ch. 66, p. 153 (also *S.P. Benedicti Regula, cum commentariis* 66, in Migne, *PL* 66. 900).

176. Blaine, "Application," p. 11, n. 31.

177. Luckhurst, *Monastic Watermills,* p. 19.

178. The importance of the Cistercians to the diffusion of water power in the medieval period is mentioned by Pacey, *Maze of Ingenuity,* pp. 36–37; Gille, "Medieval Age," pp. 559–60; and Luckhurst, *Monastic Watermills,* pp. 5–6.

179. Berman, "Cistercians in County of Toulouse," pp. 286 and table 15, p. 392.

180. David Williams, *The Welsh Cistercians: Aspects of their Economic History* (Pontypool, Wales, 1969), pp. 57–59.

181. Blaine, "Application," pp. 50–51.

182. Gimpel, *Medieval Machine*, p. 10.

183. Constance Berman, "Silvanès: A Study of the Economic Innovations of a Cistercian Monastery in Southern France" (M.A. Thesis, University of Wisconsin, 1972), pp. 63–68, 86–89.

184. Williams, *Welsh Cistercians*, pp. 57–58.

185. Berman, "Cistercians in County of Toulouse," table 15, p. 392.

186. R. A. Donkin, *The Cistercians: Studies in the Geography of Medieval England and Wales* (Toronto, 1978), pp. 138, 188.

187. Blaine, "Application," pp. 117–32 passim; Gimpel, *Medieval Machine*, pp. 67–69; Forbes, "Power," p. 610; Gille, "Medieval Age," p. 560.

188. Gille, "Machines," p. 650.

189. Gille, "Medieval Age," p. 560.

190. *Sancti Bernardi abbatis Clarae-Vallensis vita et res gestae libris septum comprehensae* 2. 5, in Migne, *PL* 185. 285; translation from Luckhurst, *Monastic Watermills*, p. 6.

191. *Descriptio positionis seu situationis monasterii Clarae-Vallensis*, in Migne *PL* 185. 570–71.

192. Bloch, *Life and Work*, p. 152, also suspects this.

193. Bennett Hill, *English Cistercian Monasteries and Their Patrons in the Twelfth Century* (Urbana, Ill., 1968), pp. 75–77.

194. The idea that Christianity improved the "cultural climate" for technological development has been developed most extensively by Lynn White. See in particular his "Cultural Climates and Technological Advance in the Middle Ages," *Viator* 2 (1971): 171–201. This and other of White's essays have been reprinted in White, *Medieval Religion and Technology: Collected Essays*. See also Forbes, "Power," pp. 605–6.

195. Gregory of Tours *Vitae patrum* 18. 2, in Migne, *PL* 71. 1085; translated in Bennett and Elton, *History of Corn Milling*, 2:71.

196. *Fontana Economic History of Europe*, 1:36. The massive drop in Europe's population is strongly emphasized by Renée Doehaerd, *The Early Middle Ages in the West: Economy and Society*, transl. W. G. Deakin (Amsterdam, 1978), pp. 22–44, 61–63. Doehaerd believes that early medieval Europe's extreme poverty was in large part due to a scarcity of agricultural labor and that this scarcity could explain many of the wars of the period (attempts to gain captive labor), the slow diffusion of three-crop rotation (it required too much labor), and the upsurge of slavery in the early centuries of the medieval period (an attempt to hold what labor one had). S. Lilley, *Men, Machines, and History* (London, 1948), pp. 36–37, stresses the labor shortage as the important factor in the emergence of water power in the West to the exclusion of all other factors.

197. Thrupp, "Medieval Industry," p. 234.

198. Carus-Wilson, "Industrial Revolution," p. 51. For a discussion of the importance of the soke in the spread of the water wheel, see Bloch, *Land and Work*, pp. 151–60.

199. Needham, *Science and Civilisation*, 4, pt. 2:396–97.

200. Ibid., pp. 390–96.

201. Ibid., pp. 398, 400.

202. Ibid., pp. 378, 393, 405, 481 f.

203. Ibid., p. 393.

204. Ibid., pp. 404, 398.

205. Ibid., p. 404 and p. 241a.

206. Ibid., pp. 404, 394, 405.

207. Sung Ying-Hsing, *T'ien-Kung K'ai-Wu: Chinese Technology in the Seventeenth Century*, transl. E-tu Zen Sun and Shiou-Chuan Sun (University Park, Pa., 1966), p. 94.

208. Needham, *Science and Civilisation*, 4, pt. 2: 400–401, and note *f* on p. 401.

209. Guy LeStrange, *Baghdad during the Abbasid Caliphate from Contemporary Arabic and*

*Persian Sources* (Oxford, 1900), pp. 142–44. One of those who gave such an account was Ya'Kubi, who wrote around 891.

210. Ibid., p. 145.

211. Al-Idrisi, *Géographia*, 2:63–64; Imamuddin, *Muslim Spain*, pp. 104–5; and Imamuddin, *The Economic History of Spain under the Umayyads, 711–1031 A.C.* (Dacca, 1963), pp. 181–86.

212. Guy LeStrange, *Palestine under the Moslem: A Description of Syria and the Holy Land from A.D. 650 to 1500*, translated from the works of medieval Arab geographers (1890; rpt. Beirut, 1965), pp. 222, 511. The reports are those of Shams ad Din Suyuti, writing about 1470, and Muqaddasi, from around 985. For Muqaddasi's report see Muhammad ibn Ahmed al-Muqaddasi, *Ahsanu-t-taqasim-fi-ma'rifati-l-aqalim*, transl. and ed. G.S.A. Ranking and R. F. Azoo, 1 (Calcutta, 1897): 271–86. Shmuel Avitsur, "Watermills in Eretz Israel," p. 389, claims that these and many other Near Eastern mills were of the horizontal type using Aruba penstocks.

213. LeStrange, *Palestine*, pp. 69 (Abu-l Finda, 1321), 240, 296 (al-Idrisi, 1154), 375 (Ibn Butlan, 1051), 398 (Idrisi), 457 (Yakut, 1225). See also Ibn Jubayr, *The Travels of Ibn Jubayr*, transl. R. J. C. Broadhurst (London, 1952), p. 263.

214. LeStrange, *Baghdad*, pp. 73, 142–45. His sources were Ya'Kubi, fl. late ninth century; Khatib, who wrote a history of Baghdad around 1058; and Tabari, who wrote about 865. See also al-Muqaddasi, *Ahsanu*, 1:201, and Adam Mez, *The Renaissance of Islam*, transl. from the German by Salahuddin Khuda Bakhsh and D. S. Margoliouth (Patna, 1937), p. 466.

215. LeStrange, *The Lands of the Eastern Caliphate: Mesopotamia, Persia, and Central Asia from the Moslem Conquest to the Time of Timur* (Cambridge, 1905), p. 178 (Ibn Hawkal in 978 described the city of Baylakan as watered by streams with many mills).

216. Ibid., pp. 222 (Kazvini, 1275) and 239. For Moslem watermills in southern Iran see pp. 271, 277.

217. Ibid., pp. 315 (Ibn Hawkal, 978), 337 (Ibn Hawkal), 360 (Nasir-i-Khusrau, 1052), 385, 386 (Mustawfi, 1340), 388 (Mustawfi), 392 (Mustawfi), 400.

218. Ibid., p. 465 (Ibn Batutah, 1355).

219. Ibid., p. 420 (Ibn Hawkal, 978).

220. Al-Idrisi, *India and the Neighboring Territories in the Kitab Nuzhat al-Mushtag Fi'Khtiraq al-'Afaq of Al-Sharif al-Idrisi*, transl. S. Maqbul Ahmud (Leiden, 1960), p. 51.

221. Ibn Jubayr, *Travels*, pp. 250, 265; Imamuddin, *Muslim Spain*, pp. 76–77; LeStrange, *Palestine*, pp. 59 (Yakut, 1225), 296 (Idrisi, 1154), 357–60 (Nasir-i-Khusrau, 1047; Dimashki, 1300; Abu-l'Fida, 1321; Ibn Batutah, 1355).

222. LeStrange, *Baghdad*, p. 73 (Istakhri, 951).

223. LeStrange, *Palestine*, p. 240 (al-Idrisi, 1154).

224. Ibid., p. 375 (Ibn Butlan, 1051).

225. Ibid., p. 457.

226. LeStrange, *Eastern Caliphate*, p. 315 (Ibn Hawkal, 978).

227. Ibid., p. 420 (Ibn Hawkal).

228. Ibid., p. 465 (Ibn Batutah, 1355).

229. Ibid., pp. 386, 388.

230. Ibid., p. 385.

231. Ibn al-Razzaz al-Jazari, *The Book of Knowledge of Ingenious Mechanical Devices*, transl. and annotated by Donald R. Hill (Dordrecht, Holland, 1974), 1. 6. 3, and 1. 6. 4, depict spoon-shaped buckets on an overshot wheel; 2. 1. 2, and 2. 2. 2, describe horizontal vaned wheels; while 5. 5. 1, shows a vertical paddle or undershot wheel. See also Hill's remarks on p. 275.

232. E. Wiedemann, "Über ein arabisches, eigentümliches Wasserrad und eine kohlenwasserhaltige Höhle auf Majorka nach al-Qazwînî," *Mitteilungen zur Geschichte der Medizin und der Naturwissenschaften* 15 (1916): 368–70; Forbes, *Studies in Ancient Technology*, 2:91–92.

233. Al-Jazari, *Ingenious Mechanical Devices*, Hill ed., p. 275.

234. Al-Idrisi, *Géographie*, Jaubert ed., 2:63–64.

235. Al-Muqaddasi, *Ahsanu*, 1:201: "The flux and reflux of the water at al-Basrah is a standing miracle and a real blessing to its inhabitants, as the water visits them twice in every day and night, entering the canals and irrigating the gardens and carrying boats to the villages; and when it ebbs it also is of use in the working of mills which stand at the mouths of the canals, so that when the water flows out they are set in motion."

236. LeStrange, *Eastern Caliphate*, p. 277.

237. Norman Smith, *Man and Water: A History of Hydro-Technology* (London, 1976), p. 142.

238. Smith, *History of Dams*, p. 81.

239. Water power was also used to power some automata. See Hill's edition of al-Jazari, *Ingenious Mechanical Devices*.

240. Smith, *History of Dams*, p. 85. Shemuel Avitsur, "Water Power in Traditional Sugar and Olive Oil Production in the Land of Israel," International Symposium on Molinology, *Transactions* 3 (1973): 176–83, claims that water-powered sugar mills were in use in Israel by the sixth to the eighth centuries A.D., and water-powered oil mills sometime after that. Unfortunately, his article does not document these claims.

241. W. Willcocks, *Egyptian Irrigation* (London, 1889), p. 209.

242. See notes 229 and 230, above.

243. Mez, *Renaissance of Islam*, p. 466.

244. Brittain, *Rivers*, pp. 251–53, discusses these factors. White, "Cultural Climates," p. 175, mentions the aridity of the lands of medieval Islam as an important factor.

245. A. G. Keller, "A Byzantine Admirer of 'Western' Progress: Cardinal Bessarion," *Cambridge Historical Journal* 11 (1955): 343–48.

## CHAPTER 3: CONTINUITY

1. John U. Nef, *The Rise of the British Coal Industry*, 2 vols. (London, 1932); Nef, *War and Human Progress: An Essay on the Rise of Industrial Civilization* (Cambridge, Mass., 1950); Nef, *The Conquest of the Material World* (Chicago, 1964); and elsewhere.

2. A. E. Musson, *The Growth of British Industry* (London, 1978), pp. 26–58, reviews the present state of scholarship on Nef's "early industrial revolution" in a very even-handed manner.

3. Kenneth Major, "The Further Contribution of England and Wales to the Molinological Map of Europe," International Symposium on Molinology, *Transactions* 2 (1969): 111, estimates that there were not less than 10,000 watermills in Britain. C. E. Bennett, "The Watermills of Kent, East of the Medway," *Industrial Archaeology Review* 1 (1977): 231, estimates that Britain had about 20,000 watermills.

4. Frank Nixon, *The Industrial Archaeology of Derbyshire* (Newton Abbot, 1969), p. 101.

5. W. Stukeley, *Itinerarium Curiosum*, p. 58, cited by Paul Mantoux, *The Industrial Revolution of the Eighteenth Century*, rev. ed. (London, 1961), p. 247.

6. In addition to Archibald Allison, "The Water Wheels of Sheffield," *Engineering* 165 (1948): 165–68, whose map is reproduced here, see Allison, "Water Power as the Foundation of Sheffield's Industries," Newcomen Society, *Transactions* 27 (1949–51): 221–24, and W. T. Miller, *The Water-Mills of Sheffield* (Sheffield, 1949).

7. D. G. Watts, "Water-Power and the Industrial Revolution," Cumberland and Westmorland Antiquarian and Archaeological Society, *Transactions*, n.s. 67 (1967): 195.

8. Stanley D. Chapman, "The Cost of Power in the Industrial Revolution in Britain: The Case of the Textile Industry," *Midland History* 1, no. 2 (Autumn 1971): 8.

9. Gimpel, *Medieval Machine*, p. 23.

10. "Projet de capitation présénté par M. de Vauban (1694)," in A. M. de Boislisle, ed., *Correspondance des contrôleurs généraux des finances avec les intendants des provinces...* [1683–99], vol. 1 (Paris, 1874), pp. 561–65. Charles L. Ducrest, *Vues nouvelles sur les courans d'eau, la navigation intérieure et la marine* (Paris, 1803) p. 10, estimated that

France had only 20,000 watermills. Later, Charles Dupin, *Forces productives et commerciales de la France*, 1 (Bruselles, 1828): 36, estimated 66,000 watermills. Ferrendier, "Anciennes utilisations," 5:785, using Vauban's figures, estimated 70,000 water-powered flour mills and 15,000 water-powered industrial mills, or 85,000 total for France c1700.

11. Nef, *War and Human Progress*, pp. 273–301 passim.

12. Bernard Forest de Bélidor, *Architecture hydraulique* ... , 1 (Paris, 1737): 282–83, warned his readers about placing mills too close together. Mill crowding had become a problem also in the Low Countries by the mid-sixteenth century. J. Schoonhoven, "Sketches of Mills in Noord-Brabant," International Symposium on Molinology, *Transactions* 3 (1973): 115, cites rules to govern mill crowding laid down by Charles V in 1545.

13. Richard L. Hills, "Water, Stampers, and Paper in the Auvergne: A Medieval Tradition," *History of Technology*, 1980, p. 143.

14. Ferrendier, "Anciennes utilisations," 5:776.

15. Ibid., p. 784.

16. Ibid., 3:334, and 5:780–81; Rivals, "Floating Mills," p. 151.

17. Kenneth Adams, *A Guide to the Industrial Archaeology of Europe* (Bath, 1971), p. 10.

18. Irimie, "Floating Mills in România," p. 437; Irimie and Bucur, "Watermills in România," p. 429.

19. Braudel, *Capitalism and Material Life*, p. 262.

20. Danilevskii, *Hydroengineering in Russia*, pp. 9–10, 19.

21. There were a large variety of options for siting mills in relation to dams and power canals (mill leats). A substantial number are reviewed by Jennifer Tann, "Some Problems of Water Power: A Study of Mill Siting in Gloucestershire," Bristol and Gloucestershire Archaeological Society, *Transactions* 85 (1965): 53–77, esp. 65.

22. Anthony Fitzherbert, *Surveying* (London, 1539), reprinted in [Robert Vansittart, ed.], *Certain Ancient Tracts concerning the Management of Landed Property Reprinted* (London, 1767), p. 91.

23. According to Tann, "Problems of Water Power," p. 62, it was common practice to use a diversion dam in conjunction with a mill leat.

24. William Camden, *Britannia*, transl. Edmund Gibson, 2nd ed., vol. 1 (London, 1722), col. 196. The work was originally published in 1594.

25. Nixon, *Industrial Archaeology of Derbyshire*, p. 99.

26. J[acques] Levainville, *L'industrie du fer en France* (Paris, 1922), p. 25; Ferrendier, "Anciennes utilisations," 5:773.

27. Hills, "Water, Stampers, and Paper in the Auvergne," pp. 145–49, 151.

28. Henri Pitot, "Nouvelle methode pour connoître & déterminer l'effort de toutes sortes de machines muës par un courant, ou une chûte d'eau," Académie des Sciences, Paris, *Mémoires*, 1725 (publ. 1727), p. 86.

29. Agricola, *De re metallica*, Hoover ed., p. 319.

30. Further information on early hydropower dams may be found in Danilevskii, *Hydroengineering in Russia*, pp. 108–72; J. A. Garcia-Diego, "The Chapter on Weirs in the Codex of Juanelo Turriano: A Question of Authorship," *Technology and Culture* 17 (1976): 217–34; Johan Matthias Beyer, *Theatrum machinarum molarium* ... (Dresden, 1767), pp. 16–23, table 6; James Leffel, *Construction of Mill Dams* (Springfield, Ohio, 1881); Edward Wegmann, *The Design and Construction of Dams*, 4th ed. (New York, 1899), pp. 111–50; John Nicholson, *The Millwright's Guide* (London, 1830), pp. 107–9; Fitzherbert, *Surveying*, pp. 91–92; *Miscellaneous Dissertations on Rural Subjects* (London, 1775), pp. 420–25; [Jean Antoine] Fabre, *Essai sur la manière le plus avantageuse de contruire les machines hydrauliques* ... (Paris, 1783), pp. 290–94; [Charles] Bossut and Vaillet, *Unterschungen über die beste construction der Deiche*, transl. from French to German by C. Kröncke (Frankfurt-on-Main, 1798); Ferrendier, "Anciennes utilisations," 5:770–74; "Water," in Abraham Rees, ed., *The Cyclopaedia; or, Universal Dictionary of Arts, Sciences, and Literature* (henceforth *Rees's Cyclopaedia*), vol. 38 (London, 1819); William C. Hughes, *The American Miller, and Millwright's Assistant* (Philadelphia, 1853), pp. 111–14; and David Craik, *The Practical American Millwright and Miller* (Philadelphia, 1882), pp. 156–76.

31. A. M. Héron de Villefosse, *De la richesse minérale*, 3 (Paris, 1819): 22–35, plate 32.

32. Ibid.

33. Arthur Raistrick, *Industrial Archaeology: An Historical Survey* (London, 1972), pp. 235–38; Nixon, *Industrial Archaeology of Derbyshire*, p. 247.

34. Smith, *History of Dams*, pp. 151–52.

35. Gabriel Jars, *Voyages métallurgiques; ou, Recherches et observations sur les mines & forges de fer...*, 1 (Lyons, 1774): 306–7; Smith, *History of Dams*, p. 157; and Smith, *Man and Water*, p. 149.

36. Smith, *History of Dams*, pp. 117–20; A. Del Aguila, "Unas presas antiguas españolas de contrafuertes," *Las Ciencias* 14 (1949): 185–202.

37. Smith, *History of Dams*, pp. 120–21, and Smith, *Man and Water*, p. 149.

38. Jose A. Garcia-Diego, "Old Dams in Extremadura," *History of Technology*, 1977, pp. 107, 119.

39. Danilevskii, *Hydroengineering in Russia*, pp. 27, 46, 51–52, 54, n. 24, 56, 173.

40. Ibid., pp. 95–103, 138–45; Smith, *Man and Water*, p. 149.

41. Additional material on the construction of the leats, head races, or power canals which led water to water wheels can be found in Nicholson, *Millwright's Guide*, pp. 105–6; *Miscellaneous Dissertations*, pp. 348–50, 425–31; Beyer, *Theatrum machinarum molarium*, esp. pp. 5–6, 13–15, 42–44, tables 2 and 14; Fabre, *Essai*, pp. 114–56, 280–90; Oliver Evans, *The Young Mill-Wright and Miller's Guide* (Philadelphia, 1795), pp. 119, 300–301; Thomas Telford, *Life of Thomas Telford...*, ed. John Rickman (London, 1838), p. 668; "Forges," *Encyclopédie*, 7 (1757): 144–45; "Meunier," *Encyclopédie méthodique*, 5 (1788): 50–51; J. Lermier, "Mémoire sur l'hydraulique...," Académie royale des sciences, belles-lettres, et arts de Bordeaux, *Séance publique*, 1825, pp. 117–25; "Water," *Rees's Cyclopaedia*.

42. Tann, "Problems of Water Power," pp. 60, 62, 67.

43. Garcia-Diego, "Dams in Extremadura," pp. 107, 119.

44. *Fontana Economic History*, 2:201–2.

45. For example, Robert L. Galloway, *A History of Coal Mining in Great Britain* (London, 1882), p. 57; Raistrick, *Industrial Archaeology*, p. 202; Arthur Raistrick and Bernard Jennings, *A History of Lead Mining in the Pennines* (London, 1965), pp. 139–40.

46. The best description of the Harz system is that of Héron de Villefosse, *De la richesse minérale*, 3:1–35. Jars, *Voyages métallurgiques*, 1:303–7, also has some material on the system.

47. W. E. Rudolph, "The Lakes of Potosi," *Geographical Review* 26 (1936): 529–54.

48. Jars, *Voyages métallurgiques*, 2:110–11, 161–64.

49. J[ean] F. d'Aubuisson de Voisins, *A Treatise on Hydraulics for the Use of Engineers*, transl. Joseph Bennett (Boston, 1852), pp. 406–17, describes both types of horizontal wheels, reviews tests made on them, and notes that they were primarily used in southern Europe.

50. Leonardo da Vinci *Codex Atlanticus* 289r.

51. *Fontana Economic History*, 2:217.

52. Ibid., p. 399.

53. Mantoux, *Industrial Revolution*, pp. 194–95.

54. The quotation is from M. J. T. Lewis, "Industrial Archaeology," in *Fontana Economic History*, 3 (Glasgow, 1973): 586. For more detailed information on the mechanization of the linen industry in Ireland see W. A. McCutcheon, "Water Power in the North of Ireland," Newcomen Society, *Transactions* 39 (1966–67): 73–83; Enid Gauldie, "Water-Powered Beetling Machines," ibid., pp. 125–28; and H. D. Gribbon, *The History of Water Power in Ulster* (Newton Abbot, 1969), pp. 81–90, 102–9.

55. Reti, "Postscript," pp. 436–37, figs. 5 and 7 (fols. 256r and 332r of Turriano's manuscript).

56. Beyer, *Theatrum machinarum molarium*, pp. 116–17, table 42, of the 1767 edition. This volume was originally published in 1735 at Leipzig and Rudelstadt. For another water-powered threshing machine see Andrew Gray, *The Experienced Millwright...*, 2nd ed. (Edinburgh, 1806), plate 14.

57. "Meunier," *Encyclopédie méthodique*, 5 (1788): 24 f.; Malouin, "L'art de meunier,"

*Descriptions des arts et métiers* . . . , new ed., 1 (Neuchatel, 1771): 74–81; R. J. Forbes, "Food and Drink," in Singer, *History of Technology*, 3:18–19.

58. Ladislao Reti, "The Leonardo da Vinci Codices in the Biblioteca Nacional of Madrid," *Technology and Culture* 8 (1967): 443–44, fig. 5.

59. Agostino Ramelli, *Le diverse et artificiose machine* (Paris, 1588), ch. 119.

60. For Howell's statement see Charles Howell, "Colonial Watermills in the Wooden Age," in Brooke Hindle, ed., *America's Wooden Age: Aspects of Its Early Technology* (Tarrytown, N.Y., 1975), p. 156; for the French mills in the Nord see Malouin, "Meunier," p. 71.

61. Nef, *Conquest of the Material World*, p. 205.

62. Alfred H. Shorter, *Paper Mills and Paper Makers in England, 1495–1800* (Hilversum, Holland, 1957), pp. 29, 38, 71–72.

63. *Fontana Economic History*, 2:181.

64. D. C. Coleman, *The British Paper Industry, 1495–1860* (Oxford, 1958), p. 32.

65. Nef, *Conquest of the Material World*, p. 125.

66. Hills, "Water, Stampers, and Paper in the Auvergne," p. 150.

67. For a description of this paper factory see "Papeterie," *Encyclopédie*, 11 (1765): 834–35, and ibid., *Planches*, vol. 5 (1767).

68. G. Hollister-Short, "Leads and Lags in Late Seventeenth-Century English Technology," *History of Technology*, 1976, p. 163.

69. *Fontana Economic History*, 2:201–2.

70. Agricola, *De re metallica*, Hoover ed., pp. 172–200 passim.

71. *Cambridge Economic History of Europe*, 5 (Cambridge, 1977): 473.

72. *Fontana Economic History*, 2:200–202.

73. Raistrick, *Industrial Archaeology*, p. 203.

74. Raistrick and Jennings, *Lead Mining in the Pennines*, pp. 139–40.

75. Galloway, *History of Coal Mining*, pp. 56, 115.

76. G. Hollister-Short, "The Vocabulary of Technology," *History of Technology*, 1977, pp. 125–55, reviews the variety of names used for the combination of water wheel, field rods, and mine pump or drawing engine at various times. This article is also the best single source for information on the stagenkunst.

77. Ibid. See also Robert P. Multhauf, "Mine Pumping in Agricola's Time and Later," United States National Museum, *Bulletin* 218 (1959): 113–20; Jars, *Voyages métallurgiques*, vol. 2 (Lyons, 1774), plate 10; and Héron de Villefosse, *De la richesse minérale*, 3 (Paris, 1819): 36 f. and plate 33.

78. Otto Vogel, "Christopher Polhem und seine Beziehungen zur Harzer Bergbau," *Beiträge zur Geschichte der Technik und Industrie* 5 (1913): 306 f.; Herman Sundholm, "Polhem, the Mining Engineer," in William A. Johnson, transl., *Christopher Polhem: The Father of Swedish Technology* (Hartford, Conn., 1963), pp. 163–82.

79. Sundholm, "Polhem, the Mining Engineer," pp. 165–70.

80. Hermann Kellenbenz, *The Rise of the European Economy: An Economic History of Continental Europe from the Fifteenth to the Eighteenth Century* (London, 1976), p. 80. See also Nef, *Conquest of the Material World*, pp. 3–64.

81. Charles K. Hyde, *Technological Change and the British Iron Industry, 1700–1870* (Princeton, 1977), p. 21.

82. *Fontana Economic History*, 2:209–10.

83. R. F. Tylecote, *A History of Metallurgy* (London, 1976), pp. 97–101; Nef, *Conquest of the Material World*, p. 131.

84. Nef, *Conquest of the Material World*, p. 131; *Fontana Economic History*, 2:253–55.

85. A. S. Britkin, *The Craftsmen of Tula: Pioneer Builders of Water-Driven Machinery*, transl. from the Russian (Jerusalem, 1967), pp. 12–13; "Forage des canons de fusil... ," *Encyclopédie*, suppl., 3 (1777): 84–86; "Fabrique des Armes," ibid., *Planches*, suppl. (1772), plate 3: "Arquebusier," *Encyclopédie méthodique*, 1 (1782): 90–92.

86. A. S. Britkin and S. S. Vidonov, *A. K. Nartov: An Outstanding Machine Builder of the Eighteenth Century*, transl. from the Russian (Jerusalem, 1964), p. 100.

87. "Forges," *Encyclopédie*, 7 (1757): 142–43, and ibid., *Planches*, vol. 4 (1765), sec. 1, plate 7, and last sec., plates 1 and 2. Also "Fer," *Encyclopédie méthodique*, 2 (1783): 576.

88. Agricola, *De re metallica*, Hoover ed., pp. 295, 297–99.

89. "Laminoir," *Encyclopédie*, 9 (1765): 230–32; "Plomb, Laminage du," ibid., *Planches*, vol. 8 (1771), plate 3; "Laminage," *Encyclopédie méthodique*, (1785):.202–13; Tylecote, *History of Metallurgy*, pp. 97–100.

90. For early methods of manufacturing gunpowder see William Anderson, *Sketch of the Mode of Manufacturing Gunpowder at the Ishapore Mills in Bengal...*, with notes and additions by Lieut.-Col. Parlley (London, 1862); Paul N. Wilson, "The Gunpowder Mills of Westmorland and Furness," Newcomen Society, *Transactions* 36 (1963–64): 48–51; Bélidor, *Architecture hydraulique*, 1:348–59; Zonca, *Novo teatro*, pp. 82–85; "Poudre à canon," *Encyclopédie méthodique*, 6 (1789): 615–41.

91. Oscar Guttmann, *The Manufacture of Explosives*, 1 (London, 1895): 17.

92. Ibid.

93. The French mill at Essonnes is illustrated in "Minéralogie, Poudre à canon," *Encyclopédie, Planches*, vol. 6 (1767); see Bennett, "Watermills of Kent," p. 230, for the Royal Gunpowder Factory.

94. Danilevskii, *Hydroengineering in Russia*, p. 21.

95. Warren C. Scoville, *Capitalism and French Glassmaking, 1640–1789* (Berkeley and Los Angeles, 1950), p. 45.

96. Kellenbenz, *Rise of the European Economy*, p. 258.

97. "Verrerie," *Encyclopédie*, 17 (1765): 102–56; "Glaces," ibid., *Planches*, vol. 4 (1765), plates 43–46; "Glacerie," *Encyclopédie méthodique*, 3 (1784): 215–16; R. J. Charleston and L. M. Angus-Butterworth, "Glass," in Singer, *History of Technology*, 3:206–44; Angus-Butterworth, "Glass," ibid., 4:365–73.

98. Scoville, *French Glassmaking*, p. 40.

99. Ibid., p. 48, n. 65.

100. Ibid., p. 45.

101. John Somervell, "Water Power and Industries in Westmorland," Newcomen Society, *Transactions* 18 (1937–38): 239–40; H. S. L. Dewar, "The Windmills, Watermills, and Horse-Mills of Dorset," Dorset Natural History and Archaeological Society, *Proceedings* 82 (1960): 118.

102. Somervell, "Water Power in Westmorland," p. 243; Dewar, "Mills of Dorset," p. 119.

103. Dewar, "Mills of Dorset," p. 119.

104. Rex Wailes, "Water-Driven Mills for Grinding Stone," Newcomen Society, *Transactions* 39 (1966–67): 95–120.

105. [Nicolas] Grollier de Servière, *Recueil d'ouvrages curieux de mathématique et de mécanique...* (Lyons, 1719), p. 59, fig. 89. Grollier exhibited a model water-driven pile driver. Whether it was ever used in practice, I do not know.

106. Danilevskii, *Hydroengineering in Russia*, p. 26; Forbes, "Food and Drink," p. 14, says the churn-mill for the manufacture of butter was first mentioned in 1600.

107. Danilevskii, *Hydroengineering in Russia*, pp. 21, 26.

108. Gustaf Sellergren, "Polhem's Contributions to Applied Mechanics," in Johnson, *Christopher Polhem*, pp. 152–53.

109. A number of water-powered automata are depicted in Isaak de Caus, *New and Rare Inventions of Water-Works...*, transl. John Leak (London, 1659), plates 14–16, 18–19, and in Robert Fludd, *Tractatus secundus, De natvrae simia seu Technica macrocosmi historia...*, 2nd ed. (Frankfurt, 1624), pp. 480, 483, for example.

110. In addition to plate 130 of Georg Andreas Böckler, *Theatrum machinarum novum...* (Nuremberg, 1661), reproduced in Fig. 3–17, see Fludd, *Tractatus secundus*, p. 463.

111. Académie des Sciences, Paris, *Machines et inventions approuvées par l'Académie royale des Sciences, depuis son établissement jusqu'à present...*, ed. M. Gallon, 2 (Paris, 1735): 31–33, 177–79; 4 (1735): 43–44, 203–7, 209–11, 213–16; 5 (1735): 95–96.

112. Biringuccio, *Pirotechnia*, Smith and Gnudi ed., p. 306.

113. Ibid., p. 22.

114. Ibid., p. 147.

115. Jacob Leupold, *Theatrum Machinarum Generale* (Leipzig, 1724), p. 239.

116. J[ohn] Mortimer, *The Whole Art of Husbandry; or, The Way of Managing and Improving of Land*... , 2nd ed. (London, 1708), p. 306.

117. [John Robison], "Mechanics," *Encyclopaedia Britannica*, 3rd ed., 10 (Edinburgh, 1797): 763.

118. Hunter, *Waterpower*, pp. 6–50 passim, as well as other points in this work.

119. Ibid., p. 29.

120. Charles B. Kuhlmann, *The Development of the Flour-Milling Industry in the United States*... (Boston, 1929), p. 30.

121. Ibid., pp. 19, 27; Hunter, *Waterpower*, pp. 29–30.

122. Kuhlmann, *Development of Flour-Milling Industry*, p. 5.

123. Smith, *History of Dams*, p. 146. South Windham, however, is on the Presumpscot River, not the Piscataqua.

124. First Iron Works Association, *The First Iron Works Restoration* (New York, 1951); E. N. Hartley, *Iron Works on the Saugus* (Norman, Okla., 1957).

125. James Walton, *Water-Mills, Windmills, and Horse-Mills of South Africa* (Cape Town, 1974), pp. 27–33; also James Walton, "South African Mills," International Symposium on Molinology, *Transactions* 3 (1973): 81–93.

126. Richard C. Harris, *The Seigneurial System in Early Canada* (Madison, Wis., 1968), pp. 72–73. By the end of the French regime in Canada there were few settled seigneuries without a mill, and some had four or five (ibid., p. 75).

127. Rudolph, "Lakes of Potosi."

128. Danilevskii, *Hydroengineering in Russia*, pp. 18–33, 46–107.

129. Witt Bowden, Michael Karpovich, and Abbott P. Usher, *An Economic History of Europe since 1750* (New York, 1937), p. 108.

130. First Iron Works Association, *First Iron Works Restoration*, pp. 6, 11.

131. Paul N. Wilson, "The Waterwheels of John Smeaton," Newcomen Society, *Transactions* 30 (1955–57): 33.

132. Craik, *American Millwright*, p. vii.

133. Zachariah Allen, *The Science of Mechanics*... (Providence, R. I., 1829), p. 207; Julius Weisbach, *A Manual of the Mechanics of Engineering and of the Construction of Machines*... , vol. 2, transl. from the 4th Ger. ed. by Jay Du Bois (New York, 1877), p. 240.

134. Craik, *American Millwright*, p. 112.

135. [Edme] Beguillet, *Manuel de meunier*... (Paris, 1775), p. 25.

136. James F. Hobart, *Millwrighting* (New York, 1909), p. 376; Fitz Water Wheel Company, *Fitz Steel Overshoot Water Wheels*, Bulletin no. 70 (Hanover, Pa., 1928), p. 35; Howell, "Colonial Watermills," p. 134.

137. Robertson Buchanan, *Practical Essays on Mill Work and Other Machinery*, with notes by Thomas Tredgold, revised by George Rennie, 3rd ed. (London, 1841), p. 202, gives the weights of several early-nineteenth-century water wheels, which were presumably somewhat larger and heavier than wheels a century or so earlier.

138. For additional information on the axles and gudgeons of water wheels, see, in particular, Evans, *Mill-Wright and Miller's Guide*, pp. 47n, 198–99, 317–19; Fitzherbert, *Surveying*, pp. 92–93; Joseph P. Frizell, "The Old-Time Water-Wheels of America," American Society of Civil Engineers, *Transactions* 28 (1893): 239; Buchanan, *Practical Essays on Mill Work*, pp. 178–79; and Weisbach, *Manual*, 2:241–42.

139. Fitzherbert, *Surveying*, p. 92.

140. Agricola, *De re metallica*, Hoover ed., p. 188. See Buchanan, *Practical Essays on Mill Work*, pp. 344–45, for a discussion of bearings.

141. Evans, *Mill-Wright and Miller's Guide*, p. 198 and 198n; Craik, *American Millwright*, pp. 63–64.

142. Evans, *Mill-Wright and Miller's Guide*, p. 344; Robert Grimshaw, *The Miller, Mill-wright, and Millfurnisher* (New York, 1882), p. 505.

143. Anders Jespersen, "Portugese Mills... ," *International Symposium on Molinology, Transactions* 2 (1969): 69.

144. Frizell, "Old-Time Water-Wheels," p. 238. Clasp-arm wheels may have surpassed compass-arm wheels in popularity in the late eighteenth and early nineteenth centuries in some regions, for certain treatises on millwrighting recommend or discuss only clasp-arm wooden wheels—for example, [David Scott], *The Engineer and Machinist's Assistant*, 2nd ed., 1 (Glasgow, 1856): 213–14; and "Water," *Rees's Cyclopaedia*.

145. In detailing the options available in the assembly of the different parts of a traditional water wheel, I utilized the bulk of the material listed in Table 3–3. Particularly helpful, however, were Frizell, "Old-Time Water-Wheels," pp. 238–41; Evans, *Mill-Wright and Miller's Guide*, pp. 315–16; Fitzherbert, *Surveying*, pp. 91–94; Bélidor, *Architecture hydraulique*, 1:227 f., and 2:132–235 passim; John T. Desaguliers, *A Course of Experimental Philosophy*, 2 (London, 1744): 422–64, 520–32 passim; "Pompe," *Encyclopédie*, 13 (1765): 10–11; Leupold, *Theatrum machinarum generale*, pp. 155–66; Beyer, *Theatrum machinarum molarium;* Scott, *Engineer and Machinist's Assistant*, 1:212–16; and Danilevskii, *Hydroengineering in Russia*, pp. 173–80.

146. Fitzherbert, *Surveying*, p. 93. Allen, *Science of Mechanics*, p. 207, recommended that two blades always be immersed.

147. Evans, *Mill-Wright and Miller's Guide*, p. 101.

148. Leonardo da Vinci *Codex Atlanticus* 207v; also MS B, 33v–34r.

149. Leonardo da Vinci, *I Manoscritti e i Disegni...* , vol. 6 (Rome, 1949), plate 229.

150. For additional information on the tail race see, in particular, Nicholson, *Millwright's Guide*, p. 106; Fabre, *Essai*, pp. 114–36; Fitzherbert, *Surveying*, p. 93; "Meunier," *Encyclopédie méthodique*, 5 (1788): 50–51; *Miscellaneous Dissertations*, pp. 392–95; and Lermier, "Mémoire sur l'hydraulique," pp. 126–35.

151. Bélidor, *Architecture hydraulique*, 1:330 f., and plate 3, fig. 3, described a sawmill which made use of a steep Alpine-like chute to deliver water to the wheel. See also plate 5, fig. 2, for a pipe-boring mill with a steep delivery chute.

152. About the only other example of this type of delivery I could find before 1750 was in Beyer, *Theatrum machinarum molarium*, plate 20.

153. For example, "Meunier," *Encyclopédie méthodique*, 5 (1788): 101; John Bate, *Mysteries of Nature and Art*, 2nd ed. (London, 1635), p. 62; *Miscellaneous Dissertations*, pp. 344–45, 347–48; and Bélidor, *Architecture hydraulique*, 1:283–84.

154. "Forges," *Encyclopédie*, 7 (1757): 145.

155. Aitchison, *History of Metals*, 2:306–8, 308–10.

156. Bern Dibner, *Agricola on Metals* (Norwalk, Conn., 1958), p. 49.

157. Ernest Straker, *Wealden Iron* (London, 1931), pp. 66, 73.

158. Paul N. Wilson, "Water-Driven Prime Movers," in *Engineering Hertage: Highlights from the History of Mechanical Engineering*, 1 (London, 1963): 29.

159. Paul Gille, "The Production of Power," in Daumas, *History of Technology*, 2:438; Usher, *Mechanical Inventions*, p. 335; Ferrendier, "Anciennes utilisations," 5: 785–86; Carlo Cipolla, *Before the Industrial Revolution: European Society and Economy, 1000–1700*, 2nd ed. (New York, 1980), p. 172.

160. See, for example, Aubuisson de Voisins, *Treatise on Hydraulics*, pp. 447–51, or Thomas Fenwick, *Essays in Practical Mechanics*, 3rd ed. (Durham, 1822), pp. 59–70. Jay M. Whitham, *Water Rights Determination from an Engineering Standpoint* (New York, 1918), pp. 2–14, found that it took anywhere from 1 to 2 hp per bushel of wheat per hour, depending on which authority one used.

161. Usher, *Mechanical Inventions*, p. 335.

162. H. Dircks, *A Biographical Memoir of Samuel Hartlib... and a Reprint of his Pamphlet, entitled "An Invention of Engines of Motion"* [1651] (London, [1865]), p. 111.

163. Agricola, *De re metallica*, Hoover ed., p. 188.

164. Ibid., pp. 196, 198–200 (pp. 196–98 for a description of the type of brake used).

Wheels of this type (i.e., with two sets of buckets for reversible action) were later reported by Jars, *Voyages métallurgiques,* 2:300–302 and plate 25, and Héron de Villefosse, *De la richesse minérale,* 3:44–45 and plate 34, among others.

165. William Pryce, *Mineralogia Cornubiensis: A Treatise on Minerals, Mines, and Mining...* (London, 1778), p. 307.

166. John Smeaton, "Description of the Statical Hydraulic Engine, invented and made by the late Mr. William Westgarth...," Society of Arts, Manufactures, and Commerce, *Transactions* 5 (1787): 189–190n.

167. Danilevskii, *Hydroengineering in Russia,* p. 38. Information on Polhem's pumping engines can be found also in Herman Sundholm, "Polhem, the Mining Engineer," in Johnson, *Christopher Polhem,* pp. 165–82, and Sten Lindroth, *Christopher Polhem och Stora Kopparberget...* (Uppsala, 1951).

168. Danilevskii, *Hydroengineering in Russia,* pp. 100–101.

169. [Robison], "Water-Works," in *Encyclopaedia Britannica,* 3rd ed., 18 (Edinburgh, 1797): 905.

170. Ibid.

171. Smeaton, "Statical Hydraulic Engine," p. 189.

172. H. Beighton, "A Description of the Water-Works at London-Bridge, explaining the Draught of Tab. I," Royal Society of London, *Philosophical Transactions* 37 (1731–32): 5–12.

173. Bélidor, *Architecture hydraulique,* 2:170–86, 204–33, describes the engines.

174. Nicholson, *Millwright's Guide,* p. 96, mentions a tide mill with a wheel 26 feet (7.9 m) wide. But this may have been an iron wheel instead of a traditional wooden wheel.

175. Smeaton, "Statical Hydraulic Engine," p. 190n.

176. Raistrick and Jennings, *Lead Mining in the Pennines,* p. 140.

177. A number of works describe the Marly installation in some detail, among them Bélidor, *Architecture hydraulique,* 2:195–203; Leupold, *Theatri machinarum hydraulicarum,* 2 (Leipzig, 1725): 38–45; Desaguliers, *Experimental Philosophy,* 2: 442–49, 530–31; and Johan F. Weidler, *Tractatus de machinis hydraulics toro terrarvm obre maximis Marlyensi et Londinensi...* (Wittenberg, 1728), pp. 4–41. See also Carl Ergang, "Die Maschine von Marly," *Beiträge zur Geschichte der Technik und Industrie* 3 (1911): 131–46.

178. Jones, "Moulin Pendant," pp. 169–70.

179. Irimie and Bucur, "Watermills in Românĭa," pp. 431–32.

180. *Miscellaneous Dissertations,* pp. 367–69.

181. Bélidor, *Architecture hydraulique,* 2:308–9.

182. Caspar Schott, *Mechanica hydraulico pneumatica* (Würzburg, 1657), pp. 365–69. Another early water-bucket engine appears in Fludd, *Tractatus secundus,* p. 467. See also Thomas Ewbank, *A Descriptive and Historical Account of Hydraulic and Other Machines for Raising Water...,* 12th ed. (New York, 1851), pp. 127–28.

183. Bélidor, *Architecture hydraulique,* 2:254–56; Ewbank, *Hydraulic Machines,* pp. 127–28.

184. Bélidor, *Architecture hydraulique,* 2:256–63; Desaguliers, *Experimental Philosophy,* 1 (London, 1734): 75–76; 2:454–64.

185. Leupold, *Theatri machinarum hydraulicarum,* 2:157, and plate 54, fig. 4.

186. [Robison], "Water-Works," p. 905. The best review of the history of water-bucket engines is G. Downs-Rose and W. S. Harvey, "Water-Bucket Pumps and the Wanlockhead Engine," *Industrial Archaeology* 10 (1973): 129–47. See also G. White, "Water Bucket Engine at Elmdon, Warwickshire," Newcomen Society, *Transactions* 16 (1935–36): 55–56.

187. This fact is also lamented by Jennifer Tann, *The Development of the Factory* (London, 1970), p. 95.

188. This is noted by Frizell, "Old-Time Water-Wheels," pp. 237–38, and Craik, *American Millwright,* p. viii, among others.

189. *A General Description of All trades...* (London, 1747).

190. William Fairbairn, *Treatise on Mills and Millwork,* 3rd ed., 1 (London, 1871): ix. The first edition of this work was published in 1863–64.

191. Ibid., p. x.

192. Ibid., pp. xii–xiii.

193. Tann, *Development of the Factory,* pp. 95–105, reviews what is known about the late-eighteenth- and early-nineteenth-century millwright.

194. Desaguliers, *Experimental Philosophy,* 2:415 (cf. p. 414).

195. Ibid., p. 521.

196. *Miscellaneous Dissertations,* p. 339.

197. Fabre, *Essai,* p. xiv. Craik, *American Millwright,* p. viii, argued in 1882 that millwrights still found it better to gather information from studying working machinery and experimenting themselves than to rely on "theories and formulae."

198. Antoine Parent, "Sur la plus grande perfection possible des machines ... ," Académie des Sciences, *Mémoires,* 1704 (1745 ed.), pp. 323–24.

199. See D. S. L. Cardwell, "The Academic Study of the History of Technology," *History of Science,* 1968, p. 118, and Cardwell, *Turning Points in Western Technology* (New York, 1972), p. 37.

200. Ramelli, *Diverse machine,* chs. 23 and 24, for example.

201. Fitzherbert, *Surveying,* pp. 93–94.

202. Ibid., p. 94. See also Venterus Mandey, *Mechanick Powers; or, The Mystery of Nature and Art unvail'd...* (London, 1702), p. 70.

203. R. D'Acres, *The Art of Water-Drawing...* (London, 1660), p. 29; see also Mandey, *Mechanick Powers,* pp. 78–79.

CHAPTER 4: ANALYSIS

1. This chapter is, in part, a considerably condensed and somewhat revised version of my dissertation, "Science and the Water Wheel: The Development and Diffusion of Theoretical and Experimental Doctrines Relating to the Vertical Water Wheel, c. 1550–c. 1850" (University of Kansas, 1973), 719 pp. Portions of this chapter have also appeared in my article, "Scientific Influences on Technology: The Case of the Overshot Water Wheel, 1752–1754," *Technology and Culture* 20 (1979): 270–95.

2. *Leic.* 15v, 22v; *MS F* 24v; *B.M.* 45r, 122r. *Leic.* 15v gives an outline of the proposed work. Book 13 was to cover "machines turned by water." See also Edward MacCurdy, ed., *The Notebooks of Leonardo da Vinci,* 2 (New York, 1956): 106.

3. *MS A* 24r; *MS H* 30v.

4. *MS H* 73v; cf. *Codex Madrid I* 69r.

5. *MS H* 79v.

6. *Codex Atlanticus* 209r, a.

7. *MS F* 8v (cf. Leonardo, *Notebooks,* MacCurdy ed., 2:37).

8. *MS F* 9r (Leonardo, *Notebooks,* MacCurdy ed., 2:38). Leonardo did use mathematics to attack problems involving the water wheel several other times, but these involved only the calculation of gearing rations. See, e.g., *Codex Atlanticus* 2r, a, and 337v, c.

9. *Codex Atlanticus* 5r, a.

10. *MS Forster III* 46v, 47r.

11. Clifford A. Truesdell, *Essays in the History of Mechanics* (New York, 1968), p. 9, found this to be true of Leonardo's work in hydraulics generally.

12. Leonardo was not completely alone. According to A. G. Keller, the Hispano-Italian engineer Juanelo Turriano (1500–1585) attempted to analyze geometrically the correct shape for water wheel blades and the best angle for water to strike them. See A. G. Keller, "Renaissance Waterworks and Hydromechanics," *Endeavour* 25 (1966): 143. Turriano's work remains in manuscript, although Keller and Ladislao Reti were reported to be preparing a critical translation for publication (Reti, "Filarete Discussion," p. 433).

13. Galileo Galilei, *On Motion and On Mechanics, De Motu* (c1590) transl. I. E. Drabkin with introduction and notes; *Le Meccaniche* (c1600) transl. Stillman Drake with introduction and notes (Madison, Wis., 1960). The text of *Le Meccaniche* forms pp. 148–86; Drake's introduction is on pp. 135–45.

14. Galileo, *On Mechanics,* pp. 155–56. See also Richard S. Westfall, "The Problem of

Force in Galileo's Physics," in Carlo L. Golino, ed., *Galileo Reappraised* (Berkeley and Los Angeles, 1966) esp. p. 71, and D. S. L. Cardwell, "Some Factors in the Early Development of the Concepts of Power, Work, and Energy," *British Journal for the History of Science* 3 (1966–67): 210.

15. Leonardo *MS A* 57v; *Codex Atlanticus* 80r, 81v, 287r, 361v; *Leic.* 6v.

16. Hunter Rouse and Simon Ince, *History of Hydraulics* ([Ames, Iowa], 1957), pp. 59–61.

17. Ibid., p. 60, and Leonardo *MS F* 112.

18. René Dugas, *A History of Mechanics*, transl. J. R. Maddox (London, 1957), p. 147; William F. Magie, ed., *A Source Book in Physics* (Cambridge, Mass., 1963), pp. 111–13; and Evangelista Torricelli, *Opere*, ed. Gino Loria and Giuseppe Vassura, 2 (Rome, 1919): 185.

19. The later evolution of the efflux law is partially covered in Rouse and Ince, *History of Hydraulics*, pp. 46–48, 60, 62, 69, 72, 83–85, 89, 95–100, 113, 124–26, 131.

20. Roger Hahn, *The Anatomy of a Scientific Institution: The Paris Academy of Sciences, 1666–1803* (Berkeley and Los Angeles, 1971), p. 119.

21. Dugas, *History of Mechanics*, p. 286.

22. Christian Huygens, *Oeuvres complètes de Christiaan Huygens*, 19 (The Hague, 1937): 166–72 (for the work on the Torricellian relationship), 120–43 (for the work on the determination of the force of impulse). For Mariotte's work see *Recueil des mémoires de l'Académie royale des sciences depuis 1666 jusqu'à 1699*, vol. 1, *Histoire de l'Académie royale des sciences, depuis son établissement en 1666, jusqu'à 1686* (Paris, 1733), p. 109.

23. Huygens, *Oeuvres*, 19:140.

24. See Edme Mariotte, *The Motion of Water and other Fluids . . .* , transl. J. T. Desaguliers (London, 1718), pp. 116–49, for his derivation of the basic laws guiding his hydraulic investigations. The original work was published under Mariotte's name in Paris in 1686 by Philippe de La Hire with the title *Traité du mouvement des eaux et des autres corps fluides. . . .*

25. Mariotte, *Motion of Water*, pp. 139–43. The dimensions and weights given by French writers *before* 1800 I have left in French feet (*pieds*) and French pounds (*livres*); 1 French foot = 1.07 English feet; 1 French pound = 1.08 English pounds.

26. Ibid., pp. 148–49.

27. Philippe de La Hire, "Examen de la force de l'homme, pour mouvoir des fardeaux absoluement et par comparison à celle des animaux qui portent et qui tirent, comme les chevaux," Académie des Sciences, Paris, *Mémoirs*, 1699 (3rd ed., 1732), pp. 153–162, and *Histoire*, 1699, pp. 96–98 (henceforth *AS–M* and *AS–H*, respectively); Guillaume Amontons, "Moyen de substituer commodement l'action du feu,`a la force des hommes et des chevaux pour mouvoir les machines," *AS–M*, 1699, pp. 112–26, and *AS–H*, pp. 101–3; Antoine Parent, "Sur les centres de conversion & sur les frotemens," *AS–H*, 1700 (2nd ed., 1761), pp. 149–53, and "Sur la position de l'axe des moulins à vent à l'égard du vent," *AS–H*, 1701 (2nd ed., 1743), pp. 138–41, and "Sur la réduction des mouvements des animaux aux lois de la méchanique," *AS–H*, 1702 (1743 ed.), pp. 95–102.

28. La Hire, "Examen de la force de l'homme," p. 157.

29. Amontons, "Moyen de substituer," pp. 121, 125.

30. Philippe de La Hire, "Examen de la force nécessaire pour faire mouvoir les bateaux tant dans l'eau dormante qui courante, soit avec une corde qui y est attachée & que l'on tire, soit avec des rames, ou par le moyen de quelque machine," *AS–M*, 1702, pp. 254–80, esp. pp. 259–61 (*AS–H*, pp. 126–34).

31. For biographical information on Parent see Benjamin Martin, *Biographia Philosophica . . .* (London, 1764), pp. 356–58.

32. Antoine Parent, "Sur la plus grande perfection possible des machines etant donné une machine qui ait pout puissance motrice quelque corps fluide que ce soit, comme, par exemple, l'eau, le vent, la flame, & c. . . . ," *AS–M*, 1704 (1745 ed.), pp. 323–38 (*AS–H*, pp. 116–23).

33. Ibid., p. 333.

34. Henri Pitot, "Nouvelle methode pour connoître & déterminer l'effort de toutes sortes de machines muës par un courant, ou une chûte d'eau . . . ," *AS–M*, 1725 (publ.

1727), pp. 78–107 (*AS–H*, pp. 80–87); "Remarques sur les aubes ou pallettes des moulins, & autres machines mûës par le courant des rivières," *AS–M*, 1729 (publ. 1731), pp. 253–58 (*AS–H*, pp. 81–87); "Comparison entre quelques machines mûës par les courants des fluides ... ," ibid., pp. 385–92 (*AS–H*, pp. 81–87); "Réfléxions sur le mouvement des eaux," *AS–M*, 1730 (publ. 1732), pp. 536–44 (*AS–H*, pp. 110–15).

35. See Bélidor, *Architecture hydraulique*, 1:245–51, for the initial exposition. For the application of Parent's theory to specific wheels see, e.g., ibid., 1:287–88, 293–94, and 2:170–86.

36. Jean Lerond d'Alembert, "Aube," *Encyclopédie*, 1 (Paris, 1751): 864; Willem Jacob Storm van s'Gravesande, *Mathematical Elements of Natural Philosophy... ,* transl. John T. Desaguliers, 6th ed. (London, 1747) 1:417–18, 434–35. Examples of Leonhard Euler's use of Parent's theory can be found in his "Discussion plus particulière de diverses manières d'élever de l'eau par le moyen des pompes... ," *Mémoires de l'Académie des Sciences de Berlin* 8 (1752; publ. 1754): 149–84, and "Maximes pour arranger le plus avantageusement les machines destinées à élever de l'eau par le moyen des pompes," ibid., pp. 185–232. Both of these papers are contained in Euler's *Opera Omnia*, 2nd ser., vol. 15, ed. Jakob Ackeret (Lausanne, 1962), pp. 251–80, 281–318.

37. John Muller, *A Mathematical Treatise: Containing a System of Conic Sections; with the Doctrine of Fluxions and Fluents Applied to Various Subjects ...* (London, 1736), pp. 172–74.

38. Colin MacLaurin, *A Treatise of Fluxions*, 2 (Edinburgh, 1742): 727–28, and Mac-Laurin, *An Account of Sir Isaac Newton's Philosophical Discoveries* (London, 1748), pp. 172–73.

39. Desaguliers, *Experimental Philosophy*, 2:424; Benjamin Martin, *Philosophia Britannica; or, A New and Comprehensive System of the Newtonian Philosophy, Astronomy, and Geography... ,* 1 (Reading, 1747): 126–28n; also Martin's *New and Comprehensive System of Mathematical Institutions... ,* 2 (London, 1764): 38–40.

40. Pitot, "Remarques sur les aubes."

41. Ignace Pardies, *Oeuvres* (Lyons, 1725), pp. 164–69. The proposition was originally published in Pardies' *Discours de mouvement local* (1670).

42. Pitot, "Remarques sur les aubes," pp. 255–56.

43. Ibid., pp. 256–57. Pitot's "Comparison" deals in more detail with this problem.

44. Bélidor, *Architecture hydraulique*, 2:182–83.

45. Ibid., 2:178.

46. D'Alembert, "Aube," pp. 863–65.

47. Desaguliers, *Experimental Philosophy*, 2:425–27.

48. The flaws in Parent's theory are also discussed by Cardwell, "Some Factors," p. 212.

49. Richard S. Westfall, *Force in Newton's Physics: The Science of Dynamics in the Seventeenth Century* (New York, 1971), p. 330.

50. Richard S. Westfall, *The Construction of Modern Science: Mechanisms and Mechanics* (New York, 1971), pp. 122–23.

51. A drawing by Newton in one of his early notebooks illustrates how seventeenth-century conceptions of gravity might have led to the assumption that gravity wheels were merely another form of impulse wheel, though powered by the impulse of aethereal matter. The drawing showed a vertical wheel suspended on a horizontal axle. It was placed so that half of it was covered by a gravitational shield which projected over the top of the wheel; the other half of the wheel extended beyond the shield and was exposed to "yᵉ rays of gravity" which would perpetually turn the wheel. Newton's perpetual motion wheel at first glance looks like an overshot water wheel, and the gravity shield, like a penstock. The basic idea could fairly easily have been employed to explain the operation of overshot water wheels. See Westfall, *Force in Newton's Physics*, p. 331 and fig. 29.

52. See Isaac Newton, *Philosophiae naturalis Principia mathematica* (London, 1687), pp. 330–32, and Newton, *Principia*, 2nd ed., (Cambridge, 1713), pp. 303 f.

53. Daniel Bernoulli, *Hydrodynamics*, transl. from the Latin by Thomas Carmody and Helmut Kobus (New York, 1968), pp. 218–20, 224.

54. Ibid., p. 327.

55. Robert-Xavier Ansart DuPetit-Vandin, "Mémoire sur l'hydraulique," *Mémoirs de mathématique et de physique presentés à l'Académie Royale des Sciences, par divers Sçavans...*, 1 (1750): 261–82.

56. Bélidor, *Architecture hydraulique*, 1:284–86.

57. Beighton's observations on the Nun-Eaton wheel were published by Desaguliers, *Experimental Philosophy*, 2:453.

58. Leonhard Euler, *Opera Omnia*, 2nd ser., 15:291–92 ("Maximes pour arranger...").

59. Stephen Switzer, *An Introduction to a General System of Hydrostaticks and Hydraulicks, Philosophical and Practical...*, 2 (London, 1729): 293 (cf. ibid., 1:273–74). The wheel that Switzer used to support his contention that overshot wheels were superior was not an overshot-gravity wheel, but an *overshot-impulse* wheel. It was a wheel with no buckets or shrouds, a wheel which differed from an undershot wheel only with respect to the point at which the water was received (see ibid., 2:319 and plate 19, fig. 2, facing p. 320). Switzer's assertion that a smaller quantity of water would drive this wheel than an undershot wheel was probably correct, but if it used less water, it was only because the water had a higher head than on most undershot wheels. It was not necessarily more efficient.

60. Desaguliers, *Experimental Philosophy*, 2:531–32.

61. Martin, *Philosophia Britannica*, 1:149.

62. Desaguliers, *Experimental Philosophy*, 2:532.

63. Ibid., pp. 453–59.

64. Paul N. Wilson, "Early Water Turbines in the United Kingdom," Newcomen Society, *Transactions* 31 (1957–59): 221.

65. D'Alembert, "Aube," p. 865.

66. "Éloge de M. De Parcieux," *AS–H*, 1765, pp. 155–65.

67. Antoine de Parcieux, "Mémoire dans lequel on démontre que l'eau d'une chûte destinée à faire mouvoir quelque machine, moulin ou autre, peut toûjours produire beaucoup plus d'effet en agissant par son poids qu'en agissant par son choc, & que les roues à pots qui tournent lentement, produisent plus d'effet que celles qui tournement vîte, relativement aux chûtes & aux dépenses," *AS–M*, 1754 (publ. 1759), pp. 606–7.

68. Ibid., p. 607.

69. Ibid., pp. 608–10.

70. Ibid., pp. 610–13. De Parcieux's work is discussed in Cardwell, *From Watt to Clausius: The Rise of Thermodynamics in the Early Industrial Age* (London, 1971), p. 69.

71. De Parcieux, "Mémoire sur une expérience qui montre qu'a dépense égale, plus une roue à augets tourne lentement, plus elle faite d'effet," *AS–M*, 1754 (publ. 1759), pp. 671–78 (*AS–H*, pp. 134–38).

72. De Parcieux, "Mémoire dans lequel on démontre," pp. 613–14.

73. De Parcieux, "Mémoire dans lequel on prouve que les aubes des roues mûes par les courans de grandes rivières, feroient beaucoup plus d'effet si elles etoient inclinées aux rayons, qu'elles ne font étant appliquées contre les rayons mêmes, comme elles le sont aux moulins pendans & aux moulins sur bateaux qui sont sur les rivières de Seine, Marne, Loire, &c.," *AS–M*, 1759 (publ. 1765), pp. 288–99 (*AS–H*, pp. 223–27).

74. No full-sized scholarly monograph has been written on Smeaton. The second volume of Samuel Smiles, *Lives of the Engineers...* (London, 1862), pp. 1–90, contains the most information among published materials. Shorter sketches can be found in John Smeaton, *Reports of the late John Smeaton, F.R.S., made on various occasions in the course of his employment as a civil engineer*, 1 (London, 1812): xv–xxx; G. Bowman, "John Smeaton: Consulting Engineer," in E. G. Semler, ed., *Engineering Heritage*, 2 (London, 1966): 8–12; and Trevor Turner, "John Smeaton, FRS (1724–1792)," *Endeavour* 33 (1974): 29–33.

75. This, at least, is the speculation of Turner, "John Smeaton," p. 31.

76. John Smeaton, "An Experimental Examination of the Quantity and Proportion of Mechanic Power necessary to be employed in giving different degrees of Velocity to Heavy Bodies from a State of Rest," Royal Society of London, *Philosophical Transactions* 66 (1776): 454–55; see also pp. 454n–455n.

77. John Smeaton, "An experimental Enquiry concerning the natural Powers of Water

and Wind to turn Mills, and other Machines, depending on a circular Motion," ibid. 51 (1759): 100–101.

78. See ibid., pp. 105–6, for a description of the experimental apparatus.

79. Ibid., pp. 101–24.

80. Ibid., pp. 124–38.

81. Ibid., p. 130. Several of Smeaton's later papers were devoted to the loss of "mechanic power" in inelastic collision, notably his "New Fundamental Experiments upon the Collision of Bodies," Royal Society of London, *Philosophical Transactions* 72 (1782): 337–54, and his "Experimental Examination" (note 76, above). The mechanical power lost was, of course, converted into heat, as Joule was to demonstrate in the 1840s. Smeaton was unable to satisfactorily explain its disappearance. Smeaton also did not come up with a satisfactory explanation of why overshot wheels had a higher efficiency at low velocities. He thought that slow motion was more effective because the water in the buckets exerted more pressure than it did when the wheel moved fast. Johann Euler's work (described below) provided a much better explanation.

82. Charles Bossut, *Traité élementaire d'hydrodynamique*, 2 (Paris, 1771): 372–98.

83. Alessandro Vittorio Papacino d'Antoni, *Institutions physico-méchaniques à l'usage des écoles royales d'artillerie et du génie de Turin*, transl. from the Italian, 2 (Strasbourg, 1777): 224–69.

84. John Banks, *A Treatise on Mills, in Four Parts...* (London, 1795), pp. 131–65.

85. For biographical data on Polhem see William A. Johnson, trans., *Christopher Polhem: The Father of Swedish Technology* (Hartford, Conn., 1963) (originally published in Swedish in 1911); Lindroth, *Polhem;* and John G. A. Rhodin, "Christofer Polhammar, ennobled Polhem: The Archimedes of the North, 1661–1751," Newcomen Society, *Transactions* 7 (1926–27): 17–23. For Polhem's work in the area of hydrodynamics see Friedrich Neumeyer, "Christopher Polhem och hydrodynamiken," *Archiv for matematik, astronomi, och fysik* 28A, no. 15 (1942):16 pp. (separately numbered).

86. Lindroth, *Polhem*, pp. 83–85.

87. Ibid., pp. 85–87. See also the article cited in note 91 below, pp. 183, 195–96, for Polhem's recognition of the superior efficiency of the gravity or overshot wheel.

88. Lindroth, *Polhem*, p. 85.

89. Ibid., p. 86.

90. Polhem to Emanuel Swedenborg, end of September 1716, in Emanuel Swedenborg, *The Letters and Memorials of Emanuel Swedenborg*, transl. and ed. Alfred Acton, 1 (Bryth Athyn, Pa., 1948): 117. See also Polhem to Swedenborg, 19 December 1715, ibid., p. 78.

91. Christopher Polhem, "Fortsetzung von der Berbindung der Theorie und Practik ben der Mechanik," Königl. Schwedischen Akademie der Wissenschaften, *Abhandlungen*, 1742, transl. from Swedish into German by Abraham Gotthelf Kästner, 4 (Hamburg, 1750): 183–96.

92. Bern Dibner, *Moving the Obelisks* (Cambridge, Mass., 1970), p. 24.

93. L. T. C. Rolt, *From Sea to Sea: The Canal du Midi* (London, 1973), pp. 37–38; Jacob Ackeret, "Vorrede," to Euler, *Opera Omnia*, 2nd ser., 15:xii.

94. Smiles, *Lives of the Engineers*, 1:311, 2:6–7, 8, 120–21.

95. For example, A. E. Musson and Eric Robinson, *Science and Technology in the Industrial Revolution* (Manchester, 1969), pp. 73–74; Musson, "Editor's Introduction," in *Science, Technology, and Economic Growth in the Eighteenth Century* (London, 1972), pp. 62–63, 67; Stephen F. Mason, *A History of the Sciences*, rev. ed. (New York, 1962), p. 276.

96. De Parcieux, "Mémoire dans lequel on démontre," p. 603 (cf. de Parcieux, "Mémoire dans lequel on prouve," p. 289). See also the comments of the perpetual secretary of the Academy of Sciences, Jean-Paul Grandjean de Fouchy, in *AS–H*, 1754, p. 134, and *AS–H*, 1768, p. 162.

97. Smeaton, "Experimental Enquiry," pp. 124–25.

98. Smeaton, "New Fundamental Experiments," p. 340.

99. Cardwell, "Academic Study," p. 117, and Cardwell, *Turning Points in Western Technology* (New York, 1972), p. 40.

100. Walter G. Vincenti, "The Air-Propeller Tests of W. F. Durand and E. P. Lesley: A Case Study in Technological Methodology," *Technology and Culture* 20 (1979): 713–17 et passim.

101. For biographical information on Johann Euler see *Biographie universelle, ancienne et moderne*, ed. Joseph and Louis Gabriel Michaud, 13 (Paris, 1855): 186–87.

102. Johann Albrecht Euler, *Enodatio Qvaestionis: Qvomodo Vis Aqvae Alivsve Flvidi cvm Maximo Lvcro Ad Molas Circvm Agendas Aliave Opera Perficienda Impendi Possit?* (Göttingen, 1754), 70 pp.

103. See ibid., pp. 7–28, for Euler's analysis of the impulse wheel.

104. For information on the vis viva controversy see Carolyn Iltis, "Leibniz and the Vis Viva Controversy," *Isis* 62 (1971): 21–35; Thomas L. Haskins, "Eighteenth-Century Attempts to Resolve the Vis Viva Controversy," ibid. 56 (1965): 281–97; Laurens Laudan, "The Vis Viva Controversy a Post-Mortem," ibid. 59 (1968): 131–43; René Dugas, *Mechanics in the Seventeenth Century*, transl. Freda Jacquot (Neuchatel, 1958), pp. 296–302.

105. Bernoulli, *Hydrodynamics*, p. 12. See p. 184 for Bernoulli's use of the product of weight or force and distance to measure the "natural effect" of a stream.

106. For Segner see Karl Keller, "Johann Andreas Segner," *Beiträge zur Geschichte der Technik und Industrie* 5 (1913): 54–72, esp. 64–69.

107. Leonhard Euler's papers on the Segner wheel are collected in his *Opera Omnia*, 2nd ser., vol. 15, ed. Jakob Ackeret.

108. For example, L. Euler, "Applications de la machine hydraulique... ," in *Opera Omnia*, 2nd ser., 15:105–33; "Discussion plus particulière... ," ibid., pp. 251–80; and "Maximes pour arranger... ," ibid., pp. 281–318.

109. L. Euler, "Recherches sur l'effet d'une machine hydraulique proposée par M. Segner, professeur à Goettingue," *Mémoires de l'Académie des Sciences de Berlin* 6 (1750; publ. 1752): 311–54, in *Opera Omnia*, 2nd ser., vol. 15, esp. pp. 31–32, 38.

110. Polhem, "Fortsetzung," pp. 183, 195–96.

111. Pehr Elvius, *Mathematisk traktat om effecter af vatndrifter...* (Stockholm, 1742).

112. Pehr Elvius, "Auszug aus einem Buche, das die königl. schwed. Akademie der Wissensch. hat drucken lassen, unter dem Titel: *Mechanik; oder, Mathematische Abhandlung von Wasserwerken...* ," Königl. Schwedischen Akademie der Wissenschaften, *Abhandlungen*, 1742, transl. and ed. Abraham Gotthelf Kästner, 4 (Hamburg, 1750): 92–100, esp. 94–95.

113. Abraham Gotthelf Kästner, *Anfangsgründe der Hydrodynamik...* (Göttingen, 1797), pp. 348–49.

114. The earliest references I have been able to find to Elvius's or Polhem's work with water wheels in British or French technical literature are Thomas Young, *Course of Lectures on Natural Philosophy and the Mechanic Arts*, 1 (London, 1807): 282, and Robison, "Water-Works," p. 907, in the 1797 edition of the *Encyclopaedia Britannica*. Young says: "Before this time [Smeaton's], the best essay on the subject of water wheels was that of Elvius, published in 1742; his calculations are accurate and extensive; but they are founded, in great measure, on the imperfect suppositions respecting the impulse of a stream of water, which were then generally adopted." Robison drew his knowledge of Elvius from Kästner's *Hydrodynamik*.

115. Kästner, *Hydrodynamik*, p. 348.

116. J. Euler, *Enodatio Qvaestionis*, pp. 28–29.

117. De Parcieux, "Mémoire sur un expérience," p. 678n, and "Mémoire dans lequel on prouve," p. 289. The source of de Parcieux's information on Euler's work was not the *Enodatio Qvaestionis* itself, but an abstract: "Quelle est la manière la plus avantageuse d'employer la force de l'eau ou de tout autre fluide, soit pour faire tourner les moulins, soit pour produire un autre effet quelconque. Piéce qui a remporté la prix proposé par la société Royale des Goëttingue, en 1754," *Journal étranger*, December 1756, secs. 193–201.

118. Bossut, *Hydrodynamique*, 1771 ed., 2:388–90, 393–94.

119. For example, Kästner, *Hydrodynamik*, pp. 348 f., and Franz von Gerstner, "Abhandlungen über die oberschlächtigen Wasserräder," Königlichen Böhmischen Gesellschaft der Wissenschaften, *Abhandlungen*, 3rd ser. 2 (1805–9): 12.

120. See Jean Muscart, *La vie et les travaux du Chevalier Jean-Charles de Borda, 1733–1799* (Lyons, 1919), for biographical information on Borda.

121. Jean Charles Borda, "Mémoire sur l'écoulement des fluides par les orifices des vases," *AS–M*, 1766 (publ. 1769), pp. 579–607, esp. 591 (*AS–H*, pp. 143–50).

122. Jean Charles Borda, "Mémoire sur les roues hydrauliques," *AS–M*, 1767 (publ. 1770), pp. 270–87 (*AS–H*, pp. 49–53).

123. Ibid., p. 274.

124. Borda's section on undershot wheels is ibid., pp. 270–74.

125. See ibid., pp. 278–80, for Borda on the overshot or gravity wheel.

126. Cardwell discusses Borda's work and its importance in "Some Factors," pp. 212–13, and in his *From Watt to Clausius*, pp. 69–70.

127. Borda, "Roues hydrauliques," pp. 280–82, 284, 286–87.

128. Lazare N. M. Carnot, *Essai sur les machines en général*... , in Carnot, *Oeuvres mathématiques du Citoyen Carnot* (Basle, 1797), pp. 104–5. The *Essai* was originally published in Dijon in 1782.

129. Joseph Louis LaGrange, *Oeuvres de LeGrange*, 13 (Paris, 1882): 230. The letter is dated 24 February 1772. D'Alembert in his correspondence with LaGrange referred on one occasion to the "bad theory of the chevalier de Borda" (ibid., p. 233). D'Alembert in response to LaGrange's comments on Borda's work noted that Borda's memoir appeared to be "full of bad reasoning, very vague and very ungeometrical" (ibid., pp. 226–28). For other negative comments by d'Alembert on Borda's 1766 paper see ibid., pp. 202–3, 261, 276.

130. Joseph Louis LaGrange, *Mécanique analytique* (Paris, 1788), pp. 428–37; LaGrange, *Oeuvres*, 12 (Paris, 1882): 265–72.

131. Charles-Louis DuCrest, *Essai sur les machines hydrauliques*... (Paris, 1777), pp. viii, 16, 47–56, esp. 47–48.

132. The hold that conservation theories had on the majority of European scientists in the period between 1600 and 1900 is emphasized in Wilson L. Scott, *The Conflict between Atomism and Conservation Theory, 1644–1860* (London, 1970). Borda's role in this conflict is discussed mainly in ch. 6, pp. 104–35.

133. Muscart, *Borda*, pp. 104–17, discusses the neglect of Borda's work in the eighteenth century.

134. Lefèvre-Gineau simply noted that in 1767 Borda examined the most advantageous forms of water wheels and their blades. He showed no appreciation at all for the importance of the paper. Lefèvre-Gineau, "Notice historique sur Jean Charles Borda," Institut national des sciences et arts [de France], Sciences mathématique et physique, *Histoire* 4 (an. XI [1804]), p. 94.

135. J. F. Montucla, *Histoire des mathématiques*... , 3 (Paris, an. X [1802]): 753–58, 771–75.

136. Charles C. Gillispie, *Lazare Carnot, Savant* (Princeton, 1971), pp. 31–32, and Scott, *Conflict*, pp. 104–35, both note that Carnot's work was initially ignored. Gillispie says that this was partially because Carnot's *Essai* read neither like the traditional works on rational mechanics in the eighteenth century, nor was it elementary enough to have been used as a practical manual. Hence both scientists and artisans ignored it.

137. LaGrange, *Oeuvres*, 9 (Paris, 1881): 409. This is a reprint of the second edition of 1813.

138. Gillispie, *Carnot*, p. 101.

139. Pierre van [Petrus van] Mussenbroek, *Cours de physique expérimentale et mathématique*, transl. Sigaud de la Fond, 1 (Paris, 1769): 234–36; Johann Heinrich Lambert, "Expériences et remarques sur les moulins que l'eau meut par en bas dans une direction horizontale," Académie royale des sciences et belles-lettres, Berlin, *Nouveaux Mémoires*, 1775 (publ. 1777), pp. 63–64.

140. James Ferguson, *Lectures on Select Subjects*... (London, 1764), p. 49; Ferguson, *Analysis of a Course of Lectures on Mechanics*... (London, 1761), p. 11; and *The Works of James Ferguson*, ed. David Brewster, 3 (Edinburgh, 1823): 63–64. See also "Mill," *Cyclopaedia*, ed. E. Chambers, vol. 3 (London, 1781), and Robison, "Mechanics," pp. 767–68 in the *Encyclopaedia Britannica*, 3rd ed. (1797).

141. Cardwell, "Some Factors," p. 212n, and Cardwell, *From Watt to Clausius*, pp. 77–78, citing a letter in the Boulton and Watt Collection of the Birmingham Reference Library. See also the letter from Watt to Boulton quoted in H. W. Dickinson, *A Short History of the Steam Engine* (Cambridge, 1938), p. 188. This may be the letter to which Cardwell referred. An earlier attempt to apply Parent's analysis to steam engines is Francis Blake, "The Greatest Effect of Engines with uniformly accelerated Motions considered," *Philosophical Transactions* 51 (1759): 1–3.

142. For example, William H. Miller, *The Elements of Hydrostatics and Hydrodynamics* (Cambridge, 1831), pp. 47, 91–92; Thomas Jackson, *Elements of Theoretical Mechanics*... (Edinburgh, 1827), p. 231; Peter Barlow, "Hydrodynamics," *Encyclopaedia Metropolitana*..., 3 (London, 1845): 245–46; and William Grier, *The Mechanic's Calculator*..., 3rd ed. (Glasgow, 1836), pp. 226, 232.

143. D'Alembert, *Traité de l'équilibre et du mouvement des fluides*... (Paris, 1744), p. 370.

144. DuPetit-Vandin, "Mémoire sur l'hydraulique," pp. 263–65, 270–78; Jacques André Mallet, "A Memoir concerning the most advantageous Construction of Water-Wheels, etc.... Communicated by M. Maty..., translated from the French by J. Bevis...," Royal Society, *Philosophical Transactions* 57 (1767): 372–88.

145. Charles Bossut, "Détermination générale de l'effet des roues mues par le choc de l'eau," *AS–M*, 1769 (publ. 1772), pp. 477–97 (*AS–H*, pp. 121–24); also his *Traité théorique et expérimental d'hydrodynamique*, 3rd ed., 1 (Paris, 1796): 499–508.

146. Bossut, "Détermination générale," p. 491.

147. Bossut, *Hydrodynamique*, 3rd ed. (1796), 1:525 (see also 1st ed. [1771], 2:424).

148. Later authorities also rejected Bossut's equations because of their complexity and uselessness. For example, David Brewster, in summarizing the implacations of Bossut's formula for the distribution of blades, commented that the rules which could be derived from it were "too difficult to be of use to the practical mechanic" (Brewster, ed., *Works*, by Ferguson, 4:19). Navier, the French engineering theoretician, commented in 1819: "The researches of Bossut" lead to "formulas so complicated that it is almost impossible to make any use of them" (C. L. M. H. Navier, ed., *Architecture hydraulique*, by Bélidor, new ed. with notes, 1 [Paris, 1819]: 474n).

149. Bossut, "Détermination générale," pp. 493, 497, and fig. 2.

150. Bossut, *Hydrodynamique*, 1st ed. (1771), 2:325–26, 361–63.

151. Charles Bossut, [Jean Lerond] d'Alembert, and [Marie-Jean de] Condorcet, *Nouvelles expériences sur la résistance des fluides* (Paris, 1777), pp. 6, 173, 178, 192, 198.

152. Bossut, *Hydrodynamique*, 3rd ed. (1796), 1:487, 496–99, and 2:376–77.

153. Jorge Juan y Santacilia, *Examen maritime, théorique et pratque ou Traité de mécanique*..., transl. from the Spanish with additions by Pierre Levêque, 1 (Nantes, 1783): 393–96, for Juan's comments on water wheels; ibid., pp. 227–80 for his theory of impulse.

154. W[illiam] Waring, "Observations on the Theory of Water Mills, &c.," American Philosophical Society, *Transactions* 3 (1793): 144–49, 319–21.

155. Evans, *Mill-Wright and Miller's Guide*, pp. 65–72. Thomas Tredgold, ed., *Tracts on Hydraulics* (London, 1826), pp. 14n–16n, derives a similar theory.

156. Buchanan, *Essays on Mill Work*, 3rd ed., pp. 326–30.

157. Isaac Milner, "Reflections on the Communication of Motion by Impact and Gravity," *Philosophical Transactions* 68 (1778): 371n.

158. George Atwood, *A Treatise on the Rectilinear Motion and Rotation of Bodies*... (Cambridge, 1784), p. 279.

159. Ibid., pp. 380–81.

160. [Philippe Louis] Lanz and [Augustin de] Bétancourt, *Analytical Essay on the Con-*

*struction of Machines* (London, c1820), pp. 15–18. The first French edition was published under the title *Essai sur la composition des machines* (Paris, 1808).

161. P. S. Girard, transl., *Recherches expérimentales sur l'eau et le vent, considérés comme forces motrices applicables aux moulins et autres machines a mouvement circulaire, etc....*, by John Smeaton, 2nd ed. (Paris, 1827), pp. xi–xii, xv.

162. Antoine Guenyveau, *Essai sur la science des machines* (Paris, 1810), pp. 87–92, 125–28, 154–61.

163. A.-T. Petit, "Sur l'emploi du principe des forces vives dans le calcul de l'effet des machines," *Annales de chimie et de physique*, 2nd ser. 8 (1818): 298–300.

164. C. L. M. H. Navier, "Détails historiques sur l'emploi du principe des forces vives dans la théorie des machines et sur diverses roues hydrauliques," ibid. 9 (1818): 149–50.

165. Young, *Lectures*, 1:321, 361, 2:62–63; Ferguson, *Works*, ed. Brewster, 4:32; [Brewster], "Hydrodynamics," *Edinburgh Encyclopaedia*, 1st American ed. (Philadelphia, 1824), p. 897; and "Mechanics," *Encyclopaedia Londinensis*, 14 (London, 1816): 621, 764.

166. Peter Ewart, "On the Measure of Moving Force," Literary and Philosophical Society of Manchester, *Memoirs*, 2nd ser. 2 (1813): 156.

167. For example, Robert Mallet, "An Inquiry as to the Co-efficient of Labouring Force in Overshot Water-wheels...," Royal Irish Academy, *Proceedings*, 1st ser. 2 (1840–44): 262 (see also Institution of Civil Engineers, *Minutes of Proceedings* 2 [1843]: 60–66]; Lewis Gordon, "Remarks on Machines recipient of Water Power; more particularly the Turbine of Fourneyron," Institution of Civil Engineers, *Minutes of Proceedings* 2 (1842): 92; Andrew S. Hart, *An Elementary Treatise on Mechanics* (Dublin, 1844), pp. 118–19; and Hart, *An Elementary Treatise on Hydrostatics and Hydrodynamics* (Dublin, 1846), pp. 104–7.

168. Navier, ed., *Architecture hydraulique*, by Bélidor (1819), 1:417–18.

169. Gaspard Gustave Coriolis, *Du calcul de l'effet des machines...* (Paris, 1829), pp. 140–41, 171–73, 187–89.

170. Jean Victor Poncelet, *Cours de mécanique appliquée aux machines*, published by M. X. Kretz (Paris, 1876), pp. 134–45, 175–89.

171. Julius Weisbach, *Principles of the Mechanics of Machinery and Engineering*, ed. Walter R. Johnson, 2 (Philadelphia, 1849): 165–97.

172. D'Aubuisson, *Hydraulics*, pp. 347–54. The second French edition of d'Aubuisson de Voisins' treatise, on which this 1852 translation was based, was published in Paris in 1840.

173. F. M. G. Pambour, "Sur la théorie des roues hydrauliques," Académie des Sciences, Paris, *Comptes rendus* 60 (1865): 1181–85, 1283–86; 61 (1865): 30–31, 200–204, 1121–25; 62 (1866): 218–22, 787–90. See also Pambour's "Mémoire sur la théorie des roues hydrauliques," extrait par l'auteur," ibid. 64 (1867): 1272–75; "Addition au mémoire sur la théorie des roues hydrauliques: Du mode d'introduction des résistances dans le calcul," ibid. 67 (1868): 1134–38; and "Roues hydrauliques: Du calcul des effets par le méthode des coefficients," ibid. 75 (1872): 1757–61.

174. C. Bach, *Die Wasserräder* (Stuttgart, 1886), pp. 127–63; Julius Weisbach, "Theoretische Untersuchungen über den Eintritt des Wassers in die Zellen verticaler Wasserräder," *Der Civilingenieur* 4 (1858): 95–98; Gustav Zeuner, "Ueber den Arbeitsverluft im Ausgussbogen ober- und rückerschlächtiger Wasserräder," ibid., pp. 89–95; and Wilhelm Müller, *Die Eisernen Wasserräder* (Leipzig, 1899).

175. Pierre L. G. DuBuat, *Principes d'hydraulique*, 2nd ed., 2 (Paris, 1786): 350–60.

176. Bossut, *Hydrodynamique*, 3rd ed. (1797), 1:487, 496–97.

177. Navier, ed., *Architecture hydraulique*, by Bélidor (1819), 1:414n.

178. D'Aubuisson, *Hydraulics*, pp. 350, 358–59, 403–4, and elsewhere.

179. Pambour, "Roues hydrauliques: Du calcul des effets par le méthode des coefficients," pp. 1757–61.

180. Ibid., p. 1761.

181. For an example of the use of graphic methods to solve vertical wheel problems, see Theodor Seeberger, "Ableitung der Theorie der ober schlächtigen Wasserräder auf

graphischem Wegen," *Der Civilingenieur* 15 (1869): 397–416, and Seeberger, "Ableitung der Theorie der Wasserräder auf graphischem Wege," ibid. 16 (1870): 339–74.

182. In engineering handbooks coefficient equations of one sort or another tended to prevail in the sections on vertical water wheels in the post-1850 period. Some examples are Arthur Morin, *Aide-Mémoir de mécanique pratique*, 4th ed. (Paris, 1858), pp. 134–57; William J. M. Rankine, *A Manual of the Steam Engine and Other Prime Movers*, 12th ed. (London, 1885), pp. 174–77; William Kent, *The Mechanical Engineer's Pocket Book*, 7th ed. (New York, 1907), pp. 588–89; Lionel S. Marks, *Mechanical Engineers' Handbook* (New York, 1916), pp. 1070–71; Reginald Bolton, *Motive Powers and Their Practical Selection* (New York, 1895), pp. 67, 69–70; Henry H. Suplee, *Mechanical Engineer's Reference Book*, 3rd ed. (Philadelphia, c1907), pp. 581–84; H. Rojkow, "Ueber einige neuere hydraulische Untersuchungen und deren Anwendung auf die Uralischen Wasserwerke," *Archiv für wissenschaftliche Kunde von Russland* (ed. A. Erman) 8 (1850): 271–306.

183. Isolated tests on specific operative water wheels were sometimes made in the eighteenth century, but not tests in which the chief variables—load, wheel velocity, quantity of water, and so on—were manipulated to any great extent. For an example of measurements and simple tests on full-size water wheels in the eighteenth century see John Rennie, "Experiments made by the late John Rennie... on the Power of Water-Wheels," *Quarterly Papers on Engineering*, vol. 4 (1845).

184. John Sutcliffe, *A Treatise on Canals and Reservoirs...* (Rochdale, 1816), pp. 67, 69. Similar distrust of model experiments was expressed by Craik, *American Millwright*, p. 115: "No experiment with a small model, however accurately made and figured up, for a wheel of working size, and another head of water, can be relied upon."

185. [Gérard Joseph] Christian, *Traité de mécanique industrielle ...* , 1 (Paris, 1822): 328–30.

186. Robert Mallet, "An Experimental Inquiry as to the Co-efficient of Labouring Force in Overshot Water-Wheels, whose diameter is equal to, or exceeds the total descent due to the fall; and of Water-wheels moving in circular channels," Institution of Civil Engineers, *Minutes of Proceedings* 2 (1843): 60–66. See also Mallet, "Inquiry," pp. 262–66.

187. Weisbach, *Manual*, pp. 261, 305, citing Lagerhjelm, Forselles, and Kallstenius, *Hydrauliska Forsok* (Stockholm, 1822).

188. For the background of the Franklin Institute and its work on water wheels see Bruce Sinclair, *Philadelphia's Philosopher Mechanics: A History of the Franklin Institute, 1824–1865* (Baltimore, 1974), pp. 135–50.

189. Franklin Institute, Philadelphia, "Report of the Committee of the Franklin Institute of Pennsylvania, appointed May, 1829, to ascertain by experiment the value of Water as a Moving Power," Franklin Institute, *Journal* 11 (1831): 145–54; 12 (1831): 73–89, 147–53, 221–30, 296–305, 367–73; 13 (1832): 31–39, 153–57, 295–303, 370–77; 14 (1832): 10–16, 294–302, 366–71; 31 (1841): 145–54, 217–24, 289–96, 361–69; 32 (1841): 1–8.

190. See [Gaspard] de Prony, "Note sur un moyen de mesurer l'effet dynamique des machines de rotation," *Annales de chimie et de physique*, 2nd ser. 19 (1821): 165–73, for the original description. See also Prony's "Moyen de mesurer l'effet dynamique des machines de rotation...," *Bulletin des sciences technologiques* 6 (1826): 173–79.

191. Arthur Morin, *Expériences sur les roues hydrauliques à aubes planes, et sur les roues hydrauliques à augets* (Metz, 1836). Morin's initial experiments were published in *Memorial de l'artillerie*, no. 3 (1830), pp. 389 f. I have not seen this publication, but extracts from it appeared in several places, e.g., Morin, "Examen des effets des moteurs employés dans les fonderies de l'artillerie," *Bulletin des sciences technologiques* 18 (1831): 178, and Eug. Lefébure de Fourcy, "Extrait d'un mémoir de M. Morin, capitaine d'artillerie, ayant pour titre; 'Compte rendu d'un mission dans les fonderies d'artillerie ...,'" *Annales des mines*, 3rd ser. 3 (1833): 93–122, 259–79. Lefébure de Fourcy also published an abstract of Morin's experiments entitled "Expériences sur les roues hydrauliques...," *Annales des mines*, 3rd ser. 12 (1837): 3–46.

192. Morin, *Expériences*, pp. 13–67, esp. pp. 65–67.

193. Ibid., pp. 68–108, 121–23.

194. Jean Victor Poncelet, Pierre-Simon Girard, and C. L. M. H. Navier, "Rapport sur un mémoire ayant pour titre: 'Expériences sur les roues hydrauliques . . . ,'" *Comptes rendus* 3 (1836): 359–60. See Chapter 5 for a discussion of the water wheels in use in Morin's time.

195. Erik Nordwall, *Afhandling rorande Mechaniquen med tillampning i Synnerhet til Bruk och Bergverk*, vol. 1 (Stockholm, 1800), esp. pp. 320–80.

196. These are reported by Navier in his 1819 edition of Bélidor, *Architecture hydraulique*, 1:348n, 408n, 409n, 410n. The title of Boistard's publication was *Expériences sur la main d'oeuvre de différens travaux dépendans du service des ingénieurs des ponts et chaussées . . .* (Paris, 1804), 76 pp. I have not seen this work.

197. See d'Aubuisson, *Hydraulics*, pp. 392–94, for a summary of these experiments. The original account is Blavot-Duchesne and Daubuisson [*sic*], "Expériences sur les machines hydrauliques des mines de Poullaouen . . . ," *Journal des mines* 21 (1807): 161–248.

198. Christian, *Mécanique industrielle*, 1:331–35, 346–61.

199. The experiments on boat mills are reported in J. V. Poncelet, *Traité de mécanique industrielle . . .* , 2nd ed., 3 (Metz, 1844): 207. The results of these experiments may have been published earlier, but I have been unable to locate an earlier publication. For the experiments with rims see Poncelet, *Mémoire sur les roues hydrauliques à aubes courbes mues par-dessous . . .* , 2nd ed. (Metz, 1827), pp. 6n, 65n, 117–18. They are also reported in Poncelet, "Note relative au *Mémoire sur les roues verticales à aubes courbes*, inséré dans les *Annales de chimie et de physique*, tome xxx, p. 136," *Annales de chimie et de physique*, 2nd ser. 30 (1825): 388–89. For the experiments on the mills at low water see Poncelet, "Expériences sur la dépense d'eau des moulins de la ville de Metz, et sur l'eau qui était fournie par la Moselle, à cette époque où la sécheresse était extrême, et ou le produit de la rivière pouvait être considéré comme un minimum," *Bulletin des sciences technologiques* 11 (1829): 44–51, also reported in Poncelet, *Mémoire sur les roues hydrauliques*, pp. 117–18.

200. P. N. E. Egen, *Untersuchungen über den effekt einiger in Rheinland-Westphalen bestehenden Wasserwerke* (Berlin, 1831), reports on experiments on the water wheels of Rhenish Westphalia made for the Prussian government. I have not seen this work, but Egen's experiments are cited by Weisbach, *Manual*, 2:261, 283, and by d'Aubuisson, *Hydraulics*, pp. 396, 400, 450, as well as by Poncelet, Girard, and Navier, "Rapport," p. 362n.

201. For Morin's and Dieu's work see Morin, *Hydraulique*, 2nd ed. (Paris, 1858), pp. 247–50; for Caligny's work see [Antoine de] Caligny, "Note sur l'effet utile d'une roue de côté à palettes plongeantes, selon le système de MM. Coriolis et Bellanger," *Comptes rendus* 21 (1845): 249–53.

202. Marozeau, "Note sur une roue hydraulique récomment établie dans la blanchisserie, dite du Breuil, située à St.-Amarin, et appartenant à MM. Gros, Odier, Roman et Comp., de Wesserling," Société industrielle de Mulhouse, *Bulletin* 18 (1844): 49–59. This article is translated into English in the *Journal* of the Franklin Institute, 3rd ser. 9 [vol. 39] (1845): 9–16. See also Jérémie Risler-Dollfus and Henri Thierry, "Rapport fait un nom du comité de mécanique . . . sur la roue hydraulique de M. Marozeau, à Wesserling," Société industrielle de Mulhouse, *Bulletin* 18 (1844): 60–75.

203. For examples see Anthony Rouse, "Overshot Water Wheels employed in pumping water at Wheal Friendship Lead and Copper Mines, near Tavistock, in July, 1841," Institution of Civil Engineers, *Minutes of Proceedings* 2 (1842): 97–102; William Fairbairn, "On Water-Wheels with Ventilated Buckets," *Quarterly Papers on Engineering*, 6 (London, 1849): 18, 19; Shute Barrington Moody, "Description of a Water-wheel constructed by Mr. W. Fairbairn . . . and erected in Lombardy," Institution of Civil Engineers, *Minutes of Proceedings* 3 (1844): 67; [Jacques] Armengaud, *Traité théorique et pratique des moteurs hydrauliques . . .* , new ed., 1 (Paris, 1858): 87, 128–30, 171; "Roues hydrauliques à augets recevant l'eau à leur sommet construites soit en fonte et en fer, soit en bois et en fonte," *Publication industrielle*, 4th ed. 2 (1858): 454–55; "Roue hydraulique de côté des grandes dimensions à aubes planes et à coursier circulaire recevant l'eau en déversoir établis par MM. Cartier et

Armengaud aîné," *Publication industrielle*, 5th ed. 1 (1869): 41–43; Weisbach, *Manual*, 2:284; F. Redtenbacher, *Theorie und Bau der Wasserräder* (Mannheim, 1846), p. 308; and G. Zeuner, "Ueber ein eisernes Wasserräd mit Coulissenschüsse zum Betreib der Schneidemühle in Deuben bei Dresden," *Der Civilingenieur* 2 (1856): 85–87.

204. Carl Robert Weidner, "Theory and Test of an Overshot Water Wheel," University of Wisconsin, *Bulletin*, Engineering Series, 7 (1913): 219.

205. Where water was abundant with high falls, and low cost was a central object, or where very high velocities were needed from a water wheel, a type of engine sometimes referred to as a "flutter wheel" was often used in place of more conventional vertical water wheels. The flutter wheel was a bladed wheel of modest diameter (say around 6–9 ft, or 2–3 m). It was sometimes provided with a tight-fitting casing like a breast wheel, but often was not. Because of the high falls it used and the wheel's modest diameter, it rotated at very high speeds. Wheels of this type are described in Weisbach, *Manual*, 2:336–37, and d'Aubuisson, *Hydraulics*, pp. 353–54. Bélidor pictures several wheels related very closely to this species.

206. Poncelet, "Mémoire sur les roues verticales à palettes courbes mues par en dessous, suivi d'expériences sur les effets mécaniques de ces roues," *Annales de chimie et de physique*, 2nd ser. 30 (1825): 143.

207. The paper cited in the preceding note is dated Metz, 15 Decembre 1821. This is apparently a misprint. Didion in his "Notice historique sur les roues hydrauliques à aubes courbes," in Poncelet, *Cours de mécanique*, p. 323, says the idea came to Poncelet in 1823.

208. Poncelet, "Mémoire sur les roues verticales," pp. 149–53.

209. Ibid., pp. 154–63.

210. Poncelet, "Note," pp. 288–95.

211. Poncelet, *Mémoire sur les roues hydrauliques*, pp. 63-109.

212. Besides its appearance in the *Annales de chimie et de physique*, already cited, Poncelet's memoir appeared in *Annales des mines*, 1st ser. 12 (1825): 433–523, and Société d'encouragement pour l'industrie nationale, *Bulletin* 24 (1825): 335–68, 381–403. The separate publication was a 56-page booklet extracted from the *Bulletin* of the Société d'encouragement published in Paris in 1825.

213. Ernest Maindron, *Les fondations de prix à l'Académie des sciences: Les lauréats de l'Académie, 1714–1880* (Paris, 1881), p. 93.

214. Reynolds, *Windmills and Watermills*, p. 41.

215. [Lewis] Gordon, "Remarks on Machines recipient of Water Power," p. 93. Whitham, *Water Rights Determination*, p. 153, says Poncelet wheels were never extensively used in America.

216. For descriptions of Sagebien's wheel see "Sagebien's Water Wheel," *The Practical Magazine* 6 (1876): 89–92; Tresca, "Rapport fait par M. Tresca, un nom du comité des arts mécaniques, sur la système de roues hydrauliques de M. Sagebien, ingénieur à Amiens," Société d'encouragement pour l'industrie nationale, *Bulletin*, 2nd ser. 17 (1870): 193–203; Ch. Leblanc, "Mémoire sur la roue-vanne, inventée et exécutée par M. Sagebien," *Annales des ponts et chaussées*, 3rd ser. 15 (1858): 129–70; Martin Grübler, "Zur Theorie mittelschlächtiger Wasserräder und des Sagebienrades," *Der Civilingenieur* 22 (1876): 409–22; and A. Sagebien, "Roue de côté a niveau mittenu dans les aubes," in Armengand, *Moteurs hydrauliques*, 1:98–105.

217. Leblanc, "Mémoire sur la roue-vanne," p. 170, has a table of the major wheels erected by Sagebien up to 1858.

218. "Sagebien's Water Wheel," p. 89.

219. Müller, *Eisernen Wasserräder*, 1:12, and Grübler, "Theorie," pp. 425–26, for example.

220. Tresca, "Rapport," p. 198, provides a list of wheels erected by Sagebien up to 1870.

221. Maindron, *Les fondations de prix*, p. 149.

222. Evans, *Mill-Wright and Miller's Guide*, pp. 1–160.

223. Andrew Gray, *The Experienced Millwright*, 2nd ed. (Edinburgh, 1806), pp. 16,

24–27; Nicholson, *Millwright's Guide;* Frederick Overman, *Mechanics for the Millwright, Machinist, Engineer, C. Engineer, Architect and Student*... (Philadelphia, 1858), pp. 184, 190–201 passim; and Robert Brunton, *A Compendium of Mechanics; or, Text Book for Engineers, Mill-Wrights, Machine-Makers, Founders, Smiths, &c....* (Glasgow, 1828), pp. 131–35. The emergence of quantitative theory and experiment as a supplement to the traditional methods and craft rules of millwrights is discussed in conjunction with the emergence of water turbines in two articles by Edwin Layton: "Millwrights and Engineers, Science, Social Roles, and the Evolution of the Turbine in America," in Wolfgang Krohn, Edwin Layton, and Peter Weingart, eds., *The Dynamics of Science and Technology,* 2 (Dordrecht, 1978): 61–87, and "Scientific Technology, 1845–1900: The Hydraulic Turbine and the Origins of American Industrial Research," *Technology and Culture* 20 (1979): 64–89.

224. For example, R. James Abernathey, *Practical Hints on Mill Building* (Moline, Ill., and London, 1880), p. 4, noted a decline in the mechanical ability of millwrights, since they no longer built shaft or wheel as before, but merely put them up as furnished by the machine shop.

## CHAPTER 5: CHANGE

1. Brian R. Mitchell, *Abstract of British Historical Statistics* (Cambridge, 1962), p. 131.

2. Cardwell, *From Watt to Clausius,* ch. 3: "The Rival Power Technologies: Steam and Water," pp. 67–88, and Cardwell, "Power Technologies and the Advance of Science, 1700–1825," *Technology and Culture* 6 (1965): 188–207.

3. Hunter, "Waterpower in the Century of the Steam Engine," pp. 163–68. The New England manufacturer Zachariah Allen commented in 1825: "Water power, wherever found, is highly valued in England; but from the inadequate supply of waterfalls in this country to operate the enlarged mills required at the present day to manufacture... the additional power of steam engines has been resorted to." See Allen, *The Practical Tourist,* 1 (Providence, R.I., 1832): 347.

4. D. G. Watts, "Water-Power and the Industrial Revolution," Cumberland and Westmorland Antiquarian and Archaeological Society, *Transactions,* n.s. 67 (1967): 199; John Somervell, "Water Power and Industries in Westmorland," Newcomen Society, *Transactions* 18 (1937–38): 236; and Somervell, *Water-Power Mills of South Westmorland...* (Kendal, 1930).

5. Ian Donnachie, *The Industrial Archaeology of Galloway* (Newton Abbot, 1971), p. 32.

6. R. A. Pelham, "The Water-Power Crisis in Birmingham in the Eighteenth Century," *University of Birmingham Historical Journal* 9 (1963): 64–91, esp. 66–71. See also Pelham's "Corn Milling and the Industrial Revolution in England in the Eighteenth Century," ibid., 6 (1958): 161–75.

7. Owen Ashmore, *The Industrial Archaeology of Lancashire* (Newton Abbot, 1969), pp. 40–42.

8. Stanley D. Chapman, "The Cost of Power in the Industrial Revolution in Britain: The Case of the Textile Industry," *Midland History* 1, no. 2 (Autumn 1971): 2.

9. Edward Baines, *History of the Cotton Manufacture in Great Britain* (London, c1835) p. 86n.

10. "An Industrial Museum for Sheffield," *British Steelmaker,* April 1941, p. 75, and Archibald Allison, "Water Power as the Foundation of Sheffield's Industries," Newcomen Society, *Transactions* 27 (1949–51): 221.

11. Albert E. Musson and Eric Robinson, *Science and Technology in the Industrial Revolution* (Manchester, 1969), pp. 68–69.

12. Chapman, "Cost of Power," p. 5.

13. Kenneth Hudson, *The Industrial Archaeology of Southern England,* 2nd ed. (Newton Abbot, 1968), p. 141. Chapman, "Cost of Power," p. 20, says that 24 mills were concentrated on less than 5 miles (8 km) of the Frome, with an average fall of only 5.2 feet (1.6 m).

14. R. J. Spain, "The Len Water-Mills," *Archaeologia Cantiana* 82 (1967): 33; Robert H.

Goodsall, "Watermills on the River Len," ibid., 71 (1957): 106–29; Spain, "The Loose Watermills," ibid. 87 (1972): 43.

15. Frederick Braithwaite, "On the Rise and Fall of the River Wandle; Its Springs, Tributaries, and Pollution," Institution of Civil Engineers, *Minutes of Proceedings* 20 (1860–61): 191–210.

16. H. D. Gribbon, *The History of Water Power in Ulster* (Newton Abbot, 1969), p. 19.

17. John Sutcliffe, *A Treatise on Canals and Reservoirs . . .* (Rochdale, 1816), p. 246.

18. Ibid., p. 248.

19. "Water," *Rees's Cyclopaedia,* vol. 38 (London, 1819) [pages unnumbered].

20. Chapman, "Cost of Power," p. 8, agrees with this conclusion.

21. The best accounts of the development of mechanized equipment in the textile industry of late-eighteenth-century Britain are Richard Hills, *Power in the Industrial Revolution* (Manchester, 1970), and W. English, *The Textile Industry: An Account of the Early Inventions of Spinning, Weaving, and Knitting Machines* (London, 1969).

22. Mitchell, *Abstract of British Historical Statistics,* pp. 177–78, 180.

23. See, for example, Chapman, "Cost of Power," p. 3.

24. Ibid., p. 8.

25. Ibid., pp. 8–9.

26. George Unwin, *Samuel Oldknow and the Arkwrights: The Industrial Revolution at Stockport and Marple,* 2nd ed. (Manchester, 1968), p. 115, and English, *The Textile Industry,* p. 115.

27. E. Butterworth, *Historical Sketches of Oldham* (Oldham, 1856), pp. 130, 134, cited by Hills, *Power in the Industrial Revolution,* pp. 91 and 114, n. 12.

28. Chapman, "Cost of Power," p. 19.

29. Ashmore, *Industrial Archaeology of Lancashire,* p. 43, and Tann, *Development of the Factory,* p. 59.

30. Mitchell, *Abstract of British Historical Statistics,* pp. 185, 198, 203, 210; Baines, *Cotton Manufacture,* p. 394; and Andrew Ure, *The Cotton Manufacture of Great Britain . . . ,* 1 (London, 1836): 354–57. Other rapidly expanding elements in Britain's economy were also making demands on the available water, most notably canal and navigation interests. The conflict between canal promoters and mill interests over the use of water appears in the reports of Smeaton. See, for example, John Smeaton, *Reports of the Late John Smeaton . . . ,* 1 (London, 1812): 239–44, 284–86; 2 (London, 1812): 44–45, 83, 133–34, 155–56, 195, 234, and elsewhere.

31. Hills, *Power in the Industrial Revolution,* p. 94; Pacey, *Maze of Ingenuity,* pp. 250–61; W. H. K. Turner, "The Significance of Water Power in Industrial Location: Some Perthshire Examples," *Scottish Geographical Magazine* 74 (1958): 98–115.

32. Cardwell, *From Watt to Clausius,* pp. 68, 71, also emphasizes the power needs of the new textile industry as an important factor in the changes undergone by water-power technology after 1750.

33. David M. Smith, *The Industrial Archaeology of the East Midlands* (Dawlish and London, 1965), pp. 80–83, and Chapman, "Cost of Power," pp. 4–8.

34. Chapman, "Cost of Power," p. 9.

35. Ibid., pp. 16–17.

36. Hills, *Power in the Industrial Revolution,* p. 110.

37. Unwin, *Oldknow and the Arkwrights,* pp. 120–21.

38. Tann, *Development of the Factory,* p. 67.

39. Fairbairn, *Mills and Millwork,* 1:79–83; Joseph Glynn, *Treatise on the Power of Water,* 3rd ed. (New York, 1869), pp. 31–33; John F. Bateman, "Description of the Bann Reservoirs, County Down, Ireland," Institution of Civil Engineers, *Minutes of Proceedings* 1 (1837–41): 168–70; Robert Mallet, "Papers upon the Principles and Practice of the Application of Water Power," *Quarterly Papers on Engineering* 6 (1846): 37–46.

40. Smith, *History of Dams,* p. 180.

41. George Head, *A Home Tour through the Manufacturing Districts of England, in the*

*Summer of 1835*, new ed. (London, 1836), pp. 359–65; "Water-Power," *Chambers's Encyclopaedia*, 10 (London, 1868): 95–96; Glynn, *Power of Water*, pp. 31–33; "New System of Water Power," Franklin Institute, *Journal* 11 (1831): 214–15; "The Condition and Prospects of American Cotton Manufactures in 1849," *The Merchants' Magazine* 22 (1850): 32.

42. Frank Booker, *The Industrial Archaeology of the Tamar Valley* (Newton Abbot, 1967), p. 130; Glynn, *Power of Water*, p. 83; and D. B. Barton, *Essays in Cornish Mining History*, 1 (Truro, Cornwall, 1968): 183.

43. Booker, *Tamar Valley*, pp. 148–51.

44. Burton, *Essays in Cornish Mining History*, 1: 187–88.

45. Ibid., pp. 176–77.

46. Smith, *History of Dams*, p. 185.

47. Ibid., p. 186.

48. Ibid., p. 185; Raistrick, *Industrial Archaeology*, pp. 251–54; and Raistrick and Jennings, *Lead Mining in the Pennines*, pp. 214–19.

49. Baines, *Cotton Manufacture*, p. 86n.

50. Somervell, *Mills of South Westmorland*, pp. xii–xiv; Paul N. Wilson, "Water Power and the Industrial Revolution," *Water Power* 6 (1954): 314–15; and Wilson, "British Industrial Waterwheels," International Symposium on Molinology, *Transactions* 3 (1973): 27. Other examples of British schemes to develop water power are described by [John] Leslie and [Robert] Jameson, "On the Value of Water as a Moving Power for Machinery, illustrated in an extract from a Report in regard to the Water of Leith," *Edinburgh Philosophical Journal* 13 (1825): 170–72; C. Carmichael, "On Water as a Moving Power for Machinery," ibid., pp. 346–48; and Mallet, "Application of Water Power" [a report on the Dodder Reservoirs, Ireland], pp. 5–36.

51. Margaret T. Parker, *Lowell: A Study of Industrial Development* (New York, 1940), esp. pp. 59–68; James B. Francis, *Lowell Hydraulic Experiments* (Boston, 1855), pp. ix–x; and United States Census Office, "Report on the Water-Power of the Streams of Eastern New England," in *10th Census, 1880*, 16 (Washington, 1885): 30–35. The most complete account of the Lowell system is to be found in Louis Hunter's *Waterpower*, pp. 204–91.

52. "Report on the Water-Power of the Streams of Eastern New England," *10th Census*. 16: 25–30.

53. Ibid., pp. 35–39.

54. Hunter, *Waterpower*, pp. 262–64.

55. "Report on the Water-Power of the Region Tributary to Long Island Sound," *10th. Census*, 16: 51–56; Wegmann, *Dams*, pp. 147–49, plate 83.

56. In conjunction with the tenth census, 1880, a comprehensive two-volume study was made of water power in the United States under the title *Reports on the Water-Power of the United States*. This study, published in Washington in 1885 and 1887, formed vol. 16 and 17 of the 1880 census. A list of large developed water powers in the United States is given in vol. 16, pp. xxx–xxxv. See also Hunter, *Waterpower*, pp. 204–50 passim.

57. *Miscellaneous Dissertations*, p. 377.

58. Gille, "Moulin à eau," p. 4 and fig. 4.

59. Reti, "Problem of Prime Movers," p. 94, claims that Leonardo sketched a breast wheel in *MS H*, 63r. There are several crude drawings of water wheels on that sheet, but none are sufficiently clear to permit positive identification as a breast wheel. But Leonardo did sketch an overshot wheel with a casing to hold water on the wheel, a device which was essential to the successful operation of the breast wheel (see note 62, below).

60. Several of these are pictured in Reti, "Postscript," pp. 428–41; see figs. 5 and 7 of that article.

61. Fitzherbert, *Surveying*, p. 92.

62. Leupold, *Theatrum machinarum generale*, p. 165 and table 62. The idea of encasing the periphery of a vertical wheel to hold water on the wheel was not original with Leupold. It dates back at least to the time of Leonardo, for one of his sketches shows an overshot

wheel with a casing over most of its loaded arc (Leonardo da Vinci, *I Manoscritti e i Disegni*, vol. 6 [Rome, 1949], plate 239).

63. Bélidor, *Architecture hydraulique*, 1:283–84, and plate 1, figs. 3 and 8 (for vol. 1, bk. 2, ch. 1).

64. Desaguliers, *Experimental Philosophy*, 2:453.

65. Ibid., pp. 453–54, and plate 33, fig. 1. The claim by Caligny, "Sur l'origine de la roue dite de côté," Académie des Sciences, Paris, *Comptes rendus* 21 (1845): 16–18, that the water wheel with blades encased in its race was invented in 1777 by DuCrest, a Frenchman, is clearly in error.

66. Wilson, "Waterwheels of Smeaton," p. 30, and pp. 39–42 (table of technical particulars on watermills designed or erected by Smeaton). More than half of all the wheels Smeaton designed or erected were low breast wheels. He almost never used undershot wheels.

67. Smeaton, *Reports*, 1:313–16. See also the Wakefield flour mill, ibid., 2:430.

68. Peter Ewart, "On the Measure of Moving Force," Literary and Philosophical Society of Manchester, *Memoirs*, 2nd ser. 2 (1813): 161. The nearly contemporary testimony of Ewart on the importance and popularity of Smeaton's improvements is supported by contemporary scholars. See, for example, Cardwell, "Some Factors," p. 213: "My own limited experience tends to confirm this [Ewart's testimony]: water-powered textile mills like those at Cromford, Belper, Styal, etc., are certainly built in accordance with Smeatonian precepts."

69. "Water," *Rees's Cyclopaedia*.

70. John Farey, *A Treatise on the Steam Engine, Historical, Practical, and Descriptive* (London, 1827), p. 296.

71. For example, Diderot and d'Alembert, eds., *Encyclopédie: Planches*, vol. 5 (1767), "Papetterie," plate 7.

72. "Water," *Rees's Cyclopaedia*.

73. For Rennie's career see John Rennie, *Autobiography of Sir John Rennie* (London, 1875); Cyril T. G. Boucher, *John Rennie, 1761–1821* (Manchester, 1963); Boucher, "John Rennie (1761–1821)," Newcomen Society, *Transactions* 34 (1961–62): 1–13; and Wallace Reyburn, *Bridge across the Atlantic: The Story of John Rennie* (London, 1972).

74. Rennie, "Experiments," pp. 22–24; Glynn, *Power of Water*, p. 89; Boucher, *John Rennie*, pp. 79–80.

75. The operation of the sliding hatch overflow sluice is described in a number of works, among them, Fairbairn, *Mills and Millwork*, 1:143; "Water," *Rees's Cyclopaedia;* Glynn, *Power of Water*, pp. 89, 90; and Nicholson, *Millwright's Guide*, pp. 89–92.

76. Glynn, *Power of Water*, p. 90, attributes the invention of the double hatch to Rennie. The article "Water" in *Rees's Cyclopaedia* describes a double hatch used in the Strutt Mills at Belper in some detail, but does not mention who was responsible for the design.

77. See Smeaton, *Reports*, 1:364–65, and 2:426–27, for examples.

78. "Water," *Rees's Cyclopaedia*. Examples of this construction can be found in Smeaton's *Reports*, 2:426–27, 428–29.

79. Anders Jespersen, *The Lady Isabella Waterwheel of the Great Laxey Mining Company, Isle of Man, 1854–1954* (Virum, Denmark, 1954), pp. 58–59; Fairbairn, *Mills and Millwork*, 1:116, 128; [David Scott], *Engineer and Machinist's Assistant . . .* , 2nd ed. (Glasgow, 1856), p. 219.

80. John Millington, *An Epitome of the Elementary Principles of Natural and Experimental Philosophy . . .* (London, 1823), p. 222.

81. Francis, *Lowell Hydraulic Experiments*, p. 1.

82. William Fairbairn, "On Water-Wheels with Ventilated Buckets," *Quarterly Papers on Engineering* 6 (1849): 3 (see also Fairbairn, *Mills and Millwork*, 1:116); Scott, *Engineer and Machinist's Assistant*, p. 213.

83. The disadvantages of wood as a construction material for water wheels are described in some detail in several pamphlets published by the Fitz Water Wheel Company in

the early twentieth century and contained in the Eleutherian Mills Historical Library, Greenville, Delaware. See, for example, Fitz Water Wheel Co., *Fitz Steel Overshoot Water Wheels*, Bulletin no. 70 (Hanover, Pa., 1928), pp. 6, 16, 30–35. See also Hunter, *Waterpower*, pp. 88–89.

84. George Woodbury, *John Goffe's Mill* (New York, 1948), pp. 98–99; Hunter, *Waterpower*, p. 89, for the Springfield Armory; and Hobart, *Millwrighting*, p. 376.

85. Buchanan, *Essays on Mill Work*, 3rd ed., 1:178.

86. Smeaton, *Reports*, 1:410–11.

87. Wilson, "Waterwheels of Smeaton," p. 33. See Smeaton, *Reports*, 1:364 f. for some of Smeaton's papers relative to his work as a consulting engineer at Carron.

88. R. A. Mott, "The Coalbrookdale Group Horsehay Works, Part I," Newcomen Society, *Transactions* 31 (1957–59): 276, and John Smeaton, *A Catalogue of the Civil and Mechanical Engineering Designs 1741–1792 of John Smeaton, F.R.S. . . .* , ed. H. W. Dickinson and A. A. Gomme, Newcomen Society Extra Publication no. 5 (London, 1950), p. 36.

89. "Water," *Rees's Cyclopaedia;* Farey, *Steam Engine*, pp. 305n, 443. See also Smeaton, *Reports*, 1:410–11. For evidence of axle breakage see Smeaton, *Catalogue of Smeaton's Designs*, p. 21.

90. "Water," *Rees's Cyclopaedia;* Farey, *Steam Engine*, p. 305n.

91. Wilson, "Waterwheels of Smeaton," p. 33n. Wilson investigated the six-volume collection of Smeaton's drawings and designs preserved in the library of the Royal Society. This collection has been catalogued by the Newcomen Society; see note 88, above.

92. Farey, *Steam Engine*, pp. 275, 305n. For Smeaton-designed iron axles see Smeaton, *Catalogue of Designs*, pp. 17, 18, 21, 23, 30, 31, 36, 37.

93. Wilson, "Waterwheels of Smeaton," p. 33; Smeaton, *Reports*, 1:378–79.

94. Wilson, "Waterwheels of Smeaton," p. 33; Smeaton, *Catalogue of Designs*, pp. 31, 35.

95. Wilson, "Waterwheels of Smeaton," p. 32; Smeaton, *Reports*, 2:426–27.

96. Wilson, "Waterwheels of Smeaton," p. 32, and Wilson, "British Industrial Waterwheels," International Symposium on Molinology, *Transactions* 3 (1973): 19 and 30, fig. 4.

97. Smeaton, *Reports*, 2:424.

98. Smeaton, *Catalogue of Designs*, p. 36.

99. Wilson, "Waterwheels of Smeaton," p. 33; Smeaton, *Catalogue of Designs*, p. 17.

100. Rennie, "Experiments," p. 23; Boucher, *John Rennie*, p. 80.

101. Hills, *Power in the Industrial Revolution*, pp. 107–8. For a biographical sketch of Ewart see William C. Henry, "A Biographical Notice of the Late Peter Ewart, Esq.," Literary and Philosophical Society of Manchester, *Memoirs*, 2nd ser. 7 (1846): 114–35.

102. Hills, *Power in the Industrial Revolution*, pp. 107, 112.

103. Jan van der Straet, *The "New Discoveries" of Stradanus* (Norwalk, Conn., 1953), plate 10 (a flour mill with rim gearing) and plate 13 (a sugar mill with rim gearing). This collection is a facsimile of a set of 24 engravings issued in the early 1580s.

104. Wilson, "British Industrial Waterwheels," pp. 20 and 23, n. 2.

105. Hewes is credited with the invention of the suspension wheel by Thomas Baines and William Fairbairn, *Lancashire and Cheshire, Past and Present . . .* , 2 (London, 1868): cxliv; by Musson and Robinson, *Science and Technology*, pp. 70, 98, 445; and by W. H. Chaloner and A. E. Musson, *Industry and Technology* (London, 1963), fig. 12. But H. R. Johnson and A. W. Skempton, "William Strutt's Cotton Mills, 1793–1812," Newcomen Society, *Transactions* 30 (1955–57): 204, attribute the suspension wheel to William Strutt. As the Fairbairn quote in the text (at note 107, below) indicates, the idea may have been suggested by Strutt and first executed by Hewes.

106. Most of what is known of Hewes's career is contained in an obituary published in the *Manchester Guardian*, 11 February 1832, and in Musson and Robinson, *Science and Technology*, pp. 70–71, 98–99, 425, 436, 445–47.

107. Fairbairn, *Mills and Millwork*, 1:120.

108. This is described in *Rees's Cyclopaedia*, s.v. "Water." Johnson and Skempton, "Strutt's Cotton Mills," p. 191n, say that these wheels were designed by Strutt but built by T. C. Hewes sometime between 1802 and 1811.

109. Musson and Robinson, *Science and Technology,* p. 71. Johnson and Skempton, "Strutt's Cotton Mills," p. 204, suggest that Strutt was reponsible for the initial application of the governor to a water wheel.

110. Chapman, "Cost of Power," p. 6, and G. N. von Tunzelmann, *Steam Power and British Industrialization to 1860* (Oxford, 1978), p. 164.

111. "Water," *Rees's Cyclopaedia;* Glynn, *Power of Water,* pp. 91–92.

112. For a history of the centrifugal ball governor see Otto Mayr, *The Origins of Feedback Control* (Cambridge, Mass., 1970), pp. 100–115.

113. "Mill-Work," *Rees's Cyclopaedia,* vol. 23 (London, 1819); Nicholson, *Millwright's Guide,* p. 37.

114. Fairbairn, *Mills and Millwork,* 1:143–47; Nicholson, *Millwright's Guide,* pp. 38, 113–14; Buchanan, *Essays on Millwork,* pp. 312–17; "Water" and "Mill-Work," *Rees's Cyclopaedia.*

115. The most complete account of Fairbairn's career is William Fairbairn, *The Life of Sir William Fairbairn . . . ,* partly written by himself, completed and edited by William Pole (London, 1877). There are shorter accounts, including A. I. Smith, "William Fairbairn: Experimental Engineer," in *Engineering Heritage: Highlights from the History of Mechanical Engineering,* 2 (London, 1966): 20–25.

116. Fairbairn, *Mills and Millwork,* 1:12–21; Fairbairn, "Ventilated Buckets," pp. 8–9.

117. Robison, "Water-Works," p. 904; also John Robison, *System of Mechanical Philosophy,* 2 (Edinburgh, 1822): 598.

118. Robison, "Water-Works," pp. 904–5; see also Robison, *Mechanical Philosophy,* 2:599.

119. Fairbairn, *Mills and Millwork,* 1:138, see also Fairbairn, "Ventilated Buckets," p. 7.

120. Fairbairn, *Mills and Millwork,* 1:136, and Fairbairn, "Ventilated Buckets," p. 6.

121. Fairbairn, *Mills and Millwork,* 1:136, see also Fairbairn, "Ventilated Buckets," p. 6.

122. For Fairbairn's discussion of his ventilated buckets and their operation see "Ventilated Buckets," pp. 8–10, 12; his *Mills and Millwork,* 1:137–40; and Baines and Fairbairn, *Lancashire and Cheshire,* 2:cxlv–cxlvii. Fairbairn's ventilated wheel with closed soal plate was anticipated in the large breast wheels constructed around 1800 in the Strutt mills at Belper. The radial floatboards in those wheels were supported between rings and mounted so that they did not touch the combination soal-axle of the wheels. Two inches (5.1 cm) were left between the bottom of the floats and the barrel axle for the escape of air (see "Water," *Rees's Cyclopaedia*). Leonardo, in *MS Forester I,* 50v, has a sketch of a water wheel which, according to Reti ("Problem of Prime Movers," pp. 93–94, and fig. 31), has ventilated buckets. Since the sketch is rather crude and some of the buckets have no "ventilation holes," I am inclined to doubt Reti's interpretation of the drawing.

123. Information on the axles, gudgeons, and bearings of iron water wheels is available from a variety of sources. Among those I have used are Buchanan, *Essays on Millwork,* pp. 178–82, 200–211, 246; Fairbairn, *Mills and Millwork,* 1:116–19; Scott, *Engineer and Machinist's Assistant,* p. 213; "Water," *Rees's Cyclopaedia;* Weisbach, *Manual,* 2:241–43; W. H. Uhland, *Handbuch für den Praktischen Maschinen-Constructeur,* 1 (Leipzig, 1883): 136–38; C. Bach, *Die Wasserräder* (Stuttgart, 1886), pp. 170–72; "Roue hydraulique de côté des grandes dimensions . . . ," *Publication industrielle,* 5th ed., 1 (1869): 10–13; "Roues hydrauliques à augets recevant l'eau à leur summet construites soit en fonte et en fer, soit en bois et en fonte," ibid., 4th ed., 2 (1858): 427–29, 436–37.

124. For information on the arms, shrouds, and soals of iron water wheels I have used the following: Fairbairn, *Mills and Millwork,* 1:117–25, 140–41, and elsewhere; Scott, *Engineer and Machinist's Assistant,* pp. 213–15, 220; Uhland, *Handbuch,* pp. 135–37; Bach, *Die Wasserräder,* pp. 172–75; Weisbach, *Manual,* 2:238–39; "Water," *Rees's Cyclopaedia;* "Roue hydraulique de côté," pp. 13–14; "Roues hydrauliques à augets recevant l'eau," pp. 429–32, 436–38.

125. For the buckets of iron water wheels I have utilized material from Fairbairn, *Mills and Millwork,* 1:125–26; Scott, *Engineer and Machinist's Assistant,* pp. 215–18; Glynn, *Power of Water,* pp. 78, 93–96; [Julien-Napoléon] Haton de la Goupillière, *Cours de machines,* 1

(Paris, 1889): 234–42; "Roues hydrauliques à augets recevant l'eau," pp. 432–33, 438–40; "Roue hydraulique à augets avec coyaux creux en fonte pour l'échappement et l'entrée de l'air...," *Publication industrielle*, 3rd ed., 4 (1862): 393–98.

126. Scott, *Engineer and Machinist's Assistant*, p. 213.

127. "Roues hydrauliques à augets recevant l'eau," p. 426.

128. Hybrid wheels, mainly wood-iron hybrids, are briefly discussed by John Vince, *Discovering Watermills* (Tring, Hertshire, 1970), pp. 5–8. Contemporary material on them can be found in "Water," *Rees's Cyclopaedia;* Scott, *Engineer and Machinist's Assistant*, pp. 213–15; Buchanan, *Essays on Millwork*, pp. 180–81; F. Redtenbacher, *Theorie und Bau der Wasserräder* (Mannheim, 1846), pp. 192–94; "Roues hydrauliques à augets recevant l'eau," pp. 427–30, 438–40; "Roue hydraulique de côté," pp. 10–17.

129. F. R. I. Sweeny, "The Burden Water-Wheel," American Society of Civil Engineers, *Transactions* 79 (1915): 708–26.

130. Sutcliffe, *Canals and Reservoirs*, pp. 255–56 (see also p. 254).

131. Redtenbacher, *Wasserräder*, table on p. 308.

132. "Water-Power," *Chambers's Encyclopaedia*, p. 96.

133. Jespersen, *Lady Isabella Waterwheel*, p. 14.

134. Sydney B. Donkin, "Bryan Donkin, F.R.S., M.I.C.E. 1768–1855," Newcomen Society, *Transactions* 27 (1949–51): 88 and plate 22, fig. 52.

135. Glynn, *Power of Water*, p. 84n. This wheel is also mentioned by Barton, *Essays in Cornish Mining History*, 1:181, who adds the information that the wheel was 8 feet (2.4 m) wide. In the 1850s an even taller wheel was apparently planned by an American manufacturer, James Prince, for a cotton mill at Palitus, Mexico. It was to be 100 feet (30.5 m) in diameter and was to develop 140 hp under a 160-foot (48.8-m) head. Prince was, however, persuaded to use water turbines instead by Alexis Du Pont. This episode is reported by H. H. Suplee in a discussion of a paper by Samuel Webber, "Water-Power: Its Generation and Transmission," American Society of Mechanical Engineers, *Transactions* 17 (1895–96): 54.

136. Wilson, "Waterwheels of Smeaton," pp. 39–42.

137. Fairbairn, *Mills and Millwork*, 1:148–49.

138. Müller, *Eisernen Wasserräder*, 1:72–77, and 2:110–13, 156–57.

139. "Water," *Rees's Cyclopaedia*, describes a wooden breast wheel 40 feet (12.2 m) wide which was used at Strutt's Belper works. This wheel was only 12.33 feet (3.76 m) bigh. It had a hollow axle built up like a cask. This large cylinder was assembled from thirty-two planks 6 inches (0.15 m) thick and 48 feet (14.6 m) long. The float boards of the wheel were secured to it and were lent further support by ten circular rings. The floats were staggered between these rings to render the operation of the wheel smoother.

140. These wheels are described in "Water," *Rees's Cyclopaedia*.

141. Chaloner and Musson, *Industry and Technology*, plate 12. It remained in operation until 1955.

142. Wilson, "Waterwheels of Smeaton," pp. 39–42.

143. Müller, *Eisernen Wasserräder*, 1:72–77, and 2:110–13, 156–57.

144. Scott, *Engineer and Machinist's Assistant*, p. 244.

145. Fairbairn, *Mills and Millwork*, 1:129.

146. For a description of the Catrine wheels see Fairbairn, *Mills and Millwork*, 1:129–33, and Fairbairn, "Ventilated Buckets," pp. 13–14; also Wilson, "British Industrial Waterwheels," pp. 23–24, n. 4.

147. Fairbairn, *Mills and Millwork*, 1:140–41, and Fairbairn, "Ventilated Buckets," pp. 11–13.

148. Fairbairn, *Mills and Millwork*, 1:133–37: Turner, "Significance," p. 108.

149. "Water-Power," *Chambers's Encyclopaedia*, p. 96; Moritz Rühlmann, *Allgemeine Maschinenlehre...*, 1 (Berlin, n.d.): 356.

150. The best study of any single water wheel is Anders Jespersen's *The Lady Isabella Waterwheel of the Great Laxey Mining Company*, which includes a very detailed description (with photographs and drawings) not only of the wheel itself, but also of the catchment basin of its water supply, the pumps it powered, and so on. See also P. Kenyon, "The Great

Laxey Mining Company, Isle of Man: Surveying Evidence, April 1965," *Industrial Archaeology* 2 (1965): 154–57, and H. D. Varley, "'Lady Isabella' Wheel, Isle of Man," ibid. 7 (1970): 337–39.

151. F.R.I. Sweeny, *The Burden Water-Wheel,* Society for Industrial Archeology, Occasional Publications, no. 2 (April 1973). This is a reprint of the article cited in note 129, above, with additional illustrations and an introduction by Robert Vogel.

CHAPTER 6: DEMISE

1. There are a large number of works which deal with the early development of the steam engine, among them H. W. Dickinson, *Short History of the Steam Engine* (Cambridge, 1938, and later editions); L. T. C. Rolt and J. S. Allen, *The Steam Engine of Thomas Newcomen* (New York, 1977); and Farey, *Steam Engine.*

2. Farey, *Steam Engine,* pp. 212–308, deals with the early applications of the steam engine and attempts to produce rotary motion with the single-action Newcomen engine. On the latter subject see also Rolt and Allen, *Engine of Newcomen,* pp. 118–23.

3. Thomas Savery, *The Miner's Friend; or, An Engine to Raise Water by Fire* (London, 1702), pp. 28–29.

4. Denis Papin, *Nouvelle manière d'élever l'eau par la force du feu, mis en lumière* (Cassel, 1707), cited by Farey, *Steam Engine,* p. 109.

5. Farey, *Steam Engine,* p. 296.

6. See, for example, Musson and Robinson, *Science and Technology,* p. 398n, and Hills, *Power in the Industrial Revolution,* p. 134.

7. Farey, *Steam Engine,* pp. 296–308.

8. Ibid., pp. 122–24.

9. Musson and Robinson, *Science and Technology,* pp. 403–4.

10. Hills, *Power in the Industrial Revolution,* pp. 134–41; Tann, *Development of the Factory,* pp. 75–77, 91; Musson and Robinson, *Science and Technology,* pp. 398–406; and Farey, *Steam Engine.*

11. Smeaton, *Reports,* 2:378–79.

12. Von Tunzelmann, *Steam Power,* pp. 170–71.

13. Ibid., pp. 170–73; Hunter, *Waterpower,* pp. 504–8.

14. John Coolidge, *Mill and Mansion: A Study of Architecture and Society in Lowell, Massachusetts, 1820–1865* (New York, 1942), p. 167, noted, in commenting on why Boston capitalists established works at Lowell in the 1820s and 1830s, that "proximity to capital, to markets, to raw materials, to labor, was not considered. . . . Water power, and plenty of it, was all they sought" (cited by Hunter, *Waterpower,* p. 484).

15. G. F. Swain, "Statistics of Water Power Employed in the United States," American Statistical Association, *Publications* 1 (1888): 36; Hunter, *Waterpower,* pp. 482–85.

16. There is a moderately extensive literature dealing with the comparative costs of steam and water power in the late eighteenth and the nineteenth century. See Hunter, *Waterpower,* pp. 487, 516–28; von Tunzelmann, *Steam Power,* pp. 129–41, 161 63, and elsewhere; Chapman, "Cost of Power"; and Peter Temin, "Steam and Waterpower in the Early Nineteenth Century," *Journal of Economic History* 26 (1966): 187–205.

17. The most extensive and best review of the factors which led to the decline of water power is chapter 10 of Louis Hunter's *Waterpower,* pp. 481–535.

18. Musson, "Industrial Motive Power," pp. 417, 439.

19. Hunter, *Waterpower,* p. 103.

20. Swain, "Statistics," p. 35.

21. Temin, "Steam and Waterpower," p. 199.

22. Ibid., p. 203.

23. Ibid., p. 204.

24. Dupin, *Forces productives,* 1:36–38.

25. Cited by David S. Landes, *The Unbound Prometheus: Technological Change and Industrial Development in Western Europe from 1750 to the Present* (Cambridge, 1969), p. 182.

26. Cited by Arthur Louis Dunham, *The Industrial Revolution in France, 1815–1848* (New York, 1955), p. 117.

27. Hunter, "Waterpower in the Century of the Steam Engine," p. 188. See also France, Ministère de l'économie nationale, *Statistique des forces motrices en 1931* (Paris, 1936), p. 6.

28. France, *Statistique des forces*, p. 6.

29. Ibid., p. 208.

30. Chapman, "Cost of Power," esp. pp. 12–13, 18–20.

31. Farey, *Steam Engine*, p. 296.

32. Maurice Block, *Statistique de la France*, 2 (Paris, 1875): 139–41.

33. Musson, "Industrial Motive Power," pp. 415–39 passim.

34. See Cardwell, "Power Technologies," p. 195.

35. Fludd, *Tractatus secundus*, p. 467.

36. "Machine hydraulique inventée par Messieurs Denisart et de la Deuille... ," *Machines et inventions approuveés par l'Académie royale des Sciences...*, ed. M. Gallon (Paris, 1776), 1:162–63 and 3 (plates):352–57.

37. Bélidor, *Architecture hydraulique*, 2:235–53, and plates 1–5.

38. For a description of Höll's engine see Jars, *Voyages métallurgiques*, 2:152–55; Jars, "Description d'une nouvelle machine executée aux mines de Shemnitz en Hongrie, au mois de Mars 1755," *Mémoirs mathématique et physique, présenté par divers Sçavans...* 5 (1768): 67–71; and A. Guenyveau, *Essai sur la science des machines* (Paris, 1810), pp. 191–206.

39. John Smeaton, "Description of the Statical Hydraulic Engine, invented and made by the late Mr. William Westgarth... ," Society of Arts, Manufactures and Commerce, *Transactions* 5 (1787): 185–210.

40. For Trevithick's work with water-pressure engines see Francis Trevithick, *Life of Richard Trevithick...* , 1 (London, 1872): 73–89, and H. W. Dickinson and Arthur Titley, *Richard Trevithick: The Engineer and the Man* (Cambridge, 1934), pp. 39–42.

41. Joseph Glynn, "On Water-pressure Engines," British Association for the Advancement of Science, *Report* 18 (1849): 15; Glynn, *Power of Water*, pp. 103–7.

42. William L. Baker, "Description of a Water-pressure Engine at Ilsang, in Bavaria," Institution of Civil Engineers, *Minutes of Proceedings* 2 (1842): 55; d'Aubuisson, *Hydraulics*, pp. 452–53; [J.N.P.] Hachette, *Traité élémentaire des machines*, 2nd ed. (Paris, 1819), pp. 121–24.

43. Weisbach, *Manual*, 2:580–83; Glynn, "Water-pressure Engines," p. 12; William L. Baker, "Description of the Water-pressure Engine at Freyberg, Saxony," Institution of Civil Engineers, *Minutes of Proceedings* 2 (1843): 143–44.

44. C. R. Bornemann, "On a New Water-Pressure Pumping Engine at Clausthal," Institution of Civil Engineers, *Minutes of Proceedings* 45 (1876): 347–48; Weisbach, *Manual*, 2:579.

45. Weisbach, *Manual*, 2:574.

46. Guenyveau, *Science des machines*, p. 205.

47. Weisbach, *Manual*, 2:626.

48. Daniel Livermore, "Remarks upon the Employment of Pressure Engines, as a substitute for Water Wheels," Franklin Institute, *Journal*, n.s. 15 (1835): 166, estimated that 90% of the waterfalls in the United States were less than 10 feet (3 m) high.

49. Weisbach, *Manual*, 2:627. Livermore, "Remarks upon Pressure Engines," attempted to develop a water-pressure engine to compete with vertical wheels on very low falls.

50. John Whitehurst, "Account of a Machine for Raising Water, executed at Oulton, in Cheshire, in 1772," Royal Society, *Philosophical Transactions* 65 (1775): 277–79. See also Ewbank, *Hydraulic Machines*, pp. 368–69.

51. For the history of the hydraulic ram see H. W. Dickinson, "Early Years of the Hydraulic Ram," Newcomen Society, *Transactions* 17 (1936–37): 73–83, and Charles Cabanes, "Joseph de Montgolfier et la Bélier Hydraulique," ibid., pp. 85–90. See also Ewbank, *Hydraulic Machines*, pp. 369–70.

52. D'Aubuisson, *Hydraulics*, pp. 461–63.

53. Ibid., p. 464.

54. Dickinson, "Early Years of the Hydraulic Ram," p. 81.

55. D'Aubuisson, *Hydraulics*, p. 466.

56. Ibid.

57. Dickinson, "Early Years of the Hydraulic Ram," p. 83.

58. The best account of the prehistory of Fourneyron's turbine is Norman Smith, "The Origins of the Water Turbine and the Invention of Its Name," *History of Technology*, 1977, pp. 215–59.

59. [J. V. Poncelet], *Cours de mécanique appliquée aux machines* [École d'Application de l'Artillerie et du Génie], sec. 7 (n.p., 1836?), p. 45. See also Poncelet, *Cours*, 2:201.

60. Weisbach, *Manual*, 2:364–67.

61. Frederic W. Keaton, "Benoit Fourneyron (1802–1867)," *Mechanical Engineering* 61 (1939): 295–301.

62. Benoit Fourneyron, "Mémoire sur l'application en grand, dans les usines et manufactures, des turbines hydrauliques ou roues à palettes courbes de Bélidor," Société d'encouragement pour l'industrie nationale, *Bulletin* 33 (1834): 4.

63. Ibid., pp. 49–53, 85–89.

64. Ibid., pp. 53–56.

65. Ibid., pp. 56–61; also d'Aubuisson, *Hydraulics*, pp. 420–22.

66. The memoir cited in note 62, above.

67. D'Aubuisson, *Hydraulics*, p. 433.

68. Moritz Ruhlman[n], *On Horizontal Water-Wheels, especially Turbines or Whirl-Wheels...*, edited with an introduction and notes by Sir Robert Kane (Dublin, 1846), pp. 17–18. For more information on the St. Blasien turbine see Benoit Fourneyron, "Lettre de M. Fourneyron à M. Arago, datée d'Augsbourg, le 17 septembre 1837, sur une nouvelle turbine," *Comptes rendus* 5 (1837): 562, and d'Aubuisson, *Hydraulics*, pp. 434–35.

69. Marcel Crozet-Fourneyron, *Invention de la turbine* (Paris, 1924), p. 38.

70. Ellwood Morris, "Remarks on Reaction Water Wheels used in the United States; and on the Turbine of M. Fourneyron...," Franklin Institute, *Journal* 34 (1842): 217–27, 289–304.

71. The best accounts of the post-Fourneyron history of the water turbine are Louis Hunter, "Les origines de turbines Francis et Pelton: Développement de la turbine hydraulique au États Unis de 1820 à 1900," *Revue d'histoire des sciences et leur applications* 17 (1964): 209–42; Hunter, *Waterpower*, pp. 292–415; Smith, *Man and Water*, pp. 176–99; Layton, "Millwrights and Engineers" and "Scientific Technology"; and Arthur T. Safford and Edward P. Hamilton, "The American Mixed-Flow Turbine and Its Setting," American Society of Civil Engineers, *Transactions* 85 (1922): 1237–1356.

72. Livermore, "Remarks upon Pressure Engines," p. 166. Hunter, *Waterpower*, pp. 294–98, strongly emphasizes the turbine's adaptability to a wide range of falls and to varying falls as its chief advantages over the vertical water wheel.

73. A number of nineteenth-century writers on water power review the advantages of the turbine over the vertical water wheel. Among them are Morris, "Remarks," pp. 298–304; Grimshaw, *Miller, Millwright, and Millfurnisher*, pp. 83–86; Redtenbacher, *Wasserräder*, pp. 309–13; C. F. Tomkins, "The Overshot Wheel vs. the Turbine," *American Miller* 13 (1885): 513; and J. Humphrey, "Turbine and Overshot," ibid. 22 (1894): 598–99. See also the discussion by Hunter, *Waterpower*, pp. 293–98.

74. H. H. Suplee in comments appended to the paper of Samuel Webber, "Water-Power: Its Generation and Transmission," American Society of Mechanical Engineers, *Transactions* 17 (1895–96): 54.

75. John Turnbull, "On Water Wheels and Turbines," Institution of Engineers and Shipbuilders of Scotland (Glasgow), *Transactions* 26 (1883): 81–82.

76. Francis, *Lowell Hydraulic Experiments*, pp. 1–2, emphasized the low cost of turbines as the most important factor behind their adoption in America, where open streams were

plentiful but capital scarce. In European countries, where capital was more abundant but water more scarce, the turbine's high efficiency and ability to handle varying falls were probably more critical to its acceptance.

77. "Water Motors," *Encyclopaedia Britannica*, 10th ed., 33 (1902): 767n.

78. A. C. Rice, "Notes on the History of Turbine Development in America," *Engineering News* 48 (1902): 210.

79. Daniel W. Mead, *Water Power Engineering*, 2nd ed. (New York, 1915), p. 668. Mead also noted (p. 669) that, other things being equal, the cost per horsepower of developing a stream varied inversely as the head. The turbine's ability to handle high heads would thus give it a cost advantage here also.

80. Francis, *Lowell Hydraulic Experiments*, p. 1.

81. Hunter, *Waterpower*, p. 503.

82. Robert H. Thurston, "The Systematic Testing of Turbine Water-Wheels in the United States," American Society of Mechanical Engineers, *Transactions* 8 (1886–87): 382.

83. Ibid., p. 361.

84. W. C. Unwin, "Water Motors," *Van Nostrand's Engineering Magazine* 34 (1886): 3.

85. William Carter Hughes, *The American Miller*, new ed. (London and Philadelphia, 1873); Grimshaw, *Miller, Millwright, and Millfurnisher*, esp. pp. 83–85; and Abernathey, *Mill Building*.

86. Abernathey, *Mill Building*, p. 93.

87. Wilson, "Early Water Turbines," p. 219. William Morshead, "On the relative Advantages of Steam, Water, and Animal Power," Bath and West and Southern Counties Society, *Journal*, n.s. 4 (1856): 35, noted with respect to the turbine: "This machine is frequently employed in France... , but is not yet much known in this country."

88. Joseph Addison and Rex Wailes, "Dorset Watermills," Newcomen Society, *Transactions* 35 (1962–63): 193–216, esp. 194; and Addison and Wailes, "Dorset Watermills: Addendum," ibid. 36 (1963–64): 175–81, esp. 177.

89. The advantages that overshot water wheels retained over turbines are outlined by Weidner, "Test of an Overshot Water Wheel," esp. pp. 124–28. See also Fitz Water Wheel Company, *Fitz Steel Overshoot Water Wheels*, pp. 5–6, 15–16, 24–28.

90. "Nouvelles machines de Marly établies à Bougival sous la direction de M. Dufrayer, ingénieur... ," *Publication industrielle* 14 (1861): 246–61, esp. 252–53.

91. "Some Modern Breast Water Wheels," *Engineering News* 48 (1902): 346.

# Bibliography

ABBREVIATIONS

AS–H   Académie des Sciences, Paris, *Histoire*
AS–M   Académie des Sciences, Paris, *Mémoires*

Abernathey, R. James. *Practical Hints on Mill Building.* Moline, Ill., and London, 1880.

Académie des Sciences, Paris. *Histoire . . . avec les mémoires.* Paris, 1699–1792.

_____. *Machines et inventions approuvées par l'Académie royale des Sciences, depuis son établissement jusqu'à present. . . .* Ed. M. Gallon. 3 vols. Paris, 1776–78.

_____. *Recueil des mémoires de l'Académie royale des Sciences depuis 1666 jusqu'à 1699.* 11 vols. Paris, 1729–34.

Acres, R. d'. *The Art of Water-Drawing. . . .* London, 1660.

Adams, Kenneth. *A Guide to the Industrial Archaeology of Europe.* Bath, 1971.

Addison, Joseph, and Rex Wailes. "Dorset Watermills." Newcomen Society, *Transactions* 35 (1962–63): 193–216.

_____. "Dorset Watermills: Addendum." Newcomen Society, *Transactions* 36 (1963–64): 175–81.

Aebischer, Paul. "Les dénominations du 'moulin' dans les chartes italiennes du Moyen Age." *Archivvm Latinitatis Medii Aevi (Bulletin Du Cange)* 7 (1932): 49–109.

Agricola, Georgius. *De re metallica.* Translated from the first Latin edition of 1556 by Herbert Clark Hoover and Lou Henry Hoover. New York, 1950.

Aguila, A. Del. "Unas presas antiguas españolas de contrafuertes." *Las Ciencias* 14 (1949): 185–202.

Aitchison, Leslie. *A History of Metals.* 2 vols. London, 1960.

Alembert, Jean Lerond d'. "Aube." In *Encyclopédie; ou, Dictionnaire raisonné des sciences, des arts, et des métiers,* ed. Denis Diderot and Jean d'Alembert, 1:863–65. Paris, 1751.

_____. *Traité de l'équilibre et du mouvement des fluides. . . .* Paris, 1744.

Allen, Zachariah. *The Practical Tourist; or, Sketches of the State of the Useful Arts. . . .* Vol. 1. Providence, R.I., and Boston, 1832.

_____. *The Science of Mechanics, as Applied to the Present Improvements in the Useful Arts in Europe, and in the United States of America. . . .* Providence, R.I., 1829.

Allison, Archibald. "Water Power as the Foundation of Sheffield's Industries." Newcomen Society, *Transactions* 27 (1949–51): 221–24.

_____. "The Water Wheels of Sheffield." *Engineering* 165 (1948): 165–68.

Allison, Robert. "The Old and the New [in Millwrighting]." American Society of Mechanical Engineers, *Transactions* 16 (1894): 742–61.

Allodi, Leone, and G. Levi, eds. *Il registo Sublacense de secolo XI*. Rome, 1885.

Amira, Karl von, ed. *Die Dresdener Bilderhandschrift des Sachsenspiegels* [c1350]. Leipzig, 1902.

Amontons, Guillaume. "Moyen de substituer commodement l'action du feu, à la force des hommes et des chevaux pour mouvoir les machines." *AS–M,* 1699 (3rd ed., 1732), pp. 112–26 (*AS–H,* pp. 101–3).

*Ancient Laws of Ireland.* Vols. 1–3. Dublin and London, 1865–73.

Anderson, William. *Sketch of the Mode of Manufacturing Gunpowder at the Ishapore Mills in Bengal....* With notes and additions by Lieut.-Col. Parlley. London, 1862.

Antipater of Thessalonica [?]. "On a Water-mill." In *Greek Anthology,* transl. W. R. Paton, 3: 232–33. London, 1917.

Aquensis, Albertus. *Incipit Historia Hierosolymitanae Expeditionis.* In J. P. Migne, ed., *Patrologia Latina,* vol. 166, cols. 389–716. Paris, 1854.

Armengaud, [Jacques Eugéne]. *Traité théorique et pratique des moteurs hydrauliques....* New ed. 1 vol. text plus 1 vol. plates. Paris, 1858.

Arnold, John P., and Frank Penman. *History of the Brewing Industry and Brewing Science in America.* Chicago, 1933.

Ashmore, Owen. *Industrial Archaeology of Lancashire.* Newton Abbot, 1969.

Atwood, George. *A Treatise on the Rectilinear Motion and Rotation of Bodies....* Cambridge, 1784.

Aubuisson de Voisins, J. F. d'. *Traité d'hydraulique, à l'usage des ingénieurs.* 2nd ed. Paris and Strasbourg, 1840.

_____. *A Treatise on Hydraulics, for the use of Engineers.* Transl. Joseph Bennett. Boston, 1852.

[Ausonius?]. "Mosella." In *Ausonius,* transl. Hugh G. Evelyn White, pp. 224–67. London, 1919.

Avitsur, Shemuel. "Watermills in Eretz Israel, and Their Contribution to Water Power Technology." International Symposium on Molinology, *Transactions* 2 (1969): 389–407.

_____. "Water Power in Traditional Sugar and Olive Oil Production in the Land of Israel." International Symposium on Molinology, *Transactions* 3 (1973): 175–83.

Bach, C. *Die Wasserräder.* 1 vol. text plus 1 vol. plates. Stuttgart, 1886.

Baines, Edward. *History of the Cotton Manufacture in Great Britain.* London, c1835.

Baines, Thomas, and William Fairbairn. *Lancashire and Cheshire, Past and Present... with an Account of the Rise and Progress of Manufactures and Commerce, and Civil and Mechanical Engineering in These Districts.* 2 vols. London, 1868.

Baker, William. "Description of a Water-Pressure Engine at Ilsang, in Bavaria." Institution of Civil Engineers, *Minutes of Proceedings* 2 (1842): 55.

_____. "Description of the Water-Pressure Engine at Freyberg, Saxony." Institution of Civil Engineers, *Minutes of Proceedings* 2 (1843): 143–44.

Banks, John. *A Treatise on Mills, in Four Parts....* London, 1795.

Barlow, Peter. "Hydrodynamics." *Encyclopaedia Metropolitana; or, Universal Dictionary of Knowledge...* , 3:161–296. London, 1845.

Baroja, Julio Caro. "Norias, azudas, aceñas." *Revista de dialectologia y tradicion, populares* 10 (1954): 29–160.

_____. "Sobre la historia de la noria de tiro." *Revista de dialectologia y tradiciones populares* 11 (1955): 15–79.

Barton, D[enys] B. *Essays in Cornish Mining History.* Vol. 1. Truro, Cornwall, 1968.

Bassert, Helmuth T., and Willy F. Storck, eds. *Das Mittelalterlich Hausbuch* [c1475]. Leipzig, 1912.

Bate, John. *Mysteries of Nature and Art.* 2nd ed. London, 1635.

Bateman, John F. "Description of the Bann Reservoirs, County Down, Ireland." Institution of Civil Engineers, *Minutes of Proceedings* 1 (1841): 168–70.

Bautier, Anne-Marie. "Les plus anciennes mentions de moulins hydrauliques industriels et de moulins à vent." *Bulletin philologique et historique* 2 (1960): 567–626.

Beck, Theodor. *Beiträge zur Geschichte des Maschinenbaues.* Berlin, 1900.

_____. "Philon von Byzanz (etwa 260–200 v. Chr.)." *Beiträge zur Geschichte der Technik und Industrie* 2 (1910): 64–77.

Beckmann, John. *A History of Inventions and Discoveries.* Transl. from the German by William Johnston. Vol. 1. London, 1817.

Beguillet, [Edme]. *Manuel du meunier et du charpentier des moulins....* Paris, 1775.

Beighton, Henry. "A Description of the Water-Works at London-Bridge, explaining the Draught of Tab. I." Royal Society of London, *Philosophical Transactions* 37 (1731–32): 5–12.

Bélidor, Bernard Forest de. *Architecture hydraulique; ou, L'art de conduire, d'élever, et de ménager les eaux pour les différens besoins de la vie....* Vols. 1–2. Paris, 1737–39.

_____. *Architecture hydraulique; ou, L'art de conduire, d'élever, et de ménager les eaux pour les différens besoins de la vie....* Ed. with notes by C. L. M. H. Navier. New ed. Vol. 1. Paris, 1819.

Benedict, Saint. *The Rule of Saint Benedict.* Ed. and transl. Justin McCann. Westminster, Md., 1952.

_____. *S. P. Benedicti Regula, cum commentariis.* In J. P. Migne, ed., *Patrologia Latina,* vol. 66, cols. 215–932. Paris, 1847.

Bennett, C. E. "The Watermills of Kent, East of the Medway." *Industrial Archaeology Review* 1 (1977): 205–35.

Bennett, Richard, and John Elton. *History of Corn Milling.* 4 vols. London, 1898–1904.

Benoit, Fernand. "'Usine' de meunerie hydraulique à l'époque Romaine." *Annales d'histoire sociale* 1 (1939): 183–84.

_____. "L'usine de meunerie hydraulique de Barbegal (Arles)." *Revue archéologique,* 6th ser. 15 (1940): 19–80.

Berman, Constance. "The Cistercians in the County of Toulouse, 1132–1249: The Order's Foundations and Land Acquisition. Ph.D. dissertation, University of Wisconsin, 1978.

_____. "Silvanès: A Study of the Economic Innovations of a Cistercian Monastery in Southern France." M.A. thesis, University of Wisconsin, 1972.

Bernoulli, Daniel. *Hydrodynamics* (and Johann Bernoulli, *Hydraulics*). Translated from the Latin by Thomas Carmody and Helmut Kobus with a preface by Hunter Rouse. New York, 1968.

Besson, Jacques. *Theatre des instrvmens mathématiques et méchaniques.* Lyons, 1579.

Beyer, Heinrich, ed. *Urkundebuch zur Geschichte der jetzt die Preussischen Regierungsbezirke Coblenz und Trier bildenden mittelrheinischen Territorien.* Vol. 1. Coblenz, 1860.

Beyer, Johann Matthias. *Theatrum machinarum molarium...* [1735]. Dresden, 1767.

Biringuccio, Vannoccio. *Pirotechnia* [1540]. Translated from the Italian by Cyril Stanley Smith and Martha Teach Gnudi. Cambridge, Mass., 1966.

Blaine, Bradford. "The Application of Water-Power to Industry during the Middle Ages." Ph.D. dissertation, University of California, Los Angeles, 1966.

————. "The Enigmatic Water-Mill." In Bert S. Hall and Delno C. West, eds., *On Pre-Modern Technology and Science: A Volume of Studies in Honor of Lynn White, Jr.*, pp. 163–76. Malibu, Calif., 1976.

Blake, Francis. "The Greatest Effect of Engines with uniformly accelerated Motions considered." Royal Society of London, *Philosophical Transactions* 51 (1759): 1–6.

Blavot-Duchesne, ————, and [Jean F. d'] Aubuisson. "Expériences sur les machines hydrauliques des mines de Poullaouen...." *Journal des mines* 21 (1807): 161–248.

Bloch, Marc. "Avènement et conquêtes du moulin à eau." *Annales d'histoire économique et sociale* 7 (1935): 538–63. Translated as "The Advent and Triumph of the Watermill," in Bloch, *Land and Work in Mediaeval Europe* (see below), pp. 136–68.

————. *Land and Work in Mediaeval Europe: Selected Papers by Marc Bloch.* Transl. J. E. Anderson. Berkeley and Los Angeles, 1967.

Block, Maurice. *Statistique de la France.* Vol. 2. Paris, 1875.

Blum, Andre. *On the Origin of Paper.* Transl. Harry M. Lydenberg. New York, 1934.

Blümner, Hugo. *Technologie und Terminologie der Gewerbe und Künste bei Griechen und Römern.* 2nd ed. Vol. 1. Leipzig and Berlin, 1912.

Böckler, Georg Andreas. *Theatrum machinarum novus; das ist: Neu-vermehrter Schauplatz der mechanischen künsten, handelt von aller hand- wasser- wind- rossgewicht- und hand-mühlen.* Nuremberg, 1661.

Boislisle, A. M. de, *Correspondance des Contrôleurs généraux des Finances avec les Intendants des Provinces.* Vol. 1 [1683–99]. Paris, 1874.

Boissonnarde, Prosper. *Essai sur l'orginisation du travail en Poitou depuis le XI$^e$ siècle jusqu'à la révolution.* Vol. 1. Paris, 1900.

Bolton, Reginald. *Motive Powers and their Practical Selection.* New York, 1895.

Booker, Frank. *The Industrial Archaeology of the Tamar Valley.* Newton Abbot, 1967.

Boon, George, and Colin Williams. "The Dolaucothi Drainage Wheel." *Journal of Roman Studies* 56 (1966): 122–27.

Borda, Jean Charles. "Mémoire sur l'écoulement des fluides par les orifices des vases." *AS–M*, 1766 (publ. 1769), pp. 579–607 (*AS–H*, pp. 143–50).

————. "Mémoire sur les roues hydrauliques." *AS–M*, 1767 (publ. 1770), pp. 270–87 (*AS–H*, pp. 149–53).

Bornemann, C. R. "On a New Water-Pressure Pumping Engine at Clausthal." Institution of Civil Engineers, *Minutes of Proceedings* 45 (1876): 347–48.

Bossut, Charles. "Détermination générale de l'effet des roues mues par le choc de l'eau." *AS–M*, 1769 (publ. 1772), pp. 477–97 (*AS–H*, pp. 121–24).

————. *Traité élémentaire d'hydrodynamique.* 2 vols. Paris, 1771.

_____. *Traité théorique et expérimental d'hydrodynamique*. 3rd ed. 2 vols. Paris, an IV [1796].

Bossut, Charles, and _____ Vaillet. *Unterschungen über die beste construction der Deiche.* ... Transl. C. Kröncke. Frankfurt am Main, 1798.

Bossut, Charles, [Jean Lerond] d'Alembert, and [Marie-Jean de] Condorcet. *Nouvelles expériences sur la résistance des fluides*. Paris, 1777.

Boucher, Cyril T. G. "John Rennie (1761–1821)." Newcomen Society, *Transactions* 34 (1961–62): 1–13.

_____. *John Rennie, 1761–1821: The Life and Work of a Great Engineer*. Manchester, 1963.

Bowden, Witt, Michael Karpovich, and Abbot P. Usher. *An Economic History of Europe since 1750*. New York, 1937.

Bowman, Gerald. "John Smeaton—Consulting Engineer." In E. G. Semler, ed., *Engineering Heritage: Highlights from the History of Mechanical Engineering*, 2:8–12. New York and London, 1966.

Boyer, Marjorie. "Bridges and Mill Sites in Medieval France." XIIᵉ Congrés international d'histoire des sciences, Paris, 1966, *Actes* 10*B*:13–17.

_____. *Medieval French Bridges: A History*. Cambridge, Mass., 1976.

Braithwaite, Frederick. "On the Rise and Fall of the River Wandle; Its Springs, Tributaries, and Pollution." Institution of Civil Engineers, *Minutes of Proceedings* 20 (1860–61): 191–258.

Branca, Giovanni. *La Machine*. Rome, 1629.

Braudel, Fernand. *Capitalism and Material Life, 1400–1800*. Transl. Miriam Kochan. London, 1973.

Brett, G. "Byzantine Water Mill." *Antiquity* 13 (1939): 354–56.

[Brewster, David.] "Hydrodynamics." *Edinburgh Encyclopaedia*, 10:751–909. 1st American ed. Philadelphia, 1832.

Britkin, A. S. *The Craftsmen of Tula: Pioneer Builders of Water-Driven Machinery*. Translated from the Russian. Jerusalem, 1967.

Britkin, A. S., and S. S. Vidonov. *A. K. Nartov: An Outstanding Machine Builder of the Eighteenth Century*. Translated from the Russian. Jerusalem, 1964.

Brittain, Robert. *Rivers, Man, and Myths: From Fish Spears to Water Mills*. Garden City, N.Y., 1958.

Bruce, J. Collingwood. *Handbook to the Roman Wall*. Ed. Ian A. Richmond. 11th ed. Newcastle-upon-Tyne, 1957.

Brunton, Robert. *A Compendium of Mechanics; or, Text Book for Engineers, Mill-Wrights, Machine-Makers, Founders, Smiths, &c.* ... Glasgow, 1828.

Buchanan, Robertson. "On the Velocity of Water Wheels." *Philosophical Magazine*, 1st ser. 10 (1801): 278–81.

_____. *Practical Essays on Mill Work and Other Machinery* [1814]. With notes and additional articles by Thomas Tredgold, revised into a third edition with additions by George Rennie. London, 1841.

Buckler, W. H., and D. M. Robinson. "Greek and Latin Inscriptions." *Sardis: Publications of the American Society for the Excavation of Sardis* 7, pt. 1 (Leiden, 1932): 139, no. 169.

Cabanes, Charles. "Joseph de Montgolfier et le Bélier Hydraulique." Newcomen Society, *Transactions* 17 (1936–37): 85–90.

Caligny, [Antoine de]. "Note sur l'effet utile d'une roue de côté à palettes plongeantes, selon le système de MM. Coriolis et Bellanger." Académie des Sciences, Paris, *Comptes rendus* 21 (1845): 249–53.

_____. "Sur l'origine de la roue dite de côté." Académie des Sciences, Paris, *Comptes rendus* 21 (1845): 16–18.

*Cambridge Economic History of Europe.* Ed. J. H. Clapham, Eileen Power, et al. 7 vols. Cambridge, 1941–78.

Camden, William. *Britannia.* Transl. Edmund Gibson. 2nd ed. Vol. 1. London, 1722.

Cardwell, D. S. L. "The Academic Study of the History of Technology." *History of Science* 7 (1968): 112–24.

_____. *From Watt to Clausius: The Rise of Thermodynamics in the Early Industrial Age.* London, 1971.

_____. "Power Technologies and the Advance of Science, 1700–1825." *Technology and Culture* 6 (1965): 188–207.

_____. "Some Factors in the Early Development of the Concepts of Power, Work, and Energy." *British Journal for the History of Science* 3 (1966–67): 209–24.

_____. *Turning Points in Western Technology.* New York, 1972.

Carmichael, C. "On Water as a Moving Power for Machinery." *Edinburgh Philosophical Journal* 13 (1825): 346–48.

[Carnot, L. N. M.] *Essai sur les machines en général, par un officier au Corps royal du Génie.* Dijon, 1782.

Carnot, L. N. M. *Oeuvres mathématiques du Citoyen Carnot.* Basle, 1797.

_____. *Principes fondamentaux de l'équilibre et du mouvement.* Paris, 1803.

Carus-Wilson, E. M. "An Industrial Revolution of the Thirteenth Century." *Economic History Review* 11 (1941): 39–60.

Cassiodorus. *De institutione divinarum Litterarum.* In J. P. Migne, ed., *Patrologia Latina,* vol. 70, cols. 1105–50. Paris, 1847.

_____. *Variarum libri duodecim.* In J. P. Migne, ed., *Patrologia Latina,* vol. 69, cols. 501–880. Paris, 1848.

Cassius Dio Cocceianus. *Dio's Roman History.* Transl. Earnest Cary. Vol. 7. London and New York, 1924.

Caus, Isaak de. *New and Rare Inventions of Water-Works....* Transl. John Leak. London, 1659.

Cedrenus, Georgius. *Compendium Historiarum.* In J. P. Migne, ed., *Patrologiae Graeca,* vol. 121. Paris, 1894.

Chaloner, W. H., and A. E. Musson. *Industry and Technology.* London, 1963.

Chapman, Stanley D. "The Cost of Power in the Industrial Revolution in Britain: The Case of the Textile Industry." *Midland History* 1, no. 2 (Autumn 1971): 1–23.

Christian, [Gérard J.]. *Traité de mécanique industrielle....* Vol. 1. Paris, 1822.

Cipolla, Carlo. *Before the Industrial Revolution: European Society and Economy, 1000–1700.* 2nd ed. New York, 1980.

_____. *The Economic History of World Population.* Sussex and New York, 1978.

_____, ed. *The Fontana Economic History of Europe.* 6 vols. London and Glasgow, 1972–76.

Coggin, F. G. "The Ferris and Other Big Wheels." *Cassier's Magazine* 6 (1894): 215–22.

Coleman, D. C. *The British Paper Industry, 1495–1860.* Oxford, 1958.

Colin, G. S. "La noria morocaine et les machines hydraulique dans le monde arabe." *Hesperis* 14 (1932): 22–60.

_____. "L'origine des norias de Fes." *Hesperis* 16 (1933): 156–57.

Columella, Lucius Junius Moderatus. *On Agriculture.* Transl. and ed. Harrison Boyd Ash. Vol. 1. London and Cambridge, Mass., 1941.

"The Condition and Prospects of American Cotton Manufactures in 1849." *The Merchants' Magazine* 22 (1850): 26–35.

Cook, Earl. *Man, Energy, Society.* San Francisco, 1976.

Cook, Walter W. S., and Jose Gudiol Ricard. *Pintura e imagineria romanicas.* Vol. 6 of *Ars Hispaniae.* ... Madrid, 1950.

Coriolis, Gaspard G. *Du calcul de l'effet des machines; ou, Considerations sur l'emploi des moteurs et sur leur évaluation.* ... Paris, 1829.

Cottrell, Fred. *Energy and Society: The Relation between Energy, Social Change, and Economic Development.* Westport, Conn., 1970.

Craik, David. *The Practical American Millwright and Miller.* ... Philadelphia, 1882.

Crozet-Fourneyron, Marcel. *Invention de la turbine.* Paris and Liège, 1924.

Curwen, E. Cecil. "The Problem of Early Water-mills." *Antiquity* 18 (1944): 130–46.

Danilevskii, V. V. *History of Hydroengineering in Russia before the Nineteenth Century.* Translated from the Russian. Jerusalem, 1968.

Daumas, Maurice, ed. *A History of Technology and Invention.* Transl. Eileen B. Hennessy. 3 vols. New York, 1969–79.

Deerr, Noel. *The History of Sugar.* 2 vols. London, 1949–50.

Dembińska, Maria. *Przetwórstwo Zbożowe w Polsce Średniowiecznej (X–XIV Wiek).* Wroclaw, Warsaw, Crakow, and Gdansk, 1973.

De Parcieux, Antoine. "Mémoire dans lequel on démontre que l'eau d'une chûte destinée à faire mouvoir quelque machine, moulin ou autre, peut toûjours produire beaucoup plus d'effet en agissant par son poids qu'en agissant par son choc, & que les roues à pots qui tournent lentement, produisent plus d'effet que celles qui tournent vîte, relativement aux chûtes & aux dépenses." *AS–M,* 1754 (publ. 1759), pp. 603–14 (*AS–H,* pp. 134–38).

———. "Mémoire dans lequel on prouve que les aubes des roues mûes par les courans des grandes rivières, feroient beaucoup plus d'effet si elles étoient inclinées aux rayons, qu'elles ne font étant appliquées contre les rayons mêmes, comme elles le sont aux moulins pendans & aux moulins sur bateaux qui sont sur les rivières de Seine, Marne, Loire, &c." *AS–M,* 1759 (publ. 1765), pp. 288–99 (*AS–H,* pp. 223–27).

———. "Mémoire sur une expérience qui montre qu'a dépense égale, plus une roue à augets tourne lentement, plus elle fait d'effet." *AS–M,* 1754 (publ. 1759), pp. 671–78 (*AS–H,* pp. 134–38).

Desaguliers, John T. *A Course of Experimental Philosophy.* 2 vols. London, 1734–44.

"Descriptio positionis seu situationis monasterii Clarae-Vallensis." In J. P. Migne, ed., *Patrologia Latina,* vol. 185, cols. 569–74. Paris, 1855.

Dewar, H. S. L. "The Windmills, Watermills, and Horse-Mills of Dorset." Dorset Natural History and Archaeological Society, *Proceedings* 83 (1960): 109–32.

Dibner, Bern. *Agricola on Metals.* Norwalk, Conn., 1958.

———. *Moving the Obelisks.* Cambridge, Mass., and London, 1970.

Dickinson, H. W. "Early Years of the Hydraulic Ram." Newcomen Society, *Transactions* 17 (1936–37): 73–83.

———. "The Shetland Watermill." Newcomen Society, *Transactions* 12 (1932–33): 89–94.

———. *A Short History of the Steam Engine.* Cambridge, 1938.

Dickinson, H. W., and Arthur Titley. *Richard Trevithick: The Engineer and the Man.* Cambridge, 1934.

Diderot, Denis. *A Diderot Pictorial Encyclopedia of Trades and Industry.* . . . Ed. Charles C. Gillispie. 2 vols. New York, 1959.

Didion, _____. "Notice historique sur les roues hydrauliques à aubes courbes." In Jean Victor Poncelet, *Cours de mécanique appliqué aux machines,* pp. 322–29. Paris, 1876.

Diocletian. *Edictum Diocletiani de pretiis rerum venalium.* Ed. Theodor Mommsen. Berlin, 1893; reprint ed., Berlin, 1958.

Dircks, H. *A Biographical Memoir of Samuel Hartlib, Milton's Familiar Friend . . . and a Reprint of his Pamphlet, entitled "An Invention of Engines of Motion"* [1651]. London, 1865.

Doehaerd, Renée. *The Early Middle Ages in the West: Economy and Society.* Transl. W. G. Deakin. Amsterdam, N.Y., and Oxford, 1978.

Donkin, R. A. *The Cistercians: Studies in the Geography of Medieval England and Wales.* Toronto, 1978.

Donkin, Sydney B. "Bryan Donkin, F.R.S., M.I.C.E. 1768–1855." Newcomen Society, *Transactions* 27 (1949–51): 85–95.

Donnachie, Ian. *The Industrial Archaeology of Galloway.* Newton Abbot, 1971.

Downs-Rose, G., and W. S. Harvey. "Water-Bucket Pumps and the Wanlockhead Engine." *Industrial Archaeology* 10 (1973): 129–47.

Drachmann, A. G. *Ktesibios, Philon, and Heron: A Study in Ancient Pneumatics.* Copenhagen, 1948.

DuBuat, Pierre L. G. *Principes d'hydraulique, vérifiés par un grand nombre d'expériences faites par ordre du Gouvernement.* . . . New ed. 2 vols. Paris, 1786.

Duby, Georges. *The Early Growth of the European Economy.* Transl. Howard B. Clarke. London, 1974.

DuCrest, C[harles] L. *Essais sur les machines hydrauliques.* . . . Paris, 1777.

_____. *Vues nouvelles sur les courans d'eau, la navigation intérieure, et la marine.* . . . Paris, an XII [1803].

Duffy, Joseph W. *Power: Prime Mover of Technology.* Bloomington, Ill., 1964.

Dugas, René. *A History of Mechanics.* Transl. J. R. Maddox. London, 1957.

_____. *Mechanics in the Seventeenth Century.* Transl. Freda Jacquot. Neuchatel, Switzerland, 1958.

Dunham, Arthur L. *The Industrial Revolution in France, 1815–1848.* New York, 1955.

DuPetit-Vandin, Robert-Xavier A. "Mémoire sur l'hydraulique." *Mémoires de mathématique et de physique présentés à l'Académie Royale des Sciences, par divers Sçavans* . . . 1 (1750): 261–82. The paper was read in 1746.

Dupin, Charles. *Forces productives et commerciales de la France.* 2 vols. Brussels, 1828.

Elvius, Pehr. "Auszug aus einem Buche, das die königl. schwed. Akademie der Wissensch. hat drucken lassen, unter dem Titel: Mechanik, oder mathematische Abhandlung von Wasserwerken. . . ." In Königl. Schwedischen Akademie der Wissenschaften, *Abhandlungen,* 1742, transl. and ed. Abraham Gotthelf Kästner, 4:92–100. Hamburg, 1750.

_____. *Mathematisk traktat om effecter af vatndrifter.* . . . Stockholm, 1742.

*Encyclopédie; ou, Dictionaire raisonné des sciences.* Ed. Denis Diderot and Jean Lerond d'Alembert. 17 vols. Paris, 1751–65. *Supplement,* 4 vols., Amsterdam,

1776–77. *Recueil de planches,* 11 vols., Paris, 1762–72. *Suite du Recueil des planches,* 1 vol., Paris, 1777.

*Encyclopédie méthodique: Arts et métiers mécaniques.* 8 vols. Paris and Liège, 1782–91. *Recueil de planches.* 8 vols., 1783–90.

English, W. *The Textile Industry: An Account of the Early Inventions of Spinning, Weaving, and Knitting Machines.* London and Harlow, 1969.

Ergang, Carl. "Die Maschine von Marly." *Beiträge zur Geschichte der Technik und Industrie* 3 (1911): 131–46.

Euler, Johann A. *Enodatio Qvaestionis: Qvomodo Vis Aqvae Alivsve Flvidi cvm Maximo Lvcro Ad Molas Circvm Agendas Aliave Opera Perficienda Impendi Possit?* Göttingen, 1754.

———. "Quelle est la manière la plus avantageuse d'employer la force de l'eau ou de tout autre fluide, soit pour faire tourner les moulins, soit pour produire un autre effet quelconque. Piéce qui a remporté la prix proposé par la société Royale des Goëttingue, en 1754." *Journal étranger,* December 1756, secs. 193–201, pp. 723–25.

Euler, Leonhard. "Application de la machine hydraulique de M. Segner à toutes sortes d'ouvrages et de ses avantages sur les autres machines hydrauliques dont on se sert ordinairement." *Mémoires de l'Académie des Sciences de Berlin* 7 (1751; publ. 1753): 271–304. Reprinted in Euler, *Opera Omnia,* 2nd ser. 15:105–33.

———. "De motu et reactione aquae per tubos mobiles transfluentis." *Novi commentarii academiae scientiarum Petropolitanae* 6 (1756–57; publ. 1761): 312–37. Reprinted in Euler, *Opera Omnia,* 2nd ser. 15: 80–104.

———. "Détermination de l'effet d'une machine hydraulique inventée par M. Segner, professeur à Goettingue." *Opera postuma,* 2 (1862): 146–73. Reprinted in Euler, *Opera Omnia,* 2nd ser. 15:40–79.

———. "Discussion plus particulière de diverses manières d'élever de l'eau par le moyen des pompes avec le plus grand avantage." *Mémoires de l'Académie des Sciences de Berlin* 8 (1752; publ. 1754): 149–84. Reprinted in Euler, *Opera Omnia,* 2nd ser. 15:251–80.

———. "Maximes pour arranger le plus avantageusement les machines destinées à élever de l'eau par le moyen des pompes." *Mémoires de l'Académie des Sciences de Berlin* 8 (1752; publ. 1754):185–232. Reprinted in Euler, *Opera Omnia,* 2nd ser. 15:281–318.

———. *Opera Omnia.* 2nd ser. Vol. 15. Ed. Jakob Ackeret. Lausanne, 1957.

———. "Recherches sur l'effet d'une machine hydraulique proposée par M. Segner, professeur à Goettingue." *Mémoires de l'Académie des Sciences de Berlin* 6 (1750; publ. 1752): 311–54. Reprinted in Euler, *Opera Omnia,* 2nd ser. 15:1–39.

———. "Recherche sur une nouvelle manière d'élever de l'eau proposée par M. de Mour." *Mémoires de l'Académie des Sciences de Berlin* 7 (1751; publ. 1753): 305–30. Reprinted in Euler, *Opera Omnia,* 2nd ser. 15:134–56.

———. "Sur le mouvement de l'eau par des tuyaux de conduite." *Mémoires de l'Académie des Sciences de Berlin* 8 (1752; publ. 1754): 111–48. Reprinted in Euler, *Opera Omnia,* 2nd ser. 15:219–50.

———. "Théorie plus complète des machines qui sont mises en mouvement par la réaction de l'eau." *Mémoires de l'Académie des Sciences de Berlin* 10 (1754; publ. 1756): 227–95. Reprinted in Euler, *Opera Omnia,* 2nd ser. 15:157–218.

Evans, Oliver. *The Young Mill-Wright and Miller's Guide*. . . . Philadelphia, 1795.

Ewart, Peter. "On the Measure of Moving Force." Literary and Philosophical Society of Manchester, *Memoirs*, 2nd ser. 2 (1813): 105–258. The paper was read in November 1808.

Ewbank, Thomas. *A Descriptive and Historical Account of Hydraulic and Other Machines for Raising Water*. . . . 12th ed. New York, 1851.

Fabre, [Jean-Antoine]. *Essai sur la manière la plus avantageuse de construire les machines hydrauliques, et en particulier les moulins à bled*. . . . Paris, 1783.

Fahy, Edward M. "A Horizontal Mill at Mashanaglass, County Cork." Cork Historical and Archaeological Society, *Journal* 61 (1956): 13–57.

Fairbairn, William. *The Life of Sir William Fairbairn*. . . . Partly written by himself, completed and edited by William Pole. London, 1877.

―――. "On Water-Wheels with Ventilated Buckets." *Quarterly Papers on Engineering*. Vol. 6. London, 1849. 20 pages.

―――. *Treatise on Mills and Millwork*. . . . 3rd ed. 2 vols. London, 1871. The first edition was published in 1863–64.

Farey, John. *A Treatise on the Steam Engine, Historical, Practical, and Descriptive*. London, 1827.

Feldhaus, F. M. "Ahnen des Wasserrades." *Die Umschau* 40 (1936): 472–73, 476.

―――. *Die Technik der Antike und des Mittelalters*. New York, 1971.

―――. *Die Technik der Vorzeit der geschichtlichen Zeit und der Naturvölker*. Munich, 1965.

Fenwick, Thomas. *Essays on Practical Mechanics*. 3rd ed. Durham, 1822.

Ferguson, James. *Analysis of a Course of Lectures on Mechanics*. . . . London, 1761.

―――. *Lectures on Select Subjects*. . . . London, 1764.

―――. *The Works of James Ferguson*. Ed. David Brewster. Vol. 3. Edinburgh, 1823.

Ferrendier, M. "Les anciennes utilisations de l'eau." *La houille blanche* 3 (1948): 325–34, 497–508; 4 (1949): 121–33; 5 (1950): 769–87.

Figuier, Louis. *Les merveilles de l'industrie*. . . . Vol. 3. Paris, n.d.

Filarete, [Antonio di]. *Treatise on Architecture*. Transl. John R. Spencer. 2 vols. New Haven, Conn., and London, 1965.

Finley, M. I. "Technical Innovation and Economic Progress in the Ancient World." *Economic History Review*, 2nd ser. 18 (1965): 29–45.

First Iron Works Association. *The First Iron Works Restoration*. New York, 1951.

Fitz Water Wheel Company. *Fitz Steel Overshoot Water Wheels*. Bulletin no. 70. Hanover, Pa., 1928.

Fitzherbert, Anthony. *Surveying*. London, 1539. In [Robert Vansittart, ed.], *Certain Ancient Tracts concerning the Management of Landed Property Reprinted*, London, 1767.

Fludd, Robert. *Tractatus secundus: De natvrae simia seu Technica macrocosmi historis*. . . . 2nd ed. Frankfurt, 1624.

*The Fontana Economic History of Europe*. Ed. Carlo Cippola. 6 vols. London and Glasgow, 1972–76.

Fontanon, Antoine, ed. *Les edicts et ordonnances des roys de France*. . . . 2nd ed. Vol. 2. Paris, 1585.

Forbes, R. J. "Food and Drink." In Charles Singer et al., eds., *History of Technology*, 2:103–46, 3:1–26.

―――. "Power." In Charles Singer et al., eds., *History of Technology*, 2:589–622.

_____. *Studies in Ancient Technology.* Vols. 2–3. Leiden, 1955.

_____. *La technique et l'énergie au cours des siècles.* Les Conférences du Palais de la Découverte, série D, no. 42, 2 Juin 1956, Université de Paris.

Fortunatus, Venantius. *Operum Omnium Miscellanea.* In J. P. Migne, ed., *Patrologia Latina,* vol. 88, cols. 59–362. Paris, 1850.

Fourneyron, Benoit. "Lettre de M. Fourneyron à M. Arago, datée d'Augsbourg, le 17 septembre 1837, sur une nouvelle turbine." Académie des Sciences, Paris, *Comptes rendus* 5 (1837): 562.

_____. "Mémoire sur l'application en grand, dans les usines et manufactures, des turbines hydrauliques ou roues à palettes courbes de Bélidor." Société d'encouragement pour l'industrie nationale, *Bulletin* 33 (1834): 3–17, 49–61, 85–96.

France. Ministère de l'économie nationale. *Statistique des forces motrices en 1931....* Paris, 1936.

Francesco di Giorgio Martini. *Trattati di Architettura, Ingegneria, e Arte Militare....* Milan, 1967.

Francis, James B. *Lowell Hydraulic Experiments....* Boston, 1855.

Franklin Institute. "Report of the Committee of the Franklin Institute of Pennsylvania, appointed May, 1829, to ascertain by experiment the value of Water as a Moving Power." Franklin Institute, *Journal,* 2nd ser. 7 [vol. 11] (1831): 145–54; 8 [vol. 12] (1831): 73–89, 147–53, 221–30, 296–305, 367–73; 9 [vol. 13] (1832): 31–39, 153–57, 295–303, 370–77; 10 [vol. 14] (1832): 10–16, 294–302, 366–71; 3rd ser. 1 [vol. 31] (1841): 145–54, 217–24, 289–96, 361–69; 2 [vol. 32] (1841): 1–8.

Freeman, M. Diane. "Assessing Potential Milling Capacity in Hampshire, c. 1750–1914." *Industrial Archaeology Review* 1 (1976): 47–62.

Frizell, Joseph P. "The Old-Time Water-Wheels of America." American Society of Civil Engineers, *Transactions* 28 (1893): 237–49.

Gade, Daniel W. "Grist Milling with the Horizontal Waterwheel in the Central Andes." *Technology and Culture* 12 (1971): 43–51.

Galilei, Galileo. *On Motion and On Mechanics,* comprising *De Motu* (c1590), translated with introduction and notes by I. E. Drabkin, and *Le Meccaniche* (c1600), translated with introduction and notes by Stillman Drake. Madison, Wis., 1960.

Galloway, Robert L. *A History of Coal Mining in Great Britain.* London, 1882.

Garcia-Diego, J. A. "The Chapter on Weirs in the Codex of Juanelo Turriano: A Question of Authorship." *Technology and Culture* 17 (1976): 217–34.

_____. "Old Dams in Extremadura." *History of Technology,* 1977, pp. 95–124.

Gasperinetti, A. F. "Paper, Papermakers, and Paper-Mills." In *Zonghi's Watermarks,* vol. 3 of *Monumenta chartae papyraceae historiam illustrantia,* ed. E. J. Labarre, pp. 63–82. Hilversum, Holland, 1960.

Gauldie, Enid. "Water-Powered Beetling Machines." Newcomen Society, *Transactions* 39 (1966–67): 125–28.

*A General Description of All Trades, digested in alphabetical order....* London, 1747.

Geramb, Victor. "Ein Beitrag zur Geschichte der Walkerei." *Wörter und Sachen* 12 (1929): 37–46.

Gerstner, Franz von. "Abhandlungen über die oberschlächtigen Wasserräder." Königlichen Böhmischen Gesellschaft der Wissenschaften, *Abhandlungen,* 3rd ser., vol. 2 (1805–9; publ. 1811). 62 pages.

Gille, Bertrand. *Engineers of the Renaissance*. Cambridge, Mass., 1966.

———. "Machines." In Charles Singer et al., eds., *A History of Technology*, 2:629–62.

———. "The Medieval Age of the West." In Maurice Daumas, ed., *A History of Technology and Invention*, 1:422–576.

———. "Le moulin à eau: Une révolution technique médiévale." *Techniques et civilisations* 3 (1954): 1–15.

———. "Le moulin à fer et la haut-fourneau." *Métaux et civilisations* 1 (1946): 89–94.

———. "Les origines du moulin à fer." *Revue d'histoire de la sidérurgie* 1 (1960–63): 23–32.

Gille, Paul. "The Production of Power." In Maurice Daumas, ed., *A History of Technology and Invention*, 2:437–63.

Gillispie, Charles C. *Lazare Carnot, Savant*. Princeton, N.J., 1971.

Gimpel, Jean. *The Medieval Machine: The Industrial Revolution of the Middle Ages*. New York, c1976.

Girard, P. S. "Introduction" to John Smeaton, *Recherches expérimentales sur l'eau et le vent, considérés comme forces motrices applicables aux moulins et autres machines a mouvement circulaire, etc.* . . . , pp. v–xxxii. 2nd ed. Paris, 1827.

Gleisberg, Hermann. *Triebwerke in Getreidemühlen: Eine technikgeschichtliche Studie*. Düsseldorf, 1970.

Glendinning, H., and D. W. T. Glendinning. "Pliny's Water-mill." *Nature* 127 (1931): 974.

Glynn, Joseph. "On Water-Pressure Engines." British Association for the Advancement of Science, *Report* 18 (1849): 11–16.

———. *Treatise on the Power of Water.* . . . 3rd ed. New York, 1869.

Goodsall, Robert H. "Watermills on the River Len." *Archaeologia Cantiana* 71 (1957): 106–29.

Gordon, [Lewis]. "Remarks on Machines recipient of Water Power; more particularly the Turbine of Fourneyron." Institution of Civil Engineers, *Minutes of Proceedings* 2 (1842): 92–102.

Gosse, G. "Las minas y el arte minero de España en la antigüedad." *Ampurias* 4 (1942): 43–68.

Goudie, Gilbert. "On the Horizontal Water Mills of Shetland." Society of Antiquaries of Scotland, *Proceedings* 20 (1886): 257–97.

Gravesande, W. James s'. *Mathematical Elements of Natural Philosophy, confirm'd by Experiments; or, An Introduction to Sir Isaac Newton's Philosophy*. Translated into English by J. T. Desaguliers. 6th ed. 2 vols. London, 1747.

Gray, Andrew. *The Experienced Millwright.* . . . 2nd ed. Edinburgh, 1806.

Gregory of Tours. *Historiae Ecclesiasticae Francorum Libri Decem*. in J. P. Migne, ed., *Patrologia Latina*, vol. 71, cols. 161–572. Paris, 1858.

———. *Vitae Patrum.* . . . In J. P. Migne, ed., *Patrologia Latina*, vol. 71, cols. 1009–96. Paris, 1858.

Gribbon, H. D. *The History of Water Power in Ulster*. Newton Abbot, 1969.

Grier, William. *The Mechanic's Calculator.* . . . 3rd ed. Glasgow, 1836.

Grimshaw, Robert. *The Miller, Millwright, and Millfurnisher: A Practical Treatise*. New York, 1882.

Grollier de Servière, [Nicolas]. *Recueil d'ouvrages curieux de mathématique et de mécanique.* . . . Lyons, 1719.

Grübler, Martin. "Zur Theorie mittelschlächtiger Wasserräder und des Sage-bienrades." *Der Civilingenieur* 22 (1876): 409–42.

Guenyveau, A. *Essai sur la science des machines.* Paris, 1810.

Guttmann, Oscar. *The Manufacture of Explosives.* Vol. 1. London, 1895.

Hachette, [Jean Nicolas Pierre]. *Traité élémentaire des machines.* 2nd ed. Paris, 1819.

Hahn, Roger. *The Anatomy of a Scientific Institution: The Paris Academy of Sciences, 1666–1803.* Berkeley and London, 1971.

Hall, Bert S. "The So-Called 'Manuscript of the Hussite Wars Engineer' and Its Technological Milieu...." Ph.D. dissertation, University of California, Los Angeles, 1971.

Hall, John W. "The Making and Rolling of Iron." Newcomen Society, *Transactions* 8 (1927–28): 40–55.

Harris, Richard C. *The Seigneurial System in Early Canada.* Madison, Wis., and Quebec, 1968.

Hart, Andrew S. *An Elementary Treatise on Hydrostatics and Hydrodynamics.* Dublin, 1846.

_____. *An Elementary Treatise on Mechanics.* Dublin, 1844.

Hartley, E. N. *Iron-Works on the Saugus.* Norman, Okla., 1957.

Haskins, Thomas L. "Eighteenth-Century Attempts to Resolve the vis viva Controversy." *Isis* 56 (1965): 281–97.

Haton de la Goupillière, [Julien-Napoléon]. *Cours de machines.* Vol. 1. Paris, 1889.

Hay, Thomas T. "Watermills in Japan." *Industrial Archaeology* 6 (1969): 321–24, 369–70.

Head, George. *A Home Tour through the Manufacturing Districts of England, in the summer of 1835.* New ed. London, 1836.

Helmholdi Presbyteri. *Helmholdi Presbyteri Chronica Slavorum a. 800–1172.* In *Monumenta Germaniae Historica: Scriptores,* 21:1–99. Hanover, 1869.

Henry, William C. "A Biographical Notice of the late Peter Ewart, Esq." Literary and Philosophical Society of Manchester, *Memoirs,* 2nd ser. 7 (1846): 114–35.

Herodotus. *Historiae.* Transl. A. D. Godley. Vol. 1. Cambridge, Mass., 1960.

Héron de Villefosse, A. M. *De la richesse minérale: Considérations sur les mines, usines, et salines des différens états....* 3 vols. text plus atlas. Paris, 1819.

Herrad de Landsberg. *Hortus Deliciarum* [c1190]. Ed. Rosalie Green, Michael Evans, Christine Bischoff, and Michael Curschmann. London and Leiden, 1979.

Hill, Bennett. *English Cistercian Monasteries and Their Patrons in the Twelfth Century.* Urbana, Ill., 1968.

Hillier, J[ack]. *Old Surrey Water-Mills.* London, 1951.

Hills, Richard L. *Power in the Industrial Revolution.* Manchester, 1970.

_____. "Water, Stampers, and Paper in the Auvergne: A Medieval Tradition." *History of Technology,* 1980, pp. 143–56.

Hobart, James F. *Millwrighting.* New York, 1909.

Hodgen, Margaret T. "Domesday Water Mills." *Antiquity* 13 (1939): 261–79.

Hollister-Short, G. "Leads and Lags in Late Seventeenth-Century English Technology." *History of Technology,* 1976, pp. 159–83.

_____. "The Vocabulary of Technology." *History of Technology,* 1977, pp. 125–55.

Hopkins, R. Thurston. *Old Watermills and Windmills.* London, 1930.

Horn, Walter. *The Plan of St. Gall.* 3 vols. Berkeley and Los Angeles, 1979.

Horn, Walter, and Ernest Born. "The 'Dimensional Inconsistencies' of the Plan of Saint Gall and the Problem of the Scale of the Plan." *Art Bulletin* 48 (1966): 285–307.

Howell, Charles. "Colonial Watermills in the Wooden Age." In Brooke Hindle, ed., *America's Wooden Age: Aspects of Its Early Technology*, pp. 120–59. Tarrytown, N.Y., 1975.

Howell, Charles, and Allan Keller. *The Mill at Philipsburg Manor Upper Mills and a Brief History of Milling.* Tarrytown, N.Y., 1977.

Hudson, Kenneth. *The Industrial Archaeology of Southern England.* 2nd ed. Newton Abbot, 1968.

Hughes, J. Donald. *Ecology in Ancient Civilizations.* Albuquerque, 1975.

Hughes, William Carter. *The American Miller, and Millwright's Assistant.* Philadelphia, 1853.

————. *The American Miller, and Millwright's Assistant.* New ed. Philadelphia and London, 1873.

Humphrey, J. "Turbine and Overshot." *American Miller* 22 (1894): 598–99.

Hunter, Louis C. "The Living Past in the Appalachias of Europe: Water-Mills in Southern Europe." *Technology and Culture* 8 (1967): 446–66.

————. "Les origines des turbines Francis et Pelton: Développement de la turbine hydraulique aux États Unis de 1820 à 1900." *Revue d'histoire des sciences et leur applications* 17 (1964): 209–42.

————. "Waterpower in the Century of the Steam Engine." In Brooke Hindle, ed., *America's Wooden Age: Aspects of Its Early Technology*, pp. 160–92. Tarrytown, N.Y., 1975.

————. *Waterpower in the Century of the Steam Engine.* Vol. 1 of *A History of Industrial Power in the United States, 1780–1930.* Charlottesville, Va., 1979.

Hurst, D. Gillian. "Medieval Britain in 1962 and 1963, II: Post Conquest." *Medieval Archaeology* 8 (1964): 241–99.

Huygens, Christian. *Oeuvres complètes de Christiaan Huygens.* Vol. 19. The Hague, 1937.

Hyde, Charles K. *Technological Change and the British Iron Industry, 1700–1870.* Princeton, 1977.

Ibn Jubayr. *The Travels of Ibn Jubayr.* Transl. R. J. C. Broadhurst. London, 1952.

Idrisi, Abu 'Abd Alla Muhammad al-. *Géographie.* Transl. P. Amédée Jaubert. Vol. 2. Paris, 1840.

————. *India and the Neighboring Territories in the Kitab Nuzhat al-Mushtag Fi'Khtiraq al-'Afaq of Al-Sharif al-Idrisi.* Transl. S. Maqbul Ahmud. Leiden, 1960.

Iltis, Carolyn. "Leibniz and the Vis Viva Controversy." *Isis* 62 (1971): 21–35.

Imamuddin, S. M. *The Economic History of Spain under the Umayyads, 711–1031 A.C.* Dacca, 1963.

————. *Some Aspects of the Socio-Economic and Cultural History of Muslim Spain, 711–1492 A.D.* Leiden, 1965.

"An Industrial Museum for Sheffield." *British Steelmaker*, April 1941, pp. 75–77.

Irimie, Cornel. "Floating Mills on Boats in România." International Symposium on Molinology, *Transactions* 2 (1969): 437–45.

Irimie, Cornel, and Coneliu Bucur. "Typology, Distribution, and Frequency of Water Mills in România in the First Half of the Twentieth Century." International Symposium on Molinology, *Transactions* 2 (1969): 421–34.

Jackson, Thomas. *Elements of Theoretical Mechanics....* Edinburgh and London, 1827.

Jacono, Luigi. "La ruota idraulica di Venafro." *Annali dei lavori pubblici* (Rome) 77 (1939): 217–20.

_____. "La ruota idraulica di Venafro." *L'ingegnere* 12 (1938): 850–53.

James, John. *History of the Worsted Manufacture in England.* London, 1857.

Jars, Gabriel. "Description d'une nouvelle machine executée aux mines de Schemnitz in Hongrie, au mois de Mars 1755." *Mémoires de mathématique et de physique, par divers Sçavans...* 5 (1768): 67–71.

_____. *Voyages métallurgiques; ou, Recherches et observations sur les mines & forges de fer....* 3 vols. Lyons, 1774–81.

Jazari, Ibn al-Razzaz al-. *The Book of Knowledge of Ingenious Mechanical Devices.* Transl. and ed. Donald R. Hill. Dordrecht, Holland, and Boston, 1974.

Jenkins, Rhys. *Collected Papers of Rhys Jenkins.* Cambridge, 1936.

Jeon, Sang-woon. *Science and Technology in Korea: Traditional Instruments and Techniques.* Cambridge, Mass., 1974.

Jespersen, Anders. *The Lady Isabella Waterwheel of the Great Laxey Mining Company, Isle of Man, 1854–1954: A Chapter in the History of Early British Engineering.* Virum, Denmark, 1954.

_____. "Portugese Mills...." International Symposium on Molinology, *Transactions* 2 (1969): 69–87.

_____. *A Preliminary Analysis of the Development of the Gearing in Watermills in Western Europe.* Virum, Denmark, 1953.

Johnson, H. R., and A. W. Skempton. "William Strutt's Cotton Mills, 1793–1812." Newcomen Society, *Transactions* 30 (1955–57): 179–205.

Johnson, William A., transl. *Christopher Polhem: The Father of Swedish Technology.* Hartford, Conn., 1963.

Jones, D. H. "The Moulin Pendant." International Symposium on Molinology, *Transactions* 3 (1973): 169–74.

Juan y Santacilia, Jorge. *Examen maritime, théorique, et pratique; ou, Traité de mécanique, appliqué à la construction et à la manoeuvre des vaisseaux & autres bâtiments.* Translated from the Spanish with additions by [Pierre] Levêque. 2 vols. Nantes, 1783.

Justinian. *Codex Justinianus.* Ed. Denis Gothofred. In *Corpus iuris civilis,* 2: 6–492. Amsterdam, 1664.

Kästner, Abraham G. *Anfangsgründe der Hydrodynamik....* Göttingen, 1797.

Keator, Frederic. "Benoit Fourneyron (1802–1867)." *Mechanical Engineering* 61 (1939): 295–301.

Kellenbenz, Hermann. *The Rise of the European Economy: An Economic History of Continental Europe from the Fifteenth to the Eighteenth Century.* London, 1976.

Keller, A. G. "A Byzantine Admirer of 'Western' Progress: Cardinal Bessarion." *Cambridge Historical Journal* 11 (1955): 343–48.

_____. "Renaissance Waterworks and Hydromechanics." *Endeavour* 25 (1966): 141–45.

_____. *A Theatre of Machines.* London, 1964.

Keller, Karl. "Johann Andreas Segner." *Beiträge zur Geschichte der Technik und Industrie* 5 (1913): 54–72.

Kemble, John M., ed. *Codex diplomaticus aevi Saxonici.* Vol. 1. London, 1839.

Kent, William. *The Mechanical Engineer's Pocket Book.* 7th ed. New York, 1907.

Kenyon, P. "The Great Laxey Mining Company, Isle of Man: Surveying Evidence, April 1965." *Industrial Archaeology* 2 (1965): 154–57.

Klemm, Friedrich. *A History of Western Technology.* Transl. Dorthea Waley Singer. Cambridge, Mass., 1964.

Koehne, Carl. "Die Mühle im Rechte der Völker." *Beiträge zur Geschichte der Technik und Industrie* 5 (1913): 27–53.

_____. *Das Recht der Mühlen bis zum Ende der Karolingerzeit.* Untersuchungen zur Deutschen Staats- und Rechtsgeschichte, ed. by Otto Gierke, vol. 71. Breslau, 1904.

Kranzberg, Melvin, and Joseph Gies. *By the Sweat of Thy Brow: Work in the Western World.* New York, 1975.

Kuhlmann, Charles B. *The Development of the Flour-Milling Industry in the United States, with Special Reference to the Industry in Minneapolis.* Boston and New York, 1929.

Kyeser, Conrad. *Bellifortis* [c1405]. Ed. Götz Quart. Düsseldorf, 1967.

LaGrange, Joseph L. *Méchanique analytique.* Paris, 1788.

_____. *Oeuvres de Lagrange.* Publ. by J.-A. Serret. Vols. 9, 12–13. Paris, 1881–82.

La Hire, Philippe de. "Examen de la force de l'homme, pour mouvoir des fardeaux absoluement et par comparison à celle des animaux qui portent et qui tirent, comme les chevaux." *AS–M,* 1699 (3rd ed., 1732), pp. 153–63 (*AS–H,* pp. 96–98).

_____. "Examen de la force nécessaire pour faire mouvoir les bateaux tant dans l'eau dormante qui courante, soit avec une corde qui y est attachée & que l'on tire, soit avec des rames, ou par le moyen de quelque machine." *AS–M,* 1702 (1743 ed.), pp. 254–80 (*AS–H,* pp. 126–34).

Lambert, [Johann H.]. "Expériences et remarques sur les moulins que l'eau meut par en bas dans une direction horizontale." Académie royale des sciences et belles-lettres, Berlin, *Nouveaux mémoires,* 1775 (publ. 1777), pp. 49–69.

Landels, J. G. *Engineering in the Ancient World.* London, 1978.

Landes, David S. *The Unbound Prometheus: Technological Change and Industrial Development in Western Europe from 1750 to the Present.* Cambridge, 1969.

Lanz, [Philippe Louis], and [Augustin de] Bétancourt. *Analytical Essay on the Construction of Machines.* London, n.d. [c1820].

_____. *Essai sur la composition des machines.* Paris, 1808.

Laudan, Laurens. "The vis viva Controversy: A Post-Mortem." *Isis* 59 (1968): 131–43.

Laufer, B. "The Noria or Persian Wheel." In *Oriental Studies in Honour of C. E. Pavry,* pp. 238–50. Oxford, 1933.

Layton, Edwin T. "Millwrights and Engineers, Science, Social Roles, and the Evolution of the Turbine in America." In Wolfgang Krohn, Edwin T. Layton, and Peter Weingart, eds., *The Dyanmics of Science and Technology,* 2: 61–87. Dordrecht, Holland, and Boston, 1978.

_____. "Scientific Technology, 1845–1900: The Hydraulic Turbine and the Origins of American Industrial Research." *Technology and Culture* 20 (1979): 64–89.

Leblanc, Charles. "Mémoire sur la roue-vanne, inventée et exécutée par M. Sagebien." *Annales de ponts et chaussées,* 3rd ser. 15 (1858): 129–70.

Lefébre de Fourcy, Eugene. "Expériences sur les roues hydrauliques...." *Annales des mines,* 3rd ser. 12 (1837): 3–46.

_____. "Extrait d'un mémoir de M. Morin, capitaine d'artillerie, ayant pour titre; 'Compte rendu d'un mission dans les fonderies d'artillerie....'" *Annales des mines*, 3rd ser. 3 (1833): 93–122, 259–79.

Lefèvre-Gineau, _____. "Notice historique sur Jean Charles Borda." Institut national des sciences et arts [de France]—Sciences mathématiques et physiques, *Histoire* 4 (an XI [1804]): 89–104.

Leffel, James, and Company. *Leffel's Construction of Mill Dams.* Springfield, Ohio, 1881.

*Leges Visigothorum.* In *Monumenta Germaniae Historica: Leges*, sec. 1, 1:33–456. Hanover, 1892.

Lennard, Reginald. "Early English Fulling Mills: Additional Examples." *Economic History Review*, 2nd ser. 3 (1950–51): 342–43.

_____. "An Early Fulling-Mill." *Economic History Review* 17 (1947): 150.

_____. *Rural England, 1086–1135.* Oxford, 1959.

Leonardo da Vinci. *Il Codice Atlantico di Leonardo da Vinci nella Biblioteca di Milano.* Ed. Giovanni Piumati. 2 vols. text plus 2 vols. plates. Milan, 1894–1904.

_____. *The Madrid Codices.* Ed. and transl. Ladislao Reti. 5 vols. New York, c1974.

_____. *I Manoscritti e i Disegni di Leonardo da Vinci.* 6 vols. Rome, 1928–49.

_____. *Les manuscrits de Léonard de Vinci....* Published, transcribed, and translated into French by Charles Ravaisson-Moilien. 6 vols. Paris, 1881–91.

_____. *Del moto e misura dell'acqua....* Bologna, n.d.

_____. *The Notebooks of Leonardo da Vinci.* Ed. and transl. Edward MacCurdy. New York, 1956.

_____. *Problèmes de géométrie et d'hydraulique....* Unedited manuscript reproduced from the original conserved in the Forster Library, South Kensington Museum, London. Paris, 1901.

Lermier, J. "Mémoire sur l'hydraulique...." Académie royale des sciences, belles-lettres, et arts de Bordeaux, *Séance publique*, 1825, pp. 113-51.

Leslie, [John], and [Robert] Jameson. "On the Value of Water as a Moving Power for Machinery, illustrated in an extract from a Report in regard to the Water of Leith." *Edinburgh Philosophical Journal* 13 (1825): 170–72.

LeStrange, Guy. *Baghdad during the Abbasid Caliphate from Contemporary Arabic and Persian Sources.* Oxford, 1900.

_____. *The Lands of the Eastern Caliphate: Mesopotamia, Persia, and Central Asia from the Moslem Conquest to the Time of Timur.* Cambridge, 1905.

_____. *Palestine under the Moslem: A Description of Syria and the Holy Land from A.D. 650 to 1500.* Translated from the works of medieval Arab geographers. Beirut, 1965. Originally published in 1890.

Leupold, Jacob. *Theatri machinarum hydraulicarum; oder, Schau-Platz der Wasser-Künste.* Vol. 2. Leipzig, 1725.

_____. *Theatrum machinarum generale: Schau-Platz des Grundes mechanischer Wissenschaften....* Leipzig, 1774. Originally published in 1724.

Levainville, J[acques]. *L'industrie du fer en France.* Paris, 1922.

Lewis, M. J. T. "Industrial Archaeology." In *The Fontana Economic History of Europe*, ed. Carlo Cipolla, 3: 574–603. London and Glasgow, 1976.

Liberman, Saul, ed. *Tosefta Sabbath.* New York, 1962.

Lilley, Samuel. *Men, Machines, and History.* London, 1948.

Lindebrog, Frederick, ed. *Codex legum antiquarum....* Frankfurt, 1613.

Lindet, L. "Les origines du moulin à grains." *Revue archéologique*, 3rd ser. 35 (1899): 413–27; 36 (1900): 17–44.

Lindroth, Sten. *Christopher Polhem och Stora Kopparberget: Ett Bidrag till Bergsmekanikens Historia....* Uppsala, 1951.

Livermore, Daniel. "Remarks upon the Employment of Pressure Engines, as a substitute for Water Wheels." Franklin Institute, *Journal,* 2nd ser. 15 [vol. 19] (1835): 165–68.

Loring, Charles H. "The Steam Engine in Modern Civilization." American Society of Mechanical Engineers, *Transactions* 14 (1892–93): 52–58.

Lucas, A. T. "The Horizontal Mill in Ireland." Royal Society of Antiquaries of Ireland, *Journal* 83 (1953): 1–36.

Luckhurst, David. *Monastic Watermills: A Study of the Mills within English Monastic Precincts.* London, n.d. A Society for the Protection of Ancient Buildings booklet.

Lucretius. *De rerum natura.* Transl. Cyril Bailey. Oxford, 1947.

"Ludwig der Deutsche." In *Monumenta Germaniae Historica: Diplomata regum Germaniae ex stirpe Karolinorum,* 1: 1–284. Berlin, 1934.

MacAdam, Robert. "Ancient Water-Mills." *Ulster Journal of Archaeology* 4 (1856): 6–15.

MacCurdy, George G. *Human Origins: A Manual of Prehistory.* Vol. 2. New York, 1924.

McCutcheon, W. A. "The Application of Water Power to Industrial Purposes in the North of Ireland." *Industrial Archaeology* 2 (1965): 69–81.

———. "Water Power in the North of Ireland." Newcomen Society, *Transactions* 39 (1966–67): 67–94.

MacLaurin, Colin. *An Account of Sir Isaac Newton's Philosophical Discoveries.* Published by Patrick Murdoch. London, 1748.

———. *A Treatise on Fluxions.* 2 vols. Edinburgh, 1742.

Magie, William F., ed. *A Source Book in Physics.* Cambridge, Mass., 1963.

Magnus, Olaus. *Historia de Gentibus Septentrionalibvs....* Rome, 1555.

Maindron, Ernest. *Les fondations de prix à l'Académie des Sciences: Les lauréats de l'Académie, 1714–1880.* Paris, 1881.

Major, Kenneth. "The Further Contribution of England and Wales to the Molinological Map of Europe." International Symposium on Molinology, *Transactions* 2 (1969): 111–21.

Mallet, [Jacques-André]. "A Memoir concerning the most advantageous Construction of Water Wheels, etc...." Royal Society of London, *Philosophical Transactions* 57 (1767): 372–88.

Mallet, Robert. "An Experimental Inquiry as to the Co-efficient of Labouring Force in Overshot Water-Wheels, whose diameter is equal to, or exceeds the total descent due to the fall; and of Water-wheels moving in circular channels." Institution of Civil Engineers, *Minutes of Proceedings* 2 (1843): 60–66.

———. "An Inquiry as to the Coefficient of Labouring Force in Overshot Water-Wheels, whose diameter is equal to, or exceeds the total descent due to the fall; and of Water-Wheels moving in Circular Channels." Royal Irish Academy, *Proceedings,* 1st ser. 2 (1840–44): 262–66.

———. "Papers upon the Principles and Practice of the Application of Water Power [Report on the Dodder Reservoirs]." *Quarterly Papers on Engineering.* Vol. 6. London, 1849.

Malouin, ———. "L'art de meunier." In Académie des Sciences, Paris, *Description des arts et métiers....* , 1:22–119. New ed. Neuchatel, 1771.

Mandey, V. *Mechanick Powers; or, The Mystery of Nature and Art unvail'd.* ... London, 1702.

Mantoux, Paul. *The Industrial Revolution in the Eighteenth Century: An Outline of the Beginnings of the Modern Factory System in England.* London, 1961.

Mariotte, [Edme]. *The Motion of Water, and other Fluids.* ... Transl. John T. Desaguliers. London, 1718.

_____. *Traité du mouvement des eaux et des autres corps fluides.* ... Paris, 1686.

Marius d'Avenches. *Marii Episcopi Aventicensis Chronica: A. CCCCLV–DLXXXI.* In *Monumenta Germaniae Historica: Auctores Antiquissimi,* 11:227–39. Berlin, 1894.

Marks, Lionel S. *Mechanical Engineers' Handbook.* New York, 1916.

Marozeau, _____. "Notes on a Water-Wheel established in a Bleachery at St. Amarin, by M. Marozeau, of Wesserling. ..." Franklin Institute, *Journal,* 3rd ser. 9 [vol. 39] (1845): 9–16.

_____. "Note sur une roue hydraulique récomment établie dans la blanchisserie dite du Breuil, située à St.-Amarin, et appartenant à MM. Gros, Odier, Roman et Comp., de Wesserling." Société industrielle de Mulhouse, *Bulletin* 18 (1844): 49–59.

Martin, Benjamin. *Biographia Philosophica.* ... London, 1764.

_____. *New and Comprehensive System of Mathematical Institutions.* ... Vol. 2. London, 1764.

_____. *Philosophia Britannica; or, A new and comprehensive system of the Newtonian philosophy, astronomy, and geography.* ... Vol. 1. Reading, 1747.

Martin, Henry M. R. *Légende de Saint Denis.* Paris, 1908.

Martin, Saint (of Trier). "De calamitat abbatiae sancti Martini Treverensis." In *Monumenta Germaniae Historica: Scriptores,* 15:739–41. Hanover, 1887.

Mason, Stephen F. *A History of the Sciences.* Rev. ed. New York, 1962.

Mayence, F. "La trosième campagne de fouilles à Apamée." *Bulletin des Musées Royaux d'art et d'Histoire* (Brussels), 3rd ser. 5 (1933): 5–6.

Mayr, Otto. *The Origins of Feedback Control.* Cambridge, Mass., 1970.

Mead, Daniel W. *Water Power Engineering.* 2nd ed. New York, 1915.

"Mechanics." In *Encyclopaedia Londinensis; or, Universal Dictionary of Arts, Sciences, and Literature.* ... , 14: 619–716. London, 1816.

Mez, Adam. *The Renaissance of Islam.* Translated from the German by Salahuddin Khuda Bakhsh and D. S. Margoliouth. Patna, 1937.

"Mill." In *Cyclopaedia; or, An Universal Dictionary of Arts and Sciences,* ed. Ephriam Chambers, rev. Abraham Rees, vol. 3. London, 1781. Pages unnumbered.

"Mill." In *The Cyclopaedia; or, Universal Dictionary of Arts, Sciences, and Literature,* ed. Abraham Rees, vol. 23. London, 1819. Pages unnumbered.

Millar, Eric G., ed. *The Luttrell Psalter* [c1340]. London, 1932.

Miller, William H. *The Elements of Hydrostatics and Hydrodynamics.* Cambridge, 1831.

Miller, William T. *The Water-Mills of Sheffield.* Sheffield, 1949.

Millington, John. *An Epitome of the Elementary Principles of Natural and Experimental Philosophy.* ... London, 1823.

Milner, Isaac. "Reflections on the Communication of Motion by Impact and Gravity." Royal Society of London, *Philosophical Transactions* 68 (1778): 344–79.

Minchinton, W. E. "Early Tide Mills: Some Problems." *Technology and Culture* 20 (1979): 777–86.

Miscellaneous Dissertations on Rural Subjects. London, 1775.

Mitchell, B. R. Abstract of British Historical Statistics. Cambridge, 1962.

Montaigne, Michael. The Diary of Montaigne's Journey to Italy in 1580 and 1581. Transl. E. J. Trechmann. New York, 1929.

Montucla, J. F. Histoire des mathématiques.... Vol. 3. Paris, an X [1802].

Moody, Shute Barrington. "Description of a Water-wheel constructed by Mr. W. Fairbairn... and erected in Lombardy." Institution of Civil Engineers, Minutes of Proceedings 3 (1844): 66–68.

Morin, Arthur. Aide-mémoire de mécanique pratique. 4th ed. Paris, 1858.

_____. "Examen des effets des moteurs employés dans les fonderies de l'artillerie." Bulletin des sciences technologiques 18 (1831): 178.

_____. "Expériences sur les roues hydrauliques à aubes planes et à augets." Extrait par M. Eug. Lefébure de Fourcy. Annales des mines, 3rd ser. 12 (1837): 3–46.

_____. Expériences sur les roues hydrauliques à aubes planes, et sur les roues hydrauliques à augets. Metz and Paris, 1836.

_____. Hydraulique. 2nd ed. Paris, 1858.

Moritz, L. A. Grain-Mills and Flour in Classical Antiquity. Oxford, 1958.

_____. "Vitruvius' Water-Mill." Classical Review 70 (1956): 193–96.

Morris, Ellwood. "Remarks on Reaction Water Wheels used in the United States; and on the Turbine of M. Fourneyron, an Hydraulic Motor, recently used with the greatest success on the continent of Europe." Franklin Institute, Journal, 3rd ser. 4 [vol. 34] (1842): 217–27, 289–304.

Morshead, William. "On the relative Advantages of Steam, Water, and Animal Power." Bath and West and Southern Counties Society, Journal, n.s. 4 (1856): 24–52.

Mortimer, J[ohn]. The Whole Art of Husbandry; or, The Way of Managing and Improving of Land.... 2nd ed. London, 1708.

Mott, R. A. "The Coalbrookdale Group Horsehay Works, Part I." Newcomen Society, Transactions 31 (1957–59): 271–87.

Muendel, John. "The Horizontal Mills of Medieval Pistoia." Technology and Culture 15 (1974): 194–225.

Muller, John. A Mathematical Treatise: Containing a System of Conic-Sections; with the Doctrine of Fluxions and Fluents Applied to Various Subjects.... London, 1736.

Müller, Wilhelm. Die Eisernen Wasserräder.... 2 vols. Leipzig, 1899.

Multhauf, Robert P. "Mine Pumping in Agricola's Time and Later." United States National Museum, Bulletin 218 (1959): 113–20.

Mumford, Louis. Technics and Civilisation. New York, 1934.

Muqaddasi, Muhammad ibn Ahmad al-. Ahsanu-t-taqasim-fi-ma'rifati-l-aqalim. Transl. and ed. G. S. A. Ranking and R. F. Azoo. Vol. 1. Calcutta, 1897.

Muscart, Jean. La vie et les travaux du Chevalier Jean-Charles de Borda, 1733–1799: Épisodes de la vie scientifique au XVIIIᵉ siècle. Lyons and Paris, 1919.

Mussenbroek, Pierre van [Petrus van]. Cours de physique expérimentale et mathématique. Transl. Sigaud de la Fond. Vol. 1. Paris, 1769.

Musson, A. E. The Growth of British Industry. London, 1978.

_____. "Industrial Motive Power in the United Kingdom, 1800–70." Economic History Review, 2nd ser. 29 (1976): 415–39.

_____, ed. Science, Technology, and Economic Growth in the Eighteenth Century. London, 1972.

Musson, A. E., and Eric Robinson. *Science and Technology in the Industrial Revolution.* Manchester, 1969.

Navier, C. L. M. H. "Détails historiques sur l'emploi du principe des forces vives dans la théorie des machines, et sur diverses roues hydrauliques." *Annales de chimie et de physique,* 2nd ser. 9 (1818): 146–59.

Needham, Joseph. "Chinese Priorities in Cast Iron Metallurgy." *Technology and Culture* 5 (1964): 398–404.

_____. *The Development of Iron and Steel Technology in China.* London, 1958.

_____. *The Grand Titration: Science and Society in East and West.* London, 1969.

_____. "The Pre-Natal History of the Steam Engine." Newcomen Society, *Transactions* 35 (1962–63): 3–58.

_____ (with the collaboration of Wang Ling). *Science and Civilisation in China.* Vol. 4, pt. 2. Cambridge, 1965.

_____. "L'unité de la science: L'apport indispensable de l'Asie." *Archives internationales d'histoire des sciences* 7 (1949): 563–82.

Needham, Joseph, Wang Ling, and Derek J. de Solla Price. *Heavenly Clockwork: The Great Astronomical Clocks of Medieval China.* Cambridge, 1960.

Nef, John U. *The Conquest of the Material World.* Chicago, 1964.

_____. *Industry and Government in France and England, 1540–1640.* Ithaca, N.Y., 1964.

_____. *The Rise of the British Coal Industry.* 2 vols. London, 1932.

_____. *War and Human Progress: An Essay on the Rise of Industrial Civilization.* Cambridge, Mass., 1950.

Neumeyer, Freidrich. "Christopher Polhem och hydrodynamiken." *Archiv för matematik, astronomi, och fysik* 28A, no. 15 (1942): 16 pp.

"New System of Water Power." Franklin Institute, *Journal* 11 (1831): 214–15.

Newton, Isaac. *Philosophiae naturalis Principia mathematica.* London, 1687.

_____. *Philosophiae naturalis Principia mathematica.* 2nd ed. Cambridge, 1713.

Nicholson, John. *The Millwright's Guide: A Practical Treatise on the Construction of all Kinds of Mill Work, and the application of the power of Wind and Water.* London, 1830.

Niederle, Lubor. *Manuel de l'antiquité slave.* Vol. 2. Paris, 1926.

_____. *Zivot starych Slovanu.* Vol. 3. Prague, 1921.

Nixon, Frank. *The Industrial Archaeology of Derbyshire.* Newton Abbot, 1969.

Nordwall, Erik. *Afhandling rörande Mechaniquen met tillampning i Synnerhet til Bruk och Bergverk.* Vol. 1. Stockholm, 1800.

"Nouvelles machines de Marly établies à Bougival sous la direction de M. Dufrayer, ingénieur...." *Publication industrielle* 14 (1861): 246–61.

O'Reilly, Joseph P. "Some Further Notes on Ancient Horizontal Water-Mills, Native and Foreign." Royal Irish Academy, *Proceedings* 24, sec. C (1902–4): 55–84.

Ostwald, Wilhelm. "The Modern Theory of Energetics." *The Monist* 17 (1907): 481–515.

Owen, Aneurin, ed. and transl. *Ancient Laws and Institutes of Wales.* London, 1841.

Pacey, Arnold. *The Maze of Ingenuity: Ideas and Idealism in the Development of Technology.* Cambridge, Mass., 1976.

Palladius. *De re rustica.* In G. Schneider, ed., *Scriptorum rei rusticae veterum Latinorum,* 3:1–256. Leipzig, 1775.

Palmer, R. E. "Notes on Some Ancient Mine Equipments and Systems." Institution of Mining and Metallurgy, *Transactions* 36 (1926–27): 299–336.

Pambour, F. M. G. de. "Addition au mémoire sur la théorie des roues hydrauliques. Du mode d'introduction des résistances dans le calcul." Académie des Sciences, Paris, *Comptes rendus* 67 (1868): 1134–38.

_____. "Mémoire sur la théorie des roues hydrauliques." Académie des Sciences, Paris, *Comptes rendus* 64 (1867): 1272–75.

_____. "Roues hydrauliques: Du calcul des effets par la méthode des coefficients." Académie des Sciences, Paris, *Comptes rendus* 75 (1872): 1757–61.

_____. "Sur la théorie des roues hydrauliques." Académie des Sciences, Paris, *Comptes rendus* 60 (1865): 1181–85, 1283–86; 61 (1865): 30–31, 200–204, 1121–25; 62 (1866): 218–22, 787–90.

_____. "Sur le frottement additional, dû à la charge des machines." Académie des Sciences, Paris, *Comptes rendus* 74 (1872): 1459–62.

Papacino d'Antoni, Alessandro Vittorio. *Institutions physico-méchaniques à l'usage des écoles royales d'artillerie et du génie de Turin*. Translated from the Italian. Vol. 2. Strasbourg, 1777.

Pardies, Ignace Gaston. *Oeuvres*. Lyons, 1725.

Parent, Antoine. "Sur la plus grande perfection possible des machines, etant donné une machine qui ait pour puissance motrice quelque corps fluide que ce soit, comme par exemple, l'eau, le vent, la flame, &c...." *AS–M*, 1704 (2nd ed., 1722), pp. 323–38 (*AS–H*, pp. 116–23).

_____. "Sur la position de l'axe des moulins à vent à l'égard du vent." *AS–H*, 1702 (1743 ed.), pp. 138–41.

_____. "Sur la réduction des mouvements des animaux aux loix de la mécanique."*AS–H*, 1702 (1743 ed.), pp. 95–102.

_____. "Sur les centres de conversion & sur les frotemens." *AS–H*, 1700 (2nd ed., 1761), pp. 149–53.

Parker, Margaret T. *Lowell: A Study of Industrial Development*. New York, 1940.

Parsons, Arthur W. "A Roman Water-Mill in the Athenian Agora," *Hisperia* 5 (1936): 70–90.

Pelham, R. A. "Corn Milling and the Industrial Revolution in England in the Eighteenth Century." *University of Birmingham Historical Journal* 6 (1958): 161–75.

_____. *Fulling Mills: A Study in the Application of Water Power to the Woolen Industry*. London, n.d. A Society for the Protection of Ancient Buildings booklet.

_____. "The Water-Power Crisis in Birmingham in the Eighteenth Century." *University of Birmingham Historical Journal* 9 (1963): 64–91.

Petit, A[lexis]-T[hérèse]. "Sur l'emploi du principe des forces vives dans le calcul de l'effet des machines." *Annales de chimie et de physique*, 2nd ser. 8 (1818): 287–305.

Philo of Byzantium. *Le livre des appareils pneumatiques et des machines hydrauliques, par Philon de Byzance....* Ed. and transl. Baron Carra de Vaux. Paris, 1902.

_____. *Pneumatica*. Ed. and transl. Frank D. Prager. Weisbaden, 1974.

Piccolpasso, Cipriano. *The Three Books of the Potter's Art....* Transl. Bernard Rackham and Albert Van de Put. London, 1934.

Pitot, Henri. "Comparaison entre quelques machines mûës par les courants des fluides; où, L'on donne une méthode très-simple de comparer l'effet de celles dont l'arbre qui porte les aîles ou aubes est perpendiculaire au courant de l'eau, à l'effet de celles dont le meme arbre est parallele au courant." *AS–M*, 1729 (publ. 1731), pp. 385–92 (*AS–H*, pp. 81–87).

_____. "Nouvelle methode pour connoître & déterminer l'effort de toutes sortes de machines muës par un courant, ou une chûte d'eau; où, L'on déduit de la loi des méchaniques des formules générales, par le moyen desquelles on peut faire les calculs de l'effet de toutes ces machines." *AS–M*, 1725 (publ. 1727), pp. 78–102 (*AS–H*, pp. 80–87).

_____. "Réfléxions sur le mouvement des eaux." *AS–M*, 1730 (publ. 1732), pp. 536–44 (*AS–H*, pp. 110–15).

_____. "Remarques sur les aubes ou pallettes des moulins, & autres machines muës par le courant des rivières." *AS–M*, 1729 (publ. 1731): 253–58 (*AS–H*, pp. 81–87).

Pleket, H. W. "Technology and Society in the Graeco-Roman World." *Acta Historiae Neerlandica* 2 (1967): 1–24.

Pliny. *Historia naturalis*. Ed. and transl. H. Rackham et al. Vol. 5. London, 1950.

Plutarchus. *Plutarch's Lives*. Transl. Bernadotte Perrin. Vol. 5. New York and London, 1917.

Polhem, Christopher. "Fortsetzung von der Berbindung der Theorie und Practik ben der Mechanik." König. Schwedischen Akademie der Wissenschaften, *Abhandlungen*, 1742, transl. and ed. Abraham Gotthelf Kästner, 4: 183–96. Hamburg, 1750.

Poncelet, Jean Victor. *Cours de mécanique appliquée aux machines. . .* Publ. M. X. Kretz. Paris, 1876.

[_____.] *Cours de mécanique appliquée aux machines*. [École d'Application de l'Artillerie et du Génie], sec. 7. N.p., 1836?.

_____. "Expériences sur la dépense d'eau des moulins de la ville de Metz, et sur l'eau qui était fournie par la Moselle, à cette époque où la sécheresse était extrême, et où le produit de la rivière pouvait être considéré comme un minimum." *Bulletin des sciences technologiques* 11 (1829): 44–51.

_____. *Mémoire sur les roues hydrauliques à aubes courbes mues par-dessous suivi d'expériences sur les effets mécaniques de ces roues. . . .* New ed. Metz, 1827.

_____. "Mémoire sur les roues verticales à palettes courbes mues par en dessous, suivi d'expériences sur les effets mécaniques de ces roues." *Annales de chimie et de physique*, 2nd ser. 30 (1825): 136–88, 225–57.

_____. "Note relative au mémoire sur les roues verticales à aubes courbes, inséré dans les *Annales de chimie et de physique*, tome xxx, p. 136." *Annales de chimie et de physique*, 2nd ser. 30 (1825): 388–95.

_____. *Traité de mécanique industrielle. . . .* 2nd ed. Vol. 3. Metz, 1844.

Poncelet, Jean Victor, Pierre-Simon Girard, and C. L. M. H. Navier. "Rapport sur un mémoire ayant pour titre: 'Expériences sur les roues hydrauliques; présenté par M. Arthur Morin. . . .'" Académie des Sciences, Paris, *Comptes rendus* 3 (1836): 358–67.

Prager, Frank D., and Gustina Scaglia, eds. *Mariano Taccola and His Book "De Ingeneis."* Cambridge, Mass., 1972.

Price, Derek J. de Solla. "Gears from the Greeks: The Antikythera Mechanism—A Calendar Computer from ca. 80 B.C." American Philosophical Society, *Transactions*, n.s. 64, pt. 7 (1974): 70 pp.

Procopius of Caesarea. *History of the Wars*. transl. H. B. Dewing. Vol. 3. London and New York, 1919.

Prony, [Gaspard]. "Moyen de mesurer l'effet dynamique des machines de rotation. . . ." *Bulletin des sciences technologiques* 6 (1826): 173–79.

_____. "Note sur un moyen de mesurer l'effet dynamique des machines de rotation." *Annales de chimie et de physique*, 2nd ser. 19 (1821): 165–73.

Prudentius. *Contra orationem symmachi.* Ed. and transl. H. J. Thompson. Vol. 2. London and Cambridge, Mass., 1953.

Pryce, William. *Mineralogia Cornubiensis: A Treatise on Minerals, Mines, and Mining....* London, 1778.

Raistrick, Arthur. *Industrial Archaeology: An Historical Survey.* London, 1972.

Raistrick, Arthur, and Bernard Jennings. *A History of Lead Mining in the Pennines.* London, 1965.

Ramelli, Agostino. *Le diverse et artificiose machine.* Paris, 1588.

_____. *"The Various and Ingenious Machines" of Agostino Ramelli (1588).* Translated by Martha Teach Gnudi, with technical annotations by Eugene S. Ferguson. Baltimore, 1976.

Rankine, William J. M. *A Manual of the Steam Engine and Other Prime Movers.* 12th ed. London, 1885.

Redtenbacher, F. *Theorie und Bau der Wasserräder.* Mannheim, 1846.

Reece, David W. "The Technological Weakness of the Ancient World." *Greece and Rome,* 2nd ser. 16 (1969): 32–47.

Reindl, C. "Die Entwicklung der Wasserkraftnutzung und der Wasserkraftmaschinen." *Wasserkraft Jahrbuch,* 1924, pp. 1–31.

_____. "Ein Römisches Wasserrad." *Wasserkraft und Wasserwirtschaft* 34 (1939): 142–43.

Rennie, John. *Autobiography of Sir John Rennie.* London, 1875.

_____. "Experiments made by the late John Rennie... on the Power of Water-Wheels." *Quarterly Papers on Engineering.* Vol. 4. London, 1845.

Reti, Ladislao. "The Leonardo da Vinci Codices in the Biblioteca Nacional of Madrid." *Technology and Culture* 8 (1967): 437–45.

_____. "Leonardo da Vinci the Technologist: The Problem of Prime Movers." In C. D. O'Malley, ed., *Leonardo's Legacy: An International Symposium,* pp. 67–100. Berkeley and Los Angeles, 1969.

_____. "On the Efficiency of Early Horizontal Waterwheels." *Technology and Culture* 8 (1967): 388–94.

_____. "A Postscript to the Filarete Discussion: On Horizontal Waterwheels and Smelter Blowers in the Writings of Leonardo da Vinci and Juanelo Turriano." *Technology and Culture* 6 (1965): 428–41.

Reyburn, Wallace. *Bridge across the Atlantic: The Story of John Rennie.* London, 1972.

Reynolds, John. *Windmills and Watermills.* New York, 1970.

Reynolds, Terry S. "Science and the Water Wheel: The Development and Diffusion of Theoretical and Experimental Doctrines Relating to the Vertical Water Wheel, c. 1550–c. 1850." Ph.D. dissertation, University of Kansas, 1973.

_____. "Scientific Influences on Technology: The Case of the Overshot Water Wheel, 1752–1754." *Technology and Culture* 20 (1979): 270–95.

Rhodin, John G. A. "Christofer Polhammar, ennobled Polhem: The Archimedes of the North, 1661–1751." Newcomen Society, *Transactions* 7 (1926–27): 17–23.

Rice, A. S. "Notes on the History of Turbine Development in America." *Engineering News* 48 (1902): 208–10.

Rickard, T. A. *Man and Metals: A History of Mining in Relation to the Development of Civilization.* Vol. 2. New York and London, 1932.

Risler-Dollfus, Jérémie, and Henri Thierry. "Rapport fait un nom du comité de mécanique . . . sur la roue hydraulique de M. Marozeau, à Wesserling." Société industrielle de Mulhouse, *Bulletin* 18 (1844): 60–75.

Rivals, C. "Floating-Mills in France: A Few Notes on History, Technology, and the Lives of Men." International Symposium on Molinology, *Transactions* 3 (1973): 149–58.

———. "Tide-Mills in France: A Few Notes on History, Technology, and the Lives of Men." International Symposium on Molinology, *Transactions* 3 (1973): 159–68.

[Robison, John]. "Mechanics." *Encyclopaedia Britannica,* 10:727–85. 3rd ed. Edinburgh, 1797.

———. *System of Mechanical Philosophy.* With notes by David Brewster. 4 vols. Edinburgh, 1822.

[———]. "Water-Works." *Encyclopaedia Britannica,* 18:887–912. 3rd ed. Edinburgh, 1797.

Rojkow, H. "Ueber einige neuere hydraulische Untersuchungen und deren Anwendung auf die Uralischen Wasserwerke." *Archiv für Wissenschaftliche Kunde von Russland* (ed. A. Erman) 8 (1850): 271–306.

Rolt, L. T. C. *From Sea to Sea: The Canal du Midi.* London, 1973.

Rolt, L. T. C., and J. S. Allen. *The Steam Engine of Thomas Newcomen.* New York, 1977.

"Roue hydraulique à augets avec coyaux creux en fonte pour l'échappement et l'entrée de l'air établie par M. Brière, Directeur de filature et construite par M. Cranger, à Rouen." *Publication industrielle* 4 (3rd ed., 1862): 393–98.

"Roue hydraulique de côté des grandes dimensions à aubes planes et à coursier circulaire recevant l'eau en déversoir établis par MM. Cartier et Armengaud ainé." *Publication industrielle* 1 (5th ed., 1869): 1–45.

"Roues hydrauliques à augets recevant l'eau à leur sommet construites soit en fonte et en fer, soit en bois et en fonte," *Publication industrielle* 2 (4th ed., 1858): 425–68.

Rouse, Anthony. "Overshot Water Wheels employed in pumping water at Wheal Friendship Lead and Copper Mines, near Tavistock, in July 1841." Institution of Civil Engineers, *Minutes of Proceedings* 2 (1842): 97–102.

Rudolph, W. E. "The Lakes of Potosi." *Geographical Review* 26 (1936): 529–54.

Ruhemann, Martin. *Power.* London, 1946.

Rühlmann, Moritz. *Allgemeine maschinenlehre.* Vol. 1. Berlin, n.d. [c1862].

———. *On Horizontal Water-Wheels, especially Turbines or Whirl-Wheels: Their History, Construction, and Theory.* Edited with introduction and notes by Sir Robert Kane. Dublin, 1846.

Russell, Sidney. "The Water Wheel." *Traction and Transmission* 10 (1904): 337–47.

Safford, Arthur T., and Edward P. Hamilton. "The American Mixed-Flow Turbine and Its Setting." American Society of Civil Engineers, *Transactions* 85 (1922): 1237–1356.

Sagebien, A. "Roue de côté a niveau mittenu dans les aubes." In [Jacques Eugéne] Armengaud, *Traité théorique et pratique des moteurs hydrauliques. . . . ,* 1:98–105. New ed. Paris, 1858.

"Sagebien's Water Wheel." *The Practical Magazine* 6 (1876): 89–92.

Sagui, C. L. "La meunerie de Barbegal (France) et les roues hydrauliques chez les anciens et au moyen âge." *Isis* 38 (1947): 225–31.

*Sancti Bernardi abbatis Clarae-Vallensis vita et res gestae libris septum comprehensae.* In J. P. Migne, ed., *Patrologia Latina,* vol. 185, cols. 225–466. Paris, 1855.

Savery, Thomas. *The Miner's Friend; or, An Engine to Raise Water by Fire.* London, 1702.

Schiøler, Thorkild. *Roman and Islamic Water-Lifting Wheels.* Odense, Denmark, 1973.

Schoonhoven, J. "Sketches of Mills in Noord-Brabant." International Symposium on Molinology, *Transactions* 3 (1973): 110–19.

Schott, Caspar. *Mechanica hydraulico pneumatica....* Wurtzburg, 1657.

Sclafert, Theodore. *Le Haut-Dauphiné au Moyen Age.* Paris, 1926.

[Scott, David]. *The Engineer and Machinist's Assistant....* Glasgow, 1856.

Scott, Wilson L. *The Conflict Between Atomism and Conservation Theory, 1644–1860.* London and New York, 1970.

Scoville, Warren C. *Capitalism and French Glassmaking, 1640–1789.* Berkeley and Los Angeles, 1950.

Seeberger, Theodor. "Ableitung der Theorie der ober schlächtigen Wasserräder auf graphischem Wegen." *Der Civilingenieur* 15 (1869): 397–416.

_____. "Ableitung der Theorie der Wasserräder auf graphischem Wege." *Der Civilingenieur* 16 (1870): 339–74.

Sellergren, Gustaf. "Polhem's Contributions to Applied Mechanics." In William Johnson, transl., *Christopher Polhem: The Father of Swedish Technology,* pp. 109–62. Hartford, Conn., 1963.

Shaw, R. C. "Excavations at Willowford." Cumberland and Westmorland Antiquarian and Archaeological Society, *Transactions,* 2nd ser. 26 (1926): 429–506.

Shorter, Alfred H. *Paper Mills and Paper Makers in England, 1495–1800.* Hilversum, Holland, 1957.

Sicard, Germain. *Les moulins de Toulouse au Moyen Age.* Paris, 1953.

Simpson, F. Gerald. *Watermills and Military Works on Hadrian's Wall: Excavations in Northumberland, 1907–1913.* Edited by Grace Simpson with a contribution on watermills by Lord Wilson of High Wray. Kendal, 1976.

Sinclair, Bruce. *Philadelphia's Philosopher Mechanics: A History of the Franklin Institute, 1824–1865.* Baltimore and London, 1974.

Singer, Charles, et al., eds. *A History of Technology.* 5 vols. Oxford, 1956–58.

Smeaton, John. *A Catalogue of the Civil and Mechanical Engineering Designs, 1741–1792, of John Smeaton, F.R.S., Preserved in the Library of the Royal Society.* Ed. H. W. Dickinson and A. A. Gomme. London, 1950.

_____. "Description of the Statical Hydraulic Engine, invented and made by the late Mr. William Westgarth...." Society of Arts, Manufactures, and Commerce, *Transactions* 5 (1787): 185–210.

_____. "An experimental Enquiry concerning the natural Powers of Water and Wind to turn Mills, and other Machines, depending on a circular Motion." Royal Society of London, *Philosophical Transactions* 51 (1759): 100–174.

_____. "An Experimental Examination of the Quantity and Proportion of Mechanic Power necessary to be employed in giving different Degrees of Velocity to Heavy Bodies from a State of Rest." Royal Society of London, *Philosophical Transactions* 66 (1776): 450–75.

_____. "New Fundamental Experiments upon the Collison of Bodies." Royal Society of London, *Philosophical Transactions* 72 (1782): 337–54.

_____. *Recherches expérimentales sur l'eau et le vent, considérés comme forces motrices applicables aux moulins et autres machines a mouvement circulaire, etc....* Translated with an introduction by P. S. Girard. 2nd ed. Paris, 1827. The first edition was published in 1810.

_____. *Reports of the late John Smeaton, F.R.S., made on various occasions in the course of his employment as a civil engineer.* 3 vols. London, 1812.

Smiles, Samuel. *Lives of the Engineers....* Vol. 2. London, 1862.

Smith, Cyril S. "Granulating Iron in Filarete's Smelter." *Technology and Culture* 5 (1964): 386–90.

Smith, Cyril S., and R. J. Forbes. "Metallurgy and Assaying." In Singer et al., eds., *A History of Technology*, 3:27–71.

Smith, David M. *The Industrial Archaeology of the East Midlands.* Dawlish and London, 1965.

Smith, Norman. *A History of Dams.* London, 1971.

_____. *Man and Water: A History of Hydro-Technology.* London, 1976.

_____. "The Origins of the Water Turbine and the Invention of Its Name." *History of Technology*, 1977, pp. 215–59.

"Some Modern Breast Water Wheels." *Engineering News* 48 (1902): 436.

Somervell, John. "Water Power and Industries in Westmorland." Newcomen Society, *Transactions* 18 (1937–38): 235–44.

_____. *Water-Power Mills of South Westmorland....* Kendal, 1930.

Spain, R. J. "The Len Water-Mills." *Archaeologia Cantiana* 82 (1967): 32–104.

_____. "The Loose Watermills." *Archaeologia Cantiana* 87 (1972): 43–79; 88 (1973): 159–86.

Stadtmüller, H., ed. *Anthologia Graeca.* Vol. 2. Leipzig, 1906.

Starr, Chauncey. "Energy and Power." In *Energy and Power*, pp. 3–15. San Francisco, 1971.

Steensberg, Axel. *Bondehuse og Vandmøller i Danmark gennem 2000 År (Farms and Water Mills in Denmark during 2000 Years).* With a contribution by Valdemar M. Mikkelsen. Copenhagen, 1952. In Danish with an English summary, pp. 285–321.

Storck, John, and Walter Dorwin Teague. *Flour for Man's Bread: A History of Milling.* Minneapolis, 1952.

Stowers, Arthur. "Observations on the History of Water Power." Newcomen Society, *Transactions* 30 (1955–57): 239–56.

_____. "Watermills c 1500–c 1850." in Singer et al., eds., *A History of Technology*, 4:198–213.

Strabo. *Geography.* Transl. Horace Leonard Jones. Vol. 5. London, 1928.

Strada, Jacobus. *Künstliche Abriss allerhandt Wasserkunsten....* Frankfurt am Main, 1618.

Straet, Jan van der. *The "New Discoveries" of Stradanus.* Norwalk, Conn., 1953.

Straker, Ernest. *Wealden Iron.* London, 1931.

Strauss, Felix F. "'Mills without Wheels' in the Sixteenth-Century Alps." *Technology and Culture* 12 (1971): 23–42.

Suetonius. *De vita Caesarum.* Ed. and transl. J. C. Rolfe. 2 vols. London and New York, 1914–20.

Sundholm, Herman. "Polhem, the Mining Engineer." In William A. Johnson, transl., *Christopher Polhem: The Father of Swedish Technology*, pp. 163–82. Hartford, Conn., 1963.

Sung, Ying-Hsing. *T'ien-Kung K'ai-Wu: Chinese Technology in the Seventeenth Century.* Transl. E-tu Zen Sun and Shiou-Chuan Sun. University Park, Pa., and London, 1966.

Suplee, Henry H. *Mechanical Engineer's Reference Book*. 3rd ed. Philadelphia, c1907.

Sutcliffe, John. *A Treatise on Canals and Reservoirs, and the Best Mode of Designing and Executing Them....* Rochdale, 1816.

Swain, G. F. "Statistics of Water Power Employed in the United States." American Statistical Association, *Publications* 1 (1888): 5–44.

Swedenborg, Emanuel. *The Letters and Memorials of Emanuel Swedenborg*. Transl. and ed. Alfred Acton. 2 vols. Bryn Athyn, Pa., 1948–55.

Sweeny, F. R. I. "The Burden Water-Wheel." American Society of Civil Engineers, *Transactions* 79 (1915): 708–26.

Switzer, Stephen. *An Introduction to a General System of Hydrostaticks and Hydraulicks, Philosophical and Practical....* 2 vols. London, 1729.

Syson, Leslie. *British Water-Mills*. London, 1965.

Taccola, Mariano. *Liber Tertius de Ingeneis* [c1430]. Ed. J. H. Beck. Milan, 1969.

_____. *De Machinis: The Engineering Treatise of 1449*. Ed. Gustina Scaglia. Wiesbaden, 1971.

Tann, Jennifer. *The Development of the Factory*. London, 1970.

_____. "Multiple Mills." *Medieval Archaeology* 11 (1967): 253–55.

_____. "Some Problems of Water Power: A Study of Mill Siting in Gloucestershire." Bristol and Gloucestershire Archaeological Society, *Transactions* 85 (1965): 53–77.

Telford, Thomas. *Life of Thomas Telford....* Ed. John Rickman. London, 1838.

Temin, Peter. "Steam and Waterpower in the Early Nineteenth Century." *Journal of Economic History* 26 (1966): 187–205.

Tessier, Georges. *Recueil des Actes de Charles II Le Chauve, Roi de France*. Vol. 2 [861–877]. Paris, 1952.

Theodosius. *The Theodosian Code and Novels and the Sirmondian Constitutions*. Ed. and transl. Clyde Pharr. Princeton, 1952.

Thompson, E. A., ed. and transl. *A Roman Reformer and Inventor: Being a New Text of the Treatise "De rebus bellicis," with a Translation and Commentary*. Oxford, 1952.

Thrupp, Sylvia. "Medieval Industry, 1000–1500." In *The Fontana Economic History of Europe*, ed. Carlo M. Cipolla, 1:221–73. London and Glasgow, 1972.

Thurston, Robert H. "The Systematic Testing of Turbine Water-Wheels in the United States." American Society of Mechanical Engineers, *Transactions* 8 (1886–87): 359–420.

Tompkins, C. F. "The Overshot Wheel vs. the Turbine." *American Miller* 13 (1885): 513.

Torricelli, Evangelista. *Opere*. Ed. Gino Loria and Giuseppe Vassura. Vol. 2. Rome, 1919.

Tredgold, Thomas, ed. *Tracts on Hydraulics*. London, 1826.

Tresca, _____. "Rapport fait par M. Tresca, un nom du comité des arts mécaniques, sur la système de roues hydrauliques de M. Sagebien, ingénieur à Amiens." Société d'encouragement pour l'industrie nationale, *Bulletin*, 2nd ser. 17 (1870): 193–203.

Trevithick, Francis. *Life of Richard Trevithick*. London, 1872.

Troup, J. "On a Possible Origin of the Waterwheel." Asiatic Society of Japan, *Transactions*, 1st ser. 22 (1894): 109–14.

Truesdell, Clifford A. *Essays in the History of Mechanics*. New York, 1968.

Turnbull, John. "On Water Wheels and Turbines." Institution of Engineers and Shipbuilders in Scotland, Glasgow, *Transactions* 26 (1883): 65–100.

Turner, Trevor. "John Smeaton, FRS (1714–1792)." *Endeavour* 33 (1974): 29–33.

Turner, W. H. K. "The Significance of Water Power in Industrial Location: Some Perthshire Examples." *Scottish Geographical Magazine* 75 (1958): 98–115.

Tylecote, R. F. *A History of Metallurgy.* London, 1976.

Ubbelohde, A. R. *Man and Energy.* New York, 1955.

Uhland, W[ilhelm] H. *Handbuch für den Praktischen Maschinen-Constructeur.* Vol. 1. Leipzig, 1883.

United States. *Compendium of the Enumeration of the Inhabitants and Statistics of the United States.* Washington, 1841.

United States. Bureau of the Census. *Tenth Census, 1880.* Vols. 16–17, *Reports on the Water-Power of the United States.* Washington, 1885.

Unwin, George. *Samuel Oldknow and the Arkwrights: The Industrial Revolution at Stockport and Marple.* 2nd ed. Manchester, 1968.

Unwin, W. C. "Water Motors." *Van Nostrand's Engineering Magazine* 34 (1886): 1–19.

Ure, Andrew. *The Cotton Manufacture of Great Britain....* Vol. 1. London, 1836.

———. *Philosophy of Manufactures....* 3rd ed. London, 1861.

Usher, Abbot Payson. *A History of Mechanical Inventions.* Rev. ed. Cambridge, Mass., 1954.

Van Buren, Albert W., and Gorham Phillips Stevens. "The Antiquities of the Janiculum." American Academy in Rome, *Memoirs* 11 (1933): 69–79.

———. "The Aqua Traiana and the Mills on the Janiculum." American Academy in Rome, *Memoirs* 1 (1915–16): 59–62; 6 (1927): 137–46.

Varley, H. D. "'Lady Isabella' Wheel, Isle of Man." *Industrial Archaeology* 7 (1970): 337–39.

Vegetius Renatius, Flavius. *Végèce.* Ed. and transl. François Reyniers. Paris, 1948.

Veranzio, Fausto. *Machinae novae.* Venice, c1615.

Verriest, Léo, ed. *Le polyptyque illustré dit "Veil rentier" de Messire Jehan de Pamele-Audenarde (c1275).* Gembloux, Belgium, 1950.

Villard de Honnecourt. *Sketchbook of Villard de Honnecourt.* Ed. Theodore Bowie. Bloomington, Ind., 1959.

Vince, John. *Discovering Watermills.* Tring, Hertshire, 1970.

Vincenti, Walter G. "The Air-Propeller Tests of W. F. Durand and E. P. Lesley: A Case Study in Technological Methodology." *Technology and Culture* 20 (1979): 712–51.

Vitruvius. *On Architecture.* Ed. and transl. Frank Granger. Vol. 2. London, 1934.

Vogel, Otto. "Christopher Polhem und seine Beziehungen zur Harzer Bergbau." *Beiträge zur Geschichte der Technik und Industrie* 5 (1913): 298–345.

Von Tunzelmann, G. N. *Steam Power and British Industrialization to 1860.* Oxford, 1978.

Vowles, H. P. "Early Evolution of Power Engineering." *Isis* 18 (1932): 415–16.

———. "Pliny's Water-Mill." *Nature* 127 (1931): 889.

Wailes, Rex. "Tide Mills in England and Wales." Newcomen Society, *Transactions* 19 (1938–39): 1–33.

———. "Water-Driven Mills for Grinding Stone." Newcomen Society, *Transactions* 39 (1966–67): 95–120.

Walton, James. "South African Mills." International Symposium on Molinology, *Transactions* 3 (1973): 81–93.

———. *Water-Mills, Windmills, and Horse-Mills of South Africa.* Cape Town and Johannesburg, 1974.

Waring, W[illiam]. "Observations on the Theory of Water Mills, &c." American Philosophical Society, *Transactions* 3 (1793): 144–49, 319–21.

"Water." In *The Cyclopaedia; or, Universal Dictionary of Arts, Sciences, and Literature,* ed. by Abraham Rees [*Rees's Cyclopaedia*], vol. 38. London, 1819.

"Water Motors." In *Encyclopaedia Britannica*, 10th ed., 33:765–68. Edinburgh and London, 1902.

"Water-Power." In *Chambers's Encyclopaedia*, 10: 94–98. London, 1868.

"Water-Wheels." In *Appleton's Dictionary of Machines . . .* , 2:786–843. New York, 1855.

Watts, D. G. "Water-Power and the Industrial Revolution." Cumberland and Westmorland Antiquarian and Archaeological Society, *Transactions*, n.s. 67 (1967): 199–205.

Webber, Samuel. "Water-Power: Its Generation and Transmission." American Society of Mechanical Engineers, *Transactions* 17 (1895–96): 41–57.

Wegmann, Edward. *The Design and Construction of Dams. . . .* 4th ed. New York, 1899.

Weidler, Johan F. *Tractatus de machinis hydraulicis toro terrarvm obre maximis Marlyensi et Londinensi. . . .* Wittenberg, 1728.

Weidner, Carl R. "Theory and Test of an Overshot Water Wheel." University of Wisconsin, *Bulletin* (Engineering Series) 7 (1911–14): 117–252.

Weisbach, Julius. *A Manual of the Mechanics of Engineering and of the Construction of Machines.* Vol. 2. Translated from the 4th augmented and improved German edition by A. Jay du Bois. New York, 1877.

———. *Principles of the Mechanics of Machinery and Engineering.* Ed. Walter R. Johnson. 1st American edition. 2 vols. Philadelphia, 1848–49.

———. "Theoretische Untersuchungen über den Eintritt des Wassers in die Zellen verticaler Wasserrader." *Der Civilingenieur* 4 (1858): 95–98.

Wertime, Theodore. "Asian Influences on European Metallurgy." *Technology and Culture* 5 (1964): 394–97.

Westermann, William, and Casper J. Kraemer. *Greek Papyri in the Library of Cornell University.* Ithaca, N.Y., 1926.

Westfall, Richard S. *The Construction of Modern Science: Mechanisms and Mechanics.* New York, 1971.

———. *Force in Newton's Physics: The Science of Dynamics in the Seventeenth Century.* New York, 1971.

———. "The Problem of Force in Galileo's Physics." In Carlo L. Golino, ed., *Galileo Reappraised.* Berkeley and Los Angeles, 1966.

White, George. "The Water-Bucket Engine at Elmdon, Warwickshire." Newcomen Society, *Transactions* 16 (1935–36): 55–56.

White, Leslie. "The Energy Theory of Cultural Development." In Morton H. Fried, ed., *Readings in Anthropology*, 2: 139–46. New York, 1959.

———. *The Science of Culture.* New York, 1949.

White, Lynn. "Cultural Climates and Technological Advance in the Middle Ages." *Viator* 2 (1971): 171–201.

———. "The Expansion of Technology, 500–1500." In *The Fontana Economic History of Europe*, ed. Carlo M. Cipolla, 1:143–74. London and Glasgow, 1972.

_____. "The Historical Roots of Our Ecologic Crisis." *Science* 155 (1967): 1203–7.

_____. *Medieval Religion and Technology: Collected Essays.* Berkeley and Los Angeles, 1978.

_____. *Medieval Technology and Social Change.* Oxford, 1962.

_____. "Technology and Invention in the Middle Ages." *Speculum* 15 (1940): 141–59.

_____. "What Accelerated Technological Progress in the Western Middle Ages?" In A. C. Crombie, ed., *Scientific Change*, pp. 272–91. London, 1963.

Whitehurst, John. "Account of a Machine for Raising Water, executed at Oulton, in Cheshire, in 1772. . . ." Royal Society of London, *Philosophical Transactions* 65 (1775): 277–79.

Whitham, Jay M. *Water Rights Determination from an Engineering Standpoint.* New York, 1918.

Wiedemann, E. "Über ein arabisches, eigentümliches Wasserrad und eine kohlenwasserhaltige Höhle auf Majorka nach al Qazwînî." *Mitteilungen zur Geschichte der Medizin und der Naturwissenschaften* 15 (1916): 368–70.

Willcocks, W. *Egyptian Irrigation.* London, 1889.

Williams, David. *The Welsh Cistercians: Aspects of Their Economic History.* Pontypool, Wales, 1969.

Williamson, Kenneth. "Horizontal Water-Mills of the Faeroe Islands." *Antiquity* 20 (1946): 83–91.

Wilson, David M. "Medieval Britain in 1957, I: Pre-Conquest." *Medieval Archaeology* 2 (1958): 183–90.

Wilson, Paul N. "British Industrial Waterwheels." International Symposium on Molinology, *Transactions* 3 (1973): 17–31.

_____. "Early Water Turbines in the United Kingdom." Newcomen Society, *Transactions* 31 (1957–59): 219–41.

_____. "The Gunpowder Mills of Westmorland and Furness." Newcomen Society, *Transactions* 36 (1963–64): 47–65.

_____. "Origins of Water Power." *Water Power* 4 (1952): 308–13.

_____. "Water-Driven Prime Movers." In *Engineering Heritage: Highlights from the History of Mechanical Engineering*, 1:27–35. London, 1963.

_____. *Watermills: An Introduction.* Rev. ed. London, 1973. A Society for the Protection of Ancient Buildings booklet.

_____. *Watermills with Horizontal Wheels.* Kendal, 1960. A Society for the Protection of Ancient Buildings booklet.

_____. "Water Power and the Industrial Revolution." *Water Power* 6 (1954): 309–16.

_____. "The Waterwheels of John Smeaton." Newcomen Society, *Transactions* 30 (1955–57): 25–48.

Woodbury, George. *John Goffe's Mill.* New York, 1948.

Woodbury, Robert S. *History of the Lathe to 1850.* Cambridge, Mass., 1961.

Wulff, Hans E. "A Postscript to Reti's Notes on Juanelo Turriano's Water Mills." *Technology and Culture* 7 (1966): 398–401.

_____. *The Traditional Crafts of Persia: Their Development, Technology, and Influence on Eastern and Western Civilizations.* Cambridge, Mass., and London, 1966.

Young, Thomas. *A Course of Lectures on Natural Philosophy and the Mechanical Arts.* 2 vols. London, 1807.

Zeising, Heinrich. *Theatri machinarvm erster-[funffter] Theill.* Leipzig, 1612–14.

Zeuner, Gustav. "Ueber den Arbeitsverluft im Ausgussbogen ober- und rückerschlächtiger Wasserräder." *Der Civilingenieur* 4 (1858): 89–95.

_____. "Ueber ein eisernes Wasserräd mit Coulissenschüsse zum Betreib der Schneidemühle in Deuben bei Dresden." *Der Civilingenieur* 2 (1856): 85–87.

Zimiles, Martha, and Murray Zimiles. *Early American Mills*. New York, 1973.

Zonca, Vittorio. *Novo teatro di machine et edificii per varie et sicure operationi....* Padua, 1607.

Zosimus. *Historia nova: The Decline of Rome*. Transl. James J. Buchanan and Harold T. Davis. San Antonio, 1967.

# Illustration Credits

FIG. 1–1.  Arthur Morin, *Expériences sur les roues hydrauliques à aubes planes, et sur les roues hydrauliques à augets* (Metz and Paris: Thiel, 1836), pl. 1, fig. 5.

FIG. 1–2.  [Jaques] Armengaud, *Moteurs hydrauliques* (Paris: Baudry et Cie. and Armengaud Ainé, n.d.), pl. 14.

FIG. 1–3.  Charles Howell and Allan Keller, *The Mill at Philipsburg Manor Upper Mills and a Brief History of Milling* (Tarrytown, New York: Sleepy Hollow Restorations, 1977), p. 28. Courtesy of Sleepy Hollow Restorations.

FIG. 1–4.  R. J. Forbes, *Studies in Ancient Technology, 2 (Leiden: E. J. Brill, 1955): 33, fig. 6. Courtesy of E. J. Brill.*

FIG. 1–5.  John Storck and Walter Dorwin Teague, *Flour for Man's Bread: A History of Milling* (Minneapolis: University of Minnesota Press, 1952, p. 98, fig. 51. Courtesy of University of Minnesota Press.

FIG. 1–6.  Abbot Payson Usher, *A History of Mechanical Inventions,* rev. ed. (Cambridge, Mass.: Harvard University Press, 1954), p. 169, fig. 47. Courtesy of Harvard University Press.

FIG. 1–7.  Baron Carra de Vaux, *Le libre des appareils pneumatiques et des machines hydrauliques par Philon de Byzance* (Paris: Imprimerie nationale, 1902), p. 178.

FIG. 1–8.  G. Brett, "Byzantine Water Mill," *Antiquity* 13 (1939): pl. 7. Courtesy of Antiquity Publications Ltd.

FIG. 1–9.  Thorkild Schiøler, *Roman and Islamic Water-Lifting Wheels* (Odense, Denmark: Odense University Press, 1973), p. 155, fig. 110. Courtesy of Odense University Press.

FIG. 1–10.  Wang Chên, *Nung Shu* [1313 A.D.] (n.p., n.d.), 19, 4.

FIG. 1–11.  Ibid., 19, 10.

FIG. 1–12.  Fausto Veranzio, *Machinae Novae* (Venice, [1615–16]; reprint Milan: Ferro Edizioni, 1968), pl. 18.

FIG. 1–13.  Charles Singer et al., eds., *A History of Technology, 2* (Oxford: Clarendon, 1956): 598, fig. 545 (after C. Reindl, "Ein römisches Wasserrad," *Wasserkraft und Wasserwirtschaft* 34 [1939]: 143). Courtesy of Oxford University Press.

FIG. 1–14.   F. Gerald Simpson, *Watermills and Military Works on Hadrian's Wall: Excavations in Northumberland, 1907–1913,* ed. Grace Simpson, with a contribution on watermills by Lord Wilson of High Wray (Kendal: Titus Wilson & Son, 1976), Text fig. 2, facing p. 30. Courtesy of Grace Simpson.

FIG. 1–15.   John Storck and Walter Dorwin Teague, *Flour for Man's Bread: A History of Milling* (Minneapolis: University of Minnesota Press, 1952), p. 109, fig. 56. Courtesy of University of Minnesota Press.

FIG. 1–16.   Charles Singer et al., eds., *A History of Technology,* 2 (Oxford: Clarendon, 1956): 599, fig. 547 (after Fernand Benoit, "L'usine de meunerie hydralique de Barbegal (Arles)," *Revue archéologique,* 6th ser. 15 [1940]: 24–25, fig. 3). Courtesy of Oxford University Press.

FIG. 1–17.   R. J. Forbes, *Studies in Ancient Technology,* 2 (Leiden: E. J. Brill, 1955): 91, fig. 17. Courtesy of E. J. Brill.

FIG. 2–1.   Sketch by Helga Fack, Engineering Experiment Station, University of Wisconsin—Madison; based on data in Bradford Blaine, "The Application of Water-Power to Industry during the Middle Ages" (Ph.D. diss., University of California, Los Angeles, 1966).

FIG. 2–2.   Margaret T. Hodgen, "Domesday Water Mills," *Antiquity* 13 (1939): 267. Courtesy of Antiquity Publications Limited.

FIG. 2–3.   Richard Bennett and John Elton, *History of Corn Milling* 2 (London: Simpkin, Marshall and Co., 1899): 64, from British Library, Harl. MSS 4979, 4 b. Courtesy of the British Library.

FIG. 2–4.   Joseph Needham, *Science and Civilisation in China,* 4, pt. 2 (Cambridge: University Press, 1965): 409, fig. 628. By permission of Cambridge University Press.

FIG. 2–5.   Henry M. R. Martin, *Légende de Saint Denis* (Paris: H. Champion, 1908), pl. 64, from Bibliothèque Nationale, MS. franç 2092, fol. 37v. Courtesy of Honoré Champion and the Bibliothèque Nationale.

FIG. 2–6.   Bertrand Gille, "Le moulin à eau: Une révolution technique médiévale," *Techniques et civilisations* 3 (1954): 6, fig. 6.

FIG. 2–7.   Vittorio Zonca, *Novo teatro de machine et edificii per varie et sicure operationi.* . . . (Padua: P. Bertelli, 1607), p. 14. Photograph courtesy of Memorial Library Rare Books Department, University of Wisconsin—Madison.

FIG. 2–8.   Mariano Taccola, *Liber Tertius de Ingeneis* [c1430], ed. J. H. Beck (Milan: Edizioni il Polifilo, 1969), fol. 8, from *Codex Palatinus* 766, Biblioteca Nazionale Centrale di Firenze. Courtesy of Biblioteca Nazionale Centrale di Firenze.

FIG. 2–9.   Ibid., fol. 29–30r. Courtesy of Biblioteca Nazionale Centrale di Firenze.

FIG. 2–10.   Agostino Ramelli, *Le diverse et artificiose machine* (Paris: by the author, 1588), ch. 118. Photograph courtesy of Memorial Library Rare Books Department, University of Wisconsin—Madison.

FIG. 2–11.   Vittorio Zonca, *Novo teatro di machine et edificii per varie et sicure operationi* (Padua: P. Bertelli, 1607), p. 30. Photograph courtesy of Memorial Library Rare Books Department, University of Wisconsin—Madison.

FIG. 2–12.   Jan van der Straet (Stradanus), *New Discoveries: The Sciences, Inventions, and Discoveries of the Middle Ages and the Renaissance as Represented in Twenty-four Engravings Issued in the Early 1580s* (Norwalk, Conn.: Burndy Library, 1953), pl. 13. Courtesy of Burndy Library.

FIG. 2–13.   Georg Andreas Böckler, *Theatrum machinarum novum* (Nuremberg: Paulus Fürsten, 1661), pl. 16. Photograph courtesy of Memorial Library Rare Books Department, University of Wisconsin—Madison.

FIG. 2–14.   After Bertrand Gille, *Engineers of the Renaissance* (Cambridge, Mass.: MIT Press, 1966), p. 69.

FIG. 2–15.   Georg Andreas Böckler, *Theatrum machinarum novum* (Nuremberg: Paulus Fürsten, 1661), pl. 77. Photograph courtesy of Memorial Library Rare Books Department, University of Wisconsin—Madison.

FIG. 2–16.   Rhys Jenkins, *The Collected Papers of Rhys Jenkins* (Cambridge: The University Press, for the Newcomen Society, 1936), p. 18, fig. 3, from Emanuel Swedenborg, *De Ferro* (1734), tab. XXIX. Courtesy of the Newcomen Society.

FIG. 2–17.   Georg Andreas Böckler, *Theatrum machinarum novum* (Nuremberg: Paulus Fürsten, 1661), pl. 116. Photograph courtesy of Memorial Library Rare Books Department, University of Wisconsin—Madison.

FIG. 2–18.   Georgius Agricola, *De re metallica* (Basel: Froben, 1556), p. 164. Photograph courtesy of Memorial Library Rare Books Department, University of Wisconsin—Madison.

FIG. 2–19.   Ibid., p. 142. Photograph courtesy of Memorial Library Rare Books Department, University of Wisconsin—Madison.

FIG. 2–20.   Vittorio Zonca, *Novo teatro di machine et edificii per varie et sicure operationi* (Padua: P. Bertelli, 1607), p. 68. Photograph courtesy of Memorial Library Rare Books Department, University of Wisconsin—Madison.

FIG. 2–21.   Mariano Taccola, *De Machinis: The Engineering Treatise of 1449*, ed. Gustina Scaglia (Wiesbaden: Dr. Ludwig Reichert Verlag, 1971), fol. 83r, from *Codex Latinus Monacensis* 28800, Bayerische Staatsbibliothek. Courtesy of the Bayerische Staatsbibliothek.

FIG. 2–22.   Vittorio Zonca, *Novo teatro di machine et edificii per varie et sicure operationi* (Padua: P. Bertelli, 1607), p. 42. Photograph courtesy of Memorial Library Rare Books Department, University of Wisconsin—Madison.

FIG. 2–23.   Georg Andreas Böckler, *Theatrum machinarum novum* (Nuremberg: Paulus Fürsten, 1661), pl. 73. Photograph courtesy of Memorial Library Rare Books Department, University of Wisconsin—Madison.

FIG. 2–24.   Friedrich Klemm, *A History of Western Technology* (Cambridge: MIT Press, 1964), p. 104, fig. 34, from Hugo Spechtshart, *Flores musicae* (Strasbourg: Johann Prüss, 1488).

FIG. 2–25.   Mariano Taccola, *De Machinis: The Engineering Treatise of 1449*, ed. Gustina Scaglia (Wiesbaden: Dr. Ludwig Reichert Verlag, 1971), fol. 43v, from *Codex Latinus Monacensis* 28800, Bayerische Staatsbibliothek. Courtesy of the Bayerische Staatsbibliothek.

FIG. 2–26. Georgius Agricola, *De re metallica* (Basel: Froben, 1556), p. 290. Photograph courtesy of the Memorial Library Rare Books Department, University of Wisconsin—Madison.

FIG. 2–27. Ibid., p. 223. Photograph courtesy of Memorial Library Rare Books Department, University of Wisconsin—Madison.

FIG. 2–28. Bertrand Gille, "Le moulin à eau: Une révolution technique médiévale," *Techniques et civilisations* 3 (1954): 11, fig. 11.

FIG. 2–29. Isaak de Caus, *New and Rare Inventions of Water-Works* (London: John Moxon, 1659), pl. 11. Photograph courtesy of the Memorial Library Rare Books Department, University of Wisconsin—Madison.

FIG. 2–30. Agostino Ramelli, *Le diverse et artificiose machine* (Paris: by the author, 1588), ch. 137. Photograph courtesy of the Memorial Library Rare Books Department, University of Wisconsin—Madison.

FIG. 2–31. Ibid., ch. 100. Photograph courtesy of Memorial Library Rare Books Department, University of Wisconsin—Madison.

FIG. 2–32. Vannoccio Biringuccio, *De la pirotechnia* (Venice: Venturio Roffinello, 1540), p. 141r. Photograph courtesy of Memorial Library Rare Books Department, University of Wisconsin—Madison.

FIG. 2–33. Cipriano Piccolpasso, *The Three Books of the Potter's Art* [c1560], transl. Bernard Rackham and Albert Van de Put (London: Victoria and Albert Museum, 1934), fol. 37. Courtesy of the Victoria and Albert Museum.

FIG. 2–34. Herrad of Hohenbourg, *Hortus Deliciarum* [c1190], ed. Rosalie Green, Michael Evans, Christine Bischoff, and Michael Curschmann, vol. 2 (London: The Warburg Institute, University of London, 1979), fol. 112r, pl. 69, no. 148, p. 183, from the collection of the Bibliothèque Nationale et Universitaire de Strasbourg. Courtesy of the Warburg Institute and the Bibliothèque Nationale et Universitaire de Strasbourg.

FIG. 2–35. After Léo Verriest, ed., *Le polyptyque illustré dit "Veil rentier" de Messire Jehan de Pamele-Audenarde* [c1275] (Gembloux, Belgium: J. Duculot, 1950), fol. 98v.

FIG. 2–36. Charles Singer et al., eds., *A History of Technology*, 2 (Oxford: Clarendon, 1956): 648, fig. 591 (after W. W. Cook and J. F. Ricard, *Ars Hispaniae*, vol. 6 [Madrid: Editorial Plus Ultra, 1950], fig. 261). Courtesy of Oxford University Press.

FIG. 2–37. Karl von Amira, ed., *Die Dresdener Bilderhandschrift des Sachsenspiegels* (Leipzig: Karl W. Hiersemann, 1902), fol. 77b.

FIG. 2–38. Eric G. Millar, ed., *The Luttrell Psalter* [c1340] (London: Trustees of the British Museum, 1932), pl. 114, from British Library, Additional Manuscript 42130. By permission of the British Library.

FIG. 2–39. Conrad Kyeser, *Bellifortis* [c1405], ed. Götz Quart (Düsseldorf: VDI-Verlag, 1967), fol. 84, from Göttingen Universitätsbibliothek Ms. philos. 63, fol. 84. Courtesy of VDI-Verlag and Niedersächsische Staats- und Universitätsbibliothek, Göttingen.

FIG. 2–40. Francesco di Giorgio Martini, *Trattati di Architettura, Ingegneria, e Arte Militare*, ed. Corrado Maltese and Livia Maltese Degrassi, vol. 1 (Milan: Edizioni il Polifilo, 1967), fol. 35, from Biblioteca Reale di

Torino, *Codice membranaceo Saluzziano* 148. Courtesy of Biblioteca Reale di Torino.

FIG. 2–41.  Conrad Kyeser, *Bellifortis* [c1405], ed. Götz Quart (Düsseldorf: VDI-Verlag, 1967), fol. 57, from Göttingen Universitätsbibliothek Ms. philos. 63, fol. 57. Courtesy of VDI-Verlag and Niedersächsische Staats- und Universitätsbibliothek, Göttingen.

FIG. 2–42.  Ying-Hsing Sung, *T'ien Kung K'ai Wu* [1637] (Shanghai: Commercial Publishing Co., n.d.), 92.

FIG. 2–43.  Ms. Greaves 27, fol. 101r, Bodleian Library, Oxford. By permission of the Bodleian Library, Oxford.

FIG. 3–1.  Archibald Allison, "The Water Wheels of Sheffield," *Engineering* 165 (1948): 167, fig. 5. Courtesy of *Engineering* (London).

FIG. 3–2.  Ekaterinburg plant sketch based on V. V. Danilevskii, *History of Hydroengineering in Russia before the Nineteenth Century* (Jerusalem: Israel Program for Scientific Translations, 1968), p. 56.

FIG. 3–3.  Charles Bossut and Viallet, *Untersuchungen über die beste Construction der Deiche*, transl. C. Kröncke into German (Frankfurt am Main: Behrens- und Kornerischen Buchhandlung, 1798), figs. 50–54.

FIG. 3–4.  Edward Wegmann, *The Design and Construction of Dams*, 4th ed. (New York: John Wiley, 1899), pl. 82.

FIG. 3–5.  Johannes Beyer, *Theatrum machinarum molarium* [1735] (Dresden: Waltherischen Hof-Buchhandlung, 1767), pl. 42. Photograph courtesy of Memorial Library Rare Books Department, University of Wisconsin—Madison.

FIG. 3–6.  Georg Andreas Böckler, *Theatrum machinarum novum* (Nuremberg: Paulus Fürsten, 1661), pl. 45. Photograph courtesy of Memorial Library Rare Books Department, University of Wisconsin—Madison.

FIG. 3–7.  Jacob Leupold, *Theatri machinarum hydraulicarum*, vol. 2 (Leipzig: Druckts Christoph Zunckel, 1725), pl. 26, fig. 1. Photograph courtesy of Memorial Library Rare Books Department, University of Wisconsin—Madison.

FIG. 3–8.  *Encyclopédie; ou, Dictionnaire raisonné des sciences*, ed. Denis Diderot and Jean d'Alembert, *Supplement, Planches* (1772), "Arte militaire, Fabrique des armes," pl. 3. Photograph courtesy of Memorial Library Rare Books Department, University of Wisconsin—Madison.

FIG. 3–9.  *Ibid., Planches*, vol. 4 (1765), "Grosses forges," last sec., pl. 2. Photograph courtesy of Memorial Library Rare Books Department, University of Wisconsin—Madison.

FIG. 3–10.  Georges Agricola, *De re metallica* (Basel: Froben, 1556), p. 234. Photograph courtesy of Memorial Library Rare Books Department, University of Wisconsin—Madison.

FIG. 3–11.  *Encyclopédie; ou, Dictionnaire raisonné des sciences*, ed. Denis Diderot and Jean d'Alembert, *Planches*, vol. 8 (1771), "Plomb, Laminage de," pl. 12. Photograph courtesy of Memorial Library Rare Books Department, University of Wisconsin—Madison.

FIG. 3–12.  Georg Andreas Böckler, *Theatrum machinarum novum* (Nuremberg: Paulus Fürsten, 1661), pl. 70. Photograph courtesy of Memorial

FIG. 3–13.    Library Rare Books Department, University of Wisconsin—Madison.

FIG. 3–13.    *Encyclopédie; ou, Dictionnaire raisonné des sciences,* ed. Denis Diderot and Jean d'Alembert, *Planches,* vol. 6 (1768), "Minéralogie, Fabrique de la Poudre à Canon," pl. 2. Photograph courtesy of Memorial Library Rare Books Department, University of Wisconsin—Madison.

FIG. 3–14.    Ibid., pl. 7. Photograph courtesy of Memorial Library Rare Books Department, University of Wisconsin—Madison.

FIG. 3–15.    Ibid., *Planches,* vol. 4 (1765), "Glaces," pl. 46. Photograph courtesy of Memorial Library Rare Books Department, University of Wisconsin—Madison.

FIG. 3–16.    Isaak de Caus, *New and Rare Inventions of Water-Works* (London: Joseph Moxon, 1659), pl. 19. Photograph courtesy of Memorial Library Rare Books Department, University of Wisconsin—Madison.

FIG. 3–17.    Georg Andreas Böckler, *Theatrum machinarum novum* (Nuremberg: Paulus Fürsten, 1661), pl. 130. Photograph courtesy of Memorial Library Rare Books Department, University of Wisconsin—Madison.

FIG. 3–18.    Julius Weisbach, *A Manual of the Mechanics of Engineering,* transl. A. Jay du Bois from the 4th German edition, vol. 2 (New York: John Wiley & Sons, 1877), p. 241, figs. 392–94.

FIG. 3–19.    *The Cyclopaedia; or, Universal Dictionary of Arts, Sciences, and Literature,* ed. Abraham Rees, Plates, vol. 3 (London, 1820), "Mill Work," pl. 2, fig. 17.

FIG. 3–20.    Joseph P. Frizell, "The Old-Time Water-Wheels of America," American Society of Civil Engineers, *Transactions* 28 (1893): pl. 36, fig. 3.

FIG. 3–21.    *Encyclopédie; ou, Dictionnaire raisonné des sciences,* ed. Denis Diderot and Jean d'Alembert, *Planches,* vol. 1 (1762), "Oeconomie rustique, Sucrerie," pl. 2. Photograph courtesy of the Memorial Library Rare Books Department, University of Wisconsin—Madison.

FIG. 3–22.    Ibid., *Supplement, Planches* (1772), "Meunier," pl. 1. Photograph courtesy of the Memorial Library Rare Books Department, University of Wisconsin—Madison.

FIG. 3–23.    Oliver Evans, *The Young Mill-Wright and Miller's Guide* (Philadelphia: by the author, 1795), pl. 16.

FIG. 3–24.    Bernard Forest de Bélidor, *Architecture hydraulique,* vol. 1 (Paris: Charles-Antoine Jombert, 1737), bk. 2, ch. 1, pl. 1, fig. 2. Photograph courtesy of Memorial Library Rare Books Department, University of Wisconsin—Madison.

FIG. 3–25.    Ibid., vol. 2 (1739), bk. 3, ch. 4, pl. 4, fig. 5. Photograph courtesy of Memorial Library Rare Books Department, University of Wisconsin—Madison.

FIG. 3–26.    Ibid., vol. 1: bk. 2, ch. 1, pl. 1, fig. 4. Photograph courtesy of Memorial Library Rare Books Department, University of Wisconsin—Madison.

FIG. 3–27.    John T. Desaguliers, *Course of Experimental Philosophy,* vol. 2 (London: W. Innys, M. Senex, T. Longman, 1744), pl. 32.

FIG. 3–28. Vittorio Zonca, *Novo teatro di machine et edificii per varie et sicure operationi* (Padua: P. Bertelli, 1607), p. 21. Photograph courtesy of Memorial Library Rare Books Department, University of Wisconsin—Madison.

FIG. 3–29. Georgius Agricola, *De re metallica* (Basil: Froben, 1556), p. 158. Photograph courtesy of Memorial Library Rare Books Department, University of Wisconsin—Madison.

FIG. 3–30. Ibid., p. 125. Photograph courtesy of Memorial Library Rare Books Department, University of Wisconsin—Madison.

FIG. 3–31. John T. Desaguliers, *Course of Experimental Philosophy*, vol. 2 (London: W. Innys, M. Senex, T. Longman, 1744), pl. 29.

Fig. 3–32. Bernard Forest de Bélidor, *Architecture hydraulique*, vol. 2 (Paris: Charles-Antoine Jombert, 1739), bk. 3, ch. 4, pl. 8, fig. 1. Photograph courtesy of Memorial Library Rare Books Department, University of Wisconsin—Madison.

FIG. 3–33. Ibid., pl. 18, fig. 3. Photograph courtesy of Memorial Library Rare Books Department, University of Wisconsin—Madison.

FIG. 3–34. Jacob Leupold, *Theatri machinarum hydraulicarum*, vol. 2 (Leipzig: Druckts Christoph Zunckel, 1725), pl. 25. Photograph courtesy of Memorial Library Rare Books Department, University of Wisconsin—Madison.

FIG. 3–35. Louis Figuier, *Les merveilles de l'industrie*, 3 (Paris: Furne, Jouvet et Cie., n.d.): 361.

FIG. 3–36. Agostino Ramelli, *Le diverse et artificiose machine* (Paris: by the author, 1588), ch. 117. Photograph courtesy of Memorial Library Rare Books Department, University of Wisconsin—Madison.

FIG. 3–37. Bernard Forest de Bélidor, *Architecture hydraulique*, vol. 1 (Paris: Charles-Antoine Jombert, 1737), bk. 2, ch. 1, pl. 7, fig. 10. Photograph courtesy of Memorial Library Rare Books Department, University of Wisconsin—Madison.

FIG. 3–38. Ibid., vol. 2 (1739), bk. 4, ch. 1, pl. 6, figs. 1–2. Photograph courtesy of Memorial Library Rare Books Department, University of Wisconsin—Madison.

FIG. 4–1. Sketches by Helga Fack, Engineering Experiment Station, University of Wisconsin—Madison; after Leonardo da Vinci, *Codex Atlanticus*, fol. 209r.

FIG. 4–2. After Leonardo da Vinci, *Codex Atlanticus*, fols. 2r, 151r.

FIG. 4–3. Drawing by the author.

FIG. 4–4. Henri Pitot, "Remarques sur les aubes ou pallettes des moulins, & autres machines mûës par le courant des rivières," Académie des Sciences, Paris, *Mémoires*, 1729, pl. 20, fig. 3, facing p. 258.

FIG. 4–5. Henri Pitot, "Comparaison entre quelques machines mûës par les courants des fluides," ibid., pl. 21, facing p. 292.

FIG. 4–6. John T. Desaguliers, *Course of Experimental Philosophy*, vol. 2 (London: W. Innys, M. Senex, T. Longman, 1744), pl. 33, figs. 5 and 9.

FIG. 4–7. Antoine de Parcieux, "Mémoire sur une expérience qui montre qu'a dépense égale plus une roue à augets tourne lentement, plus elle fait d'effet," Académie des Sciences, Paris, *Mémoires*, 1754, pl. 21, facing p. 676.

FIG. 4–8. Antoine de Parcieux, "Mémoire dans lequel on prouve que les aubes des roues mûes par les courans des grandes rivières, feroient beaucoup plus d'effet si elles étoient inclinées aux rayons . . . ," ibid., 1759, pl. 10, facing p. 298.

FIG. 4–9. John Smeaton, "An experimental Enquiry concerning the natural Powers of Water and Wind to turn Mills, and other Machines, depending on a circular Motion," Royal Society of London, *Philosophical Transactions* 51 (1759): pl. 4, facing p. 100.

FIG. 4–10. Ibid., pl. 5, facing p. 102.

FIG. 4–11. Sten Lindroth, *Christopher Polhem och Stora Kopparberget* (Uppsala: Almqvist & Wiksell, 1951), p. 84, fig. 40. Courtesy of Almqvist & Wiksell Förlag AB.

FIG. 4–12. Bruce Sinclair, *Philadelphia's Philosopher Mechanics: A History of the Franklin Institute, 1824–1865* (Baltimore and London: The Johns Hopkins University Press, 1974), p. 161. Courtesy of The Johns Hopkins University Press.

FIG. 4–13. Ibid., p. 163. Courtesy of The Johns Hopkins University Press.

FIG. 4–14. Arthur Morin, *Expériences sur les roues hydraliques à aubes planes, et sur les roues hydrauliques à augets* (Metz and Paris: Thiel, 1836), pl. 1, fig. 1.

FIG. 4–15. Jean Victor Poncelet, *Mémoire sur les roues hydrauliques à aubes courbes, mues par-dessous,* new ed. (Metz: Me. Ve. Thiel, 1827), pl. 1, fig. 1.

FIG. 4–16. Wilhelm Müller, *Die Eisernen Wasserräder* 2 (Leipzig: Veit & Co., 1899): 117.


FIG. 5–1. *The Cyclopaedia; or, Universal Dictionary of Arts, Sciences, and Literature,* ed. Abraham Rees, plates, vol. 2 (1820), "Cotton Manufacture," pl. 14.

FIG. 5–2. Drawing by the author.

FIG. 5–3. William Fairbairn, *Treatise on Mills and Millwork,* 3rd ed., 1 (London: Longmans, Green, 1871): 82, fig. 84.

FIG. 5–4. Jacob Leupold, *Theatrum machinarum generale,* new ed. (Leipzig: Bernhard Christoph Breitkopfs und Sohn, 1774), pl. 62, fig. 4. Photograph courtesy of Memorial Library Rare Books Department, University of Wisconsin—Madison. The first edition was published in 1724.

FIG. 5–5. Bernard Forest de Bélidor, *Architecture hydraulique,* vol. 1 (Paris: Charles-Antoine Jombert, 1737), bk. 1, ch. 1, pl. 1, fig. 3. Photograph courtesy of Memorial Library Rare Books Department, University of Wisconsin—Madison.

FIG. 5–6. John Smeaton, *Reports of the late John Smeaton,* vol. 1 (London: Longman, Hurst, Rees, Orme, and Brown, 1812), pl. 27.

FIG. 5–7. *The Cyclopaedia; or, Universal Dictionary of Arts, Sciences, and Literature,* ed. Abraham Rees, plates, vol. 4 (1820), "Waterwheels," pl. 2, fig. 4.

FIG. 5–8. F. Redtenbacher, *Theorie und Bau der Wasserräder* (Mannheim: Friedrich Bassermann, 1846), pl. 1, figs. 5–6.

FIG. 5–9. Paul N. Wilson, "The Waterwheels of John Smeaton," Newcomen Society, *Transactions* 30 (1955–57): 28. Courtesy of the Newcomen Society.

FIG. 5–10. William Fairbairn, *Treatise on Mills and Millwork*, 3rd ed., 1 (London: Longmans, Green, 1871): 125, fig. 113.

FIG. 5–11. *Encyclopédie; ou, Dictionnaire raisonné des sciences*, ed. Denis Diderot and Jean d'Alembert, *Planches*, vol. 1 (1762), "Agriculture, Economie rustique, Moulin à eau," pl. 6. Photograph courtesy of Memorial Library Rare Books Department, University of Wisconsin—Madison.

FIG. 5–12. *The Cyclopaedia; or, Universal Dictionary of Arts, Sciences, and Literature*, ed. Abraham Rees, Plates, vol. 4 (1820), "Waterwheels," pl. 2, fig. 3.

FIG. 5–13. Joseph Glynn, *Treatise on the Power of Water*, 3rd ed. (New York: D. Van Nostrand, 1869), p. 91, fig. 52.

FIG. 5–14. William Fairbairn, *Treatise on Mills and Millwork*, 3rd ed., 1 (London: Longmans, Green, 1871): 120, fig. 103.

FIG. 5–15. Ibid., p. 118, fig. 101.

FIG. 5–16. Thomas Baines and William Fairbairn, *Lancashire and Cheshire: Past and Present*, 2 (London: William MacKenzie, 1868): cxlvi–cxlvii.

FIG. 5–17. Julius Weisbach, *A Manual of the Mechanics of Engineering*, transl. A. Jay du Bois, from the 4th German ed., 2 (New York: John Wiley & Sons, 1871): 242, figs. 397–98.

FIG. 5–18. Ibid., p. 243, figs. 399–401.

FIG. 5–19. William Fairbairn, *Treatise on Mills and Millwork*, 3rd ed., 1 (London: Longmans, Green, 1871): 122, figs. 107–8.

FIG. 5–20. *Publication industrielle*, 4th ed., Plates, 2 (Paris, 1858): 37.

FIG. 5–21. William Fairbairn, *Treatise on Mills and Millwork*, 3rd ed., 1 (London: Longmans, Green, 1871): 121, fig. 104; C. Bach, *Das Wasserräder* (Stuttgart: Konrad Wittwer, 1886), plates, pl. 9, fig. 1.

FIG. 5–22. Sketch by Helga Fack, Engineering Experiment Station, University of Wisconsin—Madison; based on F. Redtenbacher, *Theorie und Bau der Wasserräder* (Mannheim: F. Bassermann, 1846), pl. 7.

FIG. 5–23. Reprinted by permission of the Council of the Institution of Mechanical Engineers from Paul N. Wilson, "Water-Driven Prime Movers," in *Engineering Heritage* 1 (London: Institution of Mechanical Engineers, 1963): 31, fig. 6.

FIG. 5–24. William Fairbairn, *Treatise on Mills and Millwork*, 3rd ed., vol. 1 (London: Longmans, Green, 1871), pl. 2, facing p. 130.

FIG. 5–25. Ibid., pl. 4, facing p. 140.

FIG. 5–26. Ibid., p. 135, fig. 116.

FIG. 5–27. *Appleton's Dictionary of Machines*, 2 (New York: D. Appleton, 1855): 795.

FIG. 5–28. F. G. Coggin, "The Ferris and Other Big Wheels," *Cassier's Magazine* 6 (1894): 214.

FIG. 5–29. F. R. I. Sweeny, "The Burden Water-Wheel," American Society of Civil Engineers, *Transactions* 79 (1915): 714–15.

FIG. 6–1. Jacob Leupold, *Theatri machinarum hydraulicarum*, vol. 2 (Leipzig: Druckts Christoph Zunckel, 1725), pl. 43, fig. 1. Photograph courtesy of Memorial Library Rare Books Department, University of Wisconsin—Madison.

# Index

*Note: Most place names and river names are not indexed individually. See the country in which they are located.*

441

Deville, de la, 331
Dibner, Bern, 172–73, 174
Dickinson, H. W., 337
Diderot, Denis, 157, 282, 290
Dieu, Théodore, 257–58
Diocletian, 30–31, 33, 35
Domesday Book, watermills in, 52, 53, 64, 67, 86, 103–4, 123
Donkin, Bryan, 307–8
Donkin, R. A., 110
Donnachie, Ian, 267
Douglas, William, 269
Drachmann, A. G., 16
Drawing Engines. *See* Hoists, water-powered
*Dresdener Bilderhandschrift des Sachsenspiegels*, 99, 100
DuBuat, Pierre L. G., 249
Duby, Georges, 52
DuCrest, Charles-Louis, 242
DuPetit-Vandin, Robert-Xavier A., 221, 227; and Parent's theory, 243–44; and Pitot's theory, 213–14
Dupin, Charles, 329
DuPont, Alexis, 344
Dürer, Albrecht, 92
Dye mills, 123

Edge-runner stones: in gunpowder mills, 147, 149; in medieval water mills, 70–75
Edward I, 64
Efficiency: of breast wheels, 178, 256, 264, 307, 319; of horizontal water wheels, 14, 106; of hydraulic rams, 337; of iron water wheels, 306–7, 319; origins of concept of, 201; of overshot water wheel, 11, 106, 178, 215, 219, 226, 229, 233, 235, 236, 239, 246, 253, 255, 257, 307, 347; of Poncelet wheel, 260–61; of Sagebien wheel, 263, 264; of undershot water wheel, 11, 106, 178, 225, 233, 234, 236, 239, 246, 251, 255, 307; of water lever, 11; of water turbines, 339, 341, 342, 347; of water wheels, 207, 211, 213, 240, 245, 246
Efflux law, 201–2, 203, 205
Egen, P.N.E., 257
Egypt, 113, 120
Ekaterinberg Dam, 126, 130, 131, 132, 155
Elton, John, 20–21, 64
Elvius, Pehr, 235–36
England. *See* Britain
Entwistle Dam, 273, 275

Escher, Wyss & Co., 340
Essonnes, France, 148, 149
Euler, Johann A., 239, 243, 244, 245, 246, 282; theory of, 233–37
Euler, Leonhard, 207, 214–15, 233, 234, 235, 338
Evans, Oliver, 157, 159, 160, 165, 173–74, 248, 264; theory of, 245, 246
Ewart, Peter, 247–48, 282, 290
Experiments, water wheel, 227–28, 250–52, 257–58; of DeParcieux, 219–22; of Desaguliers, 216–17; early use of models in, 230–31; of Franklin Institute, 252–55; of Morin, 255–57; of Polhem, 228–31; of Smeaton, 223–26, 233
Eytelwein, Johann A., 337

Fabre, Jean-Antoine, 193
Factory: characteristics of, 95; influence of water power on emergence of, 5–6, 95–96
Fairbairn, William, 273, 274, 286, 308, 309, 310, 319, 320; examples of water wheels of, 310–11, 312, 314, 315, 316; on Hewes, 292–93; on millwrights, 191–92; and the suspension wheel, 295–98; and ventilated buckets, 297–300
Fang-I, 116
Fans, water-powered: in China, 116; in mines, 78, 79
Farey, John, 282, 288, 322, 330
Ferguson, James, 173, 243
Ferrendier, M., 124–25, 127, 173, 174
Fertilizer Mills, 151
Feudal system. *See* Manorial system, and watermills
Filarete, Antonio di, 64, 79, 104
Finlay, James, 299, 314
Finugio, Geronomo, 189
Fitzherbert, Anthony, 127, 128, 157, 159, 164, 194, 278
Flint mills, 151
Floatboards. *See* Blades
Floating mills. *See* Boat mills
Floods, 326; advantage of high breast wheels in, 284, 285; operation of undershot wheels in, 186, 188
Flour mills, 4, 70, 71, 102, 123, 138–39, 173–74, 329
Fludd, Robert, 331
Flumes. *See* Chutes
Flutter wheel, 323n.14
Fontana, Domenico, 230, 232

early watermill at); horizontal watermills in, 103
Greek mill. *See* Horizontal watermill
Greenock, Scotland, water power system of. *See* Shaw's Water Works
Gregory of Tours, 49
Grimshaw, Robert, 157, 347
Grinding mills. *See* Cutlery mills
Grindstones, use of water power with, 70, 74, 76
Gudgeons: of hybrid water wheels, 306; of iron water wheels, 301; of traditional wooden water wheels, 159, 160, 287, 306
Guenyveau, Antoine, 247, 333, 335
Gunpowder mills, 74, 145–49
Guttmann, Oscar, 147

Haltwhistle Burn, England, ancient watermill at, 36, 37, 41, 42
Hammer forges. *See* Iron mills
Hammers, hydraulic trip-, 70; in China, 26, 28, 29, 115, 116; medieval applications of, 70, 81–89; types of, 79, 81
Han Chi, 27, 28
Hanging mills. *See* Bridge mills
Hargreaves, James, 269
Hartlib, Samuel, 174
Harz (mining region, Germany), 271, 333; hydropower dams, canals, and reservoirs in, 129–30, 133–34, 155; power of water wheels in, 179, 311
Hemp mills, 70, 83, 94, 116
Hero of Alexandria, 79
Héron de Villefosse, A. M., 174, 179
Herrad (abbess of Landsberg), 97–98
Hewes, T. C., 273, 290, 295, 297, 298, 309, 312; develops suspension wheel, 292–94; governor used by, on water wheels, 294–95
High breast wheel, 284–86, 307. *See also* Breast wheel
Hill, Donald, 119
Hills, Richard, 289–90
Hobart, James F., 287
Hoists, water-powered: in Harz, 134; in medieval mines, 70, 79; in mines, 1500–1750, 141, 142, 175, 180
Höll, Joseph K., 331–33
Holland, 50, 369n.12
Holland (English technician), 193
Hollander, 139
Holyoke, Mass., 277, 346
Honorius, 31

Horizontal watermill, 4, 7, 9, 112, 135; advantages of, 104; in Denmark, 18; description of, 14, 15, 352n.2; distribution of, 103–4; efficiency of, 14, 106; in Islam, 118–19, 367n.212; in medieval Europe, 49, 51, 65, 97, 103–8; origins of, 14–30 passim; power of, 106
Horizontal water wheel, 108, 338; in China, 18, 22, 23, 115; definition of, 352n.2; in Renaissance Europe, 108, 135. *See also* Horizontal watermill
*Hortus Deliciarum,* 97–98
*Hou Han Shu,* 18
Howell, Charles, 138
Hubs: of ancient water wheels, 36, 42, 43; of hybrid water wheels, 304; of iron water wheels, 297, 298, 301
Hughes, William C., 157, 347
Hungary, 135, 333
Hunter, Louis, 6–7, 55, 135–36, 152, 154
"Hussite" Engineer, 75, 81, 364n.154
Huygens, Christian, 202–3, 204, 205, 206, 211, 212
Hybrid water wheels: description of, 304, 306; examples of large, 316–19
Hydraulic ram, 10, 335–38
Hydrologic cycle, 9

Ibn Hawkal, Abū al-Kāsim Muhammad, 117
Idrisi, Abu 'Abd Alla Muhammad al-, 57, 117, 119
*Image du monde* (Gautier of Metz), 97
Impulse, method of calculating: of Bernoulli, 213, 234; of Borda, 239–40; of Huygens and Mariotte, 203, 205; of Parent, 206, 211, 243, 244
Impulse wheel. *See* Undershot water wheel
India, 21–22, 31
Industrial location, influence of water power on, 5, 82, 85, 326–28
Industrial revolution: influence of, on demise of wooden water wheel, 266–71; roots of, in medieval use of water power, 69–70
Inval turbine, 341
Iran, 117, 119
Ireland: horizontal watermills in, 51, 103, 105, 107; use of water power in linen industry of, 137
Iron: use of, in traditional wooden water wheels, 158–59, 287–88; use of water power in production of, 5, 18, 27, 127, 142–43, 266–67. *See also* Iron mills;

Iron (*continued*)
Iron water wheels; Hybrid water wheels
Iron mills, 70, 85–87, 94, 96, 123
Iron water wheels: construction of,
300–306; dimensions of, 307–10,
312–13; efficiency of, 306–7, 319;
emergence of, 286–90; examples of,
307–19; life of, 319, 320; power of,
310–11, 312–13; suspension, 290–99
Irrigation, and water power, 113, 116–17
Irwell River, 276
Islam, use of water power in, 30, 113,
117–20
Istakhri, al-, 117
Italy, 94; ancient, watermills in, 18, 36;
blast furnaces in, 87; boat mills in, 125;
diffusion of water power from, 49, 50,
51; fulling mills in, 82, 94; horizontal
watermills in, 103, 135; hydraulic bellows in, 86, 94; hydropower dams in,
63–64, 130; iron mills in, 86; medieval,
watermills in, 49; paper mills in, 84, 94;
silk mills in, 79, 136; tanning mills in,
75; water turbines in, 342

Jacono, Luigi, 36
Janiculum, the, ancient watermills at, 31,
33, 34, 41, 44
Japan, 20
Jazari, Ibn al-Razzaz al-, 118–19
Jespersen, Anders, 49, 104
Jonnes, Alexandre Moreau de, 329
Juan y Santacilia, Jorge, 245, 246
Justinian, 31
Jutland. *See* Denmark

Kallstenius, G. S., 251
Kao Li-Shih, 115
Kästner, Abraham, 236
Kent River, 276
Koran, the, 120
Kyeser, Conrad, 99, 101, 103

Labor supply, influence of, on diffusion
of water power: in antiquity, 32, 34–35,
36; in Islam, 120; in medieval Europe,
112–13
Lady Isabella Wheel, 307, 313, 315–16,
318, 320
Lagerhjelm, Pehr, 251
LaGrange, Joseph Louis, 241, 242
LaHire, Philippe de, 204–6
Lambert, Johann, 243
Lanz, Philippe Louis, 247

Lathes, water-powered, 70, 76, 94, 136,
143
Lawrence, Mass., water-power system of,
277
Lead: use of water power in, metallurgy,
144–45, 146; use of water power in,
mines, 141
Leat. *See* Canals, hydropower
Lefevre-Gineau, Louis, 242
Leibnitz, Gottfried Wilhelm, 234
Leo I, 36
Leonardo da Vinci. *See* Vinci, Leonardo
da
Leupold, Jacob, 157, 173; and breast
wheel, 278–80; and horizontal water-
mill, 104; and importance of water
power, 5, 151; water-bucket engine of,
190
Levainville, Jacques, 127
Li Hsi-Yün, 116
Lillie, James, 295, 299, 312, 320
Li Yuan-Hung, 116
Linen mills, 137
Living Force. *See* Vis Viva
Lombe, Thomas, 136–37
Lomonosov, Michail V., 149
London Bridge Water Works, 175–76,
181, 212
Loring, Charles, 2
Lough Island Reavy, 273, 274, 275
Lou Shou, 28–29
Louis XI, 59
Louis XIV, 182, 230
Louis XV, 218
Lowell, Mass., water power system at, 7,
276–77, 346
Lubrication, of water wheels, 160
Luckhurst, David, 110
Lucretius, 17, 22, 23, 25
*Luttrell Psalter*, 99, 100, 101

MacCurdy, George, 2
MacLaurin, Colin, 208, 214
Magnus, Olaus, 81
Majorca, Spain, 119
Mallet, Jacques André, 243–44
Mallet, John, 308
Mallet, Robert, 251, 252, 308
Malouin, M., 157
Manchester, N.H., water power system of,
277
Manorial system, and watermills, 48, 105,
107–8, 113–14, 154
Marble mills. *See* Stone-cutting mills

131–35 passim, 155; in period 1750–1850, 271–78 passim
Reti, Ladislao, 278
Reynolds, John, 261
Rice mills, 116, 138
Rims: of ancient water wheels, 36, 42, 43; iron, 288–89; of iron water wheels, 301–2; of medieval water wheels, 98–99; of traditional wooden water wheels, 158, 161, 163
Rio Tinto Mine, Spain, 5, 42
Riquet, Pierre Paul, 230, 232
Robinson, Eric, 324
Robinson, George, 272–73, 294
Robison, John, 151, 157, 173, 180, 297–98
Rolling and slitting mills, 70, 76, 77, 94
Romain d'Alexandre, 57
Rome, use of watermills in, 30–33, 56
Royal Society of London, 212, 216, 223
Ruel de Bellisle, 144
Ruel de Chaville, 144
Rühlmann, Moritz, 342
Rumania; boat mills in, 125; horizontal watermills in, 103; medieval, watermills in, 51; suspended mills in, 185
Russia, 154; glass factories of, use of water-power in, 149; hydropower canals in, 133; hydropower dams in, 128, 130, 131–32; medieval, watermills in, 51; mines and water power in, 175–80

Sagebien, Alphonse, 259, 262–64
Sagebien wheel, 259, 262–64, 307, 347
Sagui, C. L., 40–41
Saint Blasien turbine, 342
Saint Gall, 82
Sakiyah, 24
Salic law codes, 49
Samaratine. See Pont Neuf Water Works
San Kuo Chih, 27
Sardis (Asia Minor), 31
Savery engine, 321–24
Savery, Thomas, 321–22
Sawmills, 88–90, 91, 94, 116
Scandinavia: fulling mills in, 82; iron mills in, 85, 86; medieval, watermills in, 51; paper mills in, 85; sawmills in, 90
Schemnitz, Slovakia, 135, 331, 333
Schiøler, Thorkild, 14, 16
Schott, Caspar, 189
Science, influence of, on water wheel analysis, 231–33
Scott, David, 157, 286

Scoville, Walter, 150
Scutching mills, 137
Segner, Andreas, 234, 235
Senchus Mor, 105
Shaduf, 22, 24, 27
Sharpening mills, 76
Shaw's Cotton Spinning Company, water wheel of, 307, 312, 315, 317, 320, 344–45
Shaw's Water Works (Greenock, Scotland), 273, 275, 277, 307
Shetland Islands, 107
Shrouds, 317; iron, 292; of iron water wheels, 301–2, 303, 304; of medieval water wheels, 98–99; of traditional wooden water wheels, 158, 161, 163, 164–65, 287; of undershot water wheels, 98–99, 164–65
Siberia. See Russia
Sicily, 73, 94
Siegen (region, Germany), 85
Silesia, 79, 149
Silk mills, 70, 79, 80, 116, 136–37
Simpson, F. Gerald, 36
Siphon line, 68
Slavery, influence of, on diffusion of water power, 32, 45, 120
Slavs, 50
Sliding hatch (overflow gate), 258, 306, 319–20; double, 284; experiments with, 256; invention of, 282–86; use of, with governor, 294–95, 296
Slitting mills, 76, 77
Slovakia: boat mills in, 125; water power complex at Schemnitz in, 135; water pressure engines at Schemnitz in, 331, 333
Smeaton, John, 173–74, 178, 180, 228–64 passim, 286, 290, 304, 330; breast wheel designs of, 280–82, 284, 306; diameter of water wheels of, 308, 310; experiments of, 223–26, 233; iron axles and iron parts introduced by, 288–89; overshot wheel designs of, 285; use of steam engines with water wheels by, 322–23, 325; water pressure engines of, 333; width of water wheels of, 309, 310
Smith, James, 307, 312, 314–15, 316, 317, 319
Smith, Norman, 119, 120
Smiles, Samuel, 230
Snuff mills, 74, 151
Soals: experiments on use of holes in, 252; of iron water wheels, 304, 305; of traditional wooden water wheels, 158,

The Johns Hopkins University Press

STRONGER THAN A HUNDRED MEN
*A History of the Vertical Water Wheel*

This book was composed in Baskerville text and Palatino display type by The Composing Room of Michigan from a design by Lisa S. Mirski. It was printed on 50-lb. MV Eggshell Cream paper and bound in Holliston Kingston Natural Finish cloth by The Maple Press Company.